MW00769059

Technology
Today & Tomorrow

SECOND EDITION

Technology
Today & Tomorrow

SECOND EDITION

James F. Fales, Ed.D., CMfgE
Professor and Chairman
Department of Industrial Technology
Ohio University
Athens, Ohio

Vincent F. Kuetemeyer, D.Ed.
Associate Professor
Department of Industrial and Technical Education
Louisiana State University
Baton Rouge, Louisiana

Sharon A. Brusic, Ed.D.
Technology Teacher
Kate Collins Middle School
Waynesboro, Virginia

McGraw-Hill

New York, New York Columbus, Ohio Mission Hills,California Peoria, Illinois

Editor: Susan I. Shoff

Printed in the United States of America.

Send all inquires to:
Glencoe/McGraw-Hill
3008 W. Willow Knolls Drive
Peoria, IL 61614-1083

ISBN 0-02-677103-9 (Student Edition)
ISBN 0-02-677108-X (Student Workbook)
ISBN 0-02-677105-5 (Teacher's Resource Guide)

5 6 7 8 9 10 QPH 99 98 97 96 95

Acknowledgments

The publisher gratefully acknowledges the cooperation and assistance received from many persons and companies during the development of *Technology Today and Tomorrow*. Numerous teachers contributed activities; their names are listed within this book, following their activities. Individuals and corporations who provided illustrations are listed in the credits at the back of the book. Special recognition is given to the following persons for their contributions.

Contributing Writers, Editors, and Reviewers

Eric Thompson
Technology Teacher
Onalaska Middle School
Onalaska, WI

Dr. Ralph C. Bohn
San Jose State University
San Jose, CA

Dr. Angus J. MacDonald
San Jose State University
San Jose, CA

David E. East
Richwoods High School
Peoria, IL

Sheila K. Farrell
Freelance Editor
Whitewater, WI

Cynthia Haller
Technical Writer and Freelance Editor
Troy, NY

Carole Fletcher
Technical Writer and Editor
Normal, IL

James J. Kirkwood
Department of Industry and Technology
Ball State University
Muncie, IN

Carolyn Gloeckner
Technical Writer and Freelance Editor
Orlando, FL

Greg Harrison
Social Studies Teacher
Metamora High School
Metamora, IL

Janet McGrath
Technical Writer
Peoria, IL

Marcey Runkle
Language Arts Teacher
Peoria, IL

Scott Runkle
Technical Writer
Peoria, IL

David Fleming
Freelance Editor
Peoria, IL

Marlene Weigel
Freelance Editor
Peoria, IL

Susan E. Gorman
Freelance Editor
Peoria, IL

Dr. Robert Norton
Freelance Reviewer
LaCrosse, WI

Joe Charles "Chuck" Bridge
Middle School Technology Teacher
Round Rock, TX

Cover photography courtesy of Seagate Technology © 1991 Jim Karageorge

Table of Contents

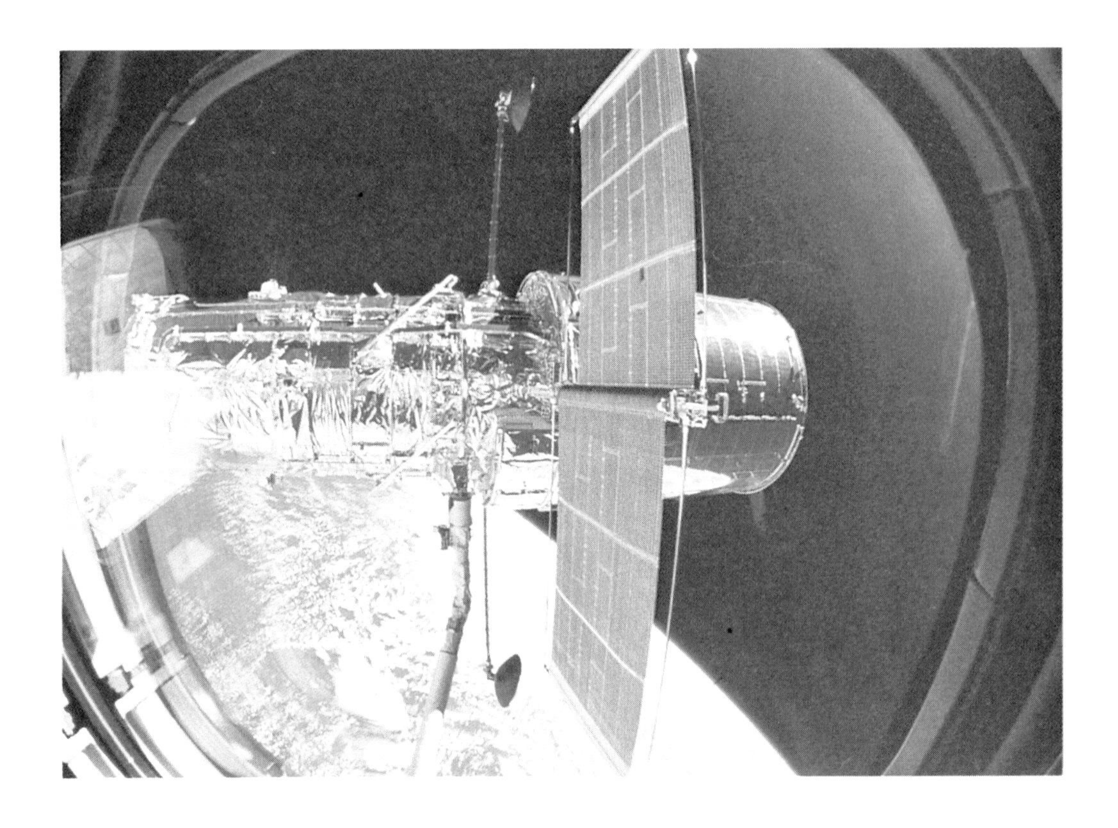

SECTION II Communication Technology......56

SECTION III Manufacturing Technology..........................168

Introduction

Technology is a part of everything that we do in life. The title of this textbook is *Technology Today and Tomorrow*. In this course, you will learn about the numerous advances being made today in the different areas of technology. Since changes are occurring very quickly in these different areas, you will also learn about the changes that are coming in technology in the near future.

People, like you, make technology what it is today. By itself, technology does not make life easier or more enjoyable. However, technology becomes vital when individuals use it to help other people, our environment, and society. Therefore, it is important that you learn all that you can about technology. Doing so will allow you to adjust better to change, become a better decision maker and more active participant in affairs that will affect your life, and enjoy life more because you understand the activities that are going on around you.

This book is about five areas of technology important to our society: communication, manufacturing, transportation, construction, and biotechnology. People all over the world use *communication* in one form or another to transfer information, increase their knowledge, and extend their potential. When you read the section about *manufacturing*, you will be learning how our natural resources are converted into industrial and consumer products. The section about *transportation* is the study of how materials/goods and people are moved from one location to another. The *construction* section will help you understand how our resources are used to build homes, office buildings, highways, and various other structures. In this book, you will learn how *biotechnology* is related to each of these other four areas of technology.

Section Content and Organization

Information in this book is presented in five sections. The first section introduces technology, defines the problem-solving process, and explains the parts of a technological system. The other four sections of the book deal with communication technology, manufacturing technology, transportation technology, and construction technology.

Section Openings

Each section opens with a list of its chapters and a photo essay that provides an overview of the main topics covered in the section. Along the bottom of the section-opening pages, there is a time line that lists major events that have occurred in the past related to the section's content.

Chapters

Following the section openers are chapters. Each major section consists of two or more chapters. Each chapter discusses a topic related to the basic idea of the section or to a family of technology. For example, in Section II, *Communication Technology*, Chapter 5 discusses telecommunication.

Section Closings

Each section closes with a problem-solving activity that relates to topics covered in the particular section. These activities are open-ended, so you can be creative and use your critical thinking skills.

Chapter Content and Organization

There are twenty-five chapters in this book.

Chapter Openings

Each chapter begins with a list of objectives and a list of vocabulary terms. The objectives, entitled "Looking Ahead," state specific facts or concepts you will learn. The vocabulary terms, entitled "New Terms," are terms used in technology. Understanding these will help you understand chapter content. These terms are printed in **boldface type** and defined where first used in the chapter. The second page of each chapter is a feature entitled "Technology Focus." These articles describe a current or future development in technology with emphasis on the people and companies that create technology.

Chapter Text

Each chapter's text is carefully organized, the information being clearly presented. The distinctive heads and subheads indicate the major divisions of information in the text. At the end of each major topic in a chapter, there are discussion questions. These questions, entitled "Apply Your Thinking Skills," are designed to get you thinking about the material you just read.

Within each chapter, there are short articles pertaining to a particular type of technology. Entitled "Fascinating Facts," these features are mostly historical items relating to a certain topic in the text.

In the book's odd-numbered chapters, there is a special feature that appears at the end of the chapter text. This special feature, entitled "Global Perspective," consists of an article that describes a particular type of technology in other countries. At the end of each article, there is an "Extend Your Knowledge" section that gives you ideas on how to learn more about the feature's topic.

In the book's even-numbered chapters, there is a special feature that appears at the end of the chapter's text. This special feature, entitled "Investigating Your Environment," consists of an article that discusses an environmental issue pertaining to the chapter's content. At the end of each article, there is a "Take Action!" section that gives you ideas on how you can get involved in solving different environmental problems.

Chapter Closings

Included in each chapter is a review. This consists of a brief chapter summary entitled "Looking Back," ten review questions, and five discussion questions. Also included in each chapter's review are cross-curricular activities. In these, you will apply your skills in language arts, social studies, science, and math to technology-related problems. Each chapter concludes with a technology activity. This hands-on activity relates to the content of the chapter and includes the step-by-step procedure for you to follow while you work.

Other Features

At the end of the book are four sections that can be helpful to you as you learn about technology—the careers handbook, the safety handbook, the glossary, and the index.

Careers Handbook

This handbook supplies you with information about a possible career in each of the different areas of technology. These areas include communication, manufacturing, transportation, construction, and biotechnology.

Safety Handbook

This handbook is in the text to help emphasize the importance of following all safety rules while working in the technology lab. You should read the safety rules listed in this handbook *before* working in the technology lab.

Glossary

If you need to know the meaning of a term, look in the glossary. Terms identified at the beginning of each chapter are listed in alphabetical order and defined in the glossary.

Index

If you would like to know where a certain subject is discussed in this book, look in the index. It will give you the numbers of pages that provide information on that subject.

Section I

Introduction to Technology

Chapter 1 **Exploring Technology**

Chapter 2 **Problem Solving: Creating Systems of Technology**

The monorail is one of the possibilities being explored in the field of mass transit. The need to conserve energy has made alternative transportation increasingly important to the future of modern cities.

A mix of technologies has contributed to the field of modern aviation. In a similar way, lessons learned in aviation have provided a foundation for technologies used in space exploration.

Technology Time Line

8000 B.C. Agriculture begins.

3600 B.C. The wheel is perfected along with many other tools.

3000 B.C. Cuneiform writing begins.

2700 B.C. Egyptians begin building pyramids.

800 Arabic numeral system we use today is developed.

8000 B.C. | 6000 B.C. | 4000 B.C. | 2000 B.C. | 1 | 500 | 1000 | 1100

Around 400 B.C. Herodutus, a Greek historian, said of messengers, "Neither snow, nor rain, nor heat, nor gloom of night stays these couriers from swift completion of their appointed rounds."

700 Waterwheels in common use in Europe.

New manufacturing techniques have begun to reduce the cost of solar cells. Solar technology represents an increasingly important source of renewable energy.

Cranes help construction workers lift large and heavy loads. For tall buildings, a crane often works from within the elevator shaft. After the job is done, the crane is removed by another crane.

Robots are doing an increasing number of jobs that were once done by production line workers. Robots can do some jobs faster and with fewer defects. Their use can protect production workers from some hazardous processes.

1232 Chinese first develop rockets.

1335 The first striking mechanical clock appears in Milan.

1454 The Latin Bible becomes the first European book printed with movable type.

1508 The first sugar mill in the Western Hemisphere is built.

1641 The English micrometer can measure a thousandth part of an inch.

1793 Eli Whitney is credited with the invention of the cotton gin.

1800 Industrial Revolution spreads to North America.

1842 Crawford Long uses ether as an anesthetic in surgery.

1862 Brown's universal milling machine made possible the quick and precise mass production of tools.

1200 | 1300 | 1400 | 1500 | 1600 | 1700 | 1800 | 1900

Shipment by container is one of the primary ways goods are shipped from home ports to overseas destinations.

Steel frame construction, pioneered in Chicago after its great fire, has made possible the modern era of skyscraper and multi-story construction.

In a short time, the laser has moved from a scientific wonder to an invaluable tool. Lasers now take the place of conventional equipment in many manufacturing processes.

Technology Time Line

1900 | 1910 | 1920 | 1930 | 1940 | 1950

- **1914** Panama Canal is opened.
- **1916** Mechanically operated windshield wipers for cars are introduced in the U.S.
- **1918** World War I ends when Germany surrenders.
- **1926** The pop-up toaster is invented in the U.S.
- **1929** The Great Depression begins.
- **1937** The Golden Gate Bridge is opened to traffic.
- **1939** NBC begins commercial television broadcasts.
- **1945** World War II ends in Europe and in the Pacific.
- **1945** The first nuclear bomb is exploded.

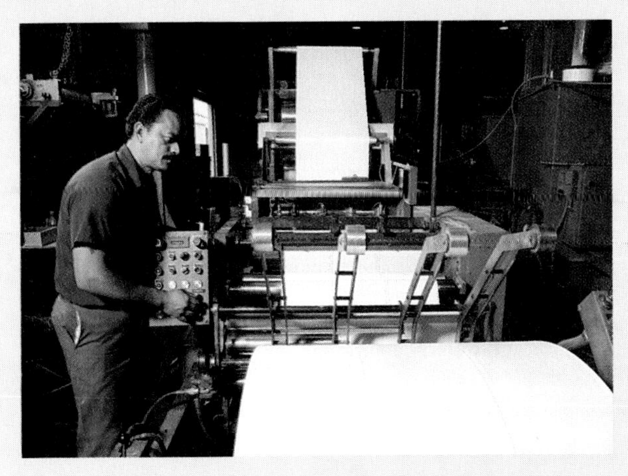

Most printed communications today are produced through offset lithography on presses like these. Some presses print as many as seven ink colors at one time.

Computers assist manufacturing companies in many ways. Most companies are installing increasingly complex networks to store data and to transmit information within the system.

The wind turbine is an ancient technology that is finding new applications today. Through creative use of some existing technologies, energy is being produced without damaging the environment. Our dependency on natural resources such as oil, coal, and natural gas is also being reduced.

1960

1962 James D. Watson and two British colleagues win Nobel prize for their research of the molecular structure of DNA.

1969 Man walks on the moon.

1970

1971 Intel of the U.S. introduces the first microprocessor, also known as a "computer on a chip."

1974 The push-button telephone makes its debut in Britain.

1976 Two amateur electronics enthusiasts build a microcomputer in their garage called Apple and revolutionize the computer field as we know it.

1979 The airplane *Gossamer Albatross* is the first human-powered aircraft to cross the English Channel.

1980

1981 NASA launches first reusable spacecraft, the space shuttle *Columbia.*

1982 Barney Clark becomes first human recipient of an artificial heart.

1983 Apple Computer introduces the "mouse" for personal computers.

1986 An accident occurs at Chernobyl's nuclear reactor in the U.S.S.R., spreading deadly radioactivity into the air and killing 23 people.

1988 Average American watches 6 1/2 hours of T.V. a day.

1990

1991 Operation Desert Storm uses the latest technological developments to end the war with Iraq in less than 2 months.

Chapter 1

Exploring Technology

Looking Ahead

In this chapter, you will discover:

- what technology is.
- how technology has been used throughout history.
- how technology is applied in different areas of human life.
- the relationship between technology and human knowledge.
- some positive and negative effects of technology.

New Terms

biotechnology
communication
communication technology
construction
construction technology
environment
manufacturing
manufacturing technology
technology
transportation
transportation technology

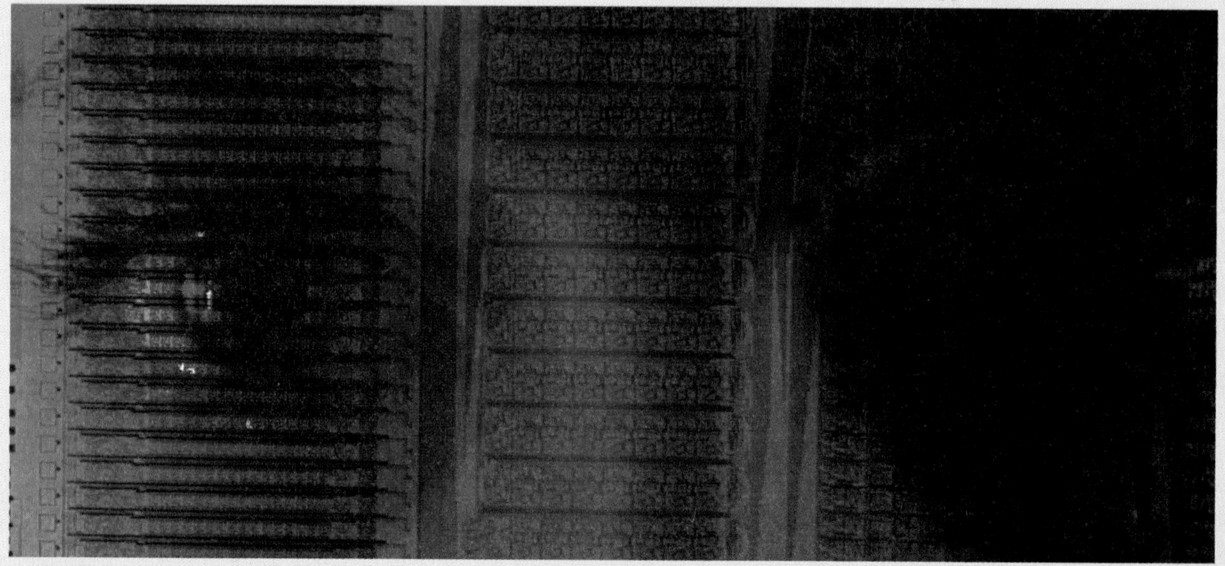

Technology Focus

Technology Surrounds Us Every Day

When you think of technology, what comes to your mind? Do you think of computers? Solar energy? High-speed planes? Robots? Technology is all these things and more. Technology is the way human beings solve problems and meet their needs. It is the way they use the resources around them creatively to do the things they want to do and create the things they want to create.

Think for a moment of something simple, like a pencil. Is this technology? Many people would say "no." However, when you understand everything that goes into making that pencil, you can see how important technology is in our daily lives. To make a pencil:

- Designers decide how the pencil will be shaped, how long it will be, what kind of eraser it will have, whether it will use soft or hard graphite, and how it will be packaged.
- Miners mine the graphite used for the writing portion of the pencil.
- Manufacturers mix the graphite with clay, form it into spaghetti-like strings, cut the strings to pencil length, and dry them in ovens.
- The lumber industry supplies the wood that encases the graphite.
- The chemical industry makes the paint to color the pencil.
- Pencil companies put wood, graphite, and erasers together according to the designer's plan to make the pencils. They also paint the pencils.

After all this has been done, the pencils are packaged. These packages are then packed in large boxes for shipping. Then trucks, boats, trains, and/or planes ship the pencils to stores, where they are sold.

As you can see, when you buy a pencil, you are not buying something so simple. The company that makes the pencils has relied on many technologies to get that pencil into your hands.

Fig. TF-1. Even simple, everyday items, such as a pencil, require the use of many technologies.

What Is Technology?

Technology is all around us. It is a part of our daily lives. **Technology** is the way people use resources to meet their wants and needs. For example, people have invented beds to meet their need for comfortable sleep. They have invented refrigerators and stoves to meet their needs for storing and cooking food. They have invented cars, buses, trains, and planes to meet their need to move from one place to another. Fig. 1-1. How many other examples of everyday technology can you think of?

A History of Technology

Technology has been around as long as the human race. When we think of modern technology, we often think of computers, complex machines, and space shuttles. However, people had to find ways of solving problems and meeting needs—in other words, develop technology—way back in history. Back in the Stone Age (beginning around two million years ago) people were using technology when they made tools out of natural materials like stone, wood, and bone. In the Bronze Age (beginning around 3000 B.C.), people learned how to make

Fig. 1-1. All technologies, whether they seem simple or complex, are ways people use resources to meet their needs.

bronze out of copper and tin. Using this technology, they could make better tools. In the Iron Age (beginning around 1200 B.C.), people learned how to mine and use iron, which is harder than bronze, and improved their tools even more.

As people began to farm land to provide their food, they developed technologies like the plow, carts and carriages, and the waterwheel to help them to plant, harvest, and prepare their food. The waterwheel, which was invented around 100 B.C., uses the power of water to help people do their work. A flour mill, for example, would be built on the bank of a stream or river. A chute directed the water from the stream into the blades on the wheel, and the weight of the water on the blades turned the wheel. The wheel was connected by shafts (horizontal and vertical) and gears to huge millstones. As the wheel turned, it caused these shafts and gears to turn, which caused one of the millstones to turn. This movement crushed the grain into flour. Because of the waterwheel, people didn't have to grind their grain by hand—a slow, hard process. They used technology to help them better meet their need to prepare their food. Later, the waterwheel was adapted (changed to make usable) for many mechanical jobs like sawing wood or running machines. Fig. 1-2.

The technologies people create change their lives. Take the example of the waterwheel. Instead of people building their own waterwheels, it made sense for one person in a community to own and operate a mill. The other people could then pay the

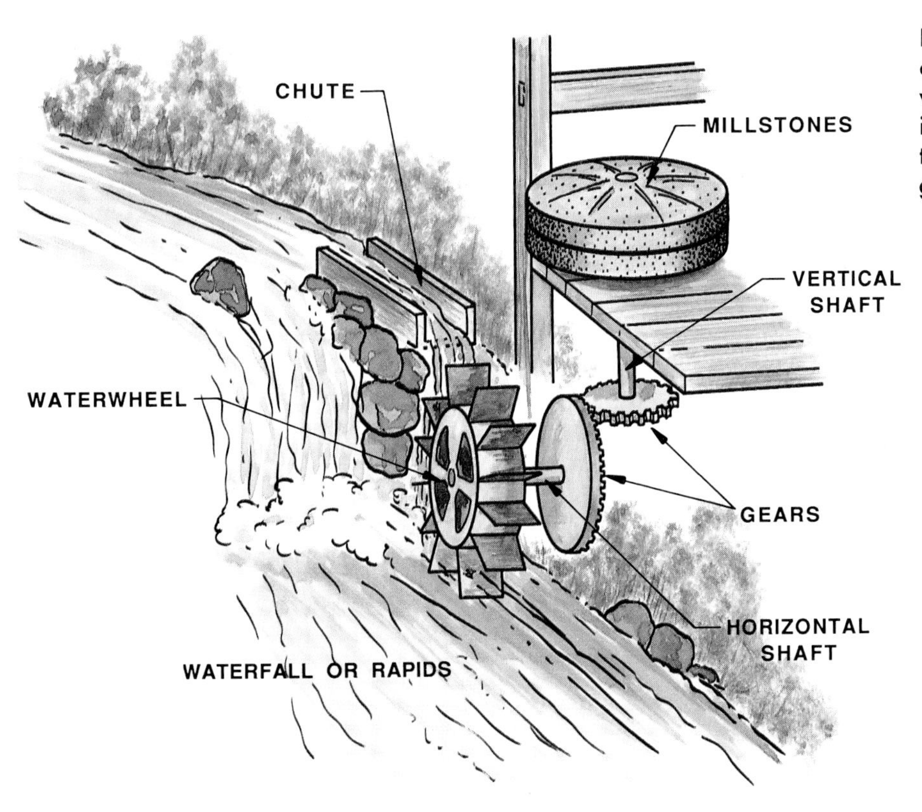

Fig. 1-2. The weight of water on the blades causes the waterwheel to turn. The wheel is connected to millstones. As the wheel turns, the millstones grind the grain.

miller money or goods in exchange for grinding their grain. The social life of the whole community changed because of the waterwheel. The waterwheel was only one of many technologies, each of which caused its own changes in the way people lived.

If technology has been around for so long, you may wonder why people are so concerned about technology in our century. The answer has to do with the amount of technology around us, how quickly it changes and grows, and how much it affects our lives. Even though people have used technology throughout history, the rate of technological change in early times was slow. New technologies came into use gradually over the course of many years. This gave people a chance to adjust to the new technologies they created. It took a long time for a technology to spread from one place to another. During the Iron Age, for example, many people were still making tools from stone because they did not know about ironmaking technology. Before the printing press was invented, information about a new technology could only be spread by the handwritten word, such as letters, or by word-of-mouth. This meant it took a long time for other people to learn about any new technology.

After the invention of the printing press, people were able to share their ideas about technology more easily and quickly. For example, someone in one place could print several copies of papers or books telling about the design of a certain machine. (The more copies made, the quicker the information could be spread.) A person in another place who knew a lot about metals and other materials could read that information and put his or her own knowledge together with the first person's to come up with a new idea. In this way, people's knowledge of technology began to spread and build on itself.

Different technologies are like different foods in a kitchen. Each time you add another food to your supply, you have a greater ability to create new and different foods. If you just have apples, for example, all you can do is eat raw or cooked apples. If you add flour, sugar, and butter to your kitchen supply, you can combine these to make something new, like an apple pie. The more technologies humans invent, the more they are able to combine these technologies to make new technologies.

Nowadays, it has become easier and easier for human beings to share their knowledge with one another. This is due to great advances in **communication**, the process of sending and receiving messages. Along with printed books and magazines, we now have computers, telephones, and satellites that make it possible for us to share ideas freely and very quickly. People can send information from one side of the globe to the other in a matter of seconds. Because of this ease of communication and the gradual buildup of technology over centuries, today we live in a time that is full of many kinds of technology that are changing rapidly each and every day. Technology, as well as its resulting impacts on our lives, has a "snowball effect"—like a snowball rolling downhill, it grows rapidly and, at times, seems out of control. Fig. 1-3. We do not

Fascinating Facts

Back in 1899, the Commissioner of the U.S. Office of Patents—Charles H. Duell—urged President William McKinley to abolish his office. His reason? "Everything that can be invented has been invented."

Fig. 1-3. Like a snowball rolling downhill, technology has developed at an ever-increasing rate and has had a greater and greater impact on our lives.

have as much time to adjust to the new technologies as people in earlier times had. Because of this, we need to be even more careful to manage our technologies well.

Apply Your Thinking Skills

1. Describe the technologies needed to make and sell a cotton blouse.
2. How has the invention of electric lighting changed people's lives?
3. Discuss a technology that is a combination of two or more other technologies.

Using Technology

In this book, you will learn about five areas of life in which technology is used to solve problems and meet needs: communication, manufacturing, transportation, construction, and biotechnology.

Communication Technology

As you just read, communication is the process of sending and receiving messages. We send messages for many reasons. Sometimes, we want to inform other people about something. For example, a newspaper informs people about important events in the world. We communicate to educate, or teach, each other about things. We may communicate to persuade others, such as when political candidates give speeches to try to persuade others to vote for them. Sometimes, communication is used to entertain, such as when people perform a play or make a movie. Fig. 1-4. People also use communication to control machines and tools. We want to tell machines how to do the work we want them to do.

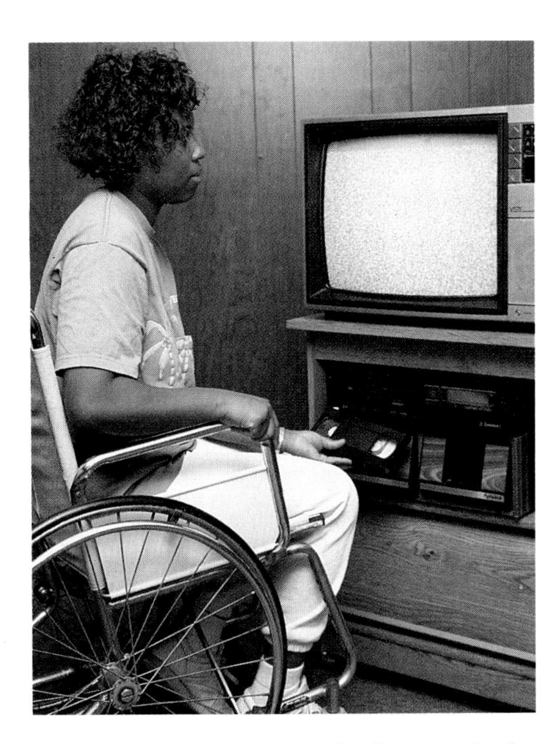

Fig. 1-4. VCRs are one example of communication technology. In this case, the purpose of the communication is to entertain.

Communication technology includes all the ways people have developed to send and receive messages. Telephones, radios, television, and computers are all examples of technologies that help us communicate with one another. In addition to communicating with other people, communication technology can be used to communicate with machines and to help machines communicate with each other, such as when a computer directs a factory cutting machine to cut a piece of metal in a certain way.

Manufacturing Technology

Another way humans use technology is to help them make the material things they need and want in their lives, such as clothing, furniture, cars, and even toothbrushes. This process of making products is called **manufacturing**. Just by looking around you, you can see the many things people manufacture. Think of your kitchen, for example. Fig. 1-5. All of the objects there, from the stove to the dishes to the plumbing, are manufactured goods.

Manufacturing technology is all the technologies people use to make the things they want and need. It includes the equipment and machines used to change the *raw materials* (materials as they occur in nature), such as trees, into finished products, such as furniture. It also includes the technologies needed to design the product, to ensure its quality, and to sell it.

Manufacturing technology doesn't have to be complex. A custom weaver, for example, may make woven cloth and blankets on a loom powered by his or her own feet. A clothing factory, by contrast, may use much electrical power to run complicated machinery to make shirts and dresses and slacks. However, both the weaver and the clothing company are using manufacturing technology.

Fig. 1-5. Your kitchen contains many examples of manufacturing technology.

Fascinating Facts

Inventors had been experimenting with electric lights at least 50 years before Thomas Alva Edison started his famous work—but Edison invented and manufactured the first practical light bulb.

For 15 months, Edison had been looking for a way to keep light bulbs glowing for a long time. In 1879, after trying 1,600 different materials as filaments, he tried carbonized (charred) cotton thread. The bulb glowed steadily for 40 hours.

Edison continued to improve the light bulb and went a step further. He designed power plants that would generate electricity and send it through wires into homes.

Transportation Technology

We also need ways to move ourselves and other things. **Transportation** is the business of carrying things, animals, and people from one place to another. Whether you fly from New York to Los Angeles, drive from home to school, or send a package to a friend in another state, you are using transportation. Manufacturers need transportation to get raw materials to their factories and their finished products to stores so people can buy them.

Transportation technology includes all the means we use to help us move through the air, in water, or over land. Fig. 1-6. Planes, boats, trains, and cars as well as the engines for those vehicles are all examples of transportation technology. Other examples include escalators, elevators, the locks that help ships move along a river, and pipeline systems that move oil from one part of the country to another.

Fig. 1-6. Subway trains are a form of transportation used to travel underground.

Construction Technology

People need buildings and structures for shelter and other purposes. **Construction** is the process of building structures. We build houses and apartment buildings to live in and skyscrapers to house business offices. Fig. 1-7. We build bridges to help us cross waterways and roads to travel on.

Construction technology is all the technology used in designing and building structures. It can range from something as simple as a hammer used to drive in nails to something as complex as a bulldozer for leveling earth or a crane for lifting heavy objects to build a tall structure.

Fig. 1-7. Construction technology is used to build the structures people live and work in.

Biotechnology

People also need to take care of their own bodies and of the living environment (surroundings) around them. If they become ill, they need to be cured. They need to know how best to grow and harvest food. They want to understand how life works.

Biotechnology is all the technology connected with plant, animal, and human life. It includes medical technologies like X-ray machines and MRI machines that help us see into our bodies. Fig. 1-8. It includes the machinery we use for plowing, sowing seed, and harvesting food. It includes the processes like canning, drying, freezing, and curing that we use to preserve our food. It also includes using living organisms to produce things we need. For example, genetic engineering uses special techniques to create improved plant varieties in order to provide more food.

In this book, you will look at biotechnology as it is related to the areas of communication, manufacturing, transportation, and construction technology.

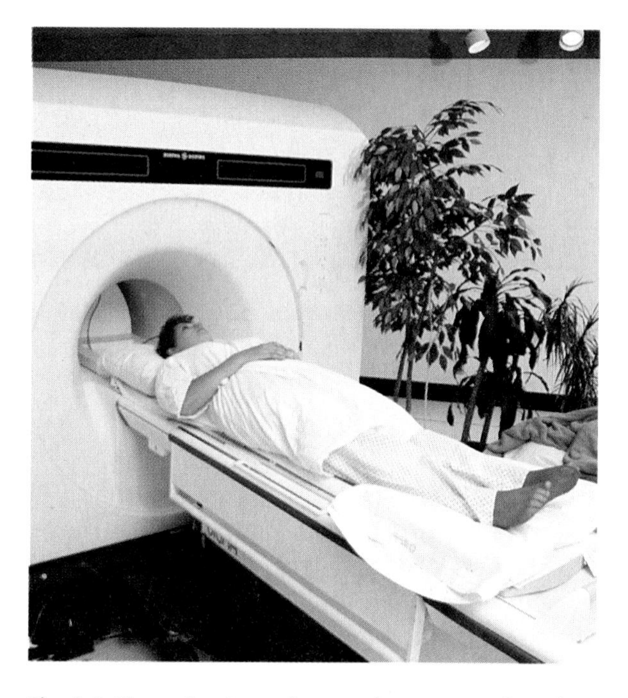

Fig. 1-8. The technology of magnetic resonance imaging (MRI) allows us to examine the human body. MRI is an example of biotechnology.

Combined Technology

We have looked briefly at five areas of life where people use technology to meet their needs. Even though we talk about these areas as if they were separate, in reality they blend together. Each area depends on the others.

Think, for example, of an airplane. In one sense an airplane is an example of transportation technology, since it is used to move people and things from one place to another. However, an airplane depends on communication technology. It has radio equipment for communicating with air traffic controllers. It has several instrument panels designed to communicate the status of the plane's systems to the pilot. A communication system allows the pilot and stewards to speak to the passengers. Some airplanes have phones so passengers can call people on the ground. On some trips, the airlines show movies for the entertainment of passengers.

An airplane is also an example of manufacturing technology. It takes more than 700,000 rivets just to hold all the parts of a medium-sized airplane together. These rivets, along with the engines, flight instruments, seats, and thousands of other parts, all need to be manufactured before the plane can be assembled. It takes the combined efforts of thousands of different companies to make all the parts needed for a single plane, as well as one company to put the parts together to make the finished aircraft.

You might think that an airplane has nothing to do with construction technology. Imagine, for a moment, however, that there were no airports, air traffic control towers, hangars, or runways. What good would it do to make airplanes without constructing the necessary buildings and structures to use those planes? An airplane would be a useless and expensive mass of parts without the structures that support its use.

How does biotechnology relate to an airplane? The meals you eat on plane trips have been processed and preserved using the knowledge provided by biotechnology. Support systems such as oxygen masks depend upon technology related to the human body and its functions.

As you can see, even though you may put an example of technology into one of the five areas of technology, it most likely includes elements of some of the other areas. Fig. 1-9.

Apply Your Thinking Skills

1. Into which area of technology would you place the following items: a car radio, a house, a leather shoe, a train? How does each of these include elements of other areas of technology?

Technology and Human Knowledge

You have seen how the areas of technology overlap with one another and are dependent on one another. In addition, technology is not separate from other forms of human knowledge. All of the subjects you study in school are related to technology, since they affect technology and technology affects them. Fig. 1-10.

Technology and Science

Technology and science are interrelated. Scientists spend their time doing research and making discoveries. A *discovery* is an observation of something in the world around us that no one has seen in quite the same way before. For example, Isaac Newton

Fig. 1-9. Most of the time, the areas of technology overlap and combine with one another. An airplane is an example of transportation technology, but it also depends on other areas of technology.

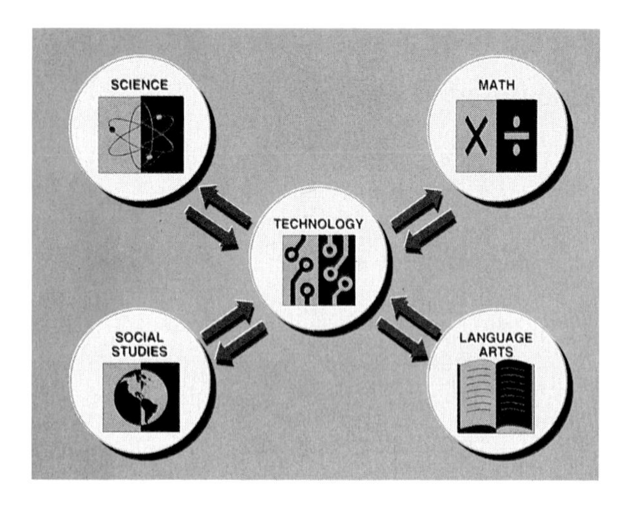

Fig. 1-10. Technology affects and is affected by other areas of human knowledge.

said he discovered gravity by watching an apple fall in his garden while he was drinking tea. Newton noticed certain patterns about the way objects fall. People had seen things fall before, but Newton figured out that the force of gravity makes them fall in certain ways.

Technologists use the knowledge discovered by scientists to develop better technologies. Knowing about gravity means technologists can design airplane engines that are able to overcome the force of gravity holding a plane to the ground.

The relationship between science and technology is greatly interwoven. Technologists use science to do their work of invention, but scientists also use technology to do their work of discovery. In 1608, for example, Hans Lippershey invented the telescope. A year later, Galileo heard about the telescope and made one of his own. Telescope technology made Galileo able to discover the moons of Jupiter and the rings of Saturn. In 1931, modern technologists invented the electron microscope, which uses beams of electrons instead of light rays to magnify objects. With it, scientists can examine objects as small as viruses and bacteria and discover more about how they work. In 1970, the electron microscope made it possible for scientists to examine and learn more about atoms. Fig. 1-11.

Technology and Math

Math is an important knowledge resource for technology. Engineers use mathematical formulas when calculating the amount of weight a structure can withstand—a very

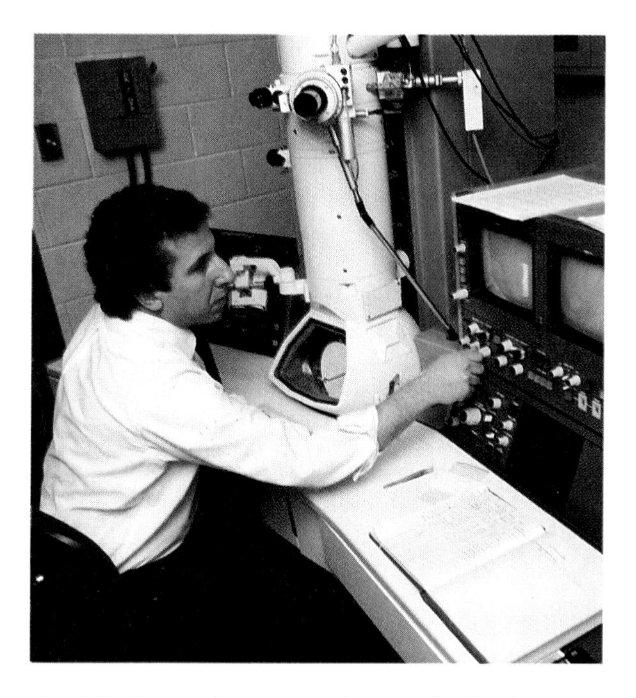

Fig. 1-11. Science helps technology, and technology helps science. The technology of the electron microscope helps scientists make important discoveries.

important piece of information when you are designing bridges or buildings. In manufacturing, measurements need to be very precise. If the part being manufactured is not extremely close to the measurement needed, it will not fit where it is supposed to fit. Imagine that you bought a do-it-yourself bicycle kit in which the factory had mismeasured and now the teeth in the gears are much too large to fit into the chain. You would not be able to assemble or use the bike.

Technology and Language Arts

Without language resources available to them, technologists would have more difficulty doing their job. Suppose a company that manufactured computers had no way of telling customers how to use them. The company would quickly go out of business if other computer companies were making instruction booklets and sending salespeople out to explain their computers to customers. Without language, companies would have no way to even let customers know their product exists.

Technology and Social Studies

Social studies helps us understand the impacts technology has on human culture. Earlier in this chapter, we talked about how the technology of the waterwheel changed the way people in communities related to one another. For a more modern example, think of how the telephone has changed human communication. People used to have to write letters or go to a telegraph office to send messages to each other. Now, they only need to pick up the phone and punch in or dial a series of numbers. They can speak directly to the other party instead of going through the long process of sending messages back and forth. Fig. 1-12.

Fig. 1-12. Technology has had many impacts on human culture. For example, before phones were invented, this person would have had to write a note or send a telegraph message to the person who is now on the other end of the line.

The Effects of Technology

Technology helps human beings solve their problems and meet their needs. We've already looked at the many benefits technology can provide—shelter; ways of producing, harvesting, and preparing food; manufactured goods; transportation; and many other useful things.

Sometimes the technologies we develop and use, however, create problems in our environment. Our **environment** includes all the conditions that surround a person, animal, or plant and affect development and growth. For example, compounds called chlorofluorocarbons (CFCs for short) are used in refrigeration to help keep food cold. The technology of CFCs thus helps human beings preserve food. However, CFCs are also used in aerosol spray cans like deodorants, hair sprays, and cleaning products. During spraying, CFCs are released into the atmosphere. Chlorofluorocarbons can contribute to the erosion (gradual wearing away) of the earth's ozone layer. The ozone layer helps prevent too many of the sun's harmful ultraviolet rays from penetrating the earth's atmosphere. People are beginning to realize that the convenience of having products in spray cans may not be worth the harm it causes to the environment. They are looking for other, less harmful ways of meeting their needs for these products.

The automobile industry also has both benefits and drawbacks. Cars help move us from one place to another and thus solve a human need. The exhaust gases from cars, however, contribute to air pollution. Is being able to drive anywhere we want worth the possible cost to our health and to the environment? Many people use public transportation or car pooling to help reduce the amount of automobile exhaust being poured into the air.

Modern technology has also provided us with many convenient products like disposable goods and foods packaged in individual portions. The advantages of these products need to be weighed against the disadvantages, such as the increasing problem of too much garbage. For example, several years ago, technologists developed disposable diapers. However, disposable diapers cannot be reused, and they take many years to decay and become part of the soil. Adding to our increasing mounds of garbage are all the containers that we throw away. Those handy individual boxes for juice, two-liter soda bottles, and squeeze bottles of detergent do not easily decompose (rot away). They just keep piling up year after year.

As you can see, technology is a mixed blessing. Technology, in solving one problem, can help people in many ways. However, technology sometimes creates other problems. It is not good or bad in itself, but becomes good or bad by the way people use it. By understanding how technology works and studying the effects it has, we can make good decisions about how to use technology to make our world a better place in which to live. Fig. 1-13.

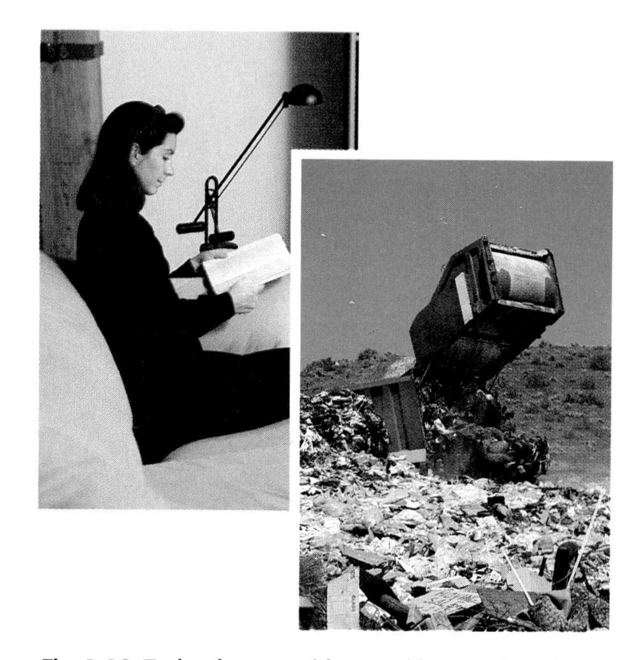

Fig. 1-13. Technology provides us with many benefits, such as electric lights. If not used wisely, however, technology can create many problems.

Apply Your Thinking Skills

1. How has human knowledge in science, math, social studies, and language arts affected the development of the airplane?
2. Most technologies have both positive and negative effects on human life. What are some positive effects of the airplane? Some negative effects?

Global Perspective

Creators of the Computer Age

Computers—these marvelous electronic machines are so common in our lives today that we tend to take them for granted. Many people played roles, both major and minor, in the development of computers. Let's look at just a sampling of the people whose contributions helped usher in "the computer age."

- *Charles Babbage* (English). In the 1800s, Babbage designed a machine called the Analytical Engine that was the ancestor of today's computers. One division of his machine was the "store" (now known as the memory) and another was the "mill" (now called the central processing unit).
- *Konrad Zuse* (German). Before and during World War II, in the 1930s and '40s, Zuse developed a computer that operated automatically and could be programmed. Much of his work was destroyed in the war. However, one surviving computer, the Z4, became the first important computer in Europe in the 1950s.
- *John V. Atanasoff* (American). The son of a Bulgarian immigrant father and American mother, Atanasoff is believed to be the inventor of the first electronic digital computer. His prototype (first of its kind and model for others) ABC computer was built in 1939.
- *Howard Aiken* (American). Aiken is known as the "builder of the first American electronic brain." His electromechanical computer, the Mark I, was completed in 1944. It weighed 5 tons and was 51 feet long and 8 feet high!
- *Grace Murray Hopper* (American). Rear Admiral Hopper, of the U.S. Navy, invented the computer "compiler," the first basic program that allowed automatic programming of computers. She is also credited with developing COBOL in the late 1950s, the first widely used computer language for business programs.

Fig. GP-1. The Mark I computer that Howard Aiken completed in 1944 weighed 5 tons and was 51 feet long and 8 feet high.

Today is indeed "the computer age." When you think of computers, think also of the many people whose vision, hard work, and creativity made this amazing age possible.

Extend Your Knowledge

1. The work of many people helped make personal computers possible, but two important ones were Steven Jobs and Stephen Wozniak. Who were these men, and what did they do that encouraged the trend towards personal computers?

Chapter 1 Review

Looking Back

Technology is the way people use resources to solve problems and meet their needs. The everyday objects that surround us as well as complicated machines and computers are all the results of technology.

Technology has been around as long as the human race. In early times, however, technology changed slowly and people had time to adjust to its effects. Today, technology is changing and growing at an ever-increasing pace.

Technology depends upon human knowledge. Science, math, social studies, and language skills are an important foundation for technology.

Modern technology has both positive and negative effects. Good management of technology means being aware of those effects and making decisions about technology that will benefit human beings.

Review Questions

1. Define technology.
2. When did people begin to use technology?
3. Define communication and give three examples of communication technology.
4. Describe five purposes of communication.
5. Define manufacturing and give three examples of manufacturing technology.
6. Define transportation and give three examples of transportation technology.
7. Define construction and give three examples of construction technology.
8. Define and give three examples of biotechnology.
9. What are four areas of human knowledge that support technology?
10. Define environment.

Discussion Questions

1. Discuss how the invention of the waterwheel changed people's lives.
2. Discuss the impact of the invention of the printing press on other technologies and on people's lives.
3. Why is technology growing and changing more quickly today than in earlier times?
4. Make a column for each of the following five technologies: communication, manufacturing, transportation, construction, and biotechnology. List the technologies you use in a normal day under the appropriate columns.
5. Pick ten items from the list you made in Question 4. Discuss the positive and negative effects of each of the ten technologies you chose.

Chapter 1 Review

Cross-Curricular Activities

Language Arts

1. Computers have become an essential part of our daily lives. Many schools are now using computers in the classroom. In an essay, discuss the positive and negative aspects of their use in the high school class.
2. Telephones are an important, if not essential, part of people's lives. What type of new telephone technology could benefit a police officer, a doctor, a student, or a teacher? In a creative essay, describe your ideas and the benefits that would be possible.

Social Studies 

1. What was the first type of manufacturing plant constructed in the United States? Where was it built and who built it? From what country did the U.S. get its idea for manufacturing? Describe ways in which manufacturing has advanced since the early 1800s.
2. What problems has NASA encountered since the establishment of the space program in the late 1950s? How were these problems solved? What were the original goals of the space program and how have these goals changed? For those goals that were changed, explain how and why they were altered.

Science 

1. You learned in this chapter that discoveries in science lead to new technology. For example, in the late 1790s, Count Alessandro Volta discovered how to create an electrical current by using stacks of silver and zinc disks separated by paper that had been wetted with saltwater. This device, known as the voltaic pile, was a battery. Name at least five technological products you use that depend on Volta's discovery.
2. Early scientists used simple means to time their experiments. Galileo, who studied gravity and acceleration, often used the human pulsebeat as a measuring device. Try this: use a clock with a second hand to count your pulsebeats for 30 seconds. Repeat this at least three times. Compare your results for each trial. How did technological developments in timekeeping aid science?

Math

Tracy and some classmates must make some paper ballots for a school election. The size of each ballot must be 2 3/4" x 4 1/2".

1. How many ballots can be cut from one piece of 8 1/2" x 11" paper?
2. Would one ream (500 sheets) of 8 1/2" x 11" paper be enough to make 3,850 ballots?

Chapter 1 Technology Activity

A Time Line of Your Life

Overview

At the beginning of this section and at the beginning of each additional section of this book, there is a time line that shows when different historical events have taken place. From the time you were born, you have been a part of this time line. You may not think that the events and actions in your life are important in comparison to what you see in these time lines, but they are, especially to your friends and family.

In this activity, you will develop a time line of your life. This time line is to include important personal events that have happened to you during your life. Then, as you develop the time line further, you will identify important world events that have happened during your life. Next, you will find events in the five technological areas (communication, manufacturing, transportation, construction, and biotechnology) that have happened in the same time frame. Plan to take several days to develop your time line.

Goal

To develop and visually present a time line of your life.

Equipment and Materials

notebook paper
drawing paper
colored pencils
straightedge
ruler or scale
reference books and periodicals

Procedure

1. On a piece of notebook paper, list the major *personal* events that have happened during your lifetime. For example: your birth, the birth of a new sister or brother, vacations, your first date, etc. Don't put down dates at this time.
2. On another piece of notebook paper, list major *world* events that you think may have happened during your lifetime. Again, do not put down the dates. Don't be afraid to list events that you are not sure about. You will check for dates later to see if they really did happen during your lifetime.
3. On a third piece of notebook paper, write a list of the *technological* events that you think may have happened during your lifetime. Write down as many as possible. You'll check the dates later.
4. Go back over the three lists and add dates to those events that did happen during your lifetime and mark out those that didn't. If you are having trouble thinking of events, go to your library and thumb through *The World Almanac* or perhaps a few encyclopedias. Be as specific as possible on the dates.
5. On a piece of drawing paper, draw a time line, to scale, that shows the years of your life so far. You may want to leave it open ended so that you can add more events as you grow older.
6. Now, using the colored pencils, place the events from the three lists on your time line in the appropriate positions. (See Fig. A1-1 for an example.) If you know the exact date of the event, try to place it in the approximate position between the years.

Chapter 1 Technology Activity

Personal, World, & Technology
Time Line

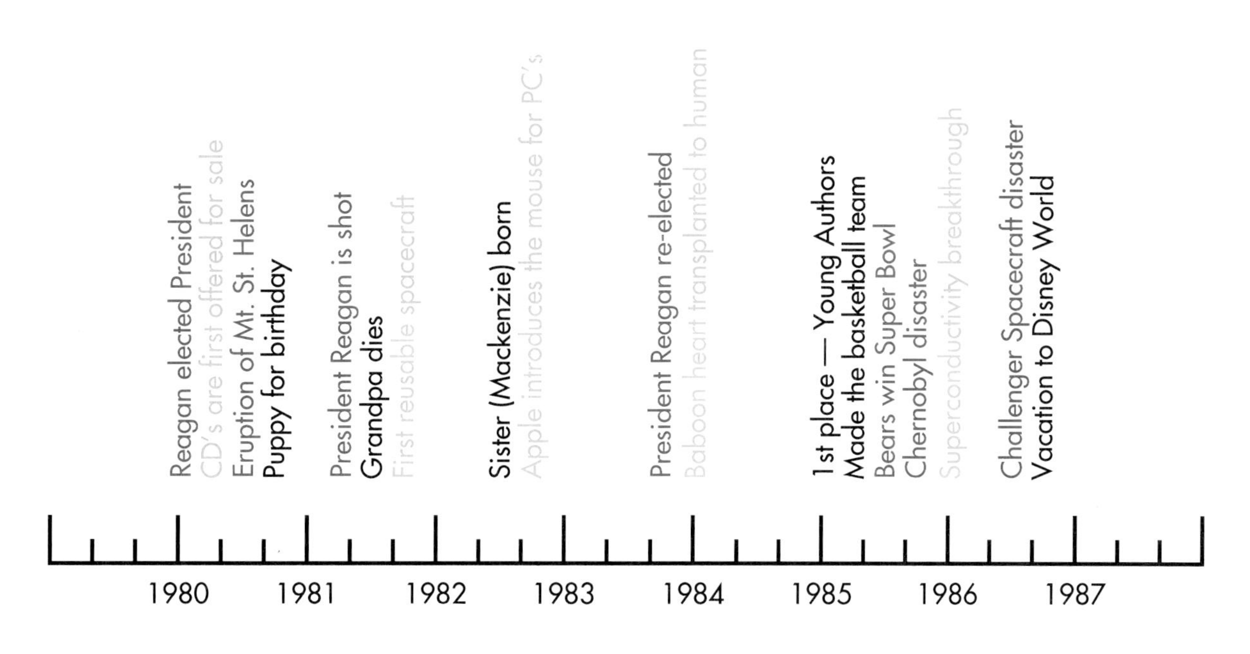

Fig. A1-1. Make your time line look something like the one shown here.

Evaluation

1. Are you surprised by how many events have actually occurred during your life? Are there any events that you would like to change if you could?
2. Ask your parents, grandparents, or an older person if they are happier because of all the technological events that have happened during their lives. Are there some aspects of technology that are not necessarily good? Are there developments in technology that have made their lives more healthy, more enjoyable, more comfortable, and more meaningful?

Chapter 2

Problem Solving: Creating Systems of Technology

Looking Ahead

In this chapter, you will discover:

- what problem solving is.
- the activities that make up the problem-solving process.
- what a system is.
- the parts of a system—input, process, output, and feedback.

New Terms

brainstorming
capital
energy
feedback
fossil fuels
input
nonrenewable energy sources
output
problem-solving process
process
renewable energy sources
resource
simulation
system

Technology Focus

The Problem: Recording Events and Ideas

The Solution: . . .

Chapter 1 described how technology is all around us. Even pencils, which we use every day, depend on many different technologies. How, though, does technology come about in the first place? It begins when people have problems they would like to solve. They develop a technology to create a means of solving those problems.

For example, people wanted to make written marks and pictures to record events and ideas. Early writers used chisels or brushes to make marks on walls, stones, and clay tablets. Around 2500 B.C., the Chinese and the Egyptians invented ink by mixing carbon and natural gum with water. This ink, however, did not work well on rough surfaces, such as stone and clay tablets. Then, around 1500 B.C., the Egyptians discovered how to make a type of paper out of papyrus, an Egyptian water plant. This paper was ideal for using ink. After this, people invented pens made out of reeds to apply ink to the papyrus.

Around 500 A.D., people learned to make pens out of goose feathers. In 1830, steelmakers discovered how to make inexpensive steel nibs, or pen points. It wasn't long before steel-nibbed pens replaced the easily damaged goose-feather pens. Then, during World War II, two brothers, Ladislao and Georg Biro, invented the modern ballpoint pen. As you know, many varieties of pens, such as rolling ball and felt-tipped, have been developed since then.

As you can see, people have been solving and re-solving the problem of how to write and make marks for thousands of years. Pens, of course, are only one type of writing tool. What about typewriters? Computer word-processing systems? Paints and brushes? People have developed many different ways to solve the problem of how to write and will continue to find new ways of communicating the written word. Technology is, in fact, an endless process of solving problems.

Fig. TF-2. Solving the problem of how to write led people to invent different types of pens.

The Problem-Solving Process

The **problem-solving process** is a multistep process used to develop workable solutions to problems. It includes the following steps:

- Stating the problem clearly
- Collecting information
- Developing possible solutions
- Selecting the best solution
- Implementing the solution
- Evaluating the solution

Stating the Problem Clearly

Solving any problem starts with knowing what that problem is. Stating the problem clearly in a sentence or two often helps to identify just what the problem is. In fact, sometimes a clear statement of the problem actually suggests a possible solution. Some people say that figuring out the problem is half of the job of solving it.

For a long time, for example, people whose legs were paralyzed couldn't drive cars. They couldn't work the brake and gas pedals. The problem was to design a car that people would not have to use their legs to operate. Once the problem is clearly stated, it is easy to see that the solution must involve developing controls that can be operated with body parts other than legs and feet. This was the first step in developing cars with hand-operated gas and brake pedals. Fig. 2-1.

Other problems are not so easy to identify. What about a situation like the problem of too much garbage? Is the problem (1) to design ways of removing the garbage, or (2) to design ways of reusing the garbage, or (3) to design ways of making products that create less garbage? A fourth possibility might be a combination of these three possibilities. Defining exactly which prob-

Fig. 2-1. Stating the problem clearly was the first step toward developing a car that could be operated by people whose legs were paralyzed.

lem to work on is important, since an environmental agency must decide where its money will be best spent.

Collecting Information

Once the problem is thoroughly understood, information that can be used to develop a good solution must be gathered. Sources of information depend, of course, on the nature of the problem, but could include libraries, museums, interviews with people who have worked in that particular area, and one's own lab or shop research. In solving a problem with toxic waste disposal, for example, a company might use a library or computer files to research information about toxic wastes. The company might also interview experts on safety in the workplace and scientists knowledgeable about toxic wastes. It could even set up a laboratory of its own to experiment with effective ways of dealing with waste.

Developing Possible Solutions

Most problems have more than one possible solution. At first, the more possibilities people can come up with, the better. That way, there are more options from which to choose.

One way of coming up with solutions is brainstorming. In **brainstorming**, people try to think of as many possible solutions as they can. They don't stop to evaluate the possible solutions at this point. The idea is just to come up with as many ideas as possible. Then all the solutions are discussed to select the ones that show the greatest promise.

Another way of developing alternative solutions is through trial and error. For example, when Thomas Edison was inventing the light bulb, he tried many types of material for the light bulb filament, including strands of red hair. He failed many times. Finally, he found that carbonized thread worked.

Sometimes a person will get really lucky and a solution to a problem will present itself accidentally. In 1974, this happened to 3M chemist Arthur Fry. As church choir director, Mr. Fry had been using slips of paper to mark songs in his hymnal, but found that they rarely stayed in place. Fry remembered an earlier discovery by a fellow 3M chemist, Spencer Silver, who had been experimenting with adhesives. Silver had discovered a "not-too-sticky" glue, but it just wasn't sticky enough for the purpose for which it was intended, and nobody could think of a good use for Silver's "invention." Fry decided to try a little of Silver's adhesive. Fry's solution to the problem of the "ever-disappearing bookmarks" paved the way for the development of 3M's Post-it™ note pads, a product that has become indispensable to office workers everywhere.

Fascinating Facts

More than 50 inventors had tried to develop a "typing machine" before the first practical typewriter appeared on the market in 1874. That typewriter—an invention that revolutionized the business world—was developed by a newspaper editor, Christopher Sholes.

Sholes tried out nearly 50 different models before inventing his machine. On the first keyboard, the letters were arranged in alphabetical order—but the "type bars" kept jamming. In 1872, after many trials and failures, Sholes tried a model with the popular letters scattered—a keyboard arranged like the letters in a typesetter's case.

E. Remington & Sons, a company that already manufactured firearms and sewing machines, bought Sholes' patent in 1873. It began selling that typewriter, at a price of $125, in 1874.

Selecting the Best Solution

In order to choose the best solution, all the possible solutions must be evaluated. Evaluating involves looking at all the advantages and disadvantages of each possible solution to determine which one best solves the problem. Many factors must be considered, and a good decision must be based on your goals and your particular situation.

As an example, suppose the problem is that a couple wants to build a house. They must look at the advantages and disadvantages of building materials and house styles and weigh these along with their goal before deciding what type of house to build. A small Cape Cod house is cozy and economical but may not be suitable for a couple starting a family. Wood is a cheaper exterior finish than brick, but it requires painting every few years. A single-level, ranch-style house would be ideal for people who have trouble climbing stairs. Ranch-style homes take up a lot of yard space, however, so people who enjoy the outdoors or like to garden may wish to build a two-story house so they will have more yard space. Fig. 2-2.

Part of good problem solving is being able to recognize which factors are most important. There is rarely a "perfect" solution to any problem.

Sometimes, more than one solution to a problem will be chosen because there is seldom one best solution that fits all circumstances. Seat belts, for example, are a good solution to the problem of passenger safety in cars. Many people, however, prefer to have air bags as well as seat belts in their cars. In addition, seat belts cannot adequately protect the small, not-yet-fully-developed bodies of infants and young children. Special car seats have been

Fig. 2-2. There can be many possible solutions to a problem. All of these examples represent solutions to the problem of needing a place to live.

designed to protect them from harm in car accidents. Thus, there are at least three solutions for protecting people in car collisions, and all of these solutions are being used. Fig. 2-3.

Implementing the Solution

After the best solution has been selected, the next step is to implement it, or put it into effect. During the implementation process, models are made and ideas are tested to make sure the solution is workable and is the best that can be made.

Often, simulations are a good way to test solutions. In a **simulation**, equipment is set up in a lab or testing area in a way that imitates as closely as possible the real-life circumstances for which the solution is designed to be used. For example, when testing seat belt and air bag designs for cars, auto manufacturers actually simulate crashes, using dummies in the cars. This way, they can see what would happen to a human being in an actual accident. Fig. 2-4.

Often a model of the proposed solution is built to aid in testing. You will learn about such models in Chapter 9.

The information people get from testing a solution helps then refine the solution. In a crash test for a seat belt, for example, perhaps the fastener designed for the belt broke loose at a certain speed. The designers would want to improve the fastener before manufacturing it.

Once the "bugs" of a new technology have been worked out, the technology can be put into effect.

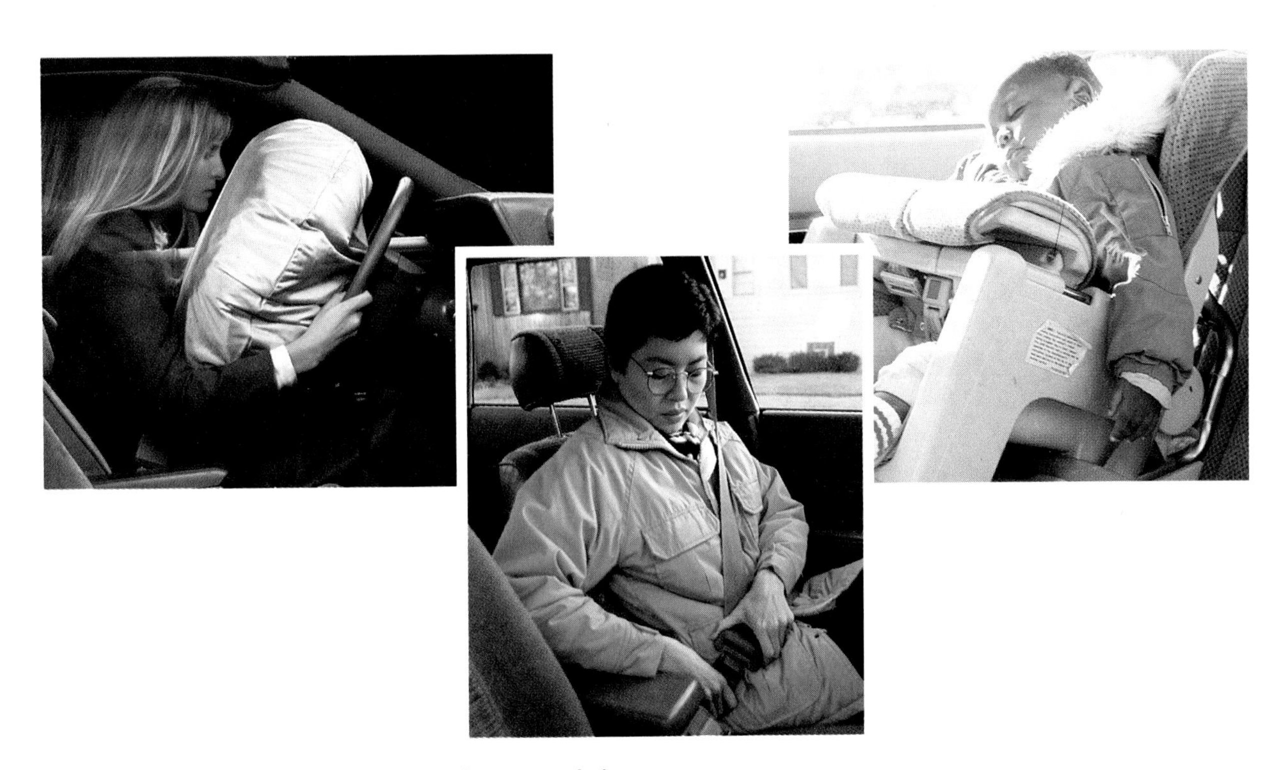

Fig. 2-3. People may decide to use more than one solution.

Fig. 2-4. Testing a solution can reveal its strengths and weaknesses. If the seat belt in a crash simulation doesn't work well enough, it can be improved.

Evaluating the Solution

Problem solving doesn't end once the solution is put into effect. Consumers may report that they are dissatisfied with the way a product works or that it wears out easily, or they may have had trouble assembling it in the first place.

Things that happen in the world may make a technological solution obsolete (out-of-date) or make a better solution possible. For example, typewriters used to be a standard piece of equipment in business offices. When personal computers were invented and became more easily available, however, people developed word-processing programs. Computers now do everything a typewriter does and more, and at a reasonable cost. Does the fact that the typewriter is being replaced by computers mean that the typewriter was a bad solution? Not at all. It's just that new information and events have made a better solution available. Problem solving is a never-ending process. Fig. 2-5.

Fig. 2-5. Sometimes one solution replaces another. In many offices, word processors have replaced typewriters.

Apply Your Thinking Skills

1. Look around you. Think about your home, community, and school. Write a list of five problems that might be solved using technology.
2. Choose one of those five problems. With a classmate, brainstorm at least five possible solutions. For each possible solution, make a list of strengths of the solution and a list of weaknesses of the solution. Which solution do you think would be the best?

Technological Systems

To solve problems, people create technology. Every technology that is invented as a solution to a problem can be thought of as a system. A **system** is a group of parts that work together to achieve a goal. These parts are: input, process, output, and feedback. Fig. 2-6.

Input

Input includes anything that is put into the system. Input comes from the system's resources. A **resource** is anything that provides support or supplies for the system. Fig. 2-7. There are seven types of resources that provide input for all technological systems:

- People
- Information
- Materials
- Tools and machines
- Energy
- Capital
- Time

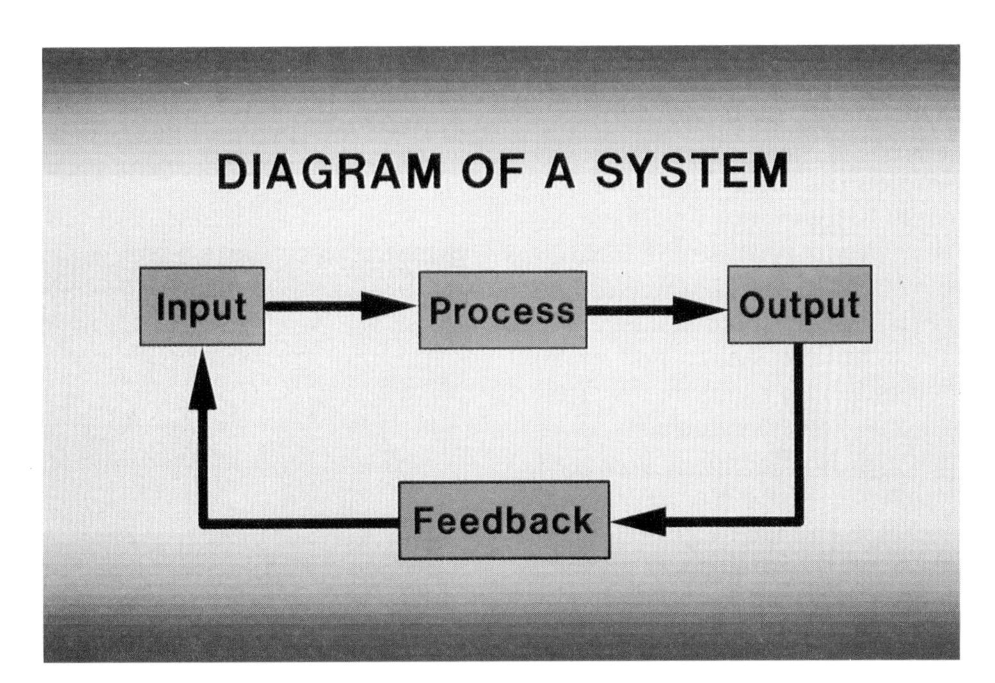

Fig. 2-6. A system is made up of input, process, output, and feedback.

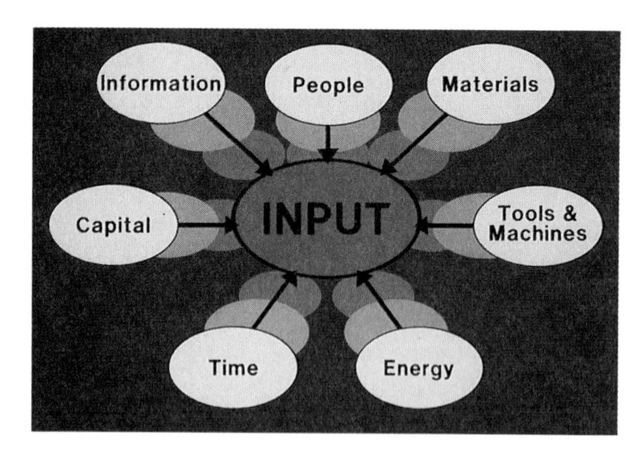

Fig. 2-7. The input to a system comes from its resources.

The most important resource is *people*. Without people, technologies wouldn't even begin to exist. People decide that problems exist and use the problem-solving process to develop workable solutions. For example, people who wanted to listen to music that had a high-quality, realistic sound spurred on the invention of CD players. If no one wanted to hear music, there would never have been a need for such a product in the first place. People created the demand, but it also took people to figure out how to make something that would provide the desired sound quality. People are also needed to actually make the players and discs. Fig. 2-8.

In order for people to design and make what is needed to solve a problem, they first need information. *Information* includes all of human knowledge. In Chapter 1, you read about how technologists need to use science, math, social studies, and language arts when they solve problems. To develop CD players, technologists used information about phonograph technology, laser technology, and digital technology. They came up with a new form of recording system where the "records" (CDs) don't wear out and can record sound more sensitively than traditional phonograph systems. Manufacturing workers need information, such as quantities, sizes, shapes, and types of parts as well as how the parts are to be assembled in order to make CD players.

In order to make a product, materials are needed. *Materials* are all those things that make up a product. A CD, for example, is made of plastic coated with a thin layer of aluminum, covered with a coat of lacquer.

Tools and machines are used to make materials take on the size and form needed to make the product. They are also used to transport materials, parts, and the final product. This category includes all the things people use to help them do their work, from the simple screwdriver and saw to a complex assembly-line robot.

Energy is needed to power the tools and machines. **Energy** can be defined as the ability to do work. Energy for the system may be provided by humans or by gas- or electrically-powered engines. In addition, energy provides the light needed to see to do the work. Energy is also needed to power the finished CD player. Energy may come from renewable or nonrenewable sources. Fig. 2-9.

Fig. 2-8. The most important input is people. It takes people to design and construct a CD player.

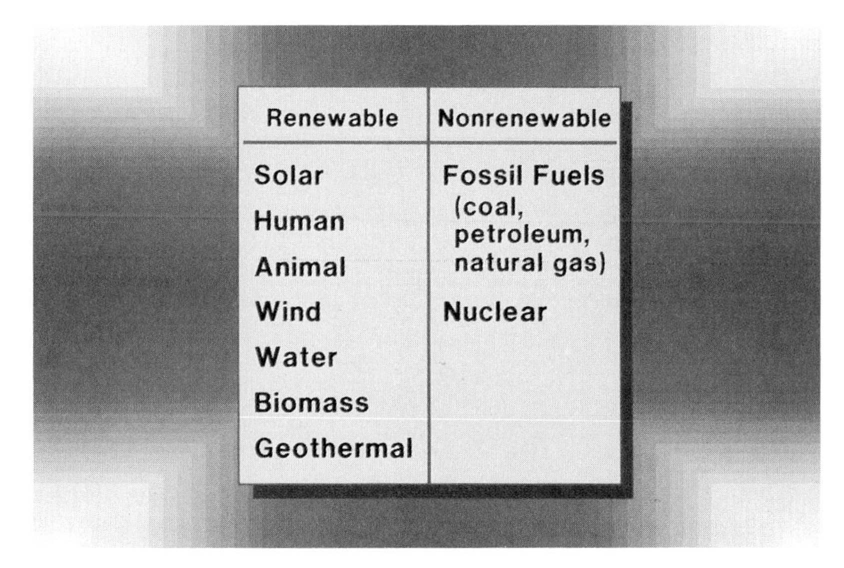

Renewable	Nonrenewable
Solar	Fossil Fuels (coal, petroleum, natural gas)
Human	Nuclear
Animal	
Wind	
Water	
Biomass	
Geothermal	

Fig. 2-9. Sources of energy.

Renewable energy sources are those that can replenish (restore) themselves regularly. Renewable energy sources include:

- The sun. The heat and light of the sun provide *solar* energy. Special solar cells, such as those in some watches and calculators, can change the sun's light energy directly into electricity. Special solar collectors can collect the heat from the sun's rays and use it to heat water and buildings.
- Humans. Human energy is used to do such things as carry materials, operate equipment, and use hand tools such as hammers.
- Animals. For centuries, animals have been used for transportation and to pull heavy loads. They are still an important source of energy in some parts of the world.
- Wind. Wind can propel sailing ships. It can also propel windmills that pump water. In some areas, generators powered by the wind produce electricity.
- Water. You learned in Chapter 1 how water is used to power waterwheels. Today, electricity can be produced by the force of water falling from a dam onto a turbine (wheel with blades) that drives an electric generator. This is called *hydroelectric energy*.
- Plants and animals. Wood from trees can be burned for energy. Plant and animal wastes as well as garbage can, with heat and pressure, be converted into fuels. This is called *biomass energy*.
- Steam and hot water from within the earth's interior. This is called *geothermal energy*. Steam below but close to the earth's surface can be piped to turbine-powered generators to produce electricity. Another form of geothermal energy involves harnessing the power from geysers, columns of hot water that shoot up from beneath the earth, to produce electricity.

Nonrenewable energy sources are those that cannot renew themselves quickly. Fossil fuels and nuclear energy are nonrenewable energy sources. **Fossil fuels** were formed from decayed plants and animals that lived millions of years ago. With pressure from the buildup of layers of the earth's crust over it, this decayed matter was gradually turned into the fossil fuels coal, petroleum, and natural gas. *Nuclear energy* is produced when the atoms of such elements as uranium are split. This splitting, or fission, produces large amounts of energy that can be used to generate electricity.

Capital is another important resource for technology. **Capital** includes the money, land, and equipment needed to set up the technological system. Without money to build factories and hire workers, CD players would remain only an interesting idea instead of an actual technology.

A final technological resource is *time*. When a company decides to make CD players, it invests time in that particular system. The company could be spending that same time making computers or furniture. Time is a valuable resource, so the ways people decide to use it are very important. Fig. 2-10.

Process

The second part of a system is the process. The **process** includes all of the activities that need to take place for the system to give the expected result. For a CD-making system, the process might include designing the product itself and its individual parts, making the various parts out of appropriate materials, and putting all the parts together to make the player. Fig. 2-11.

Fig. 2-10. All of the technological resources—people, information, capital, energy, tools and machines, materials, and time—were needed to make this CD player and disc.

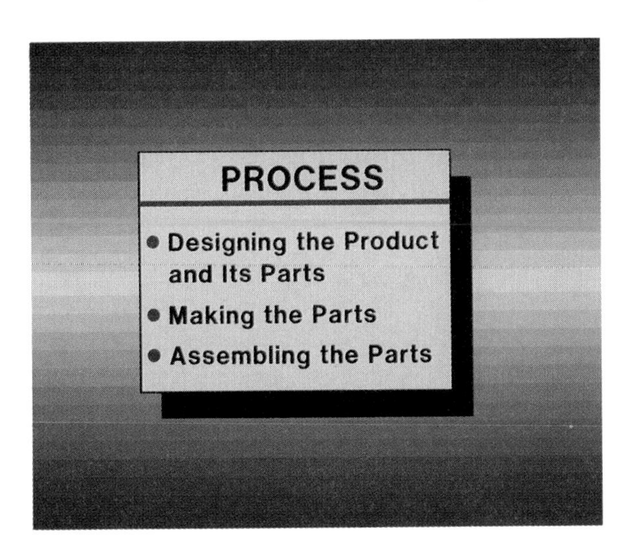

Fig. 2-11. The process part of a system includes all of the activities that are needed to solve the problem.

Output

The third part of a system is the output. The **output** includes everything that results when the input and process parts of the system go into effect. In all systems, there is, of course, the intended output, such as CD players. There are also, however, outputs that may not have been intended—such as waste that may have been created during the process, or changes in society caused by the product. Fig. 2-12.

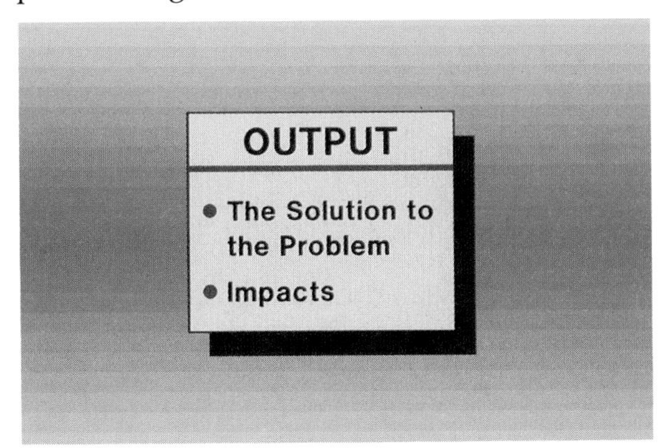

Fig. 2-12. The output of a system includes the desired goal (the solution to the problem) plus all of the impacts of the system.

CD players, for example, have had a great impact, or effect, on the recording industry. In many record stores, you no longer find vinyl phonograph records. These changes in the recording industry have an impact on other parts of the economy. Think of all the materials that were used to make vinyl records. Some of these are no longer needed. On the other hand, more of other types of materials may be needed to make compact discs.

Feedback

The outputs of a system must be closely watched to be sure that the system is solving the problem it was intended to solve and to be sure that the system is not creating new, greater problems. **Feedback** is the information about the outputs of the system that is sent back to the system to help determine whether the system is doing what it is supposed to do. Fig. 2-13.

Feedback can take many forms. Some kinds of feedback are built into a system. For example, home CD units have error checking information. Often, systems check themselves to make sure all the parts are functioning properly.

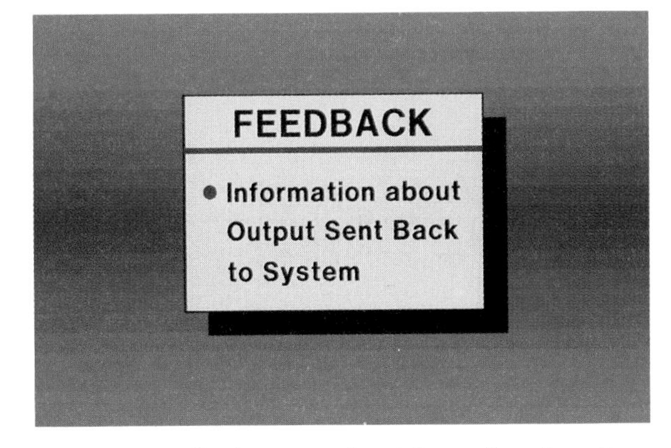

Fig. 2-13. Feedback occurs when information about the outputs of the system is sent back to the system to check to see whether the system is doing what it is supposed to do.

In addition, there is feedback from people who use the system. Suppose people liked the sound emitted from CDs, but did not like the size or bulk of the unit that played them. Manufacturers would listen to this feedback and redesign the unit so people would like it better.

Feedback becomes a form of input into the system. Fig. 2-14. It often leads to improvements in a system or to the creation of new systems. For example, the turning devices on early phonographs would often wear out and slip, distorting the sound. Feedback about this problem led manufacturers to develop belt drives and direct drives to turn the turntable. These improvements in the system lessened the sound distortion problem.

Feedback about records wearing out and sound distortion also led to the development of CD systems. CD systems have similarities to phonograph systems, but are based on such different technology that they are really a separate kind of system. They are so much better than phonograph systems that eventually they are likely to replace these systems.

Apply Your Thinking Skills

1. Describe how the seven resources of technology are used in a car manufacturing system.
2. What are some of the outputs of a car manufacturing system? Be sure to include both negative and positive impacts as well as expected output.
3. Describe how a home heating system uses feedback to keep itself working properly. (Hint: think of how the thermostat works.)

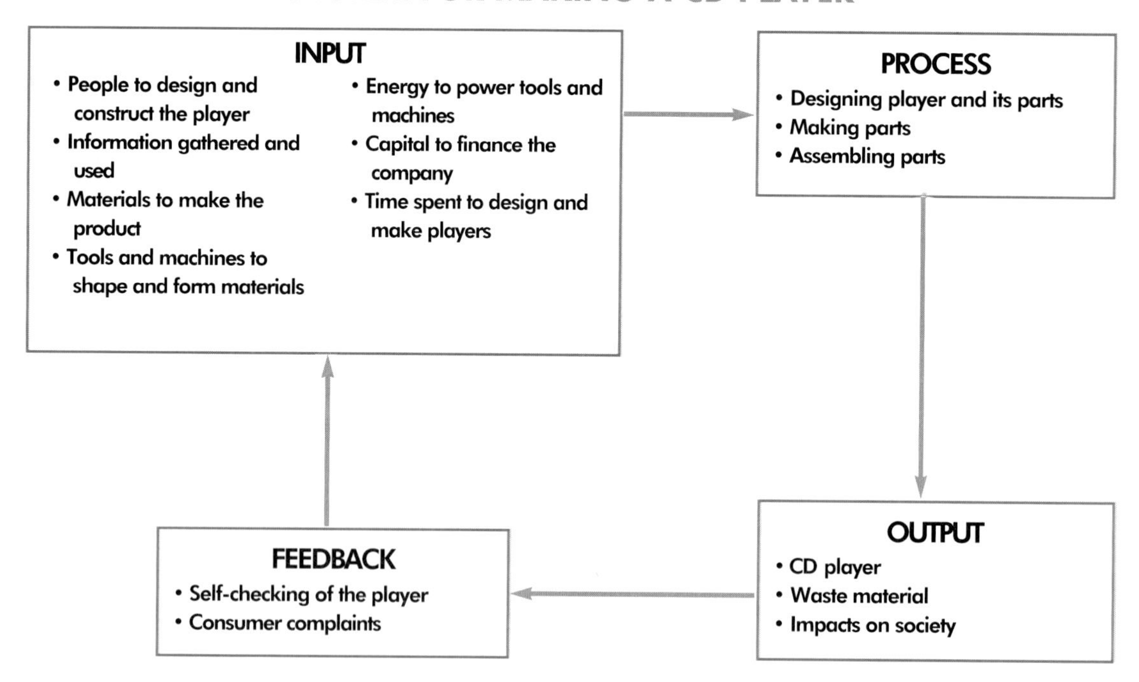

Fig. 2-14. This is the system for making a CD player. Notice that feedback becomes an input, forming a "loop"—the never-ending process of problem solving and technological change.

Investigating Your Environment
Pollution Solutions

Although technology has made our lives easier in many ways, technological systems have often produced undesirable impacts on the environment. Radioactive waste, CFCs, acid rain, noise pollution, air pollution, and the deterioration of the ozone layer are just a few. It is normal to feel helpless and upset by all the environmental problems, but there are ways you can make a difference.

Start by deciding what environmental problem you wish to solve. Eliminating oil spills would not be a problem you and your classmates could handle. However, what about paper waste in your school, or noise pollution in the community? Perhaps if your community doesn't already have a recycling program, you could find out how to get one started. Looking in the newspaper or talking to your parents may give you some good ideas. Once you have identified what problem you wish to solve, state it clearly. Stating the problem clearly helps identify exactly what the problem is and may suggest possible solutions.

Once you have identified the problem, you and your classmates need to gather all the information you can. You may wish to interview people, do research in the library, and do whatever you can to find out what other schools or communities are doing to handle similar problems.

The next step is to brainstorm with your classmates about possible solutions. Let your imagination go; even if some of the solutions seem ridiculous, they may lead you to other practical ones. Perhaps your teacher could make a list of your ideas on the chalkboard.

From the list of possible solutions, you and your classmates must choose the best solution. List and weigh the advantages and disadvantages of each possible solution. Some of your ideas may be impossible to carry out. The solution must work with your school schedule. If it is an in-school project, will the rest of the students in the school be willing to cooperate and pitch in? If you decide to gather recyclable wastes, will you have a way of transporting the waste materials to a recycling center? Questions such as these must be considered when choosing the best solution.

Fig. IYE-2. The brainstorming process is a good way to generate many ideas about how to solve an environmental problem.

Next, implement your solution. Try it out for a specified period of time. Set a date for discussing progress. Is it working? Do you need to change anything? Implement the newly revised solution.

Take Action!

1. Does your school have a recycling program? If not, as a class, use the problem-solving process to start one.
2. The storage of garbage in landfills is dangerous as well as ugly. Find out how full your local landfills are, and then find out what's being done in your community to reduce the storage of garbage in landfills. Report your findings to the class.

Chapter 2 Review

Looking Back

Technology is an endless process of solving problems. When people solve problems, they need to identify the problem by stating it clearly, collect information, develop possible solutions, choose the best solution, implement the solution, and then evaluate the solution.

It sometimes takes a lot of tries before people get a solution to work effectively. There may also be more than one good solution to a problem. New information and events often make better solutions possible, and the whole problem-solving process starts up again.

People create technology to solve problems. Every technology is a system, which is a group of parts that work together to achieve a goal. These parts include input, process, output, and feedback.

Review Questions

1. Why is it so important to state the problem clearly as the first step to problem solving?
2. What is brainstorming?
3. How do you evaluate solutions when trying to determine the best solution?
4. What is a simulation?
5. Why is it important to test solutions before putting them into effect?
6. Define input.
7. Define resource. Identify the seven resources of technology and give an example of each.
8. What does the process part of a system include?
9. What does the output of a system include?
10. What is feedback?

Discussion Questions

1. Think of five examples of technology. What are the problems that these technologies solve?
2. Suppose you lived in a tropical climate and your best friend lived in a cold, northern climate. How might your solutions to the problem of a place to live be alike? Different?
3. Describe an example from your own life where you tried one solution to a problem, failed, tried something different, and succeeded.
4. What problem does a bridge system solve? How are the seven resources of technology used in a bridge system?
5. Describe several examples of technological feedback.

Chapter 2 Review

Cross-Curricular Activities

Language Arts 

1. Divide into groups and brainstorm ways of improving the use of your school's food service, main office, library, and your classroom. Present your best ideas to the class.
2. Continue with your group to implement your ideas. Interview school personnel for input, list problem-solving possibilities, put these ideas into action, and survey other students about the results. Each group should then present the steps they followed to the class.

Social Studies 

1. How did early inventors obtain the necessary capital to build and perfect their inventions? Be specific. Choose one or two 19th century inventors and tell how they financed their inventions.
2. What kinds of problems evolved out of the development and marketing of the automobile? How were these problems dealt with? How did some of these solutions lead to other problems? How were these problems solved?

Science 

1. The problem-solving methods used for technological development are also used by scientists. Use the problem-solving process described in this chapter to find the answer to this question: Does chilling seeds for a while before planting affect their growth? List all the steps you would take to solve this problem.
2. Thomas Edison's light bulb filament was a piece of thread. Light bulb filaments today are made of the metal tungsten. A thin, coiled tungsten wire glows when electricity passes through it because of the metal's resistance. Instead of electrons flowing easily, they cause the metal to heat up. The higher the resistance, the greater the heat produced by an electrical current. Use a reference work to find the resistance of the following metals: tungsten, copper, tin, and zinc. Also find the melting points of these metals. Why is tungsten the best choice for a light bulb filament?

Math

1. Ernie obtained a part-time job inspecting bottle caps. There are too many bottle caps produced to inspect each piece individually, so he makes his inspection using a sampling method. Each production run produces 3,000 bottle caps. Ernie randomly obtains 50 samples from each run. If he finds 3 unacceptable bottle caps in each sampling of 50, what conclusion can he make about the total run of 3,000?

Chapter 2 Technology Activity

Using Problem Solving to Make a Chair

Overview

You have learned in Chapter 2 about the six steps in problem solving. You know that the problem must be stated clearly and that you must collect information. Next, you develop several possible solutions and select the best one. Finally, you put the best solution to work and evaluate its performance. Reading about problem solving, however, does not teach you all there is to know about it. The best way to learn about its usefulness is to use it in solving a real problem.

In this activity, you will use the problem-solving process to design and build a chair out of cardboard. A person will actually be able to sit on this chair. Can it be done? Let's find out. You will be competing with your classmates to see who can find the best solution to the problem.

Goal

To design and build a full-sized, usable chair of corrugated cardboard and masking tape. The chair must conform to the following specifications:

- It must be made completely of corrugated cardboard and masking tape. No glue, wood, metal, plastic, or other materials can be used.
- It must have a seat and back. The seat must be between 16" and 18" from the floor. The top of the back must be no less than 30" from the floor. It must have no fewer than 3 separate legs.
- It must support a person weighing as much as 220 pounds. The person must be able to sit in the chair in a relaxed, comfortable position.
- The chair should weigh as little as possible. In the final competition, its weight will be measured and compared to that of other chairs made in class. It should also be as attractive as possible.
- All chairs will be judged and awarded points. A total of 15 points is a perfect score. Points will be obtained as follows:
  - Following specifications—5 points
  - Weight—5 points
  - Design—5 points

Equipment and Materials

pencils and paper
measuring instruments
cutting tools, such as scissors and utility knife
corrugated cardboard
masking tape

Procedure

1. *Safety Note:* Be sure utility knives and scissors are sharp before you use them. Dull tools can cause accidents.
2. Your teacher will divide the class into teams. Each team will design and build a chair. You will find as you work that cooperation and planning are important keys to success in this activity.
3. With your teammate(s), read through the entire activity. Then clearly state the problem you must solve. You may want to copy the specifications listed under "Goal." Copying will ensure that you

understand the problem and that you will not forget anything important.

4. Collect the information you need to make your design. Study the designs of chairs you see around you. Look up chairs in an encyclopedia and learn what designs have been used in the past. Fig. A2-1. Then study your materials. How can cardboard be made strong?
5. Talk with your teammate(s) about possible chair designs. Each of you should make sketches of two or three that might work. Remember, attractiveness is also important.
6. As a team, select the design you like best. Check it against the specifications. Is anything missing?
7. Create a set of working drawings for your chair. There should be a top, front, and side view. (See Fig. 6-19 on p. 137 for an example of working drawings.)
8. Build your chair using the cardboard and masking tape.
9. Evaluate your chair. Ask someone who weighs 220 pounds to sit on it. Did the chair hold the weight? Is the person comfortable? Look at its overall design. Is it attractive? Is the tape placed neatly?
10. Submit your chair, sketches, and working drawings to the competition. Good luck.

Evaluation

1. If you were to build another chair, what would you do differently?
2. What did you learn about materials from this activity?
3. Who in real life designs chairs for people to use?

Credit: Gene Stemmann, Monroe, Oregon

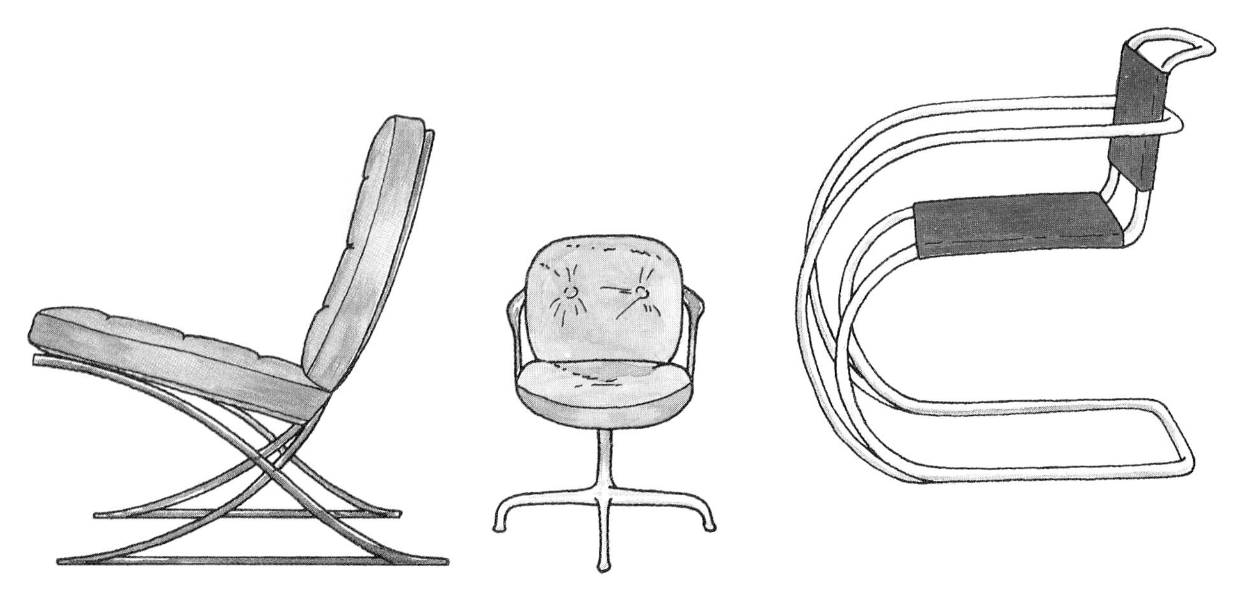

Fig. A2-1. These unusual chair designs have all been used at various times.

Section I Activity

Planning a Yearbook Advertising Campaign

Overview

Your high school yearbook will go on sale in three weeks. The yearbook staff wants the event advertised in the local media, beginning one week prior to the actual sale. Its budget for this is $200. You have been asked to create this advertising campaign.

STEP 1: State the Problem (Design Brief)

You have a variety of options in advertising. Newspaper ads are widely read and relatively inexpensive. Ads on television also have a broad audience, and their live-action quality can make them quite effective. However, they are more expensive and more energy- and time-consuming. Radio offers some of the qualities of television without all of the expense, but radio audiences tend to be sharply segmented. Careful consideration must be given to the type of station the ad is run on and the time slot in which it is played. You have a limited budget, a ten-day deadline, and a specific target audience (those buyers at whom your ad is directed). All these things must be kept in mind.

On a clean sheet of paper, type or write neatly a design brief, stating the problem to be solved (in this case, creating an advertising campaign for the yearbook), as well as some of the factors to be considered in your plan, such as time and money limitations.

STEP 2: Collect Information

Here are some ways of gathering the information you will need to design an effective advertising campaign:

- Contact the advertising departments of local newspapers, television, and radio stations to find out their rates for running a printed ad or airing a commercial.
- Contact a marketing research firm to discover the age range of television viewers and radio listeners in a given time slot. This will help you determine when to run your ad to reach the greatest number of your target audience.
- Find out what kinds of facilities and equipment you will need to produce your ad. Will a hand-drawn ad look polished enough for the newspaper, or will you need the services of a professional printer? Can you record or film your own radio or television ad, or will you need to rent time at a studio?

STEP 3: Develop Alternative Solutions

After determining the costs and resources needed for each type of ad, you will probably want to develop several different campaign strategies. For example:

- Investing the entire budget in one professional, entertaining television ad to be shown in the most desirable time slot.
- Splitting the budget among several less expensive radio ads and a newspaper ad.
- Using a less traditional or expensive medium, such as buttons or flyers.
- Having the sale advertised—for free—as a public service announcement on television or radio.

- Trying an unusual stunt to attract attention, such as a small student kazoo band marching through a public downtown square at the lunch hour.

List these alternative solutions on the same sheet of paper as your design brief. Fig. SCI-1.

STEP 4: Select the Best Solution

Consider the pros and cons of each alternative plan. Some of the things to keep in mind are:

- Does this campaign fit into the budget?
- Can it be completed on time?
- Do we have the necessary talents to do a professional job (for example, an artist to design an eye-catching newspaper ad, or an articulate, photogenic spokesperson to appear in a television ad)?
- Does it say what we want it to say about the product and the event?

STEP 5: Implement the Solution

After selecting the best plan, contact the people necessary to put it in motion: media advertising departments, artists, actors, photographers, etc. Communicate your ideas clearly, so each party knows exactly what you want.

STEP 6: Evaluate the Solution

Evaluate your advertising campaign with your teacher and classmates. Ask yourself if it is as informative and attractive as you can reasonably make it. Ask yourself: If I saw or heard this ad, would I buy a yearbook?

Fig. SCI-1. These different media are all possibilities for the advertising campaign that you're developing for your school's yearbook.

Section II

Communication Technology

Today, telecommunication instantly links people across the street or around the globe. Centers like these have evolved to coordinate the various technologies.

Technology Time Line

3000 B.C. | 1 | 1000 | 1200 | 1400 | 1600 | 1800

- **3000 B.C.** Cuneiform writing begins.
- **About 27 B.C.** First "modern" postal system is created by Augustus Caesar by using runners to relay messages from town to town.
- **780** Chinese perfect wood-block printing.
- **800** Arabs develop number system we use today.
- **1045** Movable type is invented by the Chinese printer Pi Sheng.
- **1438** Johannes Gutenberg is the first European to use movable type.
- **1454** Gutenberg prints his first book, the Latin Bible.
- **1826** Joseph Niepce produces the world's first photographic image.

Signals may be easily sent and received around the world through the use of satellite technology. Such instantaneous contact is revolutionizing the way we communicate.

Weather radar computers allow technicians to translate reflected radio signals into weather patterns.

Newspapers, once printed with ink and metal blocks, are now produced with state-of-the-art efficiency on modern printing presses.

1860 The Pony Express covers 1,966 miles using 190 relay stations to carry mail from Missouri to California.

1866 First successful telegraph cable is laid across the Atlantic Ocean.

1876 Alexander Graham Bell invents the telephone.

1877 Thomas Edison invents the phonograph.

1878 Sir William Crookes develops a vacuum tube that produces cathode rays. This tube becomes the forerunner of T.V. picture tubes.

1884 Ottmar Mergenthaler patents the Linotype typesetting machine.

1895 Guglielmo Marconi develops the wireless telegraph.

1906 The first pre-selective juke box to hold 24 discs is manufactured in Chicago.

1909 The 35mm film format is agreed as standard for the movie industry.

1919 The first dial telephones are introduced in the U.S.

1860 | 1870 | 1880 | 1890 | 1900 | 1910 | 1920

The cellular telephone is making telephone communication more accessible to even very remote locations. In the future, it may be possible to send and receive calls nearly any place that people travel.

Many job opportunities are available in radio broadcasting. A few of these include disc jockey, sports commentator, news reporter, and talk show host.

Today's era of information processing was built on the tiny silicon chip. Silicon is the same material that makes up ordinary beach sand.

Technology Time Line

1920 KDKA in Pittsburgh is the first radio station to broadcast presidential election returns.

1929 Vladimir K. Zworykin demonstrates the first completely electronic, practical T.V. system.

1934 The Federal Communications Commission (FCC) is created.

1935 Kodachrome is the first multilayer color film produced.

1939 NBC begins commercial television broadcasts.

1947 William Shockley, Walter Brattain, and John Bardeen invent the transistor.

1952 Techniques for purifying silicon are perfected, thus starting the boom in the electronics field.

1953 T.V. broadcasts in color.

1956 The "Baby Brain" is the first desk-sized computer in the world.

1920 | 1925 | 1930 | 1935 | 1940 | 1945 | 1950

In video post-production, audio and video from a variety of sources are edited and mixed to form a final master tape.

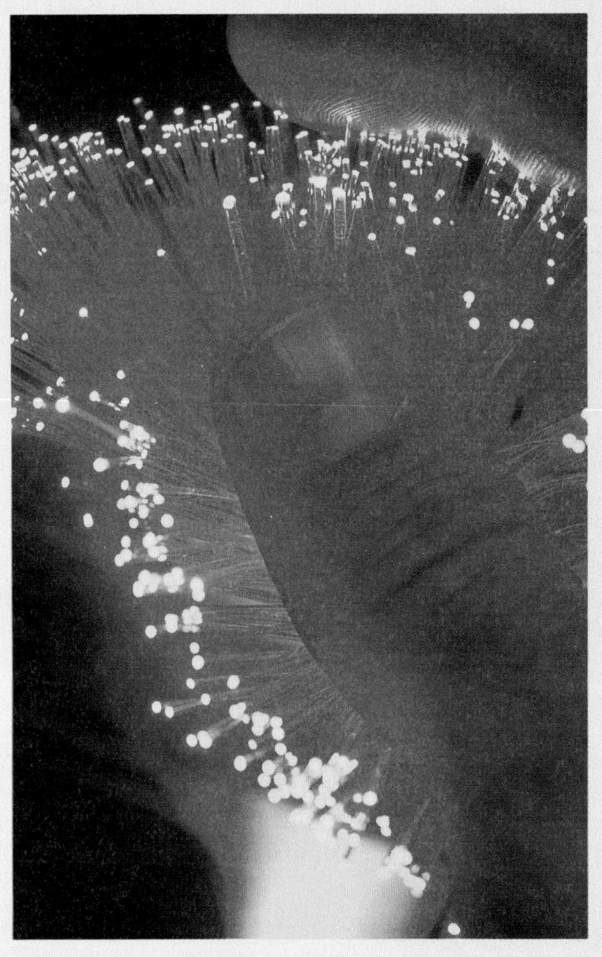

The technology of fiber optics enables light to travel through transparent glass fibers up, down, around corners, or even around the world. A cable with a diameter of no more than three-quarters of an inch can carry as many as 40,000 phone calls.

Computer-aided design (CAD) systems are replacing traditional tools such as drawing boards, pencils, and templates to produce designs faster and more accurately.

1960 | 1965 | 1970 | 1975 | 1980 | 1985 | 1990

1967 A battery-operated cordless telephone is tested in the U.S.

1974 3M's laser printer uses a beam of light to form characters from many tiny dots.

1979 A computerized laser printing method is developed in Britain.

1985 A rock concert called "Live Aid" raises $70 million for the starving people of Africa.

1988 Average American watches 2,300 hours of T.V. a year, or about 6 1/2 hours a day.

1989 The U.S. leads the world in T.V. sets with 98% of households having at least one T.V.

1990s High Definition Television (HDTV) brings incredibly sharp pictures and clear sound to homes across the world.

1991 Over 30 million T.V. sets are tuned in to the evening newscast in the U.S.

Chapter 3

What Communication Technology Is All About

Looking Ahead

In this chapter, you will discover:

- the basic reasons for communicating.
- the meaning of the Information Age.
- how microchips and computers are used in communication.
- some of the ways that communication is important in your life.

New Terms

bar code
binary digital code
bit
byte
circuit
data
data bank
facsimile (fax) system
integrated circuit (IC)
laser
microchip
modem
sensors

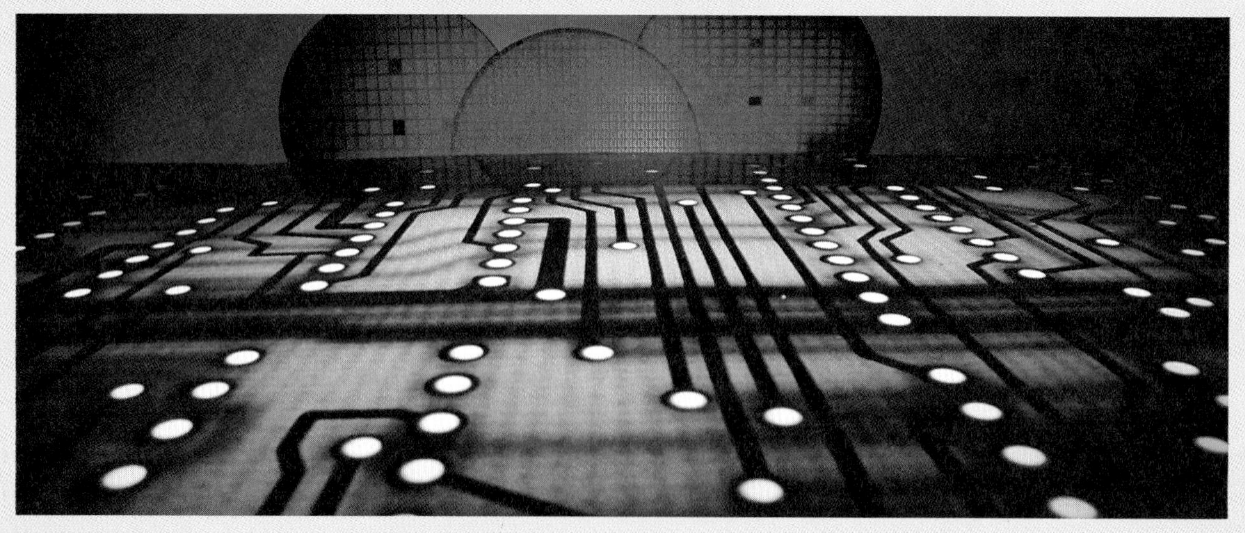

Technology Focus

Optical Cards

Credit cards contain information stored on a magnetic strip, which can be read by simple card readers. Newer cards, called "smart cards," use tiny microchips to carry much more information. However, even as smart cards are beginning to gain acceptance, an even newer pocket-size technology has been developed. This is the optical card, which can store hundreds of pages of text as well as photographs, drawings, fingerprints, and other types of information. This new card was developed by Drexler Technology of California.

A laser is used to record data (information) onto the card. The method is similar to the way music is stored on compact audio discs. Each wallet-size card can hold as much data as would fit on 1,100 typewritten pages. The cost of storing information this way is much lower than with magnetic strips or microchip smart cards.

Information is stored on the card using WORM (Write Once, Read Many) technology. Data can be updated or modified (changed), but not deleted. These rugged optical cards are not affected by magnetism or static electricity. A small part of the card's memory holds a clever program that helps protect against loss of, or damage to, the stored data.

Drexler has also developed a relatively inexpensive optical card reader/writer. It is used to transfer, store, or transmit (send) data.

Optical cards are being tested widely for personal medical records as well as automobile maintenance, personal identification, library catalogs, and stock exchange data.

Drexler has licensed Canon and other companies to manufacture and distribute the cards and readers. The optical card system marks a new generation in information exchange.

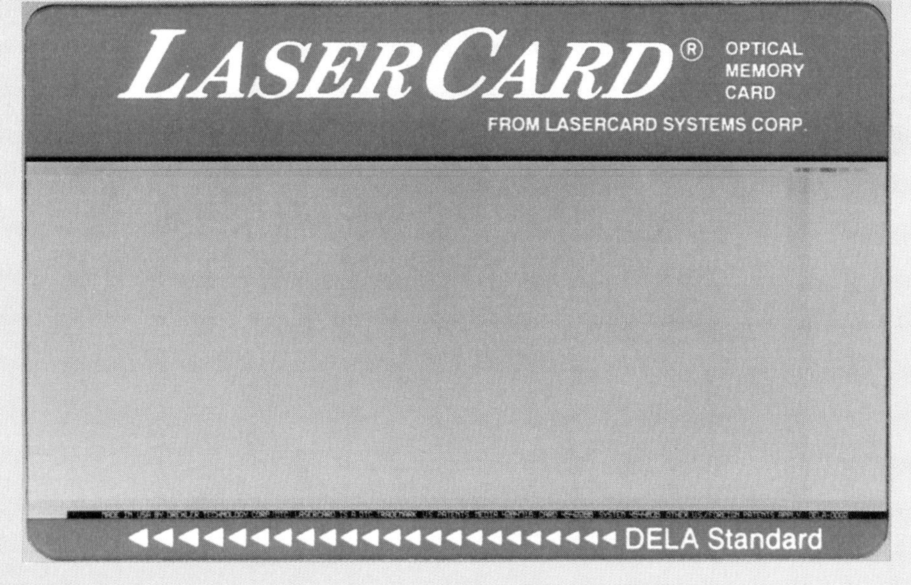

Fig. TF-3. This LaserCard® can store hundreds of pages of text as well as photographs, drawings, fingerprints, and other types of information.

What Is Communication Technology?

Imagine a world without books, signs, radios, and newspapers. What would life be like without a telephone or television? It is hard to picture yourself making it through one day without using some of these tools, yet you probably take them for granted. We depend on the tools and equipment that help us to communicate—to send and receive messages. We depend on communication technology.

As you learned in Chapter 1, communication technology is all the things people make and do to send and receive messages. It's the knowledge, tools, machines, and skills that go into communicating.

Messages come in many different forms. They reach their destinations in various ways. For example, you often send and receive messages from your friends by using the telephone. Some companies send messages to you by advertising on television or radio. A buzzer or other device in a car reminds passengers to fasten their seat belts. When we study communication technology, we are exploring the ways people use their knowledge and skills to create means for sending and receiving messages. It helps us answer questions like these: How does your voice travel 500 miles over telephone lines? How do newspapers get printed every day of the year? What inventions make it possible for you to carry a miniature television in your hand? Fig. 3-1. You will find the answers to these and other questions as we explore communication technology in this and the next four chapters.

Apply Your Thinking Skills

1. Name ten different ways you use communication technology every day.

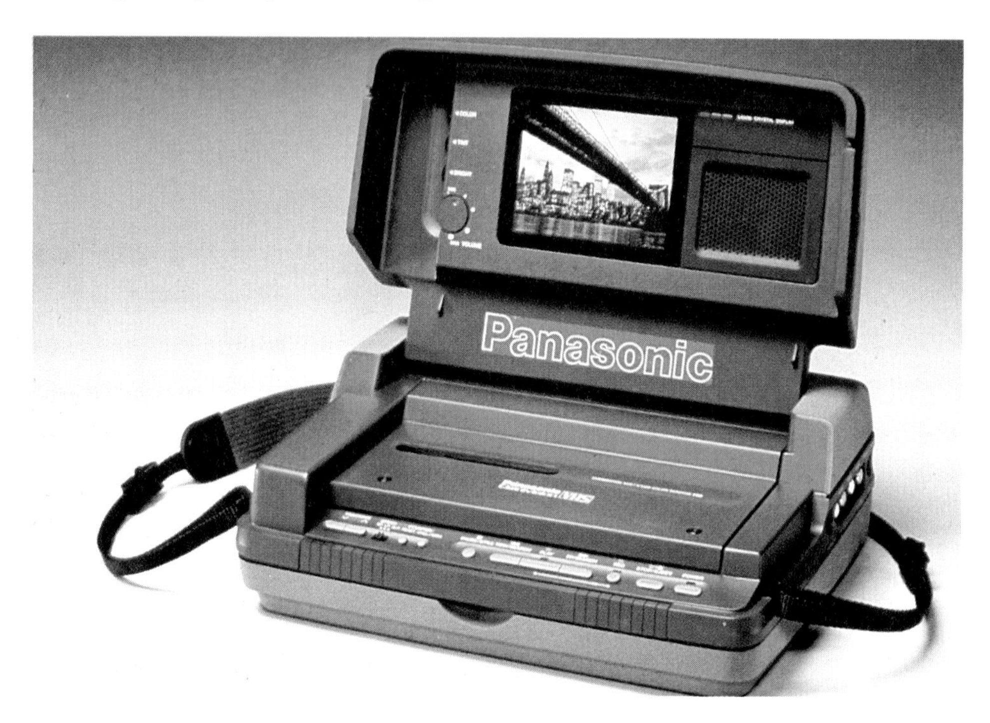

Fig. 3-1. Portable communication tools, like this television, are the result of many advancements in our ever-changing communication technology.

Why We Need Communication Technology

We send messages for many reasons Communication technology extends our ability to send and receive these messages. We all know that it's impossible to talk to someone 10 miles away without some special device. However, this task is quite easy with a telephone, television, or radio. Using these tools, we can send our voices over many, many miles. As you learned in Chapter 1, the reasons we communicate are to inform, educate, persuade, entertain, and control. Communication technology enables people to do these things faster and better. It enables them to:

- *Inform.* People read newspapers, watch television, and listen to the radio to stay informed about a wide variety of things from international politics to local sports, weather, and traffic. Salespeople use car phones to send messages to and receive messages from their offices and their customers.
- *Educate.* In addition to using textbooks, teachers use such things as video programs to help you learn about many subjects. For example, learning about another country is easier when you actually see and hear what it is like.
- *Persuade.* Advertising is an example of using communication to persuade. You can probably remember seeing a television commercial that made you think you wanted a certain product. Fig. 3-2.
- *Entertain.* You play computer games to entertain yourself. You listen to the radio to hear music and jokes. If you're like most teens, you also watch television often to relax and be entertained. The telephone serves the dual communication purposes of entertaining and informing. It's fun to talk with friends on the phone. You can also use the phone to call a store to see if they carry a certain product you need or call your parents to tell them you will be late getting home.
- *Control.* Communication technology plays an important role in controlling machines and tools. You will learn about this in the chapters on manufacturing. Traffic signals are a common example of using communication to control things. Computers and **sensors** (devices that sense things such as metal or pressure) send

Fig. 3-2. Advertisers use many different tactics to persuade you to do or buy something. What kind of strategy is being used in this ad?

messages to traffic signals, controlling when they change from red to green and back to red again. In turn the traffic signals send a message to drivers, thus controlling the flow of traffic.

Apply Your Thinking Skills

1. Think about each of the communication situations below. Determine the purpose(s) of the communication taking place in each one. Is it to inform, to educate, to persuade, to entertain, to control, or could it be a combination of two or more of these purposes? Record your ideas and then compare your ideas to a partner's ideas. Do they match? If not, is one person's answer right or wrong?

a. A street sign that indicates that there is a curve ahead.
b. A burglar alarm that is set off when movement is detected.
c. A pamphlet that discusses the AIDS epidemic.
d. An electronic bathroom scale.
e. An exercise video.
f. A radio program about the "Just Say No to Drugs" program.
g. A telephone conversation about planning a birthday party.
h. A telephone answering machine.

Fascinating Facts

Reuters, today one of the biggest news agencies in the world, once used pigeons to get their news. In 1850, bankers in Berlin, Germany, needed a faster way to get news from the stock exchange in Paris, France. There was no telegraph line from Paris to Berlin, but there was one from Paris to Brussels, Belgium. Paul Julius Reuter, a German bank clerk, trained pigeons to carry messages from Brussels to Aachen, Germany, where another telegraph line could forward the news of the Paris stock exchange to Berlin. This news was communicated by telegraph from Paris to Brussels, by pigeon from Brussels to Aachen, then by telegraph from Aachen to Berlin.

Depending on Communication Technology in the Information Age

The Stone Age was an era when people used stone tools to do things such as hunt for animals and prepare food. Their lives depended on the use of the stone. The current time period is becoming known as the *Information Age*. There has been an explosion of information as a result of our ever-changing technologies. This information affects every aspect of our lives. It has political, social, economic, and cultural impacts. Communication technology gives us access to this vast amount of information.

Our society depends on the communication of information today more than ever before. We rely on communication technology when we listen to weather reports on the television or radio, so we know how to dress for the day. When we shop, there are usually electronic communication devices that send information about the item's price to the store's cash register. When we drive, there are communication devices that tell us when to stop and when to go, how fast to go, that the lane ahead is closed, and that our gas supply is getting low. If we want to find out how to build a deck or choose a place to vacation, we can just get a videotape and put it in the VCR. When a fire occurs or a medical emergency arises, we rely on communication devices to get us help quickly. We pick up the telephone to call for help, and information about us and our plight is quickly transmitted (sent) to the nearest available units that can provide help.

It is not only in emergency situations, such as fires, that we want information to be communicated quickly. In our fast-paced and quickly changing world, we need information as fast and up-to-date as possible. New communication devices are being developed all the time to help us send and receive information faster. For example, there are many new portable communication tools that help make communication faster and easier. Fig. 3-3.

We want as much information as possible, we want it as fast as possible, and we want it to be as clear as possible. You probably take for granted that you can clearly understand people you speak with on the telephone. Do you sometimes get angry when the television picture is fuzzy? That is because you are used to having it clear. Improved technology helps us send and receive clearer messages.

The Information Age is affecting all of us. If we look at it more closely, perhaps we can better understand what this era is all about.

Apply Your Thinking Skills

1. What are some ways the Information Age has benefitted people? How has it caused problems?

The Amazing Microchip

Microchips are the basis of almost every modern communication tool you use today. A **microchip**, also called an **integrated circuit (IC)**, is a tiny piece of silicon that contains thousands of tiny, interconnected electrical circuits that work together to receive and send information. Fig. 3-4. A **circuit** is a path over which electric current or impulses flow.

You can thank the microchip for making it possible to carry a radio, tape recorder, or

Fig. 3-3. Portable communication devices like this car phone help us send and receive messages as quickly as possible.

Fig. 3-4. A microchip contains thousands of tiny, interconnected circuits that work together to send and receive information.

Fig. 3-5. Microchips in this camera control the exposure (amount of light reaching the film and length of time film is exposed to light) and automatically focus the subject.

compact disc player in your pocket. Each of these communication tools contains several microchips. Each chip has a certain job. Some may process information. Others may be assigned for memory only. For example, you might have automatic redial on your telephone. A microchip inside the phone "remembers" the last number dialed. You can simply press one button to redial the number. Fig. 3-5.

Microchips carry a great deal of information, yet they take up very little space; many can be fit into a very small electronic device. This means products can do more, yet be smaller and lighter in weight. Microchips work very fast because the electric impulses have only a very, very short distance to go. This means products work faster. Microchips cost little to operate and are cheap to make. Very little material is required and the chips can be cheaply produced in large quantities. Fig. 3-6. The use of microchips has made fast communication more affordable and available to more people.

A computer has many microchips. Computers use microchips to store information, solve problems, and run all types of programs. Fig. 3-7. Computers help us find information quickly. Computers are the very heart of the Information Age and of our modern communication system. Understanding how a computer works will help you understand how much of our modern communication system works.

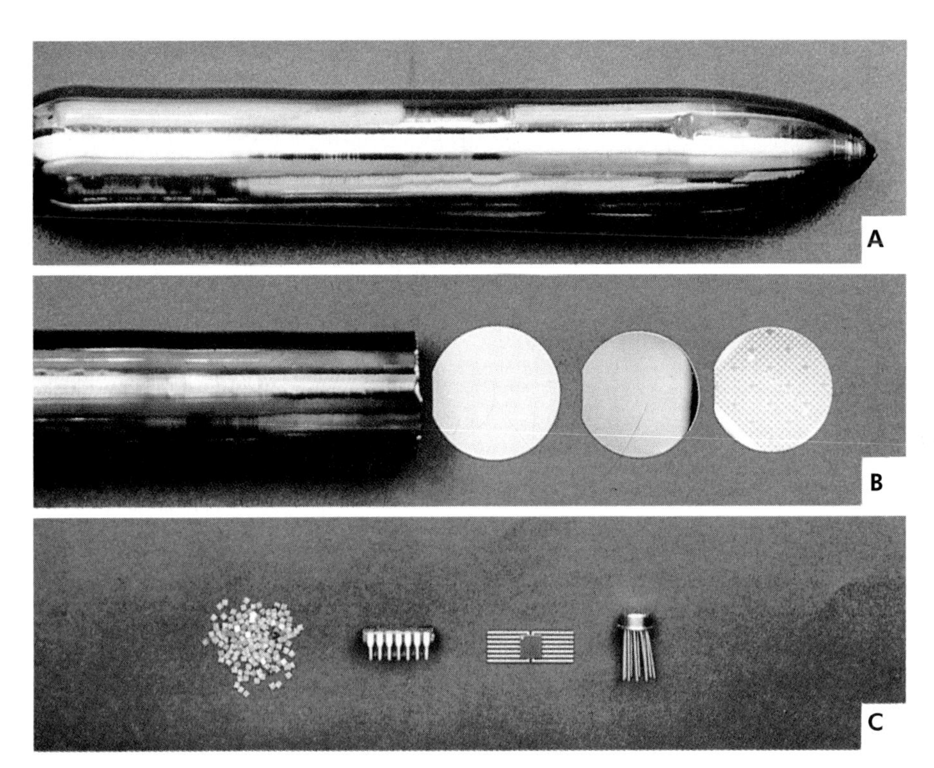

Fig. 3-6. Steps in the production of microchips.
A. A silicon ingot is grown in a vacuum oven.
B. The ingot is cut into thin slices, called wafers. The first wafer shows what the silicon looks like when it is first cut. The second wafer has been polished. The third wafer has had many integrated circuits chemically etched (engraved) into its surface.
C. A finished wafer is cut up into microchips (ICs) and the chips are put into various types of packages for easier handling.

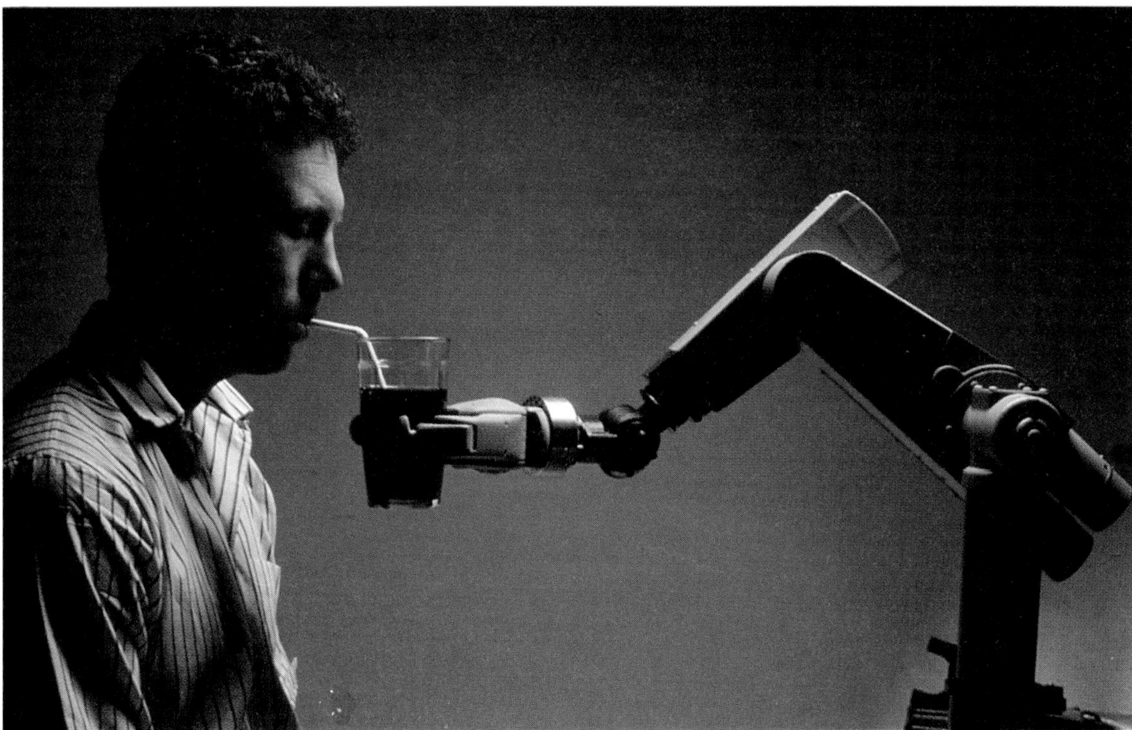

Fig. 3-7. Some computers can respond to voice commands. A person who is disabled can control robot arms and hands simply by speaking.

Apply Your Thinking Skills

1. What effect has the development of the microchip had on the size of and portability of products?

Fascinating Facts

The world's first all-electronic digital computer, designed at the University of Pennsylvania, was completed in 1946. It could perform 5,000 calculations a second. ENIAC, short for Electronic Numerical Integrator And Computer, used 18,000 vacuum tubes, weighed 30 tons, and occupied an entire room. This electronic giant began a revolution in our communication system.

Computers in Communication

Computers are used in nearly every communication system today. The main advantage of a computer is that it can handle a great deal of information quickly. Computers are not smarter than people. They are simply much faster than people. A computer can do millions or billions of calculations in seconds. However, all of its information must first be put into a form that the computer can "understand."

The Language of Computers

A computer has a language all its own. Actually, the language is a code, consisting of two factors. This is because the computer knows only two signals. It knows *on* and *off*. The factors can be written using two digits, 1 and 0. This code is called a **binary digital code**. (*Binary* means having two parts. A *digit* is any number between 0 and 9.) All information is changed electronically into this binary digital code so the computer can "understand" it. This process of changing information into a number code is called *digitizing*. Pictures, sounds, numerals, words, and letters become series of 1s and 0s. Figs. 3-8 and 3-9.

The code enters the computer as electric pulses. These turn tiny electronic switches on (1) and off (0), creating different combinations of circuits. Electricity flows through the circuits as the information is processed by the computer. Each on or off pulse of the code is called a **bit**. The term *bit* comes from *bi*nary digi*t*. Most machines combine eight bits into a **byte**. Then information can be handled in larger units.

The ability to convert everything into bits and bytes (digital code) that can be used by the computer is the key to the Information Age. Once something has been converted into digital code, it can be used in many ways. It can be stored, retrieved, sent, or altered.

Storing and Retrieving Information

Computers help us to store and retrieve data. **Data** is another word for information. The data can be pictures, words, sounds—whatever.

Digital data can be stored in several ways. You have probably seen 3 1/2" or 5 1/4" computer disks. These store digital data as a pattern of magnetized metal particles. Many computers also have hard disks. These are also magnetic media, but they hold much more data than the 3 1/2" or 5 1/4" disks.

Data can also be stored magnetically on tape. You have probably heard of digital audio tape. The sound on this tape is recorded as digital information.

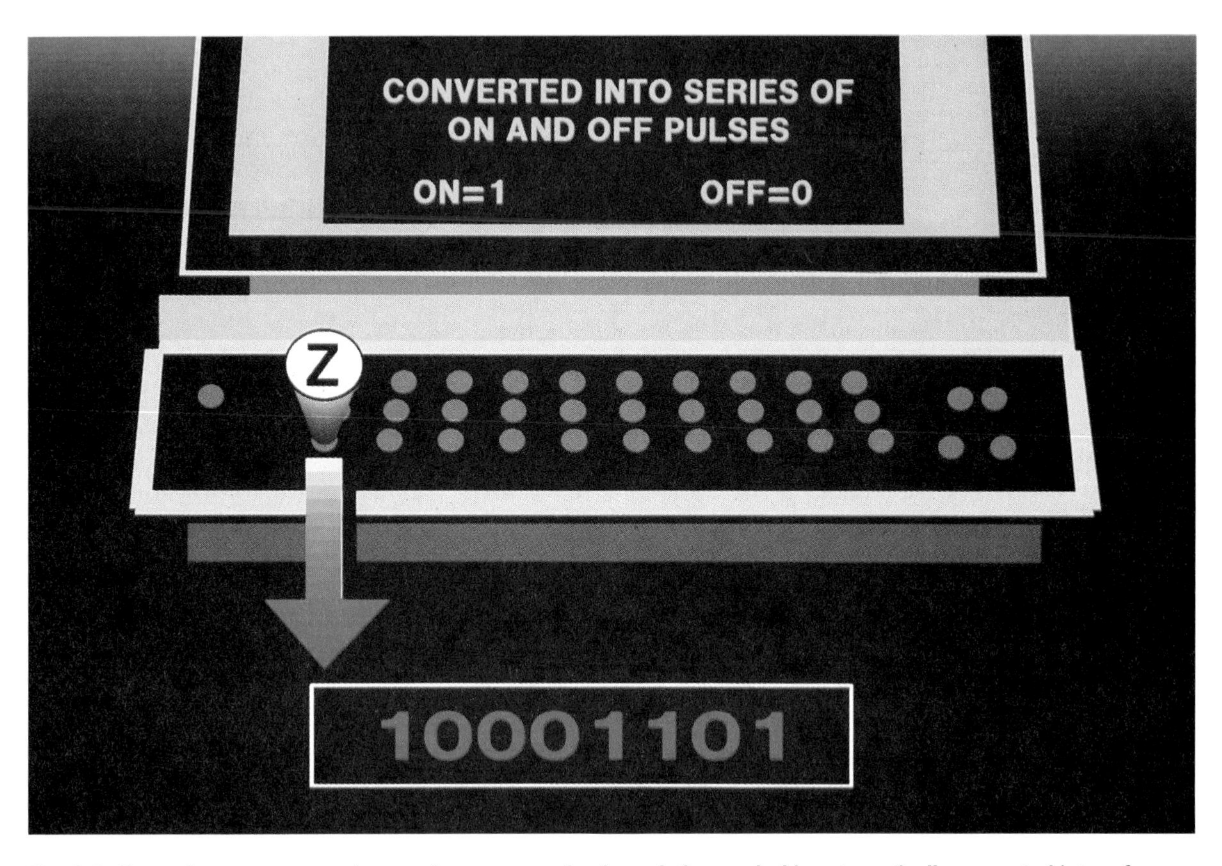

Fig. 3-8. Every time you press a key on the computer keyboard, the symbol is automatically converted into a form that the computer can understand, a series of on (1) and off (0) pulses.

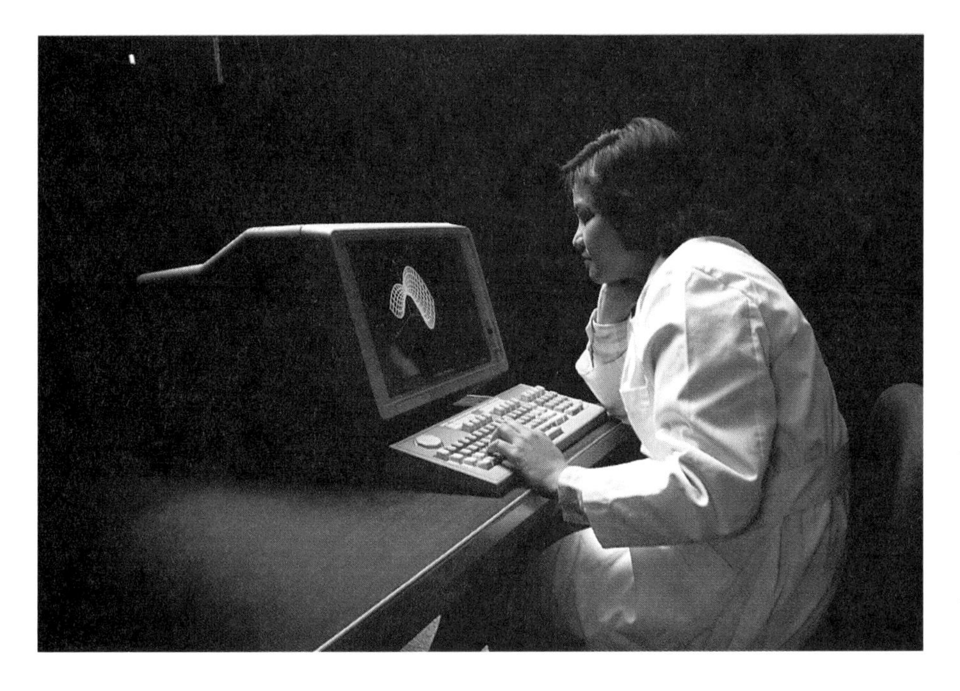

Fig. 3-9. Once the computer can "understand" what we want, it can help us do many things with our information. For example, interesting graphics can be created with the help of a computer.

Compact audio discs also store digital information, but not in magnetic form. Compact discs are optical storage media. To store data in this manner, an optical device called a laser is used to burn tiny pits into the surface of the disc. (A **laser** is a narrow, high-energy beam of light that concentrates a lot of power in a very small space.) Fig. 3-10. The pits form a digital code. When the disc is played, another laser reads reflected light from these pits. The code is converted back to sound, and we hear the music.

Videodiscs work the same way, but they store moving pictures as well as sound. CD-ROMS, which store huge amounts of text and pictures, are still another type of optical medium. Some computer systems combine CD-ROM technology with videodisc technology. The result is CD-I (compact disc-interactive). A CD-I system has sound, pictures, and video motion.

Whether the digital information is stored on computer disks, audio discs, videodiscs, or tape, computer technology is used to record the data and to retrieve it for our use.

Sending Information

There are many ways that data is transferred or moved around today. This ability to move information around quickly enables people and companies to do things that might otherwise be very difficult or time consuming.

Sending Voices

Modern telephone systems use computers to change voices into pulses which are converted into electronic signals that are sent over long distances. When someone says "hello" to you on the telephone, it is not the actual voice that you hear. Instead, you hear the sounds that are re-created from the pulses that were sent. Sometimes these pulses are electric pulses. Other times they may be light pulses.

Sending Images

Computers can change words and pictures into digital information and send it by telephone. For example, pictures are sent

Fig. 3-10. A narrow, high-energy laser beam is used to make audio discs and videodiscs.

as digital information every day. In order for newspapers to print pictures from distant nations like Iran and Russia, the pictures must be sent to the United States as digital information using a facsimile system. (*Facsimile* means copy or duplicate.) The **facsimile**, or **fax, system** turns the picture (or words) into a number code and sends this data over telephone lines to another fax machine. The fax machine at the receiving end converts this information back into pictures (or words) that can be printed in the newspaper. Fig. 3-11.

Let's look at another example. Many large companies have offices in other parts of the world. Suppose the manager of a store in New York City wants to check the supplies of a store in Japan. This could be done through the use of the computer and telephone lines.

Most major companies use data banks. A **data bank** is a central computer that stores the information from many smaller computers. The New York City manager uses her computer and modem to call the central data bank. (A **modem** is a device that sends computer data over telephone lines to other computers.) The data bank could be located in still another city, such as Dallas, Texas. The central computer sends the requested information back over the telephone lines. The New York City manager sees on a computer screen the list of supplies in Japan.

Using the Bar Code

A common data transfer method used today is the **bar code**. It is the striped code you see printed on most products. Each product has its own pattern of stripes. A computer is programmed to "read" the codes. The cashier at the store passes the code over a window. A laser scanner under the window "reads" light/no light from the

Fig. 3-11. A facsimile system can be used to send copies of documents from one city to another. The document arrives at its destination almost immediately.

Fascinating Facts

Bar codes—now seen everywhere from your neighborhood supermarket to large manufacturing plants—were first used back in the 1960s by the railroad industry. This industry, with more than two million freight cars in the United States and Canada, attached the codes to the cars to keep track of how they were being used and where they were located.

stripes. The information is sent to a computer, which locates the price of the item in its memory. The computer sends that data to the cash register. Fig. 3-12. Then the computer writes the product's code number to a file that keeps track of how many of that product have been sold.

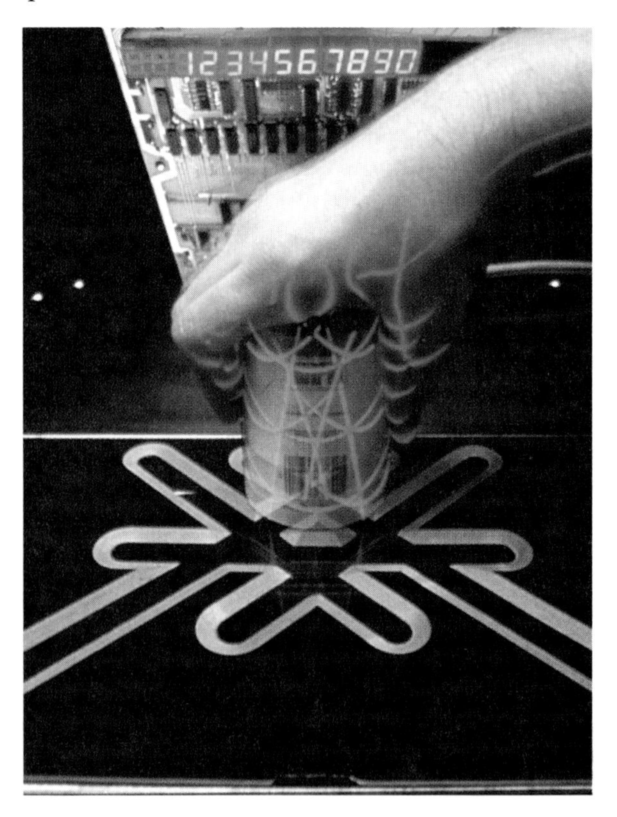

Fig. 3-12. The laser scanner under the window reads the bar code on the product and sends the coded information to a computer. The computer checks its memory for the product's cost and sends that information to the cash register.

Altering Information

When information has been reduced to binary code, it becomes fairly easy to change. Pictures, for example, can be digitized and then altered. Have you wondered how you would look with a different nose? There are computer programs that let you scan your photograph, erase your nose, select a new one from a menu of different noses, and print a picture of the new you.

Changing pictures in this way can be fun. It can also be useful. Plastic surgeons can show patients how they would look after surgery. Police departments can produce pictures of suspects based on the descriptions of witnesses. However, digital alterations also pose ethical questions. Suppose a news photographer takes a picture of a senator surrounded by his political supporters. When the picture arrives at the news bureau, the editor notices that a tree behind the senator appears to be growing out of his head. She uses her computer to remove the tree. Is this ethical? Is the picture still true? What if it wasn't a tree behind the senator, but a reputed gangster who appeared to be among the senator's supporters?

As you'll see in the next chapter, communication technology has impacts on our lives. Along with the benefits, there are problems of ethics, privacy, and health.

Apply Your Thinking Skills

1. Today there are many kinds of data banks being used. Some store personal information such as credit history and medical history. Some people fear that storing personal information in data banks will eventually destroy people's privacy. Discuss the consequences of having this information readily available.

Global Perspective

Restoring the Past/A United Future

Communication has brought the world's cultures closer together. There is less separation of "East" and "West." We have found that we all share an appreciation of beauty and great works of art. Regardless of an artwork's cultural origin, we all admire and enjoy it.

Possibly the greatest work of art in the Western Hemisphere is in the Sistine Chapel in Rome, Italy. The ceiling and upper walls of the huge chapel are covered with scenes and characters from the *Bible* painted by Michelangelo, one of the greatest artists of all time.

Michelangelo began the work in 1508 and finished four years later in 1512. During the nearly 500 years since that time, dust, soot, and other grime have combined to increasingly obscure the work's original brilliance. Conditions became so bad that, in the late 1970s, the decision was made to undertake restoration. Art historian Fabrizio Mancinelli, of the Vatican Museums, was made director and Gianluigi Colalucci was made chief restorer. Armed with the latest technology, Colalucci and his team of restorers set to work in 1980.

Restoration was carefully planned. Work was done in sections. Each section was first examined with state-of-the-art scientific instruments. The results were then entered into a computer. All information was carefully analyzed and classified and was used as work progressed. Cleaning materials were selected carefully. They were then painstakingly applied and carefully wiped away, removing 500 years' worth of grime. What emerged from under the grime was amazing! Colors that were thought to be somber and subdued blazed forth with unsuspected brilliance.

Fig. GP-3. The technique that Michelangelo used to create his paintings in the Sistine Chapel is called "fresco." Fresh plaster was placed on the surface and paint was applied while the plaster was still wet and a chemical bond was formed between the two as they dried.

Restoration of the paintings in the Sistine Chapel was completed in 1989. Today, and for years to come, people from all over the world can once again be awed by the artistic genius of Michelangelo, which has been preserved in the work he left behind.

Extend Your Knowledge

1. Another great artist of the 15th and 16th centuries was Leonardo Da Vinci. Besides being an artist, he also had ideas for machines and other inventions, including a flying machine. Find out more about his ideas for technology that did not become a reality for many, many years.

Chapter 3 Review

Looking Back

Communication technology is all the things people make and do to send and receive messages. We communicate to inform, educate, persuade, entertain, and control. Communication technology helps us do these things easier, faster, and better.

We live in what is often called the Information Age. Our society is more dependent on the communication of information today than ever before. The microchip and the computer are key technologies in the Information Age. Microchips are the basis of almost every modern communication tool we use. Computers use microchips to receive, store, send, and alter information; solve problems; and run programs. The main advantage of computers is that they can handle a great deal of information quickly, enabling people and companies to do things that might otherwise be very complicated and extremely time consuming.

A computer has a language all its own. All information is changed electronically into a digital code the computer can understand.

Review Questions

1. Name the five purposes of communication. Give at least two examples of each.
2. Briefly explain why today is becoming known as the Information Age.
3. What is a microchip? Give three examples of communication devices you commonly use that have microchips.
4. Discuss the advantages of using microchips.
5. Briefly explain what the binary digital code is and how it is used to make computers "understand."
6. Define the terms bit and byte.
7. Define data. What is digital data? What are three ways digital data can be stored?
8. Describe how a facsimile system works.
9. What is a data bank?
10. Describe how bar coding is used.

Discussion Questions

1. Select a television or radio program that you watched or listened to recently. Name the program and give a brief description of it. Then, identify which purpose(s) of communication it served.
2. Many people feel that children and young adults watch too much T.V. Present your point of view on this subject and explain your position.
3. What is the purpose of the message you send when you program a VCR to record a TV program or you play a video game? Machines can send messages, too. What is the purpose of the message you receive from a VCR? From a video game?
4. Imagine that you are an employee of a small grocery store. How would you convince the store owner that it is time to update the store's communication system by purchasing a computer and a facsimile machine?
5. Discuss the impacts of computers in the classroom. What are the advantages of computers in the classroom? Are there any disadvantages?

Chapter 3 Review

Cross-Curricular Activities

Language Arts 

1. Television is a popular form of communication for many teenagers. Keep a daily log of your television viewing for a week. In a brief report, summarize your viewing habits. Be sure to include the number of hours spent and types of programs watched. Were all of the programs for the purpose of entertainment, or were any other purposes of communication served?
2. Ask ten people of various ages to identify the most important communication device currently in use. Ask each person to give reasons for his or her opinion. Compare their answers. Write a brief report discussing your findings.

Social Studies 

1. Find three inventions that evolved out of the Industrial Revolution (which began in the 1860s) that helped to bring the world closer together. Write a one page paper explaining how these inventions helped to improve communication throughout the world.
2. Go through your home and list all the electrical appliances and gadgets you can find. Then go back and list the ones created in the last 35 years. Then see how many have been developed or greatly changed in the last 10 years. Which ones use microchips?

Science 

1. Much of our communication depends on messages in the form of sound. You can measure the speed of sound. Measure off a distance of 1500 feet outside. Have a friend stand at one end of the measured distance and bang two pan lids together overhead, so you, standing at the other end, can see when it happens. At that instant, start a stopwatch. When you hear the sound of the pan lids, stop the stopwatch. Divide the distance (1500 feet) by the time required for the sound to reach you (number of elapsed seconds). That will give you a value for the speed of sound. Use a reference work to find the speed of sound. How close was your calculation?

Math 

Obtain a copy of your family's telephone bill.

1. Check its accuracy by totaling all usage charges, special service fees, and taxes. What percentage of the bill is in the form of taxes? How much money was spent on directory assistance?
2. Analyze your family's calling habits. What changes could your family make in their calling habits to save money each month?

Chapter 3 Technology Activity

Using Graphics to Persuade

Overview

When we communicate, we pass along information. This may be done for many reasons. One reason is to teach. When your instructor communicates with you, it is to provide you with knowledge. Another reason is to persuade. When a manufacturer communicates by means of a TV commercial, it is to persuade you to buy a product. In this activity, you will use your communications skills to do both.

As you know, technological systems have had negative impacts on our environment. If you could create a design that would inform people about these negative impacts and persuade them that something must be done, what would your design look like? Let's find out.

First, you will create a design illustrating how you feel about technology's effect on our environment. You will try to persuade others to share your views. Then, when your design is finished, you will transfer it to a T-shirt or sweatshirt. Fig. A3-1.

Goal

To create a graphic design that informs others about impacts on the environment from technological systems. The design will be transferred to a T-shirt or sweatshirt.

Equipment and Materials

sketch paper and pencils
computer
CAD or other graphics software with the ability to make a mirror image
printer
heat transfer printer ribbon (color or black)
heat transfer colored pens (if using black ribbon)
T-shirt or sweatshirt
iron
ironing board

Procedure

1. *Safety Notes:*
   - When using an iron to transfer your design, do not let it rest against the fabric for too long or it will create a scorch mark.
   - When you are finished ironing, be sure to turn the iron off. Irons left turned on can cause fires.
2. Using sketch paper and pencils, make some rough sketches of possible designs. Your design may include words or it may be only a picture.
3. Select the design idea you like best. Using the computer and the proper software, create a finished drawing.
4. Flip the design image so that it appears backwards on the screen, as if seen in a

mirror. This is done so that when the image is transferred to a shirt, it will be flipped again and will appear as it should.

5. Install a heat transfer ribbon in the printer. Print your design.
6. If your ribbon was black, use heat transfer colored pens to color your design.
7. Read the instructions for the printer ribbon that tell you how to transfer the design. Follow those instructions to apply the design to your T-shirt or sweatshirt.

Evaluation

1. Why do you think others will be persuaded by your design?
2. If you could make another design for the same purpose, what would you do differently?
3. Did using the computer make your work easier or more difficult? Why?

Credit: Eric Thompson, Galesville, Wisconsin

Fig. A3-1. This is an example of a T-shirt that you're going to make in this activity.

Chapter 4

The Communication System

Looking Ahead

In this chapter, you will discover:

- that the communication system consists of many subsystems.
- the inputs, processes, outputs, and feedback in the communication system.
- some positive and negative impacts of communication systems.

New Terms

communication channel
electromagnetic radiation
subsystems
telemarketing

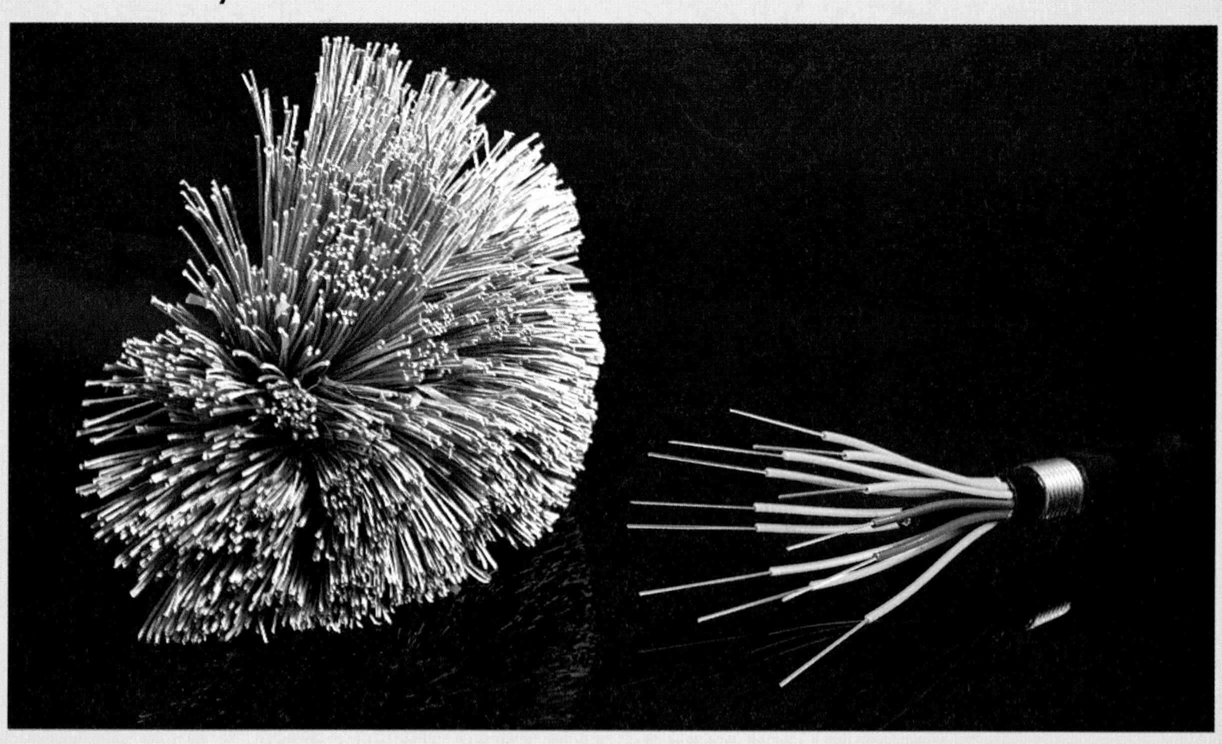

Technology Focus

Optical Computers

Imagine your TV, VCR, telephone, fax machine, and computer all assembled together in one box. Imagine that this box is connected with other similar boxes by optical fibers to form a giant network—"an intellectual power grid." All the world's art, music, books, film, and video can be brought into your home within seconds. Alan Huang, head of AT&T's Bell Laboratories' Digital Optics Research Department, and fellow scientists and engineers at Bell Labs are working on a revolution in computer and switching technology that will make this dream possible.

Huang and his coworkers are designing optical computers that will have the necessary speed to process and transmit data and images to accomplish the tasks of the future. Photons (units of light energy) replace electrons in these new computers. Photons can process information faster than electrons. They travel faster and can pass through each other without interference. Also, while electrons are limited to traveling through wires, photons can travel through free space (air) as well as through optical fibers.

S-SEEDS (Symmetric Self-Electro-optic Effect Devices) take the place of microchips in the new computers. S-SEEDS are optical switches with a potential speed of one billion operations per second. They also require very small amounts of switching energy. Huang says that computing with optics will be "1,000 times faster" than with electronic computers.

With this new intellectual power grid, you will be able to locate and communicate with any person at a minute's notice and even see them while you chat. With simple user interfaces, you will be able to dictate to your grid, or type or draw on it, or touch it. You will be able to instantly search the world's libraries to satisfy your interests.

Alan Huang says the optical processor "is an advance made possible by outstanding people with diverse talents working in a place supportive of research."

Fig. TF-4. Dr. Alan Huang, head of the Optical Computing Research Department at AT&T Bell Laboratories, in the laboratory where he and his colleagues built a digital optical processor.

Communication Subsystems and Channels

The communication system is made up of many smaller systems called **subsystems**. For example, there is the telephone system that allows personal communication over long distances. There are radio and television broadcasting systems that enable a message to be sent to millions of people at the same moment. There are computer systems that store, send, and receive large amounts of data. Each of these systems meets our demands for fast, efficient, and effective communications. Fig. 4-1. Often, many of them work together or are used in conjunction with each other so we can meet these demands.

All of the subsystems include a message, a sender, a communication channel, and a receiver. Fig. 4-2. A **communication channel** is the path over which a message must travel to get from the sender to the receiver. The channel might be a telephone line, the integrated circuit of a microchip, or sound waves.

In order to understand how the communication system works, it is useful to take a closer look at each of its basic parts. Like all systems, the communication system consists of four basic parts that work together to achieve a goal. In communication, the goal is to send or receive a message. The parts are the inputs, processes, outputs, and feedback. Fig. 4-3.

Apply Your Thinking Skills

1. What different types of communication systems does your school have?

Fig. 4-1. Many types of communication subsystems are used every working day at the New York Stock Exchange. Here, people depend on moving information quickly to process orders and make trades.

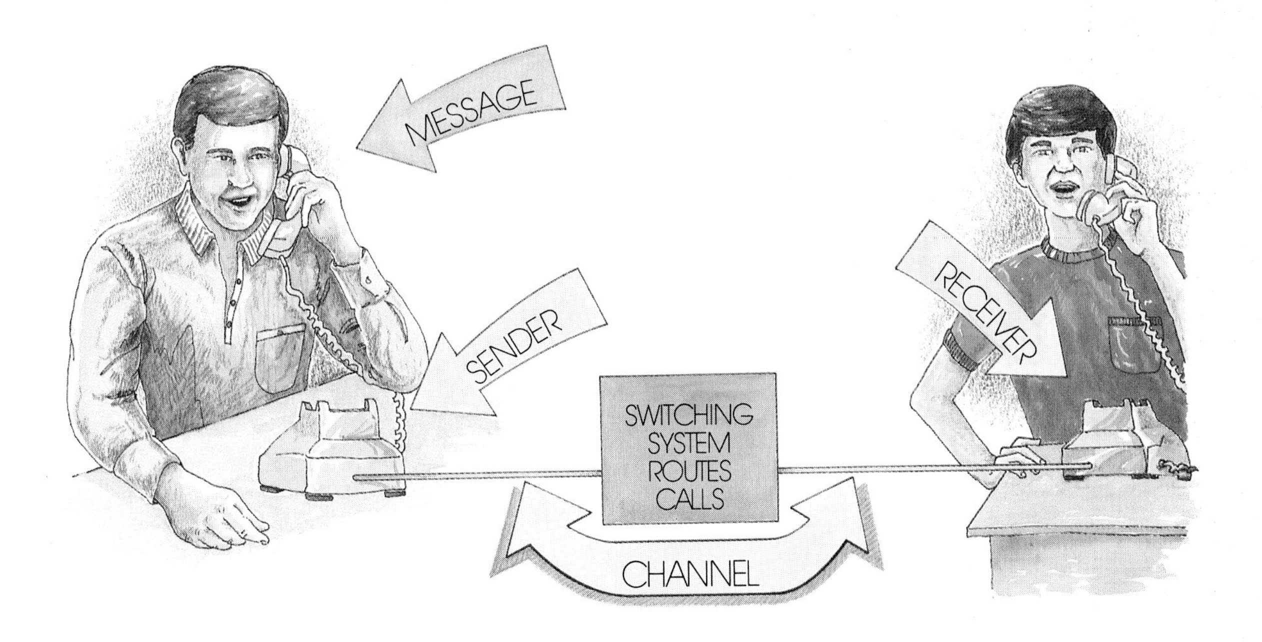

Fig. 4-2. Every communication system includes a message, sender, communication channel, and receiver.

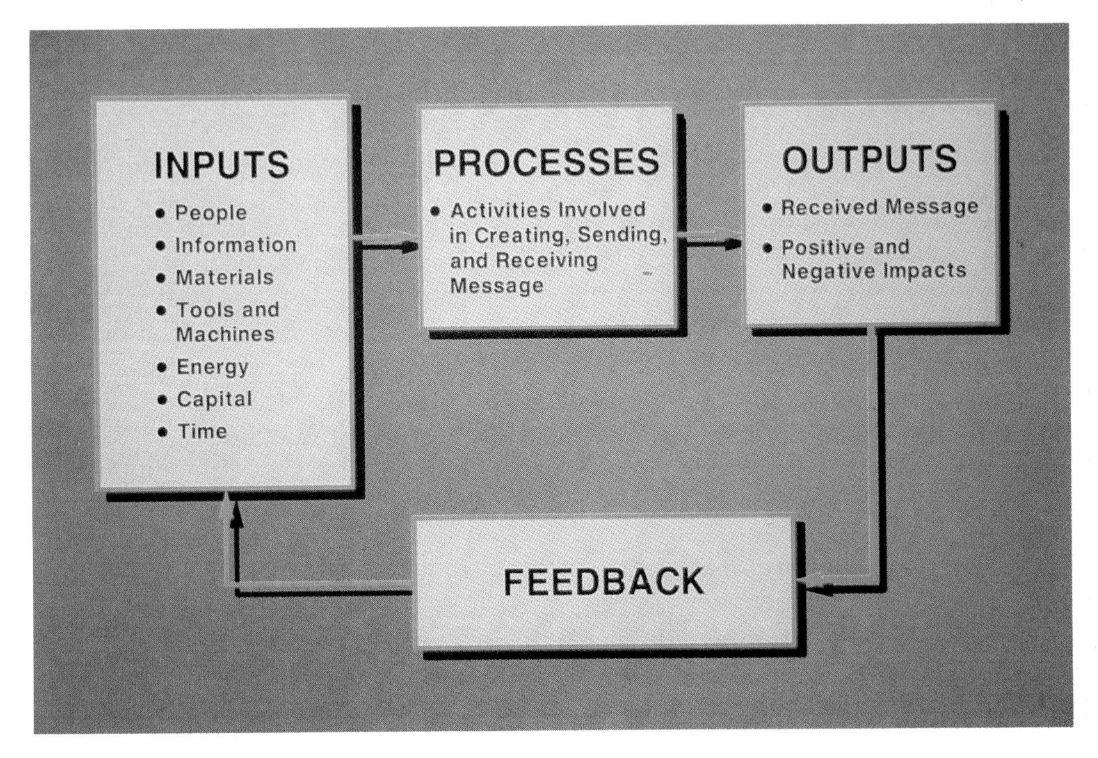

Fig. 4-3. The communication system.

Inputs

As you learned in Chapter 2, the inputs are the seven resources that provide support or supplies for the system. In the communication system, these include all the things that are needed to start or to create a message and to provide a means for the communication processes to be carried out. The communication processes we use and the outputs we get depend on the resources available.

The seven inputs that are used in communication systems include:

- People
- Information
- Materials
- Tools and machines
- Energy
- Capital
- Time

Let's look at how these resources are used in communication.

People

It is people who usually create the message that is to be sent. In most cases, it will be people who will receive and use the message. It is people who provide the knowledge, skills, and creativity needed to develop the technologies for sending and receiving messages. Many people operate communication devices as part of their jobs. Fig. 4-4. As you learned in Chapter 3, we all use and depend on communication devices and systems as part of our everyday lives.

People with a wide variety of skills and interests are needed to work in the communication industry. Many different types of careers are possible. People are needed to design and engineer communication devices. Specially trained technicians are needed to keep communication devices and systems in good operating condition. People are needed to manage communication subsystems. The communication industry also offers opportunities for work in television, radio, recording, and publishing. Computer operators, engineers, photographers, printers, and many other types of workers are needed. Fig. 4-5. If you think some of these areas are interesting, the communication industry may be for you.

Information

The people who design communication equipment and devices need information about a variety of technologies. People who operate these devices need information about how to use them to properly send or receive messages.

Designers of modern electronic communication devices must understand things about technologies such as microprocessors, digital information, photonics, electricity, and mechanical engineering. In addition to needing graphic expertise, graphic designers need to know about typesetting systems, digitization, laser scanning for color separations, and computer software and hardware. Technicians rely on their understanding of electronic circuitry, mechanical controls, signal modulation, and signal transmission to install and repair communication devices.

However, technical information is not the only source of information. People often forget about the extensive pool of general information that they draw upon each day to do their jobs. People who design, use, maintain, or manage communication systems need knowledge and skills in reading, writing, mathematics, science, and human relations. They need this information to make decisions, solve problems, access more information, relate to co-workers, and perform their jobs well.

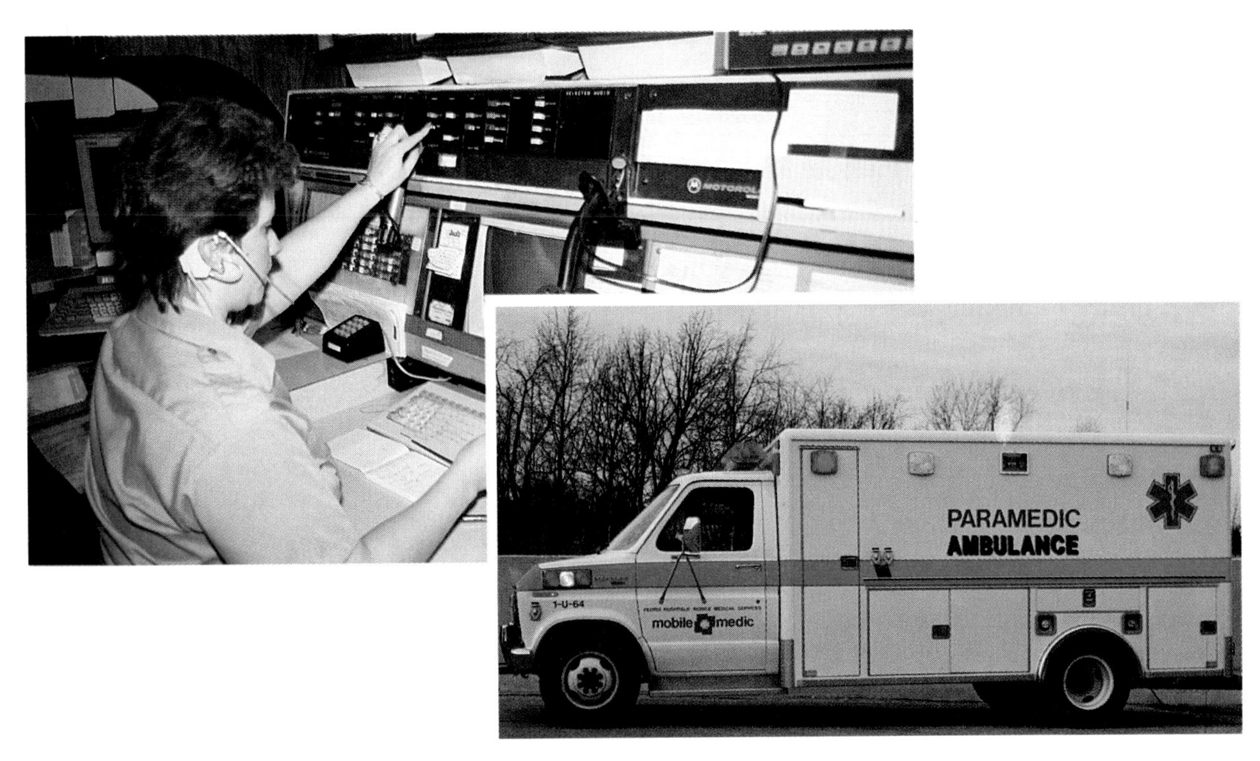

Fig. 4-4. This operator answers 911 emergency calls and then transmits needed information, such as the nature of the emergency and the address, to emergency vehicles. The needed emergency teams can then go quickly to the aid of the victim.

Fig. 4-5. This cameraperson films people and events for a local news program.

Materials

Throughout history, we have relied on access to many kinds of materials in order to create, send, and receive messages. We use natural resources such as trees, chemicals, oil, ore, and silicon to create paper, inks, plastics, metal, and microchips. In turn, we use enormous amounts of these materials in order to create the means we use to send and receive messages.

We rely on paper products for printing newspapers, pamphlets, photocopies, books, and magazines. The ink used for printing consists mainly of linseed oil and chemically manufactured pigments. Plastic materials are used in nearly all areas of communication. Video tapes, diskettes, photographic film, and compact discs are all made from plastics. Think about all the plastic that goes into making telephones, computers, and radios. Miles of metal wires and cables are laid each year to connect communication devices together. Metal is used to construct large towers that are often used to transmit (send) and receive radio signals. Fig. 4-6. Millions of microchips are integrated into useful machines that help us to control how things are made or done.

Fig. 4-6. Metal is used to construct antennas such as this tower, which sends and receives radio waves.

Fascinating Facts

For more than 5,000 years, people have been communicating by writing. In early days, they used materials that were on hand, from clay tablets to palm leaves to silk. By 2500 B.C. the Egyptians were using "writing paper" made by cutting stalks of the papyrus plant into strips and pressing them into sheets. In A.D. 105, the Chinese developed the first true paper, a material made of tree bark, hemp, rags, and fish nets! For centuries, paper was made from rags, first by hand and then by machine. In the mid-1800s, inventors developed methods of making paper from wood pulp. Wood pulp is still used today to make paper.

Tools and Machines

Consider the many kinds of tools and machines that are needed to communicate today. It takes both simple and complex tools and machines to send and receive messages. For example, simple drawing tools are often used to make very detailed technical drawings that communicate ideas about how things should be made or built. Fig. 4-7. These types of drawings can also be made using a computer and a device called a plotter. Fig. 4-8. You learned in Chapter 3 about many other ways computers are used in communication systems.

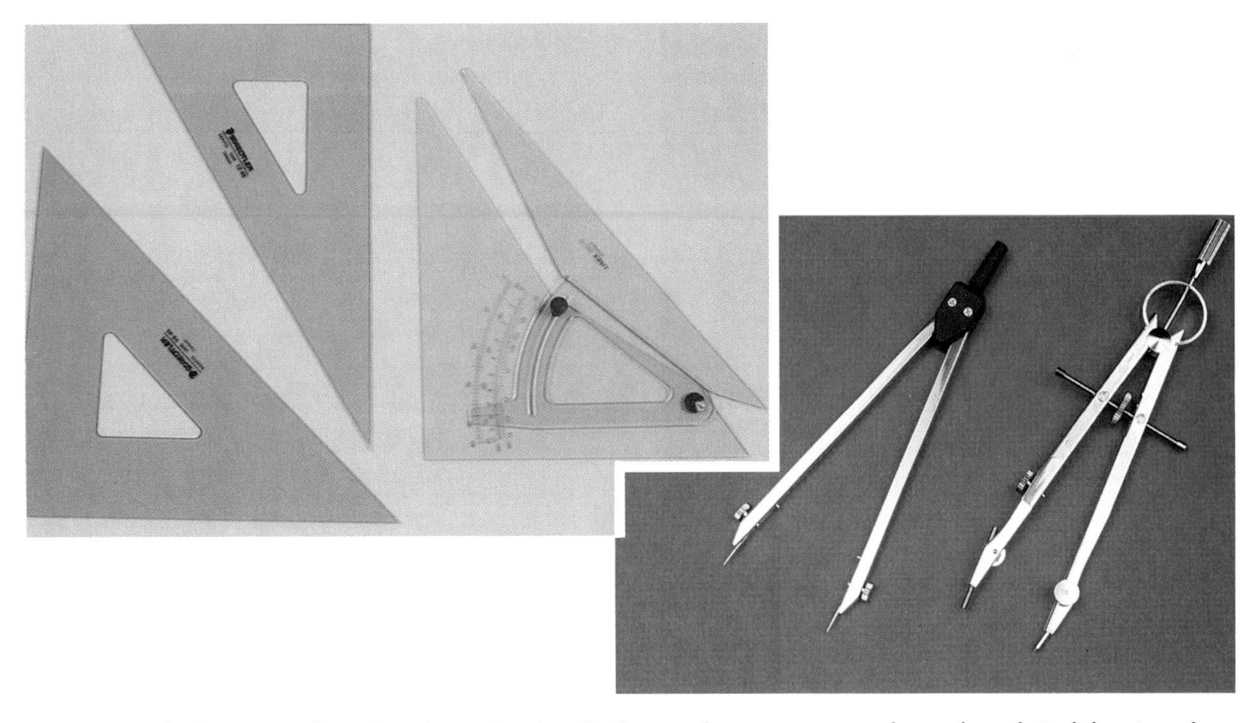

Fig. 4-7. Simple drawing tools such as these triangles, dividers, and compass are used to make technical drawings that communicate ideas and plans about how things should be made or built.

Fig. 4-8. Detailed drawings that communicate exactly how things should be made or built can also be made using computers and plotters. A drawing is created on the computer. Then the computer causes the plotter to move a pen point across a piece of paper or film to produce a hard copy drawing.

Other types of tools and machines include those used in printing such things as newspapers and magazines. A wide variety of cameras are used in communication. Some types are used to take pictures that will be used in printed matter. Other types are used for television broadcasts. Special tools and machines are needed to change voice signals into electric pulses capable of being sent over telephone lines. Radios, televisions, tape players, and VCRs are other examples of tools and machines used to carry out communication processes.

Energy

People involved in communication use their mental and physical energy to provide their special input into the system. No form of electronic communication would be able to begin without some kind of energy. These forms of energy—such as mechanical energy and light energy—must be changed into electrical energy to power the communication devices that send and receive messages.

Capital

As you learned in Chapter 2, capital includes the money, land, and equipment needed to set up and operate a technological system. Money is needed to buy the equipment, pay the workers, and pay for the energy needed to operate communication systems.

Time

Time is needed to design and develop new communication technologies. It takes time to carry out communication processes. Many of the new communication technologies were developed to save time, a very valuable resource. For example, you read in Chapter 3 that computers can make millions of math calculations in seconds. Think how long it would take humans to make *millions* of calculations!

Apply Your Thinking Skills

1. Imagine that you are a member of a social club. You want to promote an upcoming dance. The club members decided to tape a creative announcement that will be played on the school's public address system. Describe all the inputs that you will need.

Processes

Processes are all the things done to or with the inputs in a communication system in order to get the desired result, or output. Processes are all the activities involved in creating the messages, sending them, and receiving them.

This part of the communication system may include many types of activities. Using a computer to make math calculations, designing advertisements, and sketching ideas for new products are all communication processes. All the tasks involved in getting newspapers and magazines printed, from the photography to the writing to the printing, are communication processes. Turning sounds or pictures into electrical signals that can be transmitted (sent) through the air is a communication process that you will learn about in the next chapter.

Making a movie is a complex communication process. When you go to the theater to see a film, do you think you see every move that the actors are making? The truth is, you don't! The process of making a movie is actually the creation of an illusion. When you watch a movie, you are actually seeing about 24 *still* pictures flashing in front of your eye every second. Film passes through the camera at 24 frames per second. Each frame records a different position of the object or person in motion. When these frames are run through a projector at the same 24-per-second rate, your eye and your brain interpret the series of pictures as one continuous picture. Fig. 4-9.

Communication processes—and the tools and machines used with them—constantly change as new technologies are developed. Consider printing, for example. Johann Gutenberg invented movable type in the 1400s. The first book he printed was the *Bible*. It took over two years to print 200 copies. Today, using new printing processes, the same number of copies can be printed in less than a day. As another example, back in the 1800s, it often took several weeks to get a letter from one state to another. Today, letters and printed documents can be sent instantly from the East Coast to the West Coast using electronic mail and facsimile machines.

Fascinating Facts

Today's popular movies can trace their history back to the first successful photographs of motion—pictures of a race horse—taken in the 1870s with 24 still cameras. San Francisco photographer Eadweard Muybridge placed the cameras in a row. He attached strings to each shutter, and then stretched the strings across a racetrack. As the horse raced by, it broke one string after another, which tripped one shutter after another. When these successive still shots of the horse were put together in order, one could quickly flip through them and see what, to all appearances, was a race horse in action.

Fig. 4-9. Motion pictures and cartoons are actually series of still photographs shown in rapid progression. Today, computers are commonly used to help in the process of making cartoons. The computer operator creates the beginning and ending positions of a character's motion and the computer creates the different images in between.

Fascinating Facts

The Pony Express, an interesting chapter in U.S. communication history, carried mail between St. Joseph, Missouri, and Sacramento, California, from April 1860 to October 1861. The trip, which covered 1,966 miles, first took 10 days, but was later reduced to eight or nine. Sending mail by stagecoach took 12-14 days longer.

The Pony Express had about 80 riders and 400 horses. At certain stops along the routes, riders stopped only long enough to get a fresh horse. At other stops, the mail pouch was often passed on to the next rider, who would set off on a fresh horse. Riders rode day and night, in all kinds of weather, to deliver the mail as quickly as possible.

Apply Your Thinking Skills

1. Think about what it would be like if the communication processes used today were like those of the 1900s. How would your life be different? (Consider such things as security, privacy, relationships with friends and family, education, emergency situations, and recreation.)

Outputs

The message is created, coded if necessary (such as in computers), and transmitted (sent). Once the message is received, the desired output of the communication system has been achieved—information has been communicated. The received message is the output.

Outputs of a communication system come in many forms. The form depends on the inputs and processes used. The outputs can be images, words, sounds, or other forms of information. Fig. 4-10. For example, consider a telephone communication system. One person may initiate the phone call. That person is providing input to the system. The message is transmitted using communication processes. When you receive the message, you are receiving output of the system. In this case, the output is in the form of sounds. Your brain interprets these sounds. They may be voices, noises, static, or music. Everything you receive—or hear, in this case—is called output. Output may not

Fig. 4-10. The pilot and co-pilot rely on output received from the plane's instrument panel. The many types of dials, indicators, and lights are communication devices that tell them such things as fuel supply, altitude, speed, engine problems, and whether there is an obstacle ahead.

always be what you want. For example, you may hear static during your phone call. Static is an output, but not one that you like.

Impacts

The output of a system includes not only the desired result of the system, but also any effects or impacts that the system has on people and the environment. Communication systems have both positive and negative impacts.

Positive Impacts

You read in Chapter 3 about many of the impacts communication technology has had on our lives. Thanks to such things as radio, television, and computers, we now have fast access to a wealth of information. Fig. 4-11. The development of microchips has made it possible to make communication devices that are more dependable and accurate, and, at the same time, less expensive and smaller and more portable.

The ability to communicate over long distances has expanded our horizons. Today, thanks to television and satellite technology, we can see and hear what is happening on the other side of the world, even as it happens. We take it for granted today that we can quickly place a phone call and talk to a friend or relative who is almost anyplace else in the world, from across the street to across the globe. Think how the telephone has changed our world. How would business be conducted today without telephones? How many people are employed by telephone companies? How would relationships with your friends change if you couldn't call them on the phone? How have telephones affected world politics?

Consider how computers positively affect schools and learning. Computers enable administrators to manage incredible amounts of information—such as enrollment records, scheduling, grade reporting, and

Fig. 4-11. New information systems and computers are gradually replacing rows of files, records, and books. All volumes of this encyclopedia have been recorded on a single optical disc. Someday all information may only be accessible electronically. What impact might that have on your life?

lesson planning. It's difficult to imagine how all this information was handled *before* computers had been invented! Other computers linked to special compact discs enable students to search through large data bases for information related to specific topics of interest. This is especially useful when it comes time for writing reports. Computers connected with modems and telephone lines enable students to access a large amount of information outside of their school without actually leaving it.

Special telephone systems used in some schools help to improve the relationship between home and school. Computerized phone calls are placed to homes of students who fail to show up at school. In addition, parents can call a special school number during the evening to hear a tape-recorded message of their son's or daughter's homework assignments. Students won't be able to tell their parents that they have no homework!

Negative Impacts

When television was first invented, people never expected that it would become so popular. They also did not expect the problems that have come along with increased television use. Many studies have been conducted in recent years to determine the effects of television on children. Violent television shows have been highly criticized as encouraging violent or antisocial behavior. Many people feel that we depend too much on the television media. They think it affects how much people read and could therefore affect their reading ability. Is it television as a communication medium that is being questioned, or is it the way television is used? Think about it.

As you read in Chapter 3, there is a great deal of personal information regarding such things as finances and medical conditions stored in computers. This information can often be easily accessed by many businesses and agencies. This raises the question of the *right to privacy*. The increasing use of telephones in business for sales and surveys has also raised this question. This use of the telephone to sell goods and services is called **telemarketing**. Many people prefer to receive only personal calls and do not appreciate people calling and asking them questions or trying to sell them something. However, in most cases, people have to pay extra to keep their phone numbers from being published in the phone book.

Some people fear that the increased use of radios with headphones could affect hearing. Others worry that prolonged viewing of lighted screens could affect vision. Growing numbers of computer users are reporting physical problems associated with computer use. Common complaints include blurred vision, eyestrain, and headaches. Fig. 4-12.

All electric devices (computers, televisions, electric blankets) emit some amount of electromagnetic radiation. **Electromagnetic radiation** is an invisible source of energy given off by the movement of electrons. In most cases, only a very small amount of radiation is emitted. This is considered safe by most people. However, some researchers are concerned that this exposure may not be as safe as once thought. They feel that people are being exposed to greater amounts all the time. There have not been enough studies to find out how much is "safe." Some people believe that increased exposure to electromagnetic radiation contributes to a person's chances of getting different forms of cancer.

High-voltage power lines also emit electromagnetic radiation. These are the tall towers that carry electric current through wires over long distances from the power stations to where it is needed. Fig. 4-13. We depend on increased electric power to run most of our communication devices. However, some researchers feel that these

Fig. 4-12. Nonglare screens can be mounted over computer screens. This has helped reduce eyestrain from glare for many users.

Fig. 4-13. These high-voltage power lines emit electromagnetic radiation.

power lines emit enough radiation to be harmful to people and animals. They fear that some people may be living too close to them. These people are exposed to larger amounts of electromagnetic radiation for longer periods of time than other people. New studies are needed to better understand health effects from these and other communication devices.

Communication systems also affect our environment. Every year, thousands of acres of trees are cut down to make the paper for newspapers, magazines, and computer printouts.

Apply Your Thinking Skills

1. Think about the negative and positive impacts that communication systems have on your life and your community. What are some things that you can do to reduce the negative impacts?

Feedback

What happens when you receive output? Do you do anything? Do you respond? Most likely, you do. This response is called feedback. Feedback is the part of the system that checks the output to see if the system is doing what it is supposed to do and to see if there are any negative impacts. If the output is not what is desired or there are negative impacts, the inputs or the processes of the system will have to be changed.

For example, feedback may let the sender know that the message was received. On the telephone, your immediate feedback might be a statement such as, "Yes, I'll take care of that," or, you might say, "What? I cannot hear you very well. There's too much static." The person sending the message may need to repeat the message. These forms of feedback happen right away. There is also feedback that takes place later or after more time has passed. For example, you may think that the static was quite abnormal. You may decide to get your phone repaired. You may choose to buy a new phone. You may also do something drastic. You may decide to cancel or change your phone service because you feel that it is unreliable. Each of these responses are forms of feedback, too. They are delayed forms of feedback since they do not occur immediately.

Apply Your Thinking Skills

1. Think about the feedback involved when you watch television. What kinds of feedback have you given—both immediate and delayed? Consider how your feedback may differ according to the type of program or commercial you watch.

Investigating Your Environment
Communication Pollution

At first glance, communication seems like a clean technology. After all, calling a friend won't produce air pollution. Driving over to the friend's house will. We may call a television show "garbage," but it won't take up space in the landfill. However, communication technology does cause environmental problems.

One problem is noise pollution. Televisions, radios, tape players, computer printers, and many other communication devices produce noise. The noise can have many ill effects on people. Loud noise can cause hearing loss. Sometimes this happens so gradually that the victim is unaware of it until it is too late. Hearing loss is not the only danger. Noise can cause your blood vessels to constrict (become narrow). This raises blood pressure. Even noise that is not very loud can be annoying, causing tension. Noise affects you even while you are sleeping. Though you may sleep through the noise, you might wake up tired because of it. It has also been found that a loss of appetite is common in noisy surroundings.

Visual pollution can be another effect of communication technology. Some people are bothered by billboards and other forms of outdoor advertising. Large dish antennas, store signs, and political posters—all can add up to visual clutter.

Do you receive "junk mail"? This type of communication adds to our waste disposal problems. Of all the items that go into landfills, 37 percent are paper products. How many of the paper products you throw away are catalogs and advertisements you didn't request?

What can be done about communication pollution? Some communities have passed laws to regulate the volume of noise and to control the number and placement of outdoor ads. To reduce the flow of junk mail to their homes, people can contact the senders and ask to have their names removed from the mailing list. They can also notify the Mail Preference Service that they want to be left off future mailing lists. The address is: Direct Marketing Association, 6 East 43rd St., New York, NY 10017.

Take Action!

1. Brainstorm about ways you and your classmates can reduce noise pollution.
2. For two weeks, save the junk mail that arrives at your house. How much does it weigh? How much space does it take up? Sending this mail to the landfill is a poor solution. What are some better solutions?

Fig. IYE-4. How much junk mail arrives at your house?

Chapter 4 Review

Looking Back

The communication system consists of many different types of subsystems. Like all systems, the communication system, as well as its subsystems, consists of inputs, processes, output, and feedback. There are seven types of inputs. These inputs supply the things needed to create the message and provide a means for the processes to be carried out. There are a wide variety of processes used in communication systems. These are constantly changing as our technology changes. The outputs of a communication system come in many forms. They can be images, data, sounds, or other forms of information. Feedback is response to the output of a communication system. It includes all the immediate and delayed responses to the communicated message.

All communication systems have impacts. They can be positive or negative. These impacts affect our environment and our society in numerous ways. It is important to consider the impacts of communication systems as we look at communication technology today and tomorrow.

Review Questions

1. What is a communication channel?
2. List the seven inputs every communication system needs and give at least two examples of each.
3. Give three examples of careers related to the communication system.
4. Define processes, as related to the communication systems. Give three examples of communication processes.
5. What is the desired output of a communication system?
6. Describe at least two positive impacts of communication systems.
7. Describe two concerns many people have about the impacts of television.
8. How is privacy affected by communication systems?
9. What are some physical problems commonly associated with using a computer?
10. What is the concern about electromagnetic radiation?

Discussion Questions

1. In what ways can we better manage the material resources used by the communication industries?
2. How have changes in communication systems over the past fifty years affected jobs and careers?
3. Think about a situation in which you might use a computer communication system. Describe the situation, then discuss the inputs, processes, outputs, and feedback of that system.
4. Facsimile machines are becoming more widely used today. Discuss the negative and positive impacts of this communication system from your point of view.
5. "Picture phones" are telephones that display video pictures of the persons talking on the lines. Do you think these phones will ever come into common use? Why or why not? What effects do you think these phones might have on our lives?

Chapter 4 Review

Cross-Curricular Activities

Language Arts

1. Using the proper form for a business letter, write to a computer company and ask for some information on their newest home computer model.
2. Look into your local telephone directory for the nearest computer stores. List as many different outlets as you can find. Be sure to check any cross-reference listings.

Social Studies

1. Pick one of the following: television, radio, movies, or compact disc players. Do research to find out when it was invented and how long the inventor(s) worked to develop it. What technology had to be developed before this device could be developed? How has its design changed since it was first introduced? What improvements have been made?
2. Interview someone who works in some capacity with a computer in the field of medicine, business, or communication. Find out what the job entails and what training is required for such a job.

Science

1. Find out what helpful impacts communication technology has on the environment. What messages have been sent that are helping to protect earth's air, water, land, and life? ("Give a hoot, don't pollute" is one such message.)
2. Motion pictures are possible because the eye retains and "blends" successive still photographs. You can demonstrate this as follows: cut a disk of stiff cardboard four or five inches in diameter. On one half, draw a fish. On the other half, draw a fishbowl. Tape the disk on the eraser of a pencil. Twirl the disk by holding the pencil vertically between your palms and rubbing briskly back and forth. Your eye will blend the images so that the fish appears to be in the fishbowl.

Math

Number systems of base two are known as binary systems. The system uses only two digits—0 and 1. This makes it ideal for computer applications because it can be interpreted electronically as 0 for OFF and 1 for ON. The binary number has a place value that is a power of 2. (See the chart below.)

# of 16s	# of 8s	# of 4s	# of 2s	# of 1s	Place Value (Powers of 2)
				1	= 1
			1	0	= 2
			1	1	= 3
	1	0	0	1	= 9
	1	1	0	1	= 13

Sum up the numbers in each place value to find the decimal equivalent.

1. Evaluate these binary numbers: 100; 1111; 10101; 100001.
2. Write these numbers in the binary system: 6; 27; 64; 127.

Chapter 4 Technology Activity

Creating a Cyborg Communication System

Overview

As you know, communication takes place between people, between machines, and between people and machines. Have you ever seen a science fiction movie or TV show about a cyborg? A cyborg is a person who is linked to, and dependent upon, a machine. For example, a cyborg might be human from the waist up and machine from the waist down.

In this activity you will design, build, and operate a cyborg communication system. First you will design and build the electronic communication device. Next, you will design and build a helmet, mask, or set of goggles that the "cyborg" will wear. The helmet will contain the communication receiver. Then you will create a code for communicating with the cyborg by means of the sender and receiver. Finally, using your equipment and code, you will guide your cyborg through a maze.

Goal

To design, build, and operate a cyborg communication system.

Equipment and Materials

two 500-contact breadboards
red, yellow, and green LEDs (twelve total)
six 220-ohm, 1/4 watt, 5% resistors
six normally open, push-button switches
one 9-volt battery with snap connector and holder
#22 solid hook-up wire (insulated)
wire strippers
masking tape
ballpoint pen
string
jigsaw
drill press and drill bits
tin snips
25-watt soldering iron
solder for electronics
screws
scrap hardboard or paneling
wallpaper paste
strips of newspaper
chicken wire
corrugated cardboard
paint
paint brushes

Procedure

1. *Safety Note:* Follow your teacher's directions when operating tools and machines in the lab. When you guide cyborgs through the maze, one student should act as safety monitor. That person should be sure cyborgs do not trip or injure themselves as they move along.
2. Your teacher will divide the class into teams and tell you where tools and supplies are located. Read the entire activity all the way through before you begin. Discuss any questions you may have with your teacher.
3. Measure and cut 7 eight-foot-long pieces of #22 wire. Strip the ends of the wire with the wire strippers.
4. Give each piece of wire a number from 1 to 7. Using masking tape and a ballpoint pen, label the wires about 3" from each end.

5. Assemble the 7 wires into a communication cable. (A cable is a bundle of conductors or wires.) Start about 6" from one end and gently tie the wires together at 6" or 8" intervals. Stop about 6" from the other end.
6. Your teacher will show you how to use the breadboards. Connections on the boards are arranged in groups of 5. If you need more than 5 connections (to make a bus, for instance), just run a short wire between two or more groups of 5 holes.

Making the Sender and Receiver

7. *Note:* LEDs are sensitive electronic devices. Unlike flashlight bulbs, they must be protected from burnout by a resistor. Like flashlight batteries, LEDs have a positive side and a negative side. The negative side has a flat spot on the plastic or a short wire. *The negative side must go to the negative end of the battery. The positive side must go to the positive end of the battery.* If you connect them incorrectly, the LED will not work and may burn out.

   Using Fig. A4-1 as a guide, create the sending and receiving units. Start by making a negative bus on one breadboard (sending unit) and a positive bus on the other (receiving unit). Starting at the *negative* bus on the sending unit, wire one switch, one 220-ohm resistor, and one green LED (*negative* side to the resistor) in *series.* Use the drill press and soldering iron as needed. If the lead wires on the switches are not #22, solder #22 wire to the leads.
8. Connect the wire labeled "2" in your cable to the positive side of the LED.
9. Connect the other end of the number "2" wire to the second breadboard (receiver).
10. Connect the negative side of a second green LED to this wire.
11. Connect the positive side of the LED to your positive bus. Run your number "1" wire from the positive bus on the receiver through the sending unit to the red wire of the battery snap.
12. Connect the negative bus on the sending unit to the black wire of the battery snap.
13. Make sure that your board looks like Fig. A4-1. Ask your teacher to check your work and to connect the battery to the battery snap.
14. Test your device. If you wired it correctly, the LED will light up when you push the button on the switch. You have just created a one-bit digital communication circuit.
15. Build the rest of the communicator one circuit at a time:
    - Connect one lead of a switch to the negative bus.
    - Connect the other lead to a 220-ohm resistor.
    - Connect the other end of the resistor to the negative side of an LED.
    - Connect the positive side of the LED to a wire in the cable.
    - Connect the other end of the same wire to the negative side of the receiving unit's LED.
    - Connect the positive side of the receiving unit's LED to the positive bus.

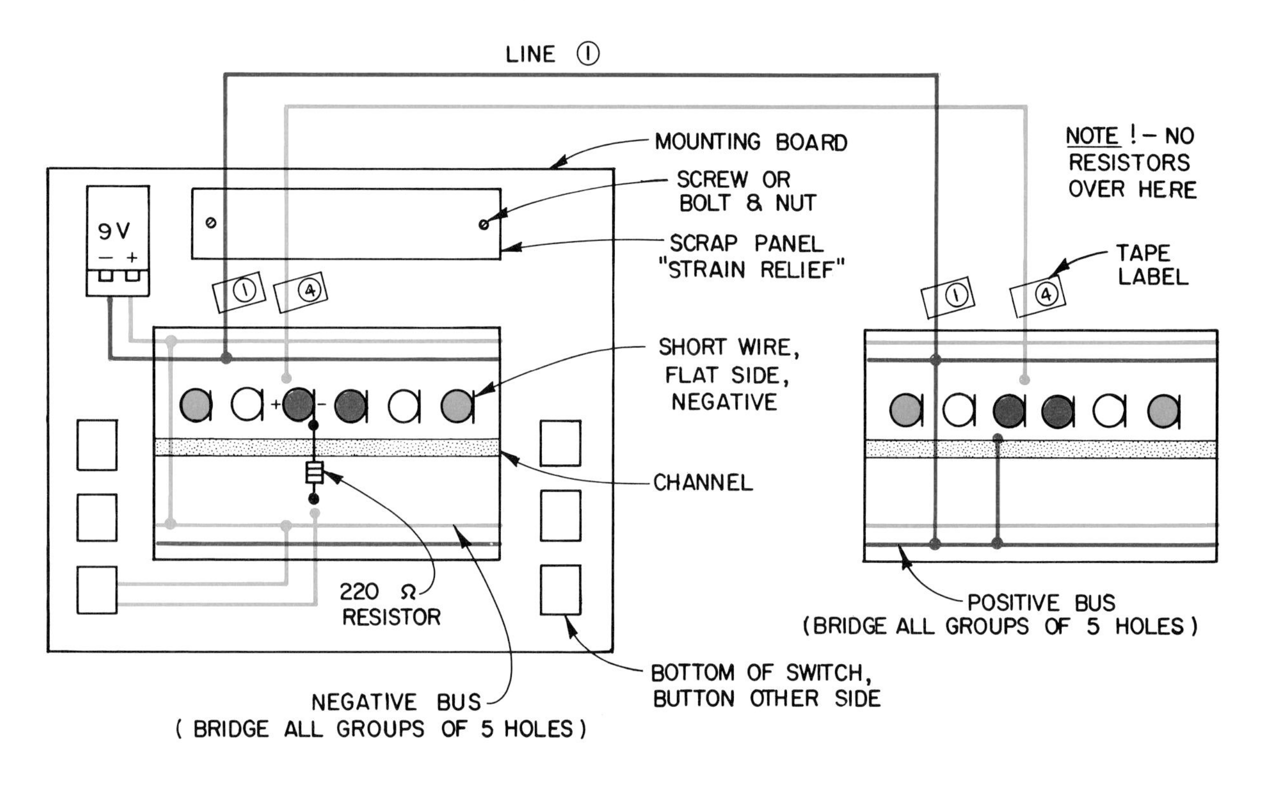

Fig. A4-1. The sending and receiving units.

- The positive bus on the receiving unit should already be connected to the number 1 wire in the cable. The cable should be connected through the sending unit to the red wire of the battery snap. Continue with these steps until all 6 circuits are complete.

16. Mount the sending unit on a piece of paneling or hardboard. Use the jigsaw to cut the board to size.

Making the Mask and Installing the Receiving Unit

17. Design a mask, helmet, or goggles that your cyborg will wear. Use the chicken wire to make a form. Cover it with papier-mâché made from the newspaper strips and paste. The mask should block the cyborg's vision so that he or she is dependent on the communicator. However, be sure that the part of the mask that covers the eyes stands at least 4" away from the face so the person can see the receiving unit. Later, you and your teammates will take turns being the cyborg.
18. Install the receiving unit in the mask.
19. Paint and decorate the mask.

Create and Memorize a Digital Code

20. You will communicate with your cyborg using a digital code. The code will be based on turning the red, yellow, and green LEDs on or off. You can use the lights one at a time, two at a time, three at a time, and so on. You can also use them in or out of sequence. This produces many combinations, but keep them simple. You will have to memorize them. Your code should be used for the following messages and any others you think necessary:
    - take a step with the left foot
    - take a step with the right foot
    - turn left
    - turn right
    - raise your right hand
    - open your right hand
    - close your right hand
21. Set up a maze in the lab using chairs and other items. Take turns guiding teammates wearing the mask through the maze.

Evaluation

1. Identify the inputs, process, and outputs in your communication system.
2. How could you add a feedback loop to your communication system?
3. How could you adapt this system for a blind person or for someone in a wheelchair?
4. How could this system operate a drill press if the sender was a computer? (Motors would be used for rotation and feed.)

Credit: Daniel B. Stout

Chapter 5

Telecommunication

Looking Ahead

In this chapter, you will discover:

- the importance of telecommunication in our society.
- how signals are transmitted using wires, cables, fiber optics, electromagnetic waves, and satellites.
- how telegraphs, telephones, radios, and televisions operate.
- many of the technical processes involved in telecommunication.

New Terms

amplitude
coaxial cable
communication satellite
downlink
earth station
electromagnet
electromagnetic waves
frequency
mass communication
microwaves
optical fibers
patent
pixels
telecommunication
uplink

Technology Focus

Telecommuting

Homework isn't just for schoolwork anymore. Today, thanks to communication technology that allows us to easily and quickly transfer information from one location to another, many people can work at home or at locations other than a typical office. Powerful personal computers, modems, and fax machines have made it easy for people working at home, in a motel, or in another remote location to quickly communicate with their company, with each other, and with clients. Working from remote offices is often called *telecommuting*.

American Express, J. C. Penney, Apple Computers, and General Electric now allow some employees to work at home. Many smaller businesses also employ telecommuters. A total of nearly 500,000 people telecommute today. Futurists say that there could be as many as seven million telecommuters by the end of the 1990s.

Who are the people who telecommute? Some are skilled workers who choose to stay at home and raise a family. By telecommuting, they can do this and still put their job-related talents and experience to good use and earn an income. Workers with physical disabilities can work out of a home office and avoid many of the difficulties they might encounter if they had to travel back and forth to work every day. President George Bush joined the ranks of telecommuters when he managed and directed the Persian Gulf war efforts from Washington D.C. and other locations. President Bush was "telecommanding."

A home office worker usually has at least one computer and a telephone modem. Many also have a fax machine. Some telecommuters work from mobile workstations to bring their expertise to clients' offices. These workers also use a cellular phone and a laptop computer.

Corporations find that telecommuters are very productive. Telecommuters also decrease the amount of office space and parking that is necessary, and they help keep a corporation's insurance rates lower.

Telecommuting workers save time by not having to drive back and forth to work every day. This also saves them the expense of fuel and of operating a car (oil, tires, maintenance, etc.). Telecommuters work on a schedule that suits their needs, and they work in more pleasant surroundings.

The environment profits too. Since telecommuters are not driving to work daily, less fuel is being used. In addition, telecommuting workers aren't crowding the highways and polluting the air with exhaust fumes.

Homework—it isn't just for schoolwork anymore.

Fig. TF-5. Telecommuting is becoming more and more common these days.

What Is Telecommunication?

Telecommunication means communication over a long distance. When we examine telecommunication, we are exploring those methods used to transmit (send) and receive messages over a distance.

Many inventions throughout history have led to the development of today's telecommunication system. However, no others stand out quite like the telegraph, telephone, radio, television, and satellite. In the next several pages, each of these technologies will be examined.

Apply Your Thinking Skills

1. Think about ways that your life would be different if you didn't have telecommunication devices like the telephone, television, and radio. How would it affect the ways you keep in touch with friends? What would you do during the hours that you usually watch television?

The Telegraph

The invention of the telegraph in the mid-1800s gave people an incredible new opportunity. For the first time, they could send messages over long distances that could be received almost instantly. Samuel Morse has been given the most credit for inventing the telegraph. Others were also successful in their experiments with devices that could send messages over wire using electricity, but it was Morse's model that caught on. Morse's telegraph marked the beginning of telecommunication in America and paved the way for future telecommunication.

Morse's telegraph was actually a very simple instrument that sent electric pulses over wire. Its operation was based on the principles of electricity and magnetism. It consisted of little more than a battery, a sending device called a "key," a wire, and a receiver called a "sounder." Fig. 5-1. The sounder consisted of an electromagnet and a metal arm that made a clicking noise. (An **electromagnet** is a soft iron core surrounded by a coil of wire that temporarily becomes a magnet when electric current flows through the wire.) The key acted as a switch that opened and closed the electrical circuit. When the key was pushed down, current flowed through the circuit to the wire-wrapped iron core and turned it into an electromagnet. The electromagnet then pulled the metal arm down, making a clicking noise. Morse developed a code to use with the telegraph based on varying the length of these clicks. The key could be held down very briefly to make a short click, which represented a dot in the code. By holding the key down a little longer, a longer click was sounded. This represented a dash. Morse assigned each letter of the alphabet, each number from one through ten, and each punctuation mark their own special sequence of dots and dashes. Fig. 5-2. Using this code, in 1844 Morse sent the message "What hath God wrought?" over 40 miles of wire, stretching from Washington, D.C., to Baltimore, Maryland.

At first, people were unsure about the ways telegraph messages could be used. Soon, however, they were able to see how quickly critical messages could be sent. With the aid of the telegraph, for example, an escaped murderer was located and captured. The telegraph also helped spread the news about national elections. Personal news was exchanged by families hundreds of miles apart. People began to realize that the telegraph could be great for long-distance

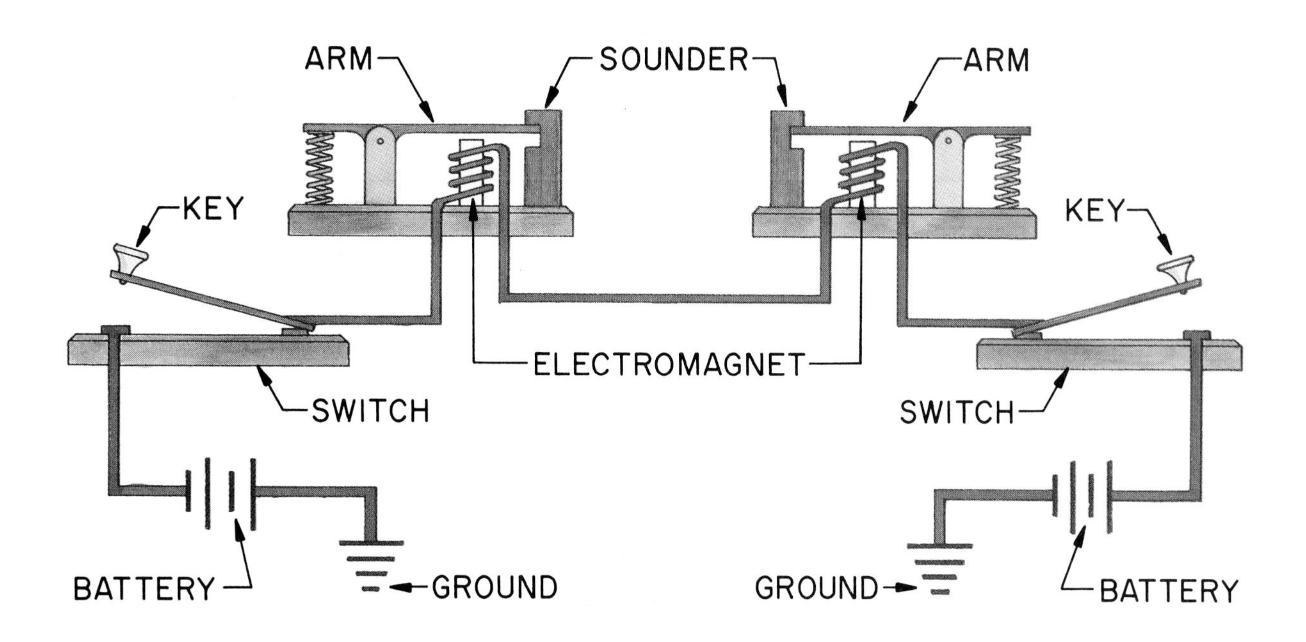

Fig. 5-1. Morse's first telegraph was similar to this diagram. When the key is pressed down, current flows through the circuit and the sounder on the other end clicks. When the key is released, the circuit is opened and there is no clicking. The clicks could be varied in length to make a code that could be sent over the telegraph wire.

INTERNATIONAL MORSE CODE						
A	.-	P	.--.	5	.....	
B	-...	Q	--.-	6	-....	
C	-.-.	R	.-.	7	--...	
D	-..	S	...	8	---..	
E	.	T	-	9	----.	
F	..-.	U	..-	10	-----	
G	--.	V	...-	'	.----.	(apostrophe)
H	....	W	.--	:	---...	
I	..	X	-..-	,	--..--	
J	.---	Y	-.--	-	-....-	(hyphen)
K	-.-	Z	--..	.	.-.-.-	
L	.-..	1	.----	( )	-.--.-	(parentheses)
M	--	2	..---	?	..--..	
N	-.	3	...--	" "	.-..-.	
O	---	4	....-	SOS	...---...	

Fig. 5-2. The Morse Code made it possible for messages to be sent via (by way of) the telegraph using electric pulses. This system has been used for international communication.

communication. By 1856, the Western Union Telegraph Company linked many parts of the United States. In 1866, a telegraph cable was successfully laid across the Atlantic Ocean, connecting America and Europe.

Improving Telegraphy

Around 1875, another young inventor began working with telegraphy. He was trying to develop the harmonic telegraph. This was a device that would make it possible to send several messages at the same time. Unfortunately for him, another person, Thomas Edison, had a more innovative idea. Edison's quadraplex telegraph could send four messages over one wire. The unsuccessful young inventor's name was Alexander Graham Bell. As you will read in the next section, he did not give up inventing.

Apply Your Thinking Skills

1. The telegraph introduced "instant" communication over a long distance. In what ways do you depend on instant communication today for emergencies? How do you rely on it for entertainment? For information?

The Telephone

After failing with the telegraph, Alexander Graham Bell began to focus his attention on another idea. He was convinced that voices could be sent over wire. Bell experimented with several different ideas. Through trial and error, Bell and his partner, Thomas Watson, developed a model telephone that became the basis of today's telephones. However, Bell was not the only one searching for a voice machine. When his patent for the telephone was filed on February 14, 1876, Bell was only two hours ahead of another inventor, Elisha Gray. (A **patent** is a government document granting the exclusive right to produce or sell an invented object or process for a specified period of time. It ensures that the inventor's idea cannot be legally copied for that specified period.)

How the Telephone Works

Although Alexander Graham Bell received the first patent for the telephone, his model did not actually work as intended. Sounds could be transmitted, but voices were not quite understandable.

Bell's telephone worked on the principle that sound waves cause vibrations. In this case, they caused a parchment (paper) drum to vibrate. Granville Woods improved Bell's design. Woods invented the carbon microphone for the telephone. Telephones today still use this same basic principle. The carbon microphone in the mouthpiece is made of a small cup filled with carbon granules (grains). A small amount of electrical current constantly flows through these granules. Next to the cup containing the carbon granules is a flexible piece of metal called a *diaphragm*. Fig. 5-3. The diaphragm vibrates when sound waves from the speaker's voice strike it. The vibrating diaphragm presses against the carbon granules. This causes more electrical current to flow. When the speaker pauses, pressure is released and less current flows. This creates a changing electrical signal. The changing signal represents the sound as it is transmitted.

The telephone receiver is located in the earpiece. The receiver works much like the transmitter in reverse. The receiver contains a coil of wire wrapped around an iron core. When electrical signals pass through this, it

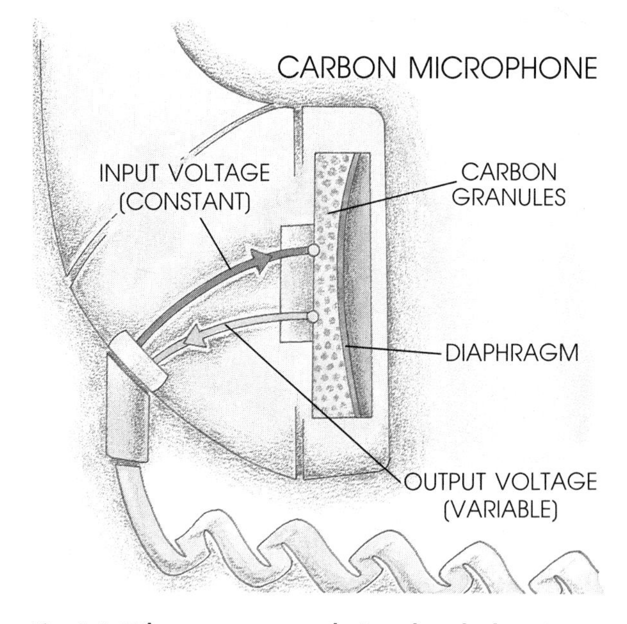

Fig. 5-3. When a person speaks into the telephone's mouthpiece, the diaphragm vibrates against the carbon granules. This creates a changing electrical signal (or voltage) that is transmitted over the telephone line to the receiver at the other end.

forms an *electromagnet*. Connected to the iron core is a flexible metal diaphragm. Fig. 5-4. When the transmitted electrical signal enters the receiver, it travels through the coil. The electricity passing through the wire magnetizes the iron core, which pulls on the metal diaphragm. The diaphragm vibrates and reproduces the sound that had originally been sent.

Transmission Channels

As you learned in Chapter 4, communication channels are the paths over which messages must travel to get from the sender to the receiver. Most telephone messages travel over wires, cables, or optical fibers. Some telephones, however, use microwaves as a transmission channel.

Copper Wire

Most local telephone messages travel over twisted-pair wire. This consists of two thin, insulated copper wires twisted around each other. Twisted-pair wires may also be bundled together to form large cables that stretch across the United States.

Coaxial Cable

Coaxial cable can carry many more messages all at once than twisted-pair wire. **Coaxial cable** consists of an outer tube made of electrical-conducting material (usually copper) that surrounds an insulated central conductor (also copper). Usually several of these cables are combined into one bundle. Fig. 5-5. This bundle is then covered in lead and plastic for protection.

Optical Fibers

Optical fibers are being increasingly used to carry telephone transmissions. **Optical fibers** are thin, flexible fibers of pure glass used to carry signals in the form of pulses of light. Fig. 5-6. Each optical fiber is surrounded by a reflective cladding (covering) material and an outside protective coating. After the sound waves from the speaker's voice have been converted to electrical signals, these signals activate a light source (a laser). The resulting light pulses enter the glass core and bounce rapidly back and forth off the reflective surface of the cladding as they pass through the glass fiber. Fig. 5-7. At the

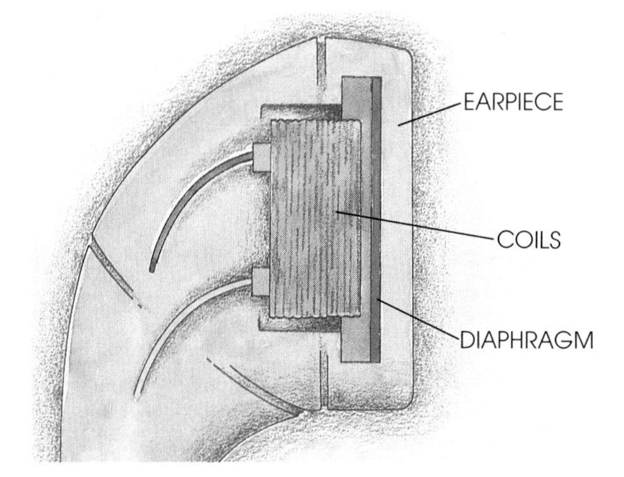

Fig. 5-4. The transmitted electrical signals pass through the coils causing the iron core within the coils to become magnetized, creating an electromagnet. The electromagnet causes the diaphragm to vibrate. This vibration reproduces the sound.

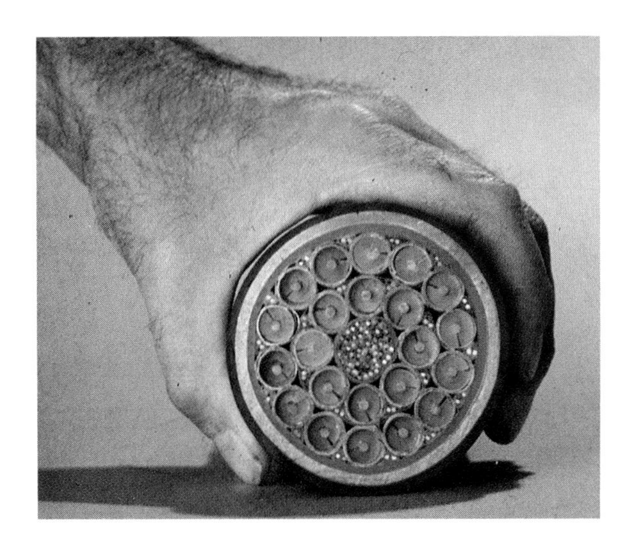

Fig. 5-5. This 22-tube coaxial cable will carry up to 108,000 two-way voice conversations.

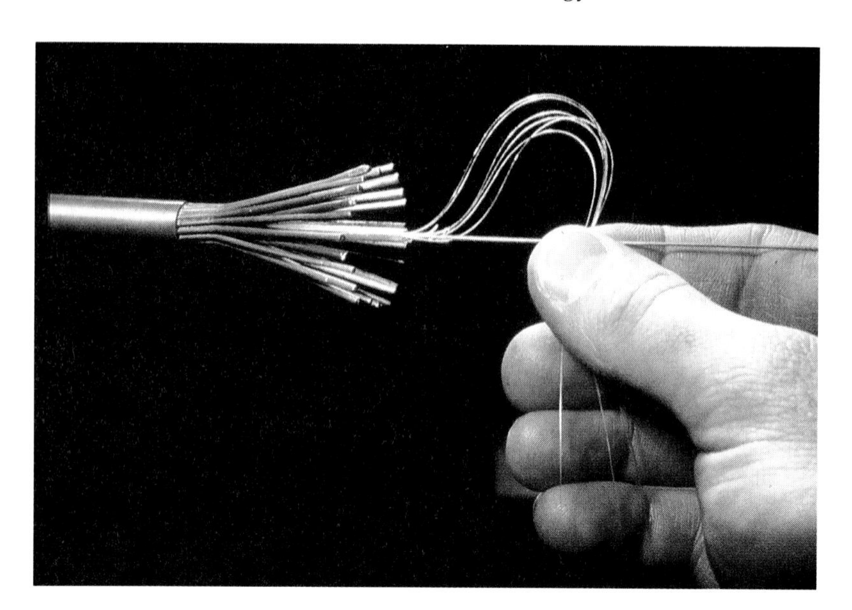

Fig. 5-6. Optical fibers are flexible strands of glass that are about as thin as human hair.

receiving end, the light pulses are changed back to an electrical signal that is changed back into the sound of the sender's voice.

Optical fibers have many advantages over copper wire or coaxial cable. A single fiber is about the thickness of a human hair, yet it can carry 6,000 voice circuits. A pair of copper wires can carry only 24 voice circuits. Since optical fiber cables are lighter and thinner than copper wires, they are ideal for communication systems in large cities or other places where the available space is limited. In addition, the light signals that travel on optical fibers do not fade as quickly as electrical signals sent on copper wire. Many of the noises you sometimes hear on the telephone, like buzzing or clicking, are also eliminated when optical fibers are used.

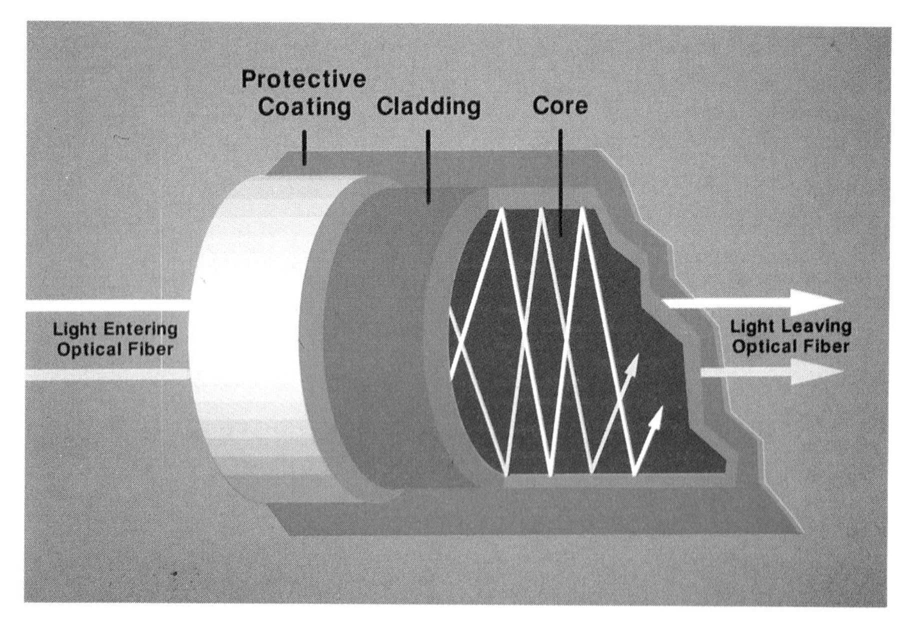

Fig. 5-7. The glass fiber core has a reflective cladding and an outside protective material. The arrows indicate how the light traveling through the fiber bounces off the reflective cladding, keeping it inside the cable. After leaving the optical fiber, the light is changed back to an electrical signal.

Microwaves

Microwaves can be used to carry phone conversations over long distances. **Microwaves** are very short electromagnetic waves. (**Electromagnetic waves** are waves created by a magnetic field. These waves travel through the atmosphere and make communication without a connecting wire possible.) Many billion microwaves can pass a given point in less than a second.

When microwaves are used in telephone communication, the sound waves are changed into microwaves. These can be sent through the air using an antenna. The microwaves travel through the air like the ripples that you see when you drop a stone into a pool of water. Another antenna is able to detect the signals and acts as the receiver. It then boosts the signal to a usable strength. The signal is then converted into sound waves that you recognize, like the voice of the person you are talking with on the telephone. Fig. 5-8.

Fig. 5-8. Signals sent through the atmosphere are transmitted and received by antennas. The messages travel as electromagnetic waves and are received and converted back to their original form, whether it be voice, data, or images.

Getting a Message to the Right Place

Whether wire, fiber, or microwave is used to send the message on the telephone, there still exists another major problem. With millions of telephones in the world, how can you get a message to the right one? When telephones were first introduced, the number of connections possible was very limited. Today, however, almost all homes and other buildings have at least one phone. The number of possible connections is enormous! A system designed to handle this great flow of information has been developed.

Early Switching Systems

The first simple telephone exchange system required human operators to connect the phone of the message sender with the phone of the expected receiver. These operators sat at stations and plugged the line of the person making the call into the right set of telephone lines. As the number of calls continued to increase, a new mechanical system had to be developed. The first new system, invented by Almon B. Strowger, used electromagnetic switches to make telephone connections. The switching system received telephone signals by "counting" the number of pulses from the dialed number. The system held the signals until the entire number was registered and then automatically (without human involvement) created a "path" through all the switches to the receiver.

Automated Telephone Exchanges

Numerous systems were used to make the line connection process possible. However, it was not until the late 1950s that electronic switching systems became a reality. Electronic switching makes connections faster and provides better signal transmission. These telephone exchange systems use computers to route a call through a series of telephone exchange centers to the intended destination.

Fascinating Facts

On Christmas Eve 1906, Canadian inventor and physicist Reginald Fessenden staged the first radio broadcast. As he spoke from his laboratory on the Massachusetts coast, his voice was heard by shipboard operators offshore in the Atlantic. That night, the operators, used to hearing the sound of the telegraph through their earphones, suddenly heard a voice. Fessenden read the Christmas story from the *Bible*, played music on the phonograph, and played "O Holy Night" on his violin. Then he wished everyone a Merry Christmas!

Apply Your Thinking Skills

1. Unlike the telegraph, the telephone was designed for individuals to have and use in their homes. Once people could afford to buy a telephone, new opportunities to communicate were possible. Discuss how this affected people's lives, their communities, and their businesses.

Radio

The telephone soon became a widely accepted means of communication. However, along with the telegraph, it still limited communication to individuals or small groups. People began to look for a way to send messages without using wires and cables. Such a development would make communication with large groups of people, or **mass communication**, a reality.

The Development of the Wireless

Radio communication evolved (developed slowly) from a series of discoveries in the field of electricity and electromagnetics. It was, however, Guglielmo Marconi who, in 1895, finally used electromagnetism to send telegraph code signals a distance of more than 1 mile (1.6 kilometers) without wire. With this wireless telegraph, the forerunner of today's radio, the dream of mass communication became a reality. Improvements on Marconi's device followed rapidly. One such improvement was the vacuum tube. Vacuum tubes greatly strengthened a weak signal so that it traveled farther and more people received it. Soon everyone wanted a radio.

Transmitting Radio Waves

Radio waves are electromagnetic waves that are longer than microwaves. All waves—sound waves, microwaves, radio waves—have both amplitude and frequency. **Amplitude** refers to the strength of the wave. Frequency refers to the wave's length. **Frequency** is the number of waves (in this case, radio waves) that pass a given point in one second. Fig. 5-9.

The process of transmitting radio waves begins with a signal source, such as a person speaking into a microphone. The microphone changes the sound waves into electrical signals. Then the low frequency sound signal is combined with a high frequency carrier signal. Fig. 5-10(A and B). These waves are then *modulated* (varied or altered). Sometimes, the amplitude, or strength, of the carrier wave is changed, or modulated. Fig. 5-10(C). This signal can only be received by *AM* (amplitude modulation) radio. Other times, the frequency is modulated. In frequency modulation, the waves are spread farther apart or crowded closer together. Fig. 5-10(D). These signals are received on *FM* (frequency modulation) radios. The modulated signal is then amplified (boosted) and sent to an antenna for transmission. The antenna sends the radio waves. These waves travel through the air in all directions.

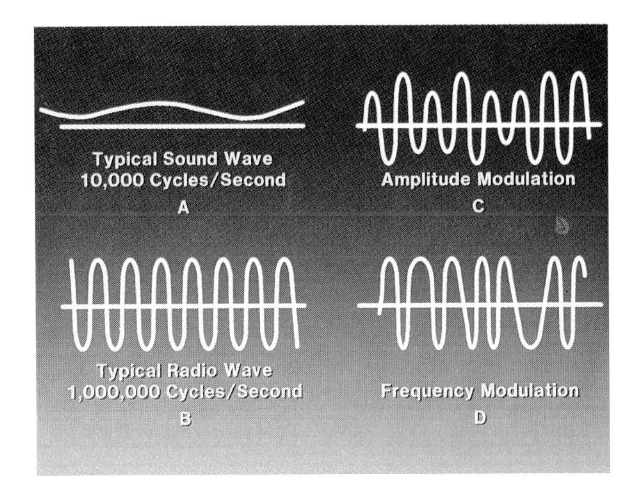

Fig. 5-10. A typical sound wave (A) is combined with a carrier wave (B). The carrier wave may receive amplitude modulation (C) or frequency modulation (D).

Tuning In

Every radio station and device that works on radio waves (such as citizens' band radios, military radios, beepers) is assigned a certain frequency. The numbers on your radio dial indicate the frequency of the different radio stations. When you set your dial at a number, you are *tuning in* the frequency of a particular radio station. The antenna or aerial on the radio picks up the waves that have the frequency that you have tuned in. The circuitry (system of wiring) in the radio separates the carrier wave from the electrical sound signal. This process is called *demodulation*. The circuitry also boosts the electrical signal and sends it to the speaker, which changes the signal back into sounds.

What is the frequency of your favorite radio station? Is it AM or FM?

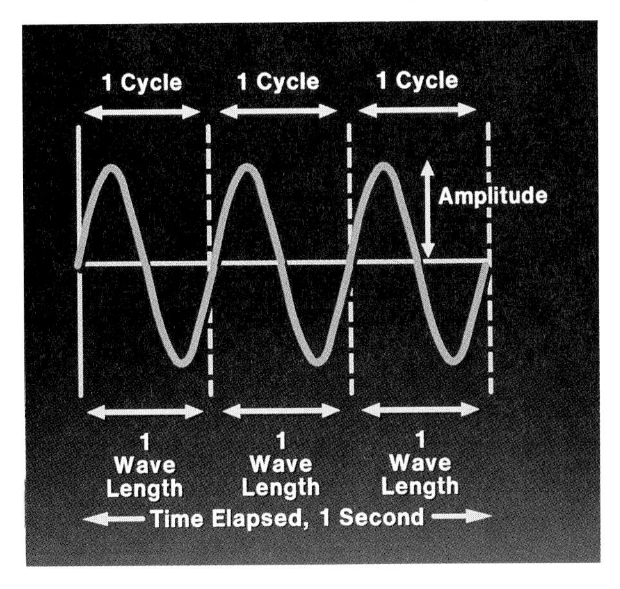

Fig. 5-9. *Frequency* refers to the number of waves that pass a given point (or leave a transmitting source) in one second. The radio signal is measured in cycles per second, or *hertz*. This radio signal has a frequency of 3 hertz. The *amplitude* is the strength (or height) of a wave. Amplitude is measured from the wave's midpoint to its peak.

Apply Your Thinking Skills

1. Consider the many ways that radio communications are used today. They help to (1) inform, (2) educate, (3) persuade, and (4) entertain large numbers of people. Identify at least one specific way that radio communications are used in each of these four categories.

Television

Television is an electronic system of transmitting pictures and sounds over a wire or through air and space. Today, television is a popular telecommunication tool with a variety of program options. Television sets come in many different sizes and are used in many more ways. Let's see how the system was developed and how it works today.

Development of the Television

Sending visual images over a long distance became the ambition of several inventors in the early 1920s. Many individuals experimented with various designs to develop a device that could send and receive pictures. Fig. 5-11.

Early televisions had poor picture quality. In addition, the screens were so small that a person had to sit very close to and directly in front of them in order to see the picture. In spite of this, the impact television had on the public was enormous. At first, most people thought television would be used solely for entertainment. However, it soon became apparent that this "picture machine" had uses in education and information distribution as well. In 1939, only one or two programs were aired daily. During the early 1940s, the war effort slowed the development and spread of television. By 1946, however, the television boom was on again.

Today, nearly every household in the United States has at least one television set—most have at least two, and some have one in almost every room. The picture is clear and sharp, and so is the sound. A complex broadcasting network is set up to transmit television programs around the world twenty-four hours a day.

How a Television Works

A television set may seem like a very complicated device. Actually, it works according to some very simple principles. To begin with, the images being televised are focused onto a special target surface within a pickup tube inside the camera. This surface

Fig. 5-11. One of the earliest televisions developed in America is this receiving device made by Charles Francis Jerkins around 1928. The motor and drum fit inside the wood case. When the motor runs, the drum spins. It projects an image up through an opening in the top of the case using a scanning motion. The mirror reflects the image through the magnifier. The viewer, looking through the magnifier, sees an orange and black image.

is divided into approximately 367,000 microscopic picture elements. The light from the televised image gives each picture element an electrical charge. The charge varies according to the amount of light falling upon each element. Each element, then, has a certain charge that equals a certain amount of light. This pattern of picture elements is scanned (examined in detail) by a beam of electrons about the diameter of a pinhead. The beam moves from left to right over the entire target surface 525 times for each picture. Thirty pictures are scanned every second. The electrons pass on through the target surface and strike a signal plate. When they hit the signal plate, an electrical current is created. This is the electrical *video* (picture) signal that will be transmitted.

If a color picture is to be broadcast, a more complex camera must be used. Color television is based on the principle that different mixtures of red, green, and blue light can produce every color in nature. Mixing light rays is not like mixing colors of paint. Paint simply reflects light. Color video cameras have three pickup tubes; one each for red, green, and blue. Light passing through the camera's lens strikes three mirrors. Each mirror is directed at a different pickup tube. In front of each tube is a filter that allows only one color to pass through. Fig. 5-12. The pickup tubes then process the images like the black-and-white pickup tube just described.

While cameras are picking up the video portion of a telecast and converting it to electrical signals, microphones are picking up the *audio* (sound) portion. The microphones are the same as those used in radio, and they convert the sound waves to an electrical signal in the same way as in radio. The audio and video signals are amplified, modulated, combined, amplified again, and then sent to an antenna where they are transmitted. (In the case of cable television, however, these signals are sent over coaxial cable to the television sets.)

Fascinating Facts

On a historic Tuesday night in July 1931, CBS launched the nation's first regular schedule of television broadcasting. On that first program, George Gershwin played the song "Lisa," the Boswell Sisters performed the song "Heebie-Jeebie Blues," and Kate Smith sang "When the Moon Comes Over the Mountain." New York mayor Jimmy Walker served as host of the telecast.

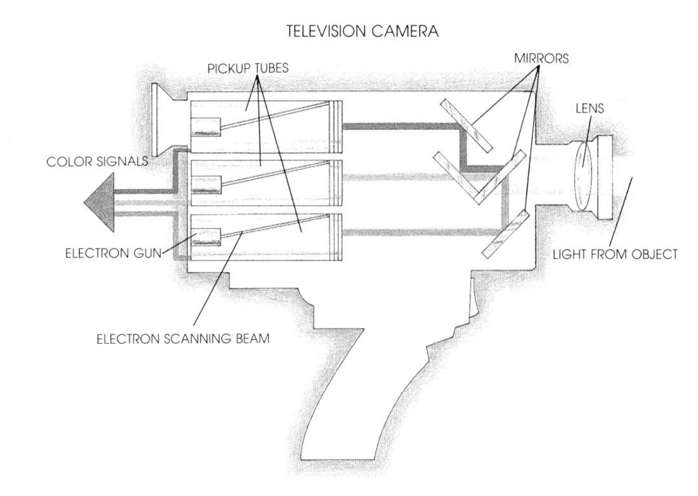

Fig. 5-12. A color video camera has three pickup tubes—one each for red, green, and blue. A black and white camera has only one pickup tube and does not need the mirrors to direct light. However, both cameras process the images in the same manner.

With the exception of cable TV, television signals are transmitted much as radio signals are transmitted. Your TV set is like a radio with a picture attached. Just as we have AM and FM radio, we have VHF and UHF television. VHF stands for *v*ery *h*igh *f*requency and UHF stands for *u*ltra *h*igh *f*requency. As in radio, each TV channel has a different frequency. The channel selector on the TV (receiver) tunes in the frequency of the channel (VHF, UHF, or cable) you select, just as the dial does on the radio.

The television receives the transmitted signals and converts them to electrical signals. The video and audio signals are then separated. The audio signals are sent to the television's speaker, which converts the signals back into sound. The video signals are sent to the picture tube.

The flat end of the picture tube (the screen) is covered with phosphor salts. (*Phosphor* is a substance that emits light when given energy.) At the back of the picture tube is an electron gun. The electron gun projects a beam onto the phosphorescent (phosphor-covered) screen. The beam goes in a left-to-right direction 525 times across the screen, 30 times every second, just as when the camera scanned the picture elements. Since each scene is replaced 30 times every second, your brain interprets these rapid changes as a continuously moving image. The beam excites the phosphor salts and makes them glow. These glowing dots, or **pixels**, create the image you see.

A color picture tube has three electron guns that sweep across the flat surface of the tube. One gun is used for each of the primary colors—red, green, and blue. The surface of the color picture tube is covered by groups of red, green, and blue phosphor dots, or pixels, that will make up the color image. Fig. 5-13. It is extremely important that the right electron gun hits the right phosphors. Most televisions have a masking guide that directs the beams to the correct phosphors. The beams excite the appropriate phosphors to create the color picture you see on the screen.

Apply Your Thinking Skills

1. About how many hours of television do you watch each day? Of this time, how much of your television viewing is for entertainment? For education? For receiving information?

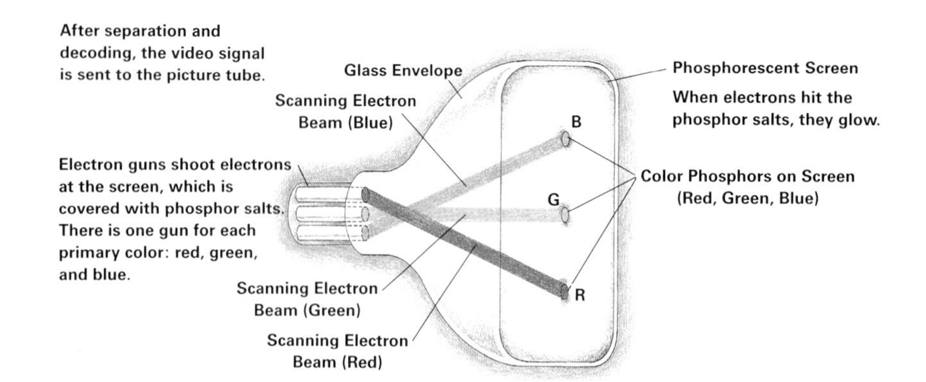

Fig. 5-13. Each of the three electron guns inside the picture tube, or cathode ray tube, scans for one of the primary colors of light.

Satellite Communication Systems

A **communication satellite** is a device placed into orbit above the earth to receive messages from one location and transmit them to another. Fig. 5-14. The satellite acts as a relay station. That is, it simply reflects signals back to earth. For that reason, a satellite is often called a "mirror in the sky." Just as a mirror reflects your image, a satellite reflects signals back to earth.

Communication satellites are placed in orbit approximately 22,300 miles above the earth. They travel at the same speed as the earth rotates. Thus, they remain above the same part of the earth at all times.

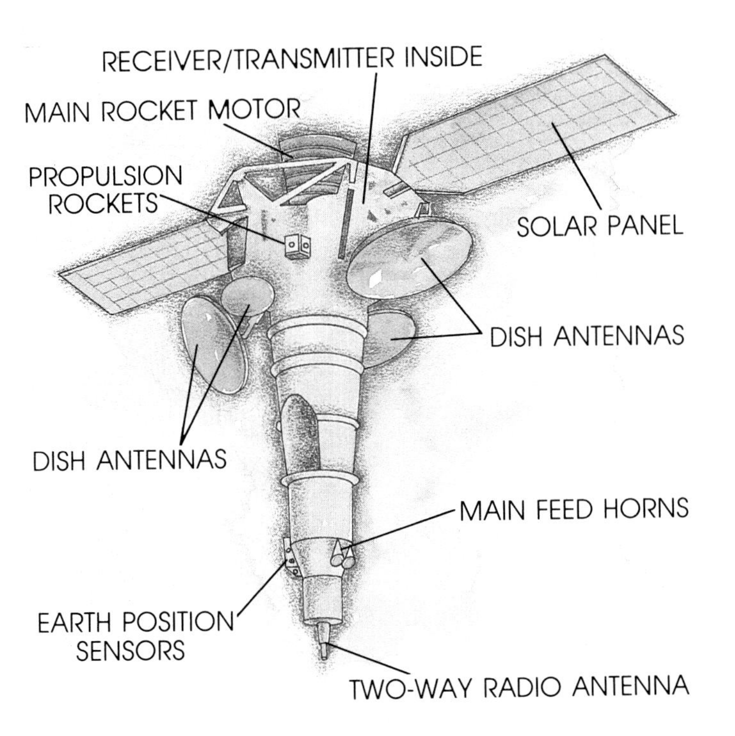

Fig. 5-14. This satellite collects electromagnetic signals with one set of antennas, amplifies (boosts) them, and then transmits them with a different set of antennas. The solar panels collect energy from the sun. This energy is used to operate the satellite.

Satellites have greatly influenced our modern communication systems. These devices are used in many aspects of today's message transfer systems. For example, satellites help transmit numerous types of messages, including telephone and television signals. Satellites are used to transmit printed information as well. For example, copy (stories and pictures) for the *Wall Street Journal* and *USA Today* newspapers is transmitted via (by way of) satellite to many printing locations throughout the country.

Satellites make it possible to communicate instantly. Live broadcasts depend on satellites to transmit messages as they are happening. Thanks to satellites, you can see things that happen far away while you sit in the comfort of your home. You, along with people all over the world, can watch the World Series games while they are being played. You can also witness exciting events, such as the live launch of a space shuttle.

Since their development in the 1960s, satellites have changed a great deal in design and efficiency. This is due to the technological developments taking place in electronics and communication. Satellite systems are being developed today with greater capacity. This means they can carry more messages. They provide better quality transmission.

How the System Works

Signals are sent to orbiting satellites through earth stations. An **earth station** (sometimes called ground station) is a large, pie-shaped antenna. It receives signals and transmits them to the satellite. This is called the **uplink**. The satellite receives the signals and transmits them to another location back on earth. This is called the **downlink**. Receiving earth stations capture the signals and send them to the desired receivers. Fig. 5-15.

Imagine how many signals are passing through the air constantly! You may wonder how the right message gets to the right earth station. When a signal is transmitted, the sender puts a certain code at the beginning of the message. The code directs the signal to the intended receiver. In addition, messages can be "scrambled." Then, only certain earth stations can pick up the message and decode ("unscramble") the information.

Apply Your Thinking Skills

1. Satellites can take pictures of earth from space. How might this help meteorologists (people who study weather patterns)?

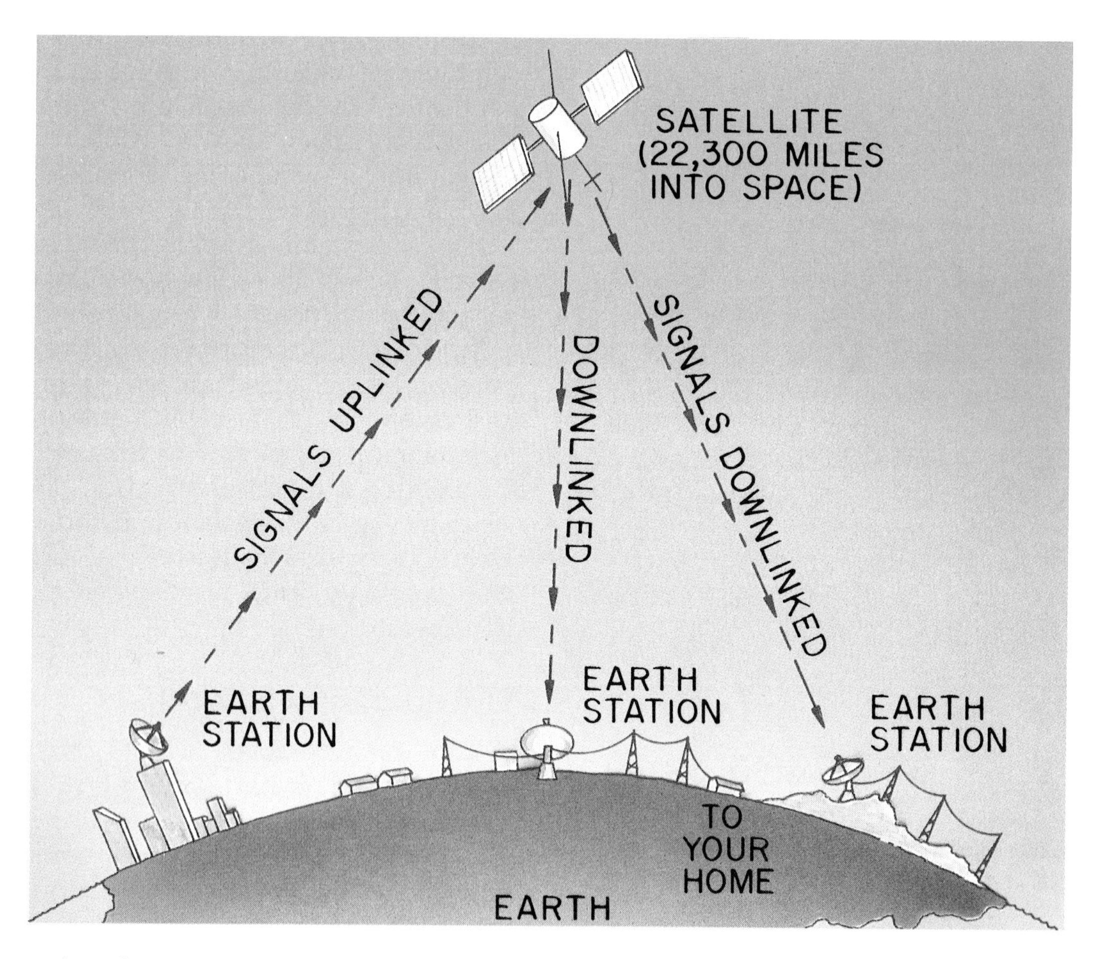

Fig. 5-15. This is how a satellite is used to transmit information from one place to another place far away.

Global Perspective

Bigger and Better

A major technological change is expected to make television bigger and better beginning in the mid-1990s. The innovation is called high definition television (HDTV).

How does HDTV differ from the television to which we are accustomed? For one thing, the shape of the screen is different. An HDTV screen is wider in proportion to its height than the screen of a present TV. The width-to-height ratio for HDTV is 16 to 9 compared with the present 4 to 3 ratio. That means, for example, that an HDTV screen 16 units (inches, for example) wide is 9 units (inches) high. A present-day TV that is 16 units wide is 12 units high.

The other major difference is that HDTV has greater picture clarity. This fact results mainly from doubling the number of scanning lines making up a TV picture. In the U.S., the number of lines will be increased from 525 to 1,050. In Europe, it will be increased from 625 to 1,250. More scanning lines means that HDTV sets can be made quite large without losing the sharpness of the picture.

A leader in the development of the new HDTV technology is Philips, a company based in the Netherlands in Europe. The company's present goal is to start commercial sales of HDTV sets in 1994, but it has several plans related to this goal.

First of all, Philips is developing technology that will allow for a gradual changeover from present TVs to HDTV. If transmission signals meant for HDTV are compatible with (agree with) present signals, then HDTV signals can be received on present sets and vice versa. This will allow people to choose for themselves the best time to buy the new equipment.

Philips also recognizes the need to resolve problems related to the variation of TV standards between countries. The company is working with others to, first of all, achieve compatibility between systems. The long-range goal, however, is for companies/countries around the world to adopt a single set of standards. In this world made ever smaller by advances in communication technology, Philips believes that the achievement of that goal would ultimately benefit all.

Extend Your Knowledge

1. Research continues on improving television. Find articles in magazines and newspapers about HDTV and other advances being made in television and share these with your class.

Fig. GP-5. The development of high definition television (HDTV) could revolutionize television in the '90s and beyond.

Chapter 5 Review

Looking Back

Telecommunication is the term given to those communication media that allow us to send messages over long distances.

The telegraph was the first device developed for telecommunication. With electricity and magnetism, coded messages could be sent over wires that stretched from one place to another.

In telephones, sound waves are converted to electrical signals that can be sent over wire. The electrical signals are converted back to sound in the receiver. Today, telephone messages may travel over wire, coaxial cable, optical fibers, or microwaves.

With radio, sounds are converted into signals that can be sent through the air. The radio waves are picked up by receivers tuned in to the right frequency.

With television, both picture and sound must be converted to signals that can be sent through the air or over cable. A TV receiver converts these signals to the pictures we see and sounds we hear on our television.

The development of radio and television made mass communication a reality.

Satellite communications is one of the main reasons why mass communication has become so easy.

Review Questions

1. What is an electromagnet?
2. Describe how the telegraph works.
3. What is a patent? Why is it important for an inventor to get a patent?
4. Describe how sound is converted into an electrical signal in a telephone.
5. What are optical fibers? Give three advantages of using optical fibers to carry telephone messages.
6. What are electromagnetic waves and how are they used for communication?
7. Define frequency.
8. Explain what happens when you tune a radio to a particular station.
9. What is the technical difference between AM and FM radio?
10. What is a communication satellite?

Discussion Questions

1. Your color TV has received the transmitted signal and sent the video signal to the picture tube. Briefly describe what happens to make the picture appear on the screen.
2. Compare the advantages and disadvantages of using radio versus TV to broadcast news information.
3. A process called *colorization* uses computer technology to turn old black-and-white movies into color ones. This process appeals to many. Others feel that this destroys an art form. They think that it intrudes on the rights of the filmmaker. What is your point of view?
4. The Federal Communications Commission (FCC) regulates communications industries. This means that it sets up specific rules and guidelines that broadcasting corporations must follow. Discuss why regulation of this industry is considered necessary by many people. Also discuss the disadvantages of such regulation.
5. In what ways do you use and/or depend on the kind of information that is transmitted via satellite?

Chapter 5 Review

Cross-Curricular Activities

Language Arts 

1. Discuss the effects of cable television versus network programming. Divide into small groups to debate the issue. You will need to support your ideas with facts.
2. Television newscasters often write their own copy, or script. Cut out an article from the newspaper and rewrite it for a television newscast.

Social Studies 

1. Write a two-page biography of Alexander Graham Bell. Include in your biography what circumstances led to Bell's invention of the telephone, as well as other significant Bell inventions.
2. You can find out a lot of information about a product by doing a patent search. This is especially true of an older product that is no longer made. Find out how to go about conducting a patent search on such a product. What kind of information would a patent search reveal?

Science 

1. A color television screen presents images in the form of combinations of pixels of red, green, and blue light. The eye blends these pixels to see the colored images. The process is similar to the way color vision works. A tiny portion of the retina at the back of the eye contains receptors that sense red, green, and blue light. These receptors are stimulated in combination to let us see all colors of the color spectrum. What receptors are stimulated so we see yellow? Use a magnifying glass to look at a patch of yellow on a color television screen and find out.

Math

A knowledge of the Morse Code is a handy way to send messages with little advanced technology. For the activities below, refer to the copy of the international Morse Code in Fig. 5-2.

1. In 1844, Samuel F. B. Morse sent a message which earned him national praise for his invention of the telegraph. Decode the following to reveal that famous message.

   .-..-. /.- -/.... /.-/-/ /.... /.-/-/.... /
   /- -. /- - -/-.. /
   /.- -/.-. /- - -/..-/- -./.... /-/..- -.. /.-..-.

2. Code a Morse Code message to one of your classmates, and then have him or her code one to you. *Note:* Using "long" for dashes and "short" for dots, you can send Morse Code messages by other means, such as by tapping or by blinking a light.

Chapter 5 Technology Activity

Video Production of a Commercial

Overview

Commercial messages have been broadcast on television almost since its invention. Most commercials are created to sell products. In this activity, you will work in small groups to write and direct a commercial. You will then produce and record a commercial message on a videocassette. The commercial can be a public service announcement, a product advertisement, an announcement of an upcoming school event, or another related message. There will be a set amount of time allowed for each commercial. You will need to edit your commercial so that it will fit this time exactly.

Your group will be required to storyboard (organize) your commercial. You will need to write a script. You may also need to prepare props, scenery, lighting, and costumes. Each group will have to appoint actors. The actors will need to read the commercial or work in front of the camera.

This activity will help you learn how a videocassette recorder (VCR), camera, and monitor work together. You will be able to see how they show a real object as an image on the monitor screen.

Goal

To produce a commercial message on a videocassette.

Equipment and Materials

VCR
video camera
color monitor
blank cassette tapes
3" x 5" cards
microphones
props as required

Procedure

1. Your teacher will assign you to a group of two or three people.
2. Your group will then select a message to produce. The messages should be original. The product that you wish to advertise must be something that can be aired on commercial television.
3. Select the time limit you will use for your commercial message. Your choices are 15 seconds, 30 seconds, or 60 seconds.
4. Prepare the storyboard cards for your commercial. Fig. A5-1. Use 3" x 5" cards. Each storyboard card should include the following information:
   - sequence number
   - action to take place
   - names of people responsible for performing the task
   - desired length of time for each action

   Organize the storyboard cards in the order in which each task is to occur. These may be reorganized as needed.
5. Write the script. Decide who is to act in front of the camera.
6. Prepare the props and scenery.
7. Rehearse the scene(s). Make changes as needed.

Chapter 5 Technology Activity

Fig. A5-1. Preparing a simple commercial.

Sample storyboarding card.

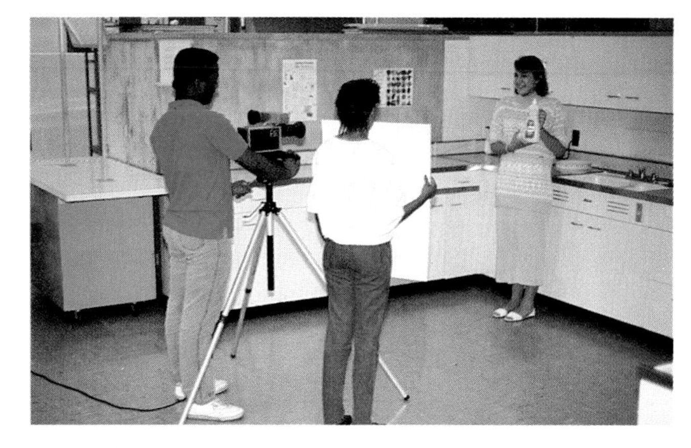

Taping the commercial.

8. Practice taping your group's commercial. At each taping session, be sure to record the VCR counter's beginning and ending numbers. This will help you preview each session. It will also help prevent accidental erasing of your commercial. Tape at least three practice takes of your commercial. Then record the final take on a separate tape.
9. All of the finished commercials are to be viewed by the class. The commercials are to be critiqued (examined critically) using the evaluation form shown in Fig. A5-2.

Evaluation

1. What did your classmates think of your group's commercial overall? What areas of the evaluation form did your group do the best and worst in?
2. What part of the production process did you enjoy the most? Why?

Ten points maximum in each category. Record the scores earned on a separate sheet of paper.*	High ⟵ ⟶ Low
Actor's Ability — knows lines, makes good eye contact, has good facial expression, moves smoothly	10 9 8 7 6 5 4 3 2 1
Set Design — neat, lettering easy to read, attractive setting	10 9 8 7 6 5 4 3 2 1
Creativity — original ideas, unique delivery, special attention-getting devices	10 9 8 7 6 5 4 3 2 1
Impact on Viewer — held your attention, made you interested in the product, convinced you to buy	10 9 8 7 6 5 4 3 2 1
Proper Length — (15, 30, or 60 seconds) — subtract 1 point for each second below or beyond the set time	10 9 8 7 6 5 4 3 2 1

*Do not write in your books   TOTAL________

Fig. A5-2. Commercial evaluation form.

Chapter 6

Graphic Communication

Looking Ahead

In this chapter, you will discover:

- the importance of graphic communication media in our society.
- what makes a good graphic design.
- those processes that help us to record and duplicate messages.
- some of the technical processes involved in graphic communication.
- how printing, photography, and drafting are used in graphic communication.

New Terms

drafting
electrostatic printing
emulsion
graphic communication
gravure printing
hologram
ink-jet printing
lithography
logos
photographic printing
planographic printing
principles of design
relief printing processes
screen printing

Technology Focus

Seeing Is Believing—or Is It?

New wave graphics—the wave of the future. In the past, a graphic artist often had to labor long and hard assembling photographs, artwork, and type for printed products. It was an important, but time-consuming, task. Today's new imaging systems make this work much easier and more fun.

The new wave graphic artist sits at a computer workstation. He or she can put together complete *layouts* (graphic versions of how the words and pictures will be arranged) for magazines and posters, all on a computer screen. A touch of a special electronic pen, and a photograph is moved from the top of the page to the bottom. Is there a tree or building in the background of a photo that is distracting? The graphic artist can electronically remove it. Would you like to combine two pictures? A photo can be altered as needed. It can even be changed from a realistic picture into a special effect.

Part of two or more photos can be combined. For example, Michael Jordan of the Chicago Bulls could appear to be playing basketball on your school playground.

The machine that makes this possible is called a *digital image processor*. It converts pictures into thousands of tiny squares called *pixels*. The computer analyzes each pixel for color and brightness. The artist chooses which pixels need to be changed and tells the computer how they should be changed. Thanks to systems like this one, printed products can be more accurate and effective . . . and more exciting to create.

Fig. TF-6. Using a computer, a graphic artist can combine two photos, such as these of a school playground and Michael Jordan on a basketball court, to make a new picture—Michael Jordan playing on a school playground.

What Is Graphic Communication?

Graphic communication is the term given to methods of sending messages using primarily visual means. The basic methods are:

- Printing
- Photography
- Drafting

Think about all the visual messages you receive every day. Photographs, pictures, product labels, signs, newspapers, magazines, books, and mail are just a few examples of graphic communication media. (*Media* refers to the *means* of communication.) Can you name others?

Apply Your Thinking Skills

1. Think about all the signs you see in one day. Consider road signs, safety signs, and signs hanging in your classroom or library. Imagine what it would be like if you did not have these signs. In what ways would it affect you and the things that you do?

Visual Design

In order to communicate a message effectively using graphic media, the visual design is extremely important. *Visual design* refers to how something looks. Do people find it appealing? Does it capture your attention? Does it communicate the idea well? Each of these questions refers to the visual design of the product. They are important questions to consider when designing a graphic message.

Certain designs or pictures seem to catch your attention better than others. Many companies and organizations develop their own symbols with which they are quickly and easily identified. These symbols are called **logos**. Think about the logos (short for logotypes) that you know. Fig. 6-1. Why do you think they appeal to you and others? A good design captures the attention of the intended reader or viewer. It effectively communicates the message and leaves a lasting impression.

Designers must consider how a design will be used. They must think about the function of the design. Does it need to appeal to young people or older people? Will it be used in a magazine or on a billboard? Will it be read and thrown away like a newspaper—or will people look at it for many years as they do a photograph?

Many factors can have an impact on the effectiveness of a design. Some of these will be discussed next.

Fig. 6-1. Logos have been developed for different companies, organizations, and services to quickly communicate their identities.

Principles of Design

When designing a graphic message, the following **principles of design** should be considered:

- Balance
- Proportion
- Contrast
- Harmony
- Unity

These factors help to determine the effectiveness of the design.

Balance

The visual weight of images on a design is referred to as *balance.* Some designs can be formally balanced. Formal balance is achieved when a line drawn down the center of the design would create two halves that are very similar to one another, or symmetrical. Others can be informally balanced. The objects in the design may look different, but they have equal weight to the eye. Fig. 6-2. The viewer must gain a sense of balance from the design—the space, type (words), and artwork should be positioned to give the viewer an impression of steadiness.

Proportion

The size relationship of the various parts of a design is important. This relationship is referred to as *proportion.*

The sizes of type, drawings, and photographs are all considered carefully. The designer wants to give the right amount of attention and space to each part of the message. A design that is out of proportion will appear awkward and displeasing to the viewer.

Contrast

Contrast refers to techniques used to call attention to certain parts of the message. It is the emphasis in a design. The designer wants the viewer to especially notice important parts of the message or idea. Contrast is often accomplished through simple techniques like underlining, adding color, changing the size of the type, or using arrows ⬅⬅.

Harmony

Whenever you look at a design or layout, your eyes will have an immediate reaction. You will either be drawn into the design or

FORMAL BALANCE

INFORMAL BALANCE

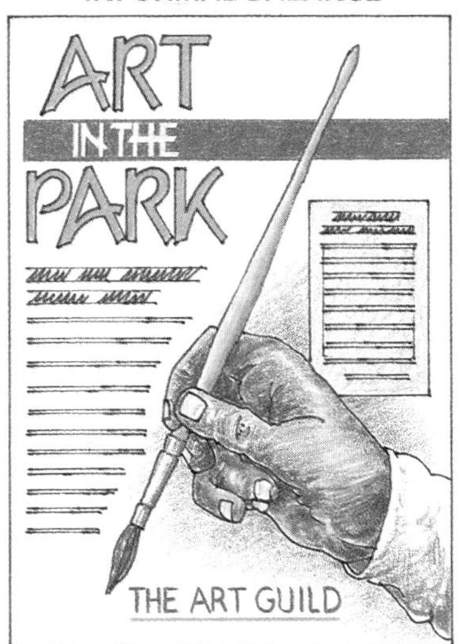

Fig. 6-2. A design with about the same amount of pictures and type on both sides of an imaginary center line is formally balanced. If there is an unequal amount of pictures and type on either side of an imaginary center line, but it appears to have even weight, it is said to be informally balanced.

be turned away. A good design keeps the viewer's attention.

Harmony refers to the aspects of a design that encourage the viewer to "look into" the design. It addresses the eye movement and flow of the design. Harmony is accomplished when the type and pictures are placed appropriately. Fig. 6-3.

Unity

Unity refers to the overall impact (effect) of the design. If all aspects of the design must appear to belong together and work well together, the design has unity.

Unity is achieved when the designer plans the kinds of type, the colors, and the shapes used in the design. These must work together to create an effective message. If any one of these is too prominent (noticeable), the effect of the design can be ruined.

Fig. 6-3. The position of the elements in this ad encourages eye movement. This is called harmony.

Apply Your Thinking Skills

1. Find an advertisement in a newspaper or magazine. Analyze how the principles of design were applied in this specific ad. Were they effective or ineffective?

The Creative Design Process

All printed products (magazines, books, labels) are designed first. Designers go through a sequence of steps to create a design that is ready for printing reproduction. The creative design process consists of four major steps:

- Thumbnail sketches
- Rough layouts
- Comprehensive layouts
- Camera-ready art

Most designs start out from thumbnail sketches. Designers will often come up with many ideas before choosing the one they like best. A rough layout and a comprehensive layout are made of the design that a designer thinks is most appealing and functional. Camera-ready art (often called a *pasteup*) is prepared after the layout stages are complete. Computers and other special tools are used to make camera-ready art.

The first three steps of the creative design process are shown in Fig. 6-4. The last step of this process, camera-ready art, shows the type, art, and lines combined together. Everything is accurate and perfectly laid into place, and the type is usually set on special machines that make neat letters.

Designing with a Computer

Today, the creative design process is aided by the use of computers. Some designers use computers to create thumbnails, rough

Fig. 6-4. Here you can see the creative design process from thumbnail sketches, to rough layout, to full-color comprehensive layout.

sketches, comprehensive layouts, or camera-ready art. The computer makes it easy to combine text with illustrations. As a matter of fact, a person could use a small desktop computer to completely lay out a publication of his or her own. This is often called *desktop publishing*.

Designing with a computer can make the creative design process much easier. This is because the computer makes changes quickly and accurately. Using a computer, a designer can turn 1-inch letters into 3-inch letters or move a picture from the top of the page to the bottom. The designer can fill a page with an illustration, or shrink it down to fit in the corner of the page. These changes can be completed within seconds or minutes. A designer doing the work by hand would need to redraw the entire layout to make these kinds of changes. It would take hours or days.

With special computer systems called *electronic pagination systems*, the designer can alter colors in photographs, too. These systems use scanners to input photographs into the computer's memory. The scanner turns the photograph into electronic information that the computer can understand. Then the designer can alter that electronic information. Fig. 6-5. For example, a yellow dress can be changed to a bright red one. A black car can be changed to blue or white within seconds. The designer can also combine photographs into one photograph.

Apply Your Thinking Skills

1. Electronic pagination systems enable designers to make changes to photographs. You cannot tell whether a photograph printed in a newspaper, book, or magazine has been electronically altered. How do you think this capability can be misused?

Printed Communication

Once the message is designed and camera-ready art is prepared, printers convert the message into a form that can be printed. Many stages are involved in this process, but we will concentrate mostly on how it can be printed.

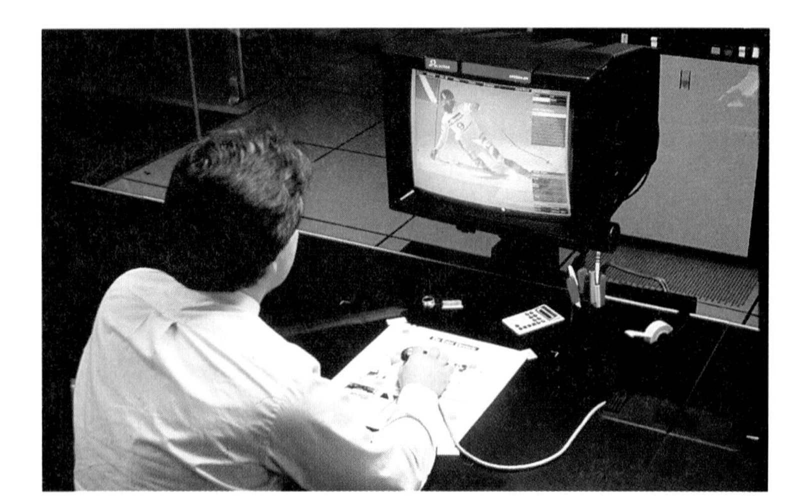

Fig. 6-5. Electronic pagination systems combine color scanners with computers.

The Importance of Printing

Most graphic messages sent and received daily involve some type of printing. Cereal boxes, candy wrappers, and wallpaper are printed. Printing is usually taken for granted. It is everywhere around us. Expressing messages without printed words and pictures would be difficult. Fig. 6-6.

Fig. 6-6. How would you know what is inside a container that had no printed information on it? These containers could contain milk, orange juice, or eggnog, for example.

Printing processes have made it possible for people to send many copies of a message in exact duplicate. In addition, printing has provided a way to preserve knowledge by recording information.

Before printing was invented, people recorded information by hand. If several copies were needed, scribes (persons employed to write by hand) would handwrite information many times over. The invention of methods that made it possible to duplicate printed information was vitally important to the advancement of our culture.

Types of Printing Processes

Printing can be done in a number of ways. The type of printing process used depends on what is being printed, the quality of reproduction desired, costs, and the speed at which it must be printed.

Generally, printing techniques are grouped according to the type of printing surface used. Major groups of printing processes are:

- Relief
- Porous
- Planographic
- Gravure
- Electrostatic

Other printing processes include photography, ink-jet printing, and laser printing.

Relief Printing Processes

Relief printing processes are methods that print from a raised surface. Fig. 6-7. Parts of the image are raised above the surface. These pick up ink and transfer it to paper. Printing processes that use raised printing surfaces include letterpress and flexography.

Letterpress is a very old form of printing. Johann Gutenberg is given credit for inventing the process upon which this technique is based. Gutenberg, the "Father of Printing," invented movable metal type. In *letterpress*, the reverse or mirror image of each character (letter of the alphabet, punctuation mark, or symbol) was made into a piece of metal type. Printers hand-set lines of metal type, letter by letter, to form sentences, paragraphs, and pages. Ink was rolled over the raised letters. Then, paper was pressed on the inked letters to transfer the message. This process was repeated until the desired number of pages were made.

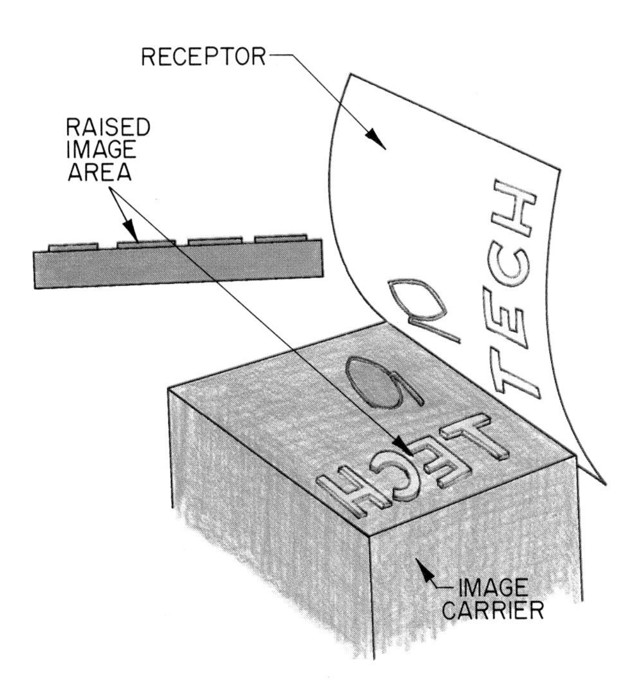

Fig. 6-7. Relief printing is printing from a raised surface.

Today, letterpress printing is being phased out of most companies. New machines can print messages more quickly and efficiently.

In *flexography*, raised letters are formed on plastic or rubber sheets. Letters do not have to be hand-set. Instead, they are produced using photographic and computer methods that are very fast.

Porous Printing Processes

A *porous* material has many small openings (pores) that allow liquids to pass through. In *porous printing processes*, ink or dye is passed through an image plate or stencil and transferred onto the material being printed. A *stencil* is a thin piece of material with holes cut out in the shape of letters or a design. Stencils can be made many different ways. The material being printed on (the receptor) may be paper, cloth, glass, or another material. The most common porous printing process is screen printing. It is sometimes referred to as silk-screen printing.

In **screen printing** a stencil is adhered (made to stick) to a porous screen. All areas of the screen are blocked out except for the

Fascinating Facts

By the end of the second century A.D., the Chinese already had paper and ink—plus the idea that pictures and word characters could be carved in a block of wood, that ink could be placed on these raised images, and that the ink could then be transferred to paper. By A.D. 700, they put these three elements together to create printed material—a Buddhist charm. This was the first relief printing. In 868, they produced the first block-printed book, an entire Buddhist tract.

area to be printed. A special tool, called a *squeegee*, is used to force the ink through the stencil and onto the product being printed. Fig. 6-8.

Screen printing is a simple and flexible printing process. This type of process can be used to print on almost any surface available. It is commonly used on fabrics, such as T-shirts and sweatshirts. Even three-dimensional products like drinking glasses and mugs or shampoo bottles are commonly screen printed.

Planographic Printing Processes

Any process that involves the transfer of a message from a flat surface is called **planographic printing**. Lithography is the most popular planographic printing method. It is also the most common printing process used today.

Lithography is based on the principle that grease and water do not mix. Using a material that has a grease base, an image is created on a flat plate. The image attracts ink because ink, too, has a grease base. However, all areas of the printing surface that do not contain the image are covered with water, which repels the ink. (Remember, grease and water do not mix.) Multiple copies of an image are made by coating the image areas with ink and the remainder of the plate with water on a continuous basis with a printing press. Fig. 6-9.

Gravure Printing Processes

Gravure, or intaglio, printing is the exact opposite of relief printing. In **gravure printing**, images are transferred from plates that have recessed (sunken) areas. The images are etched or carved into the surface. Each plate has many tiny holes, called cells, that combine to form the shape of the letters and symbols. The cells are filled with ink. Then, paper is forced against the plate. The paper absorbs the ink. The image transferred to the paper is the identical form of the image on the plate. Fig. 6-10.

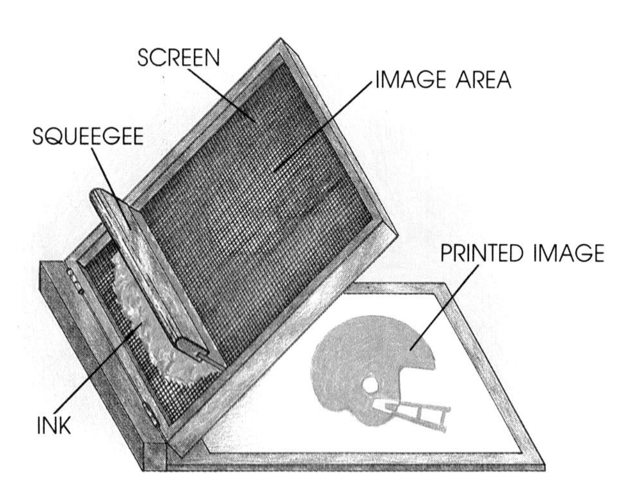

Fig. 6-8. Screen printing is a porous printing process. Ink passes through a stencil attached to a porous screen onto the printing surface below.

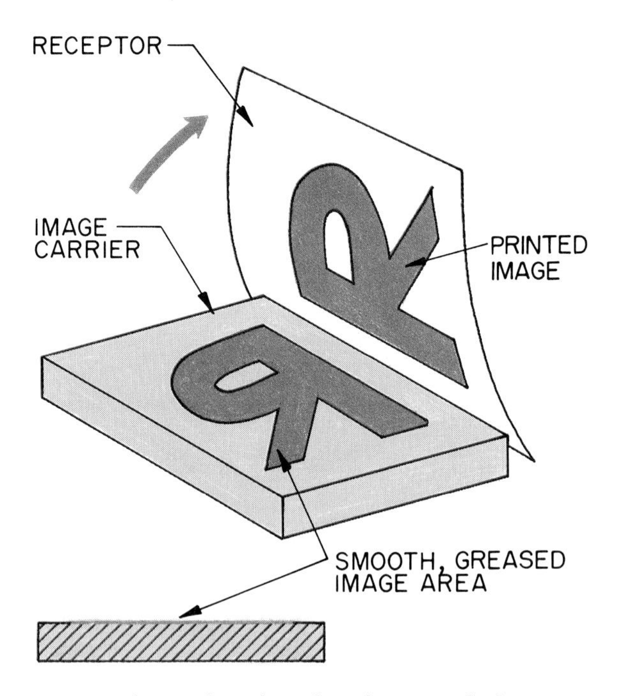

Fig. 6-9. Lithography is based on the principle that grease and water do not mix. Because the printing is done from a plane (flat) surface, this is a planographic printing process.

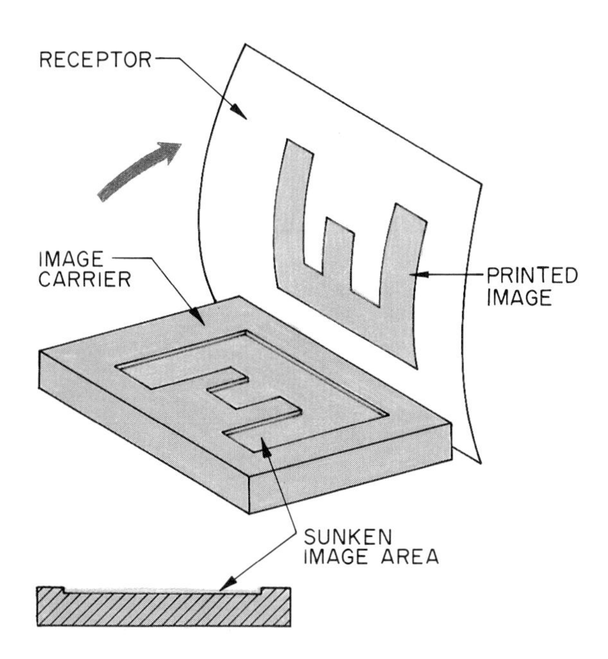

Fig. 6-10. Gravure printing transfers images from a sunken surface.

Gravure printing is an expensive printing technique, but it provides very high-quality printing results. Gravure printing plates require precise equipment. The plates are large copper drums that are etched with a message or design. Etching is done using either chemical processes or a diamond etching tool. Since the plates are so expensive to create, this process is generally used for long press runs or jobs that require high quality. In fact, all U.S. currency (paper money) is printed using this technique. Some magazines, wallpaper, stamps, and newspaper supplements are also printed using gravure.

Electrostatic Printing Processes

Many quick printing companies use electrostatic printing processes. In fact, you have probably used this technique a few times yourself. Copier machines that you use to make quick duplicates of your original fall into this type of printing category.

Electrostatic printing is based on the principle that opposite electrical charges attract, while like charges repel. Basically, when your original is placed on the glass window, it is exposed to a plate inside the copier. The plate has been given a positive charge. The light removes the charge from the non-image areas of the plate. The image area remains positively charged. A toner material is given a negative charge. It adheres to the positive image areas of the plate. A paper is given a positive charge and passed over the toner plate. Since opposite charges attract and the positive charge is stronger than the negative charge, the toner is transferred to the paper. A heating element is used to set the toner on the paper permanently. Fig. 6-11.

The electrostatic printing process makes it possible to duplicate messages quickly. However, the reproductions are generally of poorer quality than those made using other printing techniques. For long runs, this process is too slow and expensive.

Other Printing Techniques

Photography is often considered to be a printing technique. Basically, in **photographic printing**, light is projected through a plate (usually called a negative) onto a light-sensitive material. After processing, the image appears. (Photography is discussed in greater detail later in this chapter.)

Another printing technique that does not easily fit into the other categories is called **ink-jet printing**. During this process, ink jets spray ink onto paper. Fig. 6-12. This process is actually computer-controlled. Digital data control the tiny nozzles that spray the ink droplets onto the receptor material. Because there is no contact between the image carrier and the material being printed upon, the process can be used to print on a wide variety of materials.

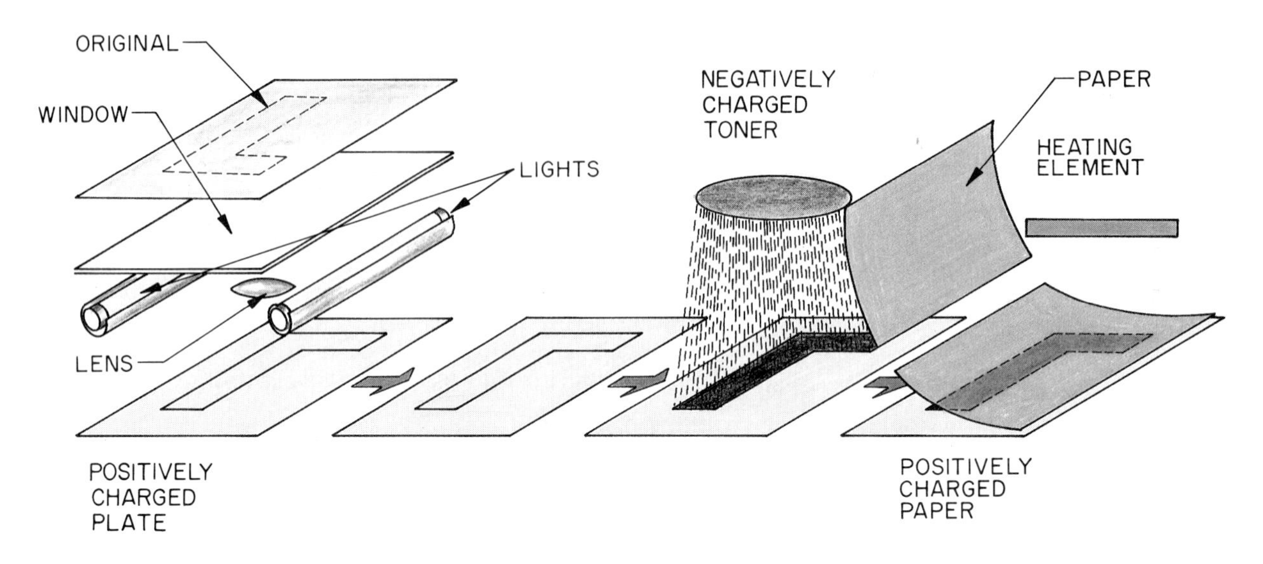

Fig. 6-11. This diagram shows the process involved in electrostatic printing.

Laser printing is one of the newest forms of printing today. As you learned in Chapter 3, *lasers* are devices that strengthen and direct light to produce a narrow, high-energy beam. The most common laser printers work much like electrostatic copiers. They print onto regular paper. Many businesses have these printers hooked up to their desktop computers. They provide good quality output, and they operate much more quietly than typical computer printers. Fig. 6-13. Other laser printers are designed to expose photographic paper. These laser printers are used to make very high quality images and type to be used for camera-ready art. They are commonly found in art departments or printing companies.

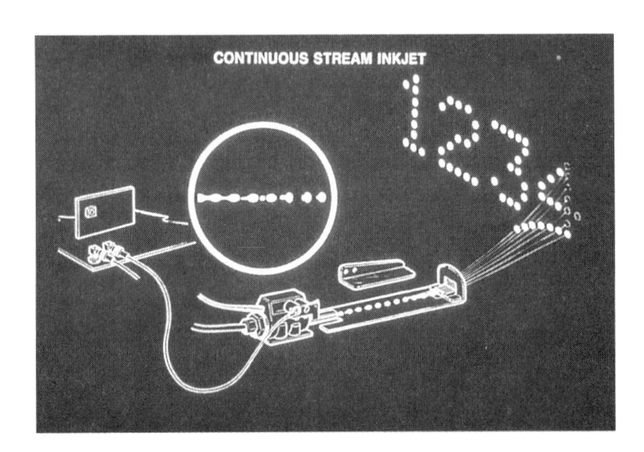

Fig. 6-12. Ink-jet printing transfers images onto a surface by spraying very thin ink.

Choosing a Printing Process

The processes used to get messages onto a surface have changed considerably over the years. Different processes are used for

Fig. 6-13. Laser printers are fast becoming the preferred printer in offices. Copies are of high quality, and the machines run quietly.

different reasons. However, the basic idea is still the same. That is, no matter how the message is printed, it must be visually effective. As computers and other devices begin to transform (change) our printing machines and methods, keep in mind that the message being communicated is what's most important.

Apply Your Thinking Skills

1. Think about each type of printing process discussed. Decide which one would be best for the following printing jobs:
   - Printing a company logo on 100 T-shirts
   - Printing ten 12-page reports
   - Printing 1,000 flyers announcing a special event

Communicating through Photography

"One picture is worth a thousand words." How many times have you heard that statement? Often, photographs can relate so many concepts (ideas) and feelings that words are hardly necessary.

Most graphic communication media depend on photography to send a message. Can you imagine magazines and newspapers that used only words? Photographs help us to capture feelings, expressions, concepts, information, and images using a camera and film.

The Photographic Process

The photographic process is much like the way you see. Everything reflects light. You see objects because your eyes pick up and focus the reflected light. In photography, light is reflected from a subject and focused through the lens of a camera onto film. When the film is developed, you see a reproduction of what the camera "saw." This image was captured when light rays passed through the camera lens and reached the film. Fig. 6-14.

Using a Camera

Many kinds of cameras are available today to help us record images. Regardless of the type or size, most have the same basic parts that work on the same basic principles. A basic camera requires six features or devices:

- Light-tight space
- Lens
- Aperture with aperture diaphragm
- Shutter
- Filmholder
- Viewfinder

In order to record an image on film, it is necessary to control the light that reaches the film. Every camera must have a completely *light-tight space* in which to store the film.

Fascinating Facts

The world's first photograph took eight hours to develop. In 1826, French scientist Joseph Niépce spread a special solution on a sheet of pewter. He put this in a "*camera obscura*" used by artists at that time and placed it on a windowsill. In eight hours, he had a likeness of a scene on his French farm.

A *lens* is needed to focus the image on the film. The lens directs the light rays so they are correctly positioned and do not appear blurred.

An *aperture diaphragm* is required to control the amount of light that can reach the film. Just as the iris of your eye adjusts to changing light, the size of the aperture (opening) must be adjusted to varying light conditions. The amount of light allowed to reach the film can be increased or decreased to get a good exposure. Fig. 6-15.

Every camera must have a shutter. A *shutter* is a device that controls the amount of time that light is allowed to reach the film. When you take a picture and hear a click,

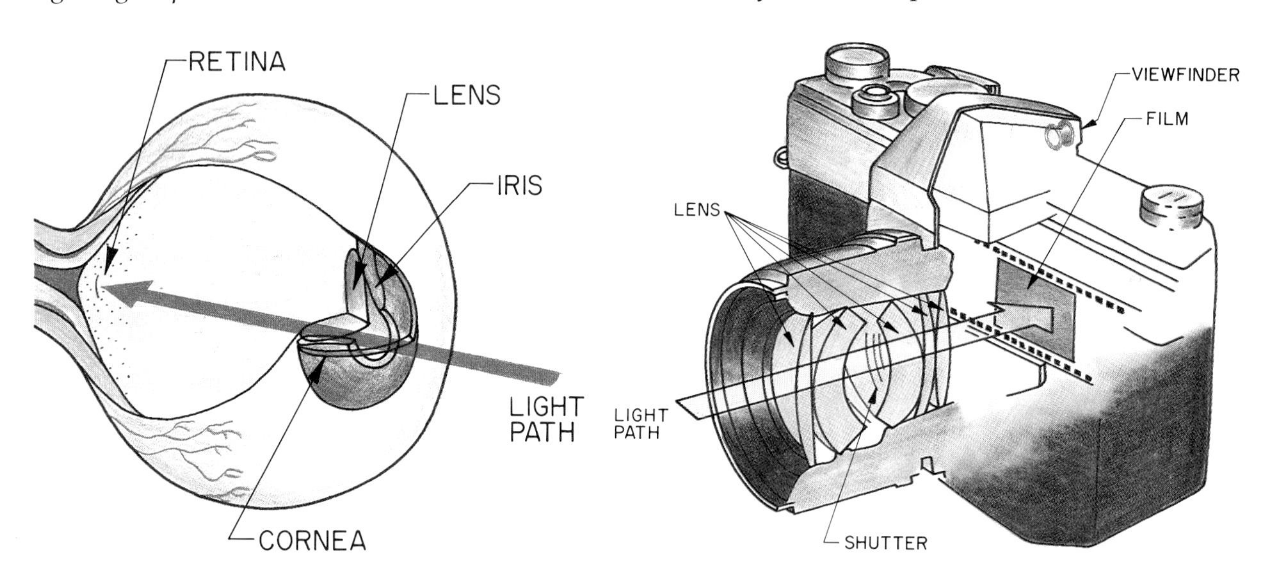

Fig. 6-14. A camera works much like an eye works, but is much less sophisticated. Your eye works because light is projected through the cornea, which bends and focuses the light to pass through a small hole called the pupil. The light then passes through the lens, which focuses the images on the retina. In a camera, light passes through the lens, which bends and focuses the light on light-sensitive film in the rear of the camera.

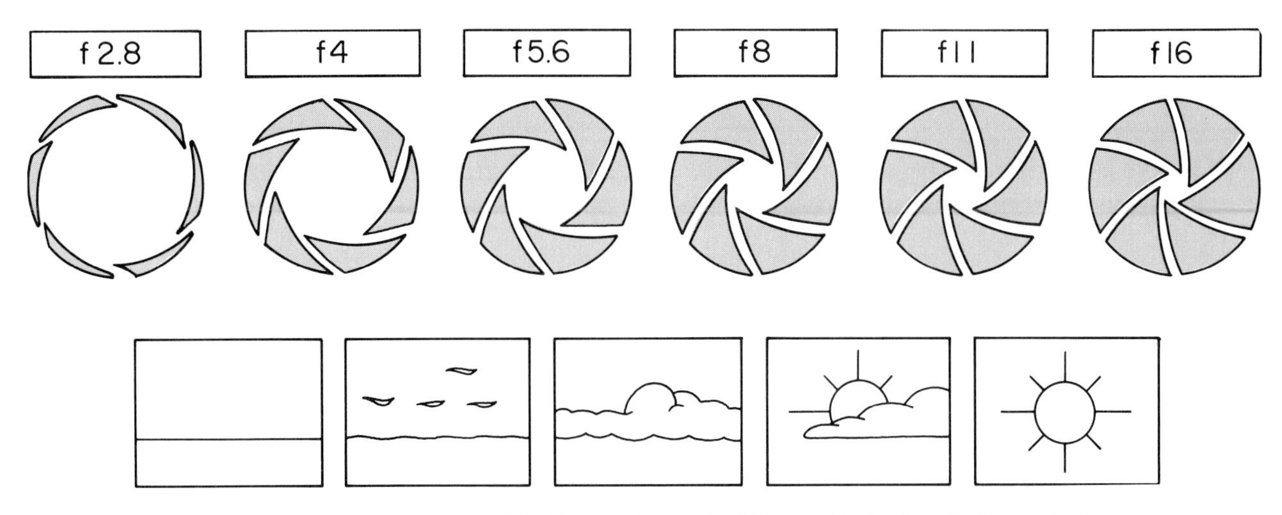

Fig. 6-15. The size of the aperture may be identified by numbers called f-stops. Notice that the larger the f-stop number, the smaller the opening. As you can see in the lower picture, small pictures are sometimes used to help you determine the correct aperture setting. The opening must be large when there is little light and small on bright, sunny days.

you are hearing the shutter open and close. It is like opening a door for a set amount of time. Usually, this set time is a fraction of a second, like 1/250 of a second or 1/30 of a second.

Next, a *filmholder* is needed. This holds the film in position so it will not move during an exposure.

Lastly, the photographer looks through a *viewfinder* to locate the subject and position the subject or the camera for the picture. The viewfinder is like a window. It allows you to see the image being photographed.

Photographic Film

Photographic film comes in a variety of formats and sizes to fit different cameras and applications. Rolled film comes in canisters or cartridges that fit inside your camera. Sheet film is most often used by professional photographers and printers who will be duplicating photographs.

Despite the size or shape of films, they have one thing in common. They contain one or more thin layers of a light-sensitive material called emulsion. The **emulsion** is the part of the film that captures the image. It must be chemically processed before the image can be seen.

Films are rated according to how sensitive their emulsion is to light. This is often called the speed of the film. A high-speed film, such as ASA/ISO 400, is very sensitive to light. It does not need much light for a good exposure. A low-speed film, such as ASA/ISO 100, needs more light for a good exposure. Photographers choose films according to the lighting conditions where they will be shooting the film. Which type of film do you think would be better to use if you were taking photographs outside around the evening campfire?

Video Photography

Technology breakthroughs in photography are improving the speed at which images can be recorded. New camera systems can record still images on computer disks instead of film. This makes it possible to instantly see the picture on a television screen or computer monitor. You don't have to take a roll of film to a store to be processed and have prints made.

Video photography works very much like regular photography. The main difference is the way the image is stored. In video photography, the image is stored as digital information. As such, it can be transmitted to other locations using telephone lines. It can also be printed out on a video printer to give you a full color photograph on paper. The paper is laminated to give it a glossy finish like a real photograph.

Photographic Composition

Photography is a very effective communication medium. Certain factors make some photographs better than others. These factors are guidelines to composition. *Composition* is the way in which all the elements in a photograph are arranged. When you compose a photograph, you plan what will be in the picture and how it will be positioned. To communicate effectively and to create a visually pleasing photo, you will want to compose it carefully. Consider some of the following tips when composing a photo.

Balance

Just as layouts should appear balanced, so too should the subjects in a photograph. You want the viewer to get a sense of balance from a photo. When composing a photograph, consider the positions of the subjects. Ask yourself questions like, "Is there too much on one side of the picture?" or "Is there appropriate space at the top and bottom of the picture?"

Photographers often follow the rule of thirds to position a subject in a photograph. The rule of thirds divides the image area into thirds both vertically and horizontally. Composition is more interesting when the subject is located at the intersections of these imaginary lines, rather than the exact center. Fig. 6-16. By positioning the subject off-center at one of these locations, the viewer gains a better sense of balance. The picture is more visually pleasing.

Framing

A photo is sometimes more effective if the photographer uses a foreground image (part of the scene nearest the viewer) to *frame* a background image. This is a good technique to use when photographing scenic landscapes. Other times a background image can be used to frame a subject. For example, the trunk and branches of a tree can be used to frame a person standing in front of the tree.

Fig. 6-16. This photo clearly shows the rule of thirds. Note that the main subject is placed off-center at the intersection of the lines that have been drawn on the photo for clarification.

Simplicity

A photographer should usually keep photos simple. This is done by concentrating on the center of interest in a picture. Background can be distracting to viewers. Sometimes, such as in portraits, no background is needed.

Leading Lines

A photographer often needs to "lead" the viewer into a picture. By positioning a subject appropriately, *leading lines* can be used to pull the attention into the photograph. This technique is especially important when photographing subjects that have apparent line structures. For example, the line structures of roads, bridges, fences, and buildings can be seen easily. Fig. 6-17.

Fig. 6-17. The lines created by the rows of flowers and the sidewalk "lead" the viewer to the building.

Visual Messages

The photograph is an important graphic communication medium. In science and technology, photos enable researchers to study microscopic forms and to explore the failure or success of experiments. Teachers and authors use them to educate and to explain information. Journalists rely on photos to inform readers about matters of current interest.

Most of us make use of photos. They provide records of times and events in our lives. We use them to recall the past.

Holography

Holography is the use of lasers to record realistic images of three-dimensional objects. As you've learned earlier, lasers are devices that strengthen and direct light to produce a narrow, high-energy beam.

In a photograph, you only see one side of the subject. In a **hologram**, which is the printed three-dimensional image produced by holography, you see the subject from different angles as you change your view of the hologram. You can see the front, back, and sides. It is as if you were walking around it! Fig. 6-18.

Holograms are a good way to record many details about an object. When viewed, they look more realistic than a photograph. It seems like the object is right there in front of you and that you can pick it up and hold it. Holograms are not used as often as photographs are used today. However, the technology to make them and reproduce them is improving all the time. One day, you will find them used in more places where you now see photographs.

Fig. 6-18. You only see one view of a person in a photograph. When you look at a hologram, you see different views of the subject, depending on the angle from which you are viewing the hologram.

Apply Your Thinking Skills

1. Think about ways that holograms could be used in schools to help students learn. Can you imagine ways that seeing holograms of plants and animals could help you in science? How could holograms of land forms help you explore geography? Would you appreciate artists' works more if you could see holograms of their sculptures instead of photographs?

Drafting and Design: The Language of Industry

Every product and structure that is made today begins as a drawing. Drawings are used to communicate ideas effectively and accurately. For example, all the design details of a building could not be verbally described to construction workers! To describe the details of a planned product or structure in a way that others can create an exact model of the planned idea, drafting is required. **Drafting** is the process of accurately representing three-dimensional (having height, width, and depth) objects and structures on a two-dimensional surface, usually paper. People who are skilled at recording and understanding drafted messages are called *drafters*.

The Process of Drafting

Drafters use many different techniques and tools to create a message. They can create messages using simple tools like pencils and rulers. However, today they often use computers in much the same way they are used in designing printed matter, which you learned about earlier in this chapter. Regardless of the materials and tools used to communicate the message, the thinking process is the same.

The first step in creating a drafted drawing is sketching. Sketches are created freehand and are drawn quickly. Sketching allows the drafter to experiment with design ideas. Sketches show the ideas, roughly but neatly. Sketches should also show the proper size relationship between parts.

Once the drafter chooses the sketch that best represents the idea, detailed drawings must be made. The goal of drafting is to describe an object or structure so accurately that someone else can use the drawings to create an exact model of the idea. To achieve this goal, the drawings must accurately describe the shape, size, dimensions, and details of the object and all of its parts.

Drafters use many different types of drawings to accurately describe the shapes of things. These drawings show the object from different views and angles. Each view communicates a different message. Each drawing has a different purpose and use.

One common type of drawing, an *orthographic,* or multiview, drawing, shows three or more different views drawn at right angles, or perpendicular, to one another. Fig. 6-19. Other drawings are used to show an object in depth. The object is tilted or rotated so that three sides are visible in one view. These drawings are called:

- *Isometric.* In this type of drawing, the object is tilted so that its edges form equal angles. (Isometric means "of equal measure.") In isometric drawings, one corner always appears closest to you. Fig. 6-20.
- *Oblique.* An oblique drawing is made from the front view, with the top and side views lying back at any angle other than 90°. The front of the object appears closest. Fig. 6-21. This method is often used in drawing irregularly shaped objects.
- *Perspective.* These drawings resemble the way something would appear in real life. Think about when you look down a

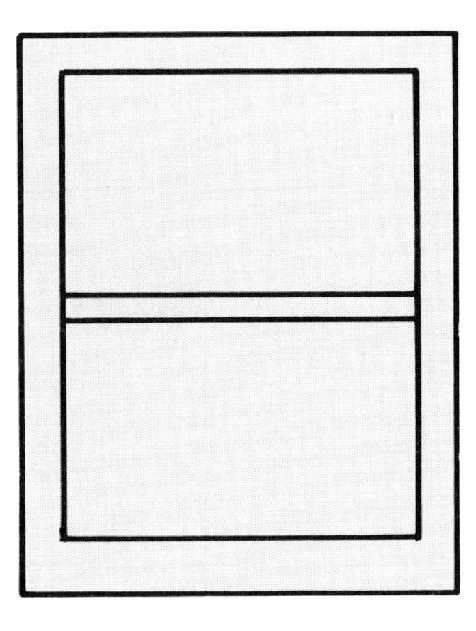

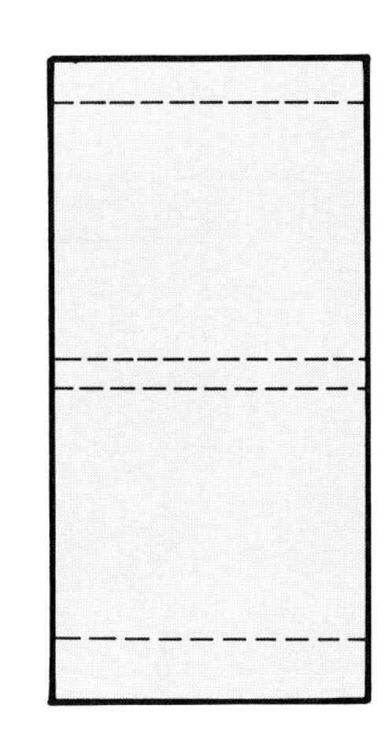

Fig. 6-19. This is an *orthographic* drawing of a cabinet. Most orthographic drawings show the view from the top, front, and right side. However, the more complex the object, the more views the drafter must show.

railroad track or long, straight highway. The track or road seems to get narrower and come together at some distant point, doesn't it? In perspective drawings, receding parallel lines appear to come together in a distance. Fig. 6-22.

Drafters add dimensions to their drawings to indicate the height, width (or length), and depth of the object (and often, of its parts). Usually, at least two dimensions are needed to describe a feature accurately. They use symbols to represent such things as materials being used, fasteners such as screws, or types of electrical outlets. They also use different kinds of lines, with each line having a different meaning. For example, a thin, dashed line is used to show a hidden feature, or one that is not normally seen. A thin line consisting of alternating long and short dashes is used to show the center of an object.

Drafting is often referred to as the universal language. A person who understands the basic symbols, lines, and rules can understand the message regardless of who communicates it. You don't need to understand the Japanese language in order to understand a drawing made by a Japanese-speaking person.

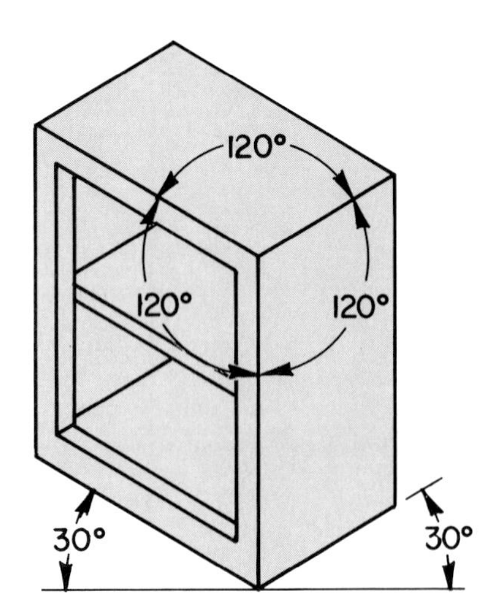

Fig. 6-20. This is an *isometric* drawing of the cabinet shown in Fig. 6-19.

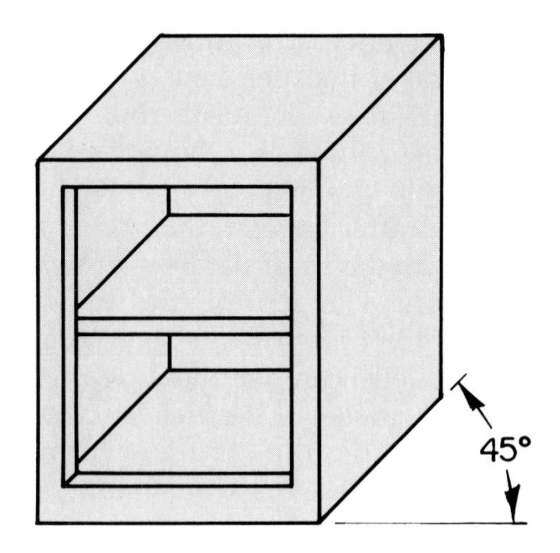

Fig. 6-21. This is an *oblique* drawing of the cabinet.

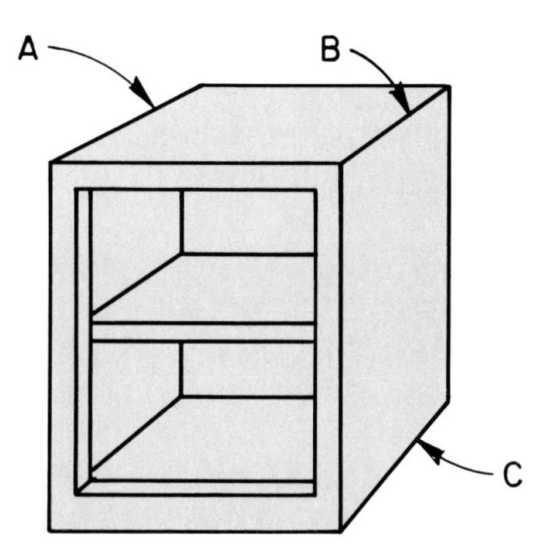

Fig. 6-22. A *perspective* drawing. If you extended lines A, B, and C, they would come together at some point in the distance. This point is called the *vanishing point.*

Apply Your Thinking Skills

1. Take a look at each of the drawings shown in Figs. 6-19 through 6-22. Think about instances in which each drawing of the cabinet would be useful. Consider which drawings best communicate how the product looks. Is one drawing better for the builder and another better for the buyer?

Investigating Your Environment
Paper Pollution

Manufacturing paper pollutes the air and the water. Fortunately, new technology has made it possible for paper manufacturers to reduce the amount of pollution they make. Many paper companies have reduced the amount of sulfur dioxide emitted during production by switching to cleaner-burning fuels. When a sulfur dioxide mixes with the nitrogen and moisture in the air, they combine to form *acid rain*. When acid rain falls to earth, it affects the growth of trees and plants and can kill all forms of life in lakes. Paper manufacturers have also worked to reduce the water pollution caused by releasing untreated production wastes, such as chemicals, into nearby streams and rivers.

Particles and chemicals are not the only environmental concerns of the paper industry. The paper they produce is a growing environmental problem. Americans use over 67 million tons of paper a year, and the majority of it ends up in our landfills. One way to keep so much paper from filling our landfills is to recycle it. A lot of paper we use is recyclable. Plain or office paper, corrugated cardboard, newspapers, and some food packaging can all be recycled. Unfortunately, a lot of things that you might think are recyclable are not. Many kinds of paper packaging contain wax or wax-coated cellophane, making them nonrecyclable. Most junk mail isn't recyclable. While newspapers are recyclable, their glossy, colored inserts are not.

Recycled paper can be used to make products such as newspapers, egg cartons, paper bags, tissues, food and clothing boxes, and corrugated boxes. In addition, recycled paper can also be used to make cardboard displays, building insulation, and roofing material.

Even the "waste" left over after recycling paper may be recycled. Manufacturing recycled paper produces a by-product called "sludge." It is expected that sludge can be used as a soil conditioner. Spreading it on farm and forest land can add important nutrients to the soil. It can be similarly used to line and cover landfills. Sludge can also be added to building and construction materials such as cement for bridges, highways, and other structures. It can even be burned and used for its energy.

In addition to all of this, the process involved in manufacturing recycled paper is better for the environment. It uses less water, fewer trees, and causes less pollution than manufacturing paper with virgin (new) pulp. Air pollution may be reduced as much as 74 percent, and water pollution as much as 35 percent.

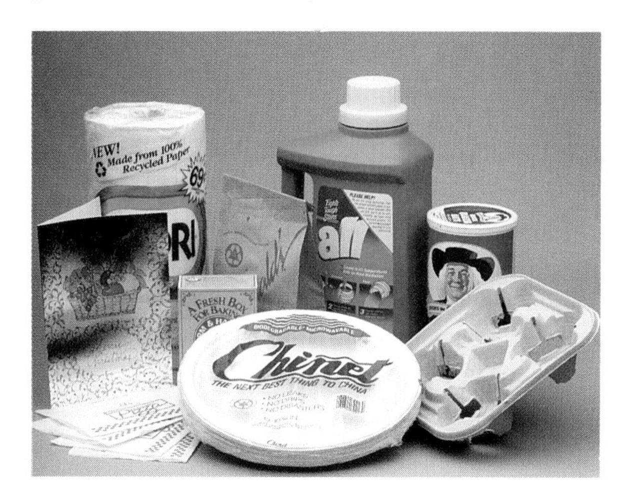

Fig. IYE-6. These days, a lot of products, including many paper products, are made from recycled material.

Take Action!

1. Is there a recycling program in your town? If not, find out if you can get one started.

Chapter 6 Review

Looking Back

Graphic communication is the term given to methods used to send visual messages. Photography, printing, and drafting are the most common examples of graphic communication media.

People are greatly affected by the visual appeal of communication media. Many factors must be taken into consideration when designing products for printing. These factors include balance, proportion, contrast, harmony, and unity.

Through a variety of printing techniques, it is possible to reproduce graphic messages in quantity, rapidly and accurately. Printing processes are divided into categories according to the surface from which the image is printed.

Photographic communication enables us to send many messages through a visual medium without words. In photography, light is reflected from a subject and focused through the lens of a camera onto film. The composition of a photograph can help communicate its message. In video photography, the image is stored as digital information.

The graphic communication medium that can accurately record information about the size, shape, and dimension of objects and structures is drafting.

Review Questions

1. What is a logo? Why do companies and organizations develop logos?
2. Name and briefly describe the five principles of graphic design.
3. Name and briefly describe the four steps in the creative design process used by graphic designers.
4. What is the main difference between relief printing and gravure printing?
5. Briefly describe screen printing.
6. Name and briefly describe the most common printing process used today.
7. Name the six basic features every camera must have and briefly tell the function of each.
8. What is photographic composition? Briefly describe four guidelines for composing a pleasing photo.
9. What is the goal of drafting?
10. Describe an orthographic drawing.

Discussion Questions

1. Analyze the visual design of this textbook's cover. Consider its appeal and function. Evaluate how each of the principles of design was applied.
2. Computers have greatly altered the creative design process in some companies. Discuss the advantages and disadvantages of such changes.
3. You encounter many forms of graphic communications during a day. Decide which ones are *not necessary* for you to effectively accomplish daily tasks such as personal hygiene or learning. Discuss ways that graphic communications media sometimes overload us with more information than we need.
4. Describe how you would compose a photograph of one or more members of your family that would best reflect the interests, lifestyle, and personality of your subject(s).
5. What school subjects would best prepare a person for a career in drafting?

Chapter 6 Review

Cross-Curricular Activities

Language Arts

1. Companies spend large amounts of money each year to advertise their products. Select a product and develop an advertising campaign. Before you begin, you must organize your thoughts and outline your strategy. Present your final promotional campaign to the class.
2. Create an ad jingle to go along with your advertising campaign.

Social Studies 

1. Choose a specific historical area of your community or a nearby community to photograph. Assemble your photos with those of your classmates to create a historical collage of the community.

Science 

1. A porous material has many small holes or spaces that let liquids or gases pass through. Use an eyedropper to drop water on various surfaces to check for porousness. Test natural materials such as leaves and human-made materials such as different kinds of paper, fabric, and ceramics. When would porousness be important in choosing materials for a printing process?
2. Electrostatic printing makes use of *static electricity*—the stationary charges on materials that have an excess of electrons (a negative charge) or a deficiency of electrons (a positive charge). You can produce static electricity by rubbing a plastic comb on a piece of nylon, fur, or wool. This builds an excess of electrons and a negative charge. Bring the charged comb near puffed rice, sawdust, or torn paper bits and it will attract them. This occurs because the tiny objects have a positive charge, and opposite charges attract.

Math

1. A circle on a CAD (computer-aided drafting) drawing has a diameter of 2". How many times must it be enlarged to become a 7" circle?
2. What percentage of reduction would be needed to shrink the 2" circle to a 3/4" circle?

Chapter 6 Technology Activity

Basic Drafting Techniques

Overview

Drafting is a form of graphic communication. It is a visual expression of information and ideas. In drafting, lines and symbols are used to graphically display ideas and data on paper as technical drawings. These drawings can be understood almost anywhere in the world. In this activity, you will learn a few of the basic techniques used in drafting, the universal language.

Goal

To learn a few of the proper techniques used in drafting by drawing a pencil holder.

Equipment and Materials

11" x 17" graph paper
drafting pencil
drawing board
masking tape
T-square
triangle
circle template
architect's scale
drafting eraser

Procedure

1. Prepare the paper.

- Hold the T-square firmly against the drawing board.
- Place the bottom of the paper against the T-square.
- Put a small piece of tape on each corner of the paper to hold it in position.
- Make sure the paper is straight. Fig. A6-1.

2. Draw lines.

- Place the T-square where the horizontal line will be drawn.
- Hold the T-square firmly against the side of the board.
- Draw the line. *Pull* the pencil, don't push it. Doing this will help you to make a smooth, even line. Fig. A6-2.
- Place the T-square *below* the point where the vertical line will end.
- Position the triangle where the line will be drawn. Hold the bottom of the triangle against the T-square.
- Keep the T-square tight against the drawing board.
- Draw the line. Pull the pencil from the bottom to the top of the paper. Fig. A6-3.

3. Draw the views.

- Most drawings are done with at least three views of the object (in this case, a pencil holder). Usually, the front, top, and right sides are shown. The front

Fig. A6-1. Placing paper on the drafting board.

Chapter 6 Technology Activity

Fig. A6-2. Drawing horizontal lines.

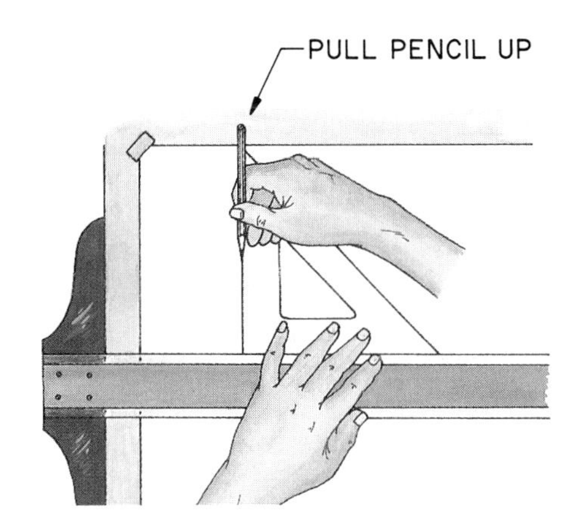

Fig. A6-3. Drawing vertical lines.

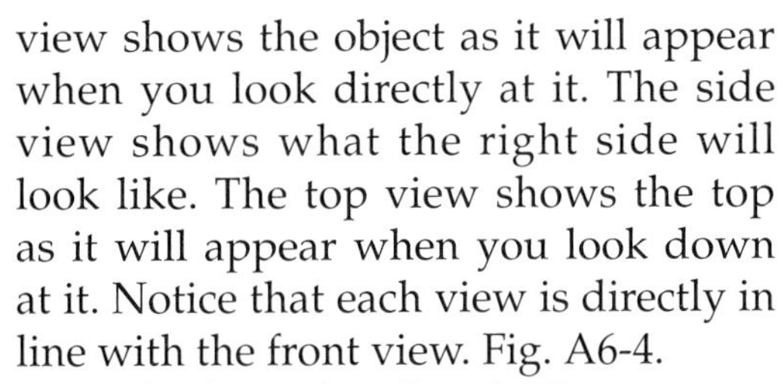

view shows the object as it will appear when you look directly at it. The side view shows what the right side will look like. The top view shows the top as it will appear when you look down at it. Notice that each view is directly in line with the front view. Fig. A6-4.

- Draw the *front view*. Fig. A6-5.
- Using the T-square, draw very light lines, called transfer lines, from each part of the front view to the side view.
- Draw the *side view* using the T-square and the triangle. Use the transfer lines to guide you. The side view must be directly in line with the front view. Fig. A6-6.
- Erase all transfer lines.
- Using the T-square and the triangle, draw very light lines from each part of the front view to the top view. Fig. A6-7.
- Draw the *top view* using the T-square and the triangle. Use the transfer lines to guide you. The top view must be directly in line with the front view. Fig. A6-8.
- Erase all transfer lines.

4. Dimension your drawings.

- Using the T-square (or triangle for vertical extension lines), draw two *extension lines* straight out from the object. Fig. A6-9.
- Extension lines should not touch the object. Make them lighter than the lines of the drawing. They must not look like part of the object.

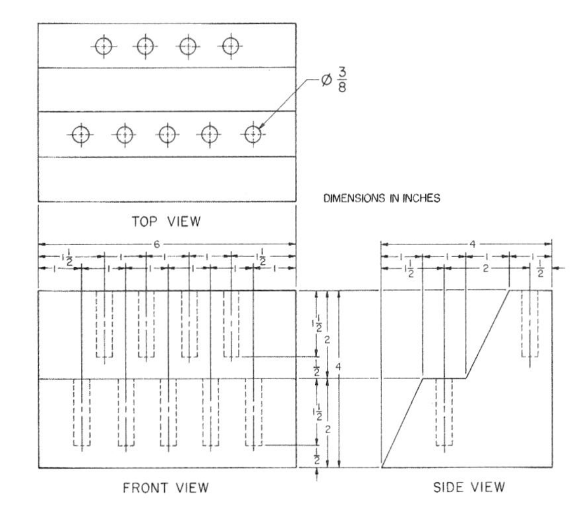

Fig. A6-4. These are three views of a pencil holder.

Chapter 6 Technology Activity

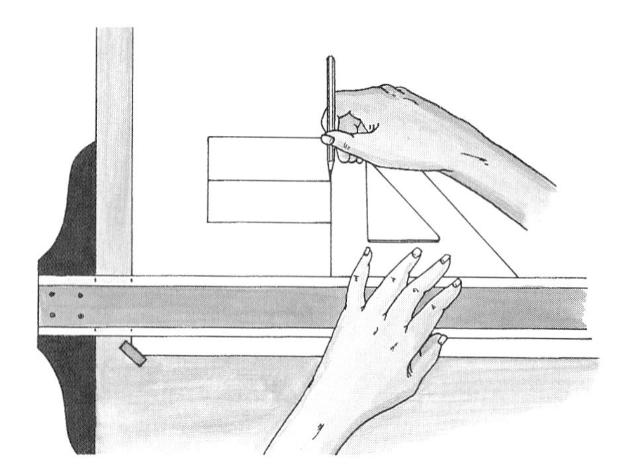

Fig. A6-5. Drawing the front view of the pencil holder.

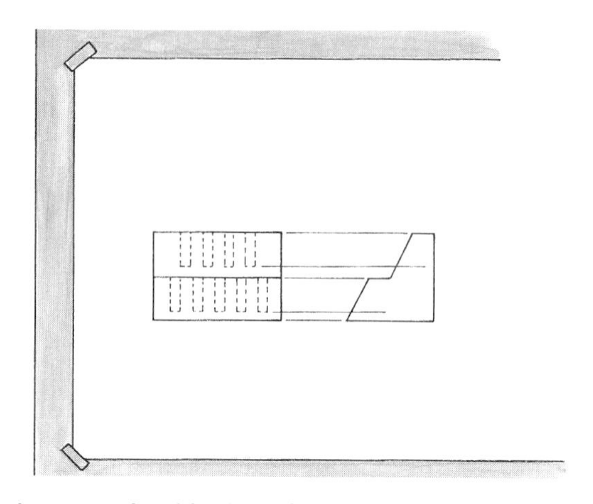

Fig. A6-6. The side view of the pencil holder.

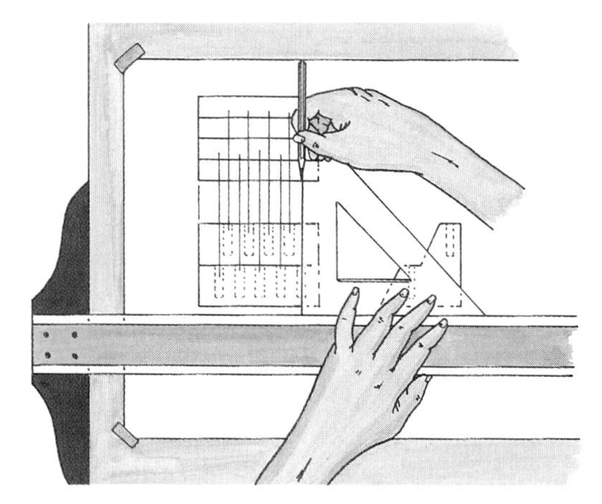

Fig. A6-7. Transferring lines for the top view.

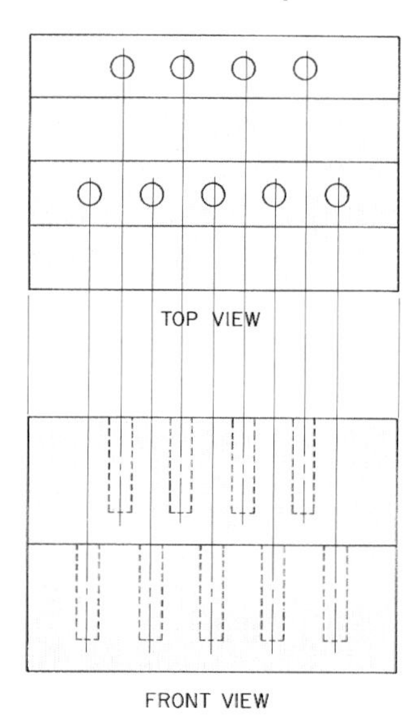

Fig. A6-8. Drawing the top view of the pencil holder.

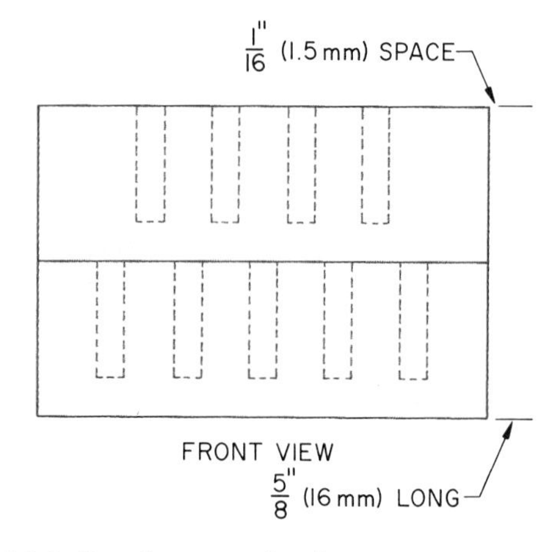

Fig. A6-9. Drawing extension lines.

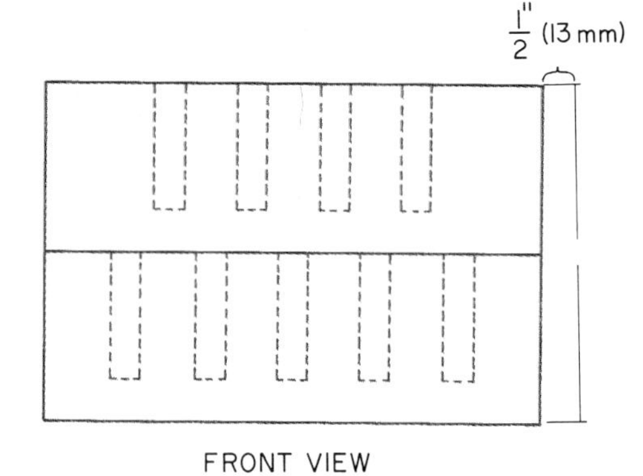

Fig. A6-10. Drawing dimension lines.

Chapter 6 Technology Activity

- Make each extension line about 3/4" (19 mm) long.
- Using the T-square and triangle (or just the T-square for horizontal lines), draw *dimension lines*. Fig. A6-10.
- Each dimension line should be about 1/2" (13 mm) away from the object.
- Leave a space in the middle of each line to put in the dimension number.
- Neatly draw arrowheads on both ends of each dimension line. Fig. A6-11.
- Arrowheads must touch the extension lines.
- Neatly write the dimension within the break of the dimension line. Put in a very neat number to show what size the object is. Fig. A6-12.
- Indicate on the top view of the object what the dimensions are of the holes. If the dimension of each hole is the same, you only need to indicate the size of one hole. However, if all sizes are not the same, you need to indicate the size of each hole. Fig. A6-13.

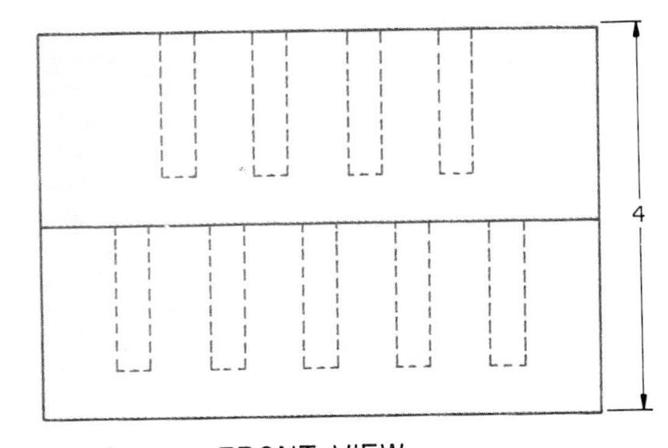

Fig. A6-12. Writing the dimension.

Evaluation

1. How long did it take you to draw the different views of this pencil holder? Now that you've learned some basic drafting techniques, would you be interested in a career as a drafter?
2. Would the use of a computer benefit you in creating these different drawings of the pencil holder? If so, how?

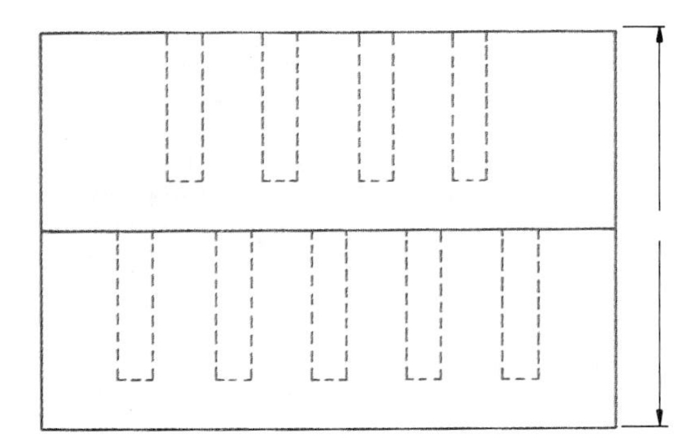

Fig. A6-11. Drawing arrowheads.

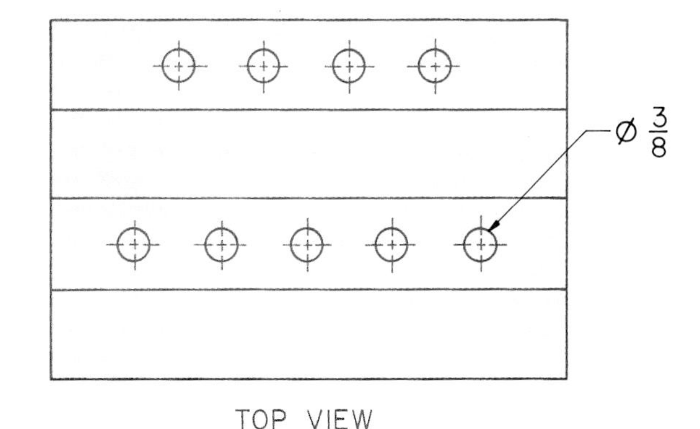

Fig. A6-13. Dimensioning the circles or holes that are in the pencil holder.

Chapter 7

Trends in Communication Technology

Looking Ahead

In this chapter, you will discover:

- ways that computer technology may revolutionize communication systems.
- how videotex systems may change the way people access information and make purchases.
- what artificial intelligence is and how expert systems are being used.
- innovative audio and video technologies.
- how advanced communication systems are used to explore space and expand our understanding of the universe.
- ways that communication technology is used in biotechnology.

New Terms

acoustics
artificial intelligence
computerized tomography (CT)
digital image processing
electronic mail
myoelectric hand
program
space observatories
teleconference
vehicle navigation system
videotex

Technology Focus

Image Processing

Powerful weapons destroyed SCUD missiles and strategic targets in Iraq during the Persian Gulf War in 1991. However, the credit for the success of these weapons should be given to electronic images, which provided the information necessary to direct "smart" weapons to their targets. (The term "smart" means that computers help humans control the functions of the weapons.)

Images of the target were obtained from many sources. The most important were images obtained by satellite cameras and infrared imagers, which "see" by heat instead of by light. A series of satellites provided overlapping images of the battle area. Images of the moving target were processed and transmitted to the Patriot missiles. (Images are said to be *processed* when they are changed, by computers, to provide new information. Powerful new computers and mathematical formulas were used to process large, complex images.) For speed of use, the images were compressed and transmitted to the site, where they were decompressed and used to make decisions. Patriot missiles were able to precisely track, intercept, and destroy incoming SCUD missiles using these images. Video footage supplied by the United States military showed the devastating accuracy achieved by using these electronic images to guide not only the Patriot missiles, but also bombs dropped from aircraft.

As complicated as these weapon systems are, even more sophisticated "launch and leave" smart weapons are now being considered. These will require extensive use of image processing technology. In addition, the next generation of "brilliant" laser-guided weapons will recognize, select, and destroy targets independent of human help. The key will be a reliable and quick image processing system that rivals human reasoning.

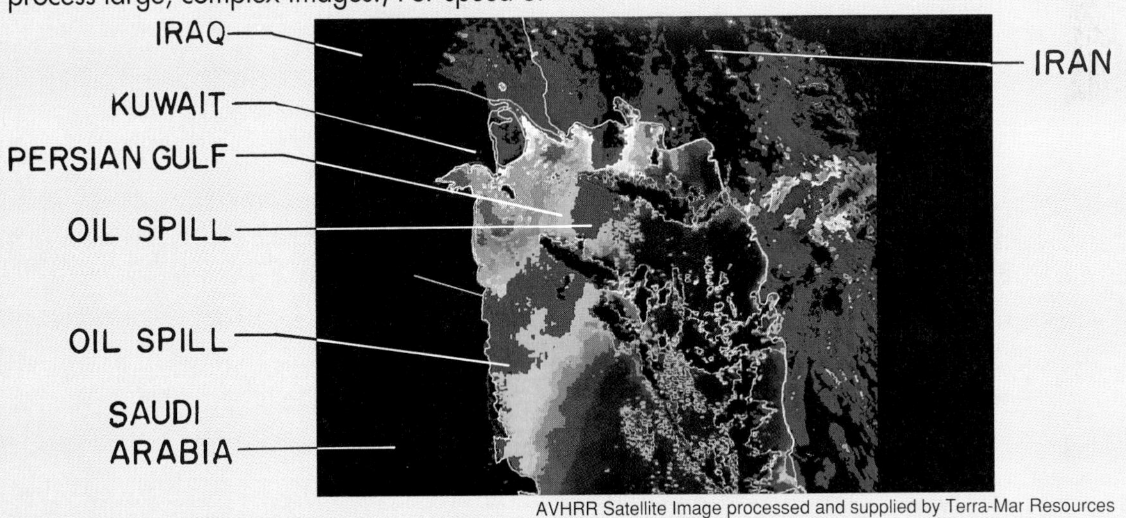

Fig. TF-7. Satellite imagery of the Persian Gulf and vicinity. The image was processed to emphasize the existence of oil spills in the Persian Gulf (labeled purple splotches).

Innovations and Trends

An *innovation* is the introduction of something new. It can be a new device, process, or idea. Portable telephones and home computers may be considered innovative devices. Using satellites to help navigate airplanes is an innovative process. Making communication tools smaller and portable is an innovative idea. Many innovative ideas lead to the development of trends.

A *trend* is a general movement or inclination toward something. It is not usually a specific device, product, or idea. Trends are often represented by several innovative devices, processes, or ideas. There are a number of trends in communication technology today. However, most of them can be categorized as being aimed at improving one of the three following areas:

- Quality of communications
- Convenience and portability of devices
- Speed and efficiency of devices and systems

These trends in communication technology are easily explained by giving examples of some of the innovations that signaled these trends. For example, compact discs have such superior sound quality and are so resistant to scratches and warpage that they have virtually replaced vinyl records. Fig. 7-1. High definition television (HDTV) demonstrates a marked improvement in television picture quality. The increased use of optical fibers in place of copper cables improves the transmission quality of sounds and data sent over telephone lines.

Convenience and portability are demonstrated by devices such as miniature televisions and cordless telephones. There are also lightweight computers that can be carried in briefcases. People have greater access to information because these devices are close at hand whether the people are in their cars, on airplanes, or away from the office on business travel. Fig. 7-2.

Advanced computer systems continue to increase the speed and efficiency of sending, receiving, and storing information. This

Fig. 7-1. Compact discs are representative of the trend toward communication devices that provide superior sound quality. This portable CD player demonstrates the trend for convenient portable communication devices.

Fig. 7-2. Even though this person is traveling on an airplane, this laptop computer allows him to have access to information and to get some work done.

trend is clearly understood when you consider how data banks are being used to centralize information in libraries and corporations.

As you can see, innovations are directly related to trends. In the rest of this chapter, we will examine some of the other important and innovative new technologies affecting trends in communication.

Apply Your Thinking Skills

1. Think about innovations in communication that you depend on or use. Do they represent one of the three communication trends listed? Do they represent more than one?

New Uses for Microchips and Computers

Computers, along with microchips, are the driving force in communication technology today. It is difficult to think of any exciting new developments in communications that do not involve the use of microchips or computers. Among the new uses of computers are videotex, vehicle navigation systems, and artificial intelligence.

Videotex

Videotex is a system that allows you to receive computer text and graphics via telephone lines. In some videotex systems, the telephone line is linked to a special video terminal, a computer-like screen designed to display the text and graphics. Some videotex systems use a regular computer and a modem. There also are systems that use a television as the display screen.

Fascinating Facts

Recent advances like digital cameras, which store images on a disk rather than on film, have contributed to the speed of today's communication system. The Associated Press used this new technology in covering George Bush's inauguration, and newspapers had photos of the ceremony 40 seconds after it occurred!

A videotex system enables users to access an incredible array of services and information. Although options vary on each system, videotex users are finding that there is a vast selection from which to choose. Using one system, users are able to:

- Buy airline tickets.
- Order groceries to be delivered to their home.
- Bring up onscreen a certain part of the daily newspaper, such as the headline news, sports, or weather forecasts.
- Look up something in the encyclopedia.
- Send flowers to a friend.
- Buy and sell stocks.
- Browse through a department store catalog and place an order.

This system also allows users to send and receive electronic mail. **Electronic mail** (or E-mail) refers to sending such things as letters, messages, and documents using computers and telephone lines instead of paper. The mail is sent electronically instead of using the traditional mail service. The mail arrives onscreen instantly instead of days later as with traditional mail. In addition, the user can send the same message to a number of people at the same time. Fig. 7-3.

This type of computer system is not exactly a new innovation. The idea started several years ago, but the technology was not sophisticated enough to make the systems appealing to a large number of people. In addition, the cost of the services was too high for the average consumer. Improved and more affordable computer systems are now making this system accessible to more people. As a result, the quality of the service is greatly improving, and users are finding the system much more appealing. In the future, it is expected that a larger number of people will find this technology beneficial to their busy lifestyles.

Vehicle Navigation Systems

By the time you are ready to buy a new car, you may have a new option on your vehicle. It's called a **vehicle navigation system**. This communication system is designed to make it much easier to get where you are going. It combines several technologies to bring you a convenient system for finding your destination quickly and efficiently. A compact disc stores map information for a particular area. The maps

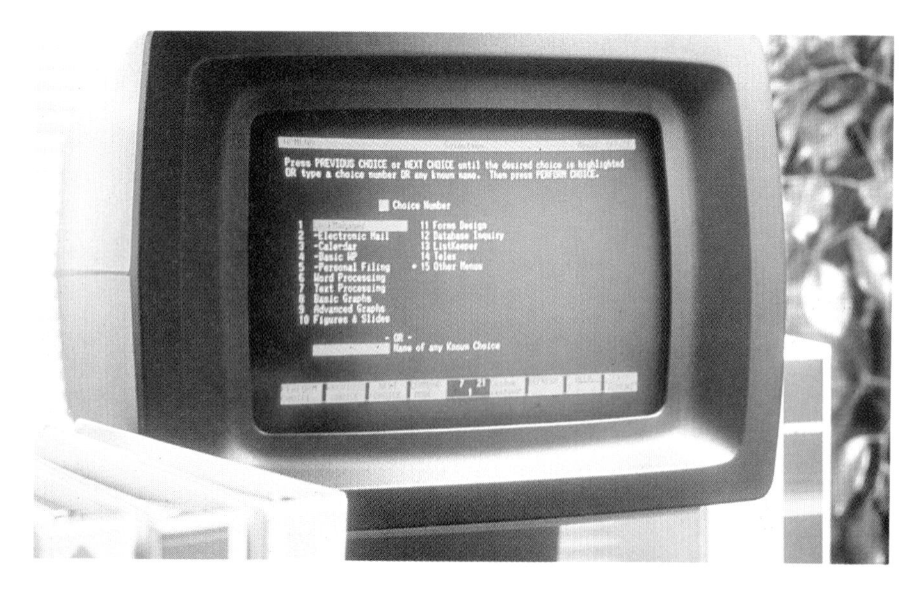

Fig. 7-3. The use of electronic mail has sped up information transfer dramatically in business. What used to take days to get delivered using the traditional mail service can now be delivered instantly using electronic mail.

can be displayed on a small screen mounted on your dashboard. Sensors on the vehicle's wheels can track the vehicle's movements, and these movements are compared to those on the computer's maps. As you drive, a cursor on the computer screen traces your path. Fig. 7-4.

One day, systems like these may be a standard feature on vehicles. The systems might also be linked with satellites. This would make it possible to expand the system's capabilities. Drivers could be warned of traffic jams ahead and be provided with alternate routes.

Artificial Intelligence

Artificial intelligence is the process computers use to solve problems and make decisions that are commonly solved or made by humans. Computers cannot think like humans. They are not intelligent. However, some people believe it is possible to "teach" computers to "think" much like people.

Teaching a Computer to Think

Computers work by following instructions devised by people. Each set of instructions is called a **program**. Basically, these instructions are logical steps to solving a problem. However, what about a problem that cannot be answered by logic?

Researchers are finding that computers can solve such problems when given the instructions in the form of reasoning processes that people use. Computers are then able to "think" like people. Sometimes a person makes a decision based on a good guess, or a solution may be based on the way an expert might solve the problem. In fact, one application of artificial intelligence is the development of expert systems.

When developing an *expert system*, developers prepare a program that has all the available facts about the topic. Experts in a particular area are interviewed. Based on the information that they provide and the facts that are available, a complex program is developed. The program instructs the

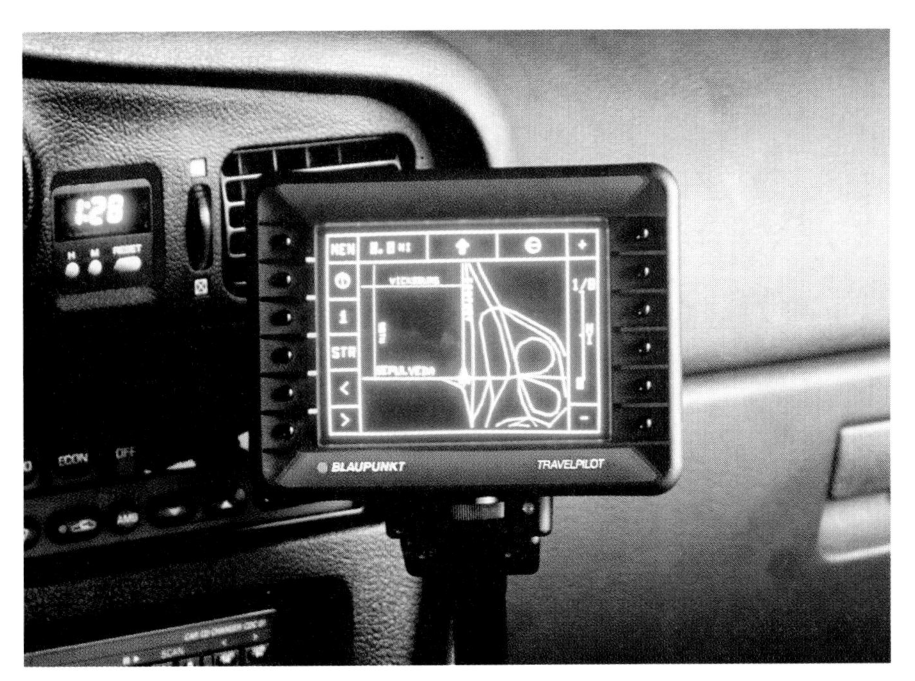

Fig. 7-4. Today, systems like this vehicle navigation system are being used on a limited basis in the U.S. Businesspeople who do a lot of traveling within a city where they need to find locations quickly really appreciate them.

Fig. 7-5. An expert system is based on a network of information from many experts on a particular subject. This particular system can help doctors diagnose illnesses.

computer how to solve a problem or reach a decision. Fig. 7-5.

Examples of Expert Systems

Various kinds of expert systems are in use today. Some repair shops use a system that figures out what is wrong with an engine. Fig. 7-6. Another system can even locate ore deposits in places where people could not.

Another expert system is used in the medical field. It is used to determine the right amount of treatment to give cancer patients who are undergoing chemotherapy. (*Chemotherapy* involves using chemicals to treat or control disease. Patients periodically receive these chemical treatments in hospitals or doctors' offices.) Since every person's body and disease is different, it can be difficult for doctors to determine the appropriate amount of chemicals to give a patient. The dose depends on the patient's body functions and on the present level of

Fig. 7-6. General Electric's expert system is programmed to diagnose what is wrong with locomotive engines. It is connected with a graphics screen that can display the parts. It is also connected to a video monitor that shows the mechanic how to repair it.

chemicals in the patient's body. Using an expert system that analyzes these factors and uses the reasoning processes of medical experts, it is possible to give patients a more accurate treatment.

Research is being done to find new ways to use expert systems in communication industries. One possible use will be in the newspaper industry. There are thousands of decisions that must be made in producing every newspaper that is published. For example, which stories should appear on the front page? Which picture would be most appealing to readers? On what page should a certain advertisement be placed? Someday these decisions might all be made by using an expert system or possibly some new form of artificial intelligence.

Apply Your Thinking Skills

1. What would be some advantages and disadvantages of using a videotex system to do your shopping?

Audio and Video Technologies

Improved quality in images and sounds can be seen in a wide variety of audio and video products. Nearly every communication product with *audio* (related to sending, receiving, or reproducing sound) and *video* (related to sending, receiving, or reproducing pictures) capabilities is affected by this trend toward increased quality. Televisions, radios, compact disc players, and recording systems reflect substantial improvement in both communication quality and efficiency over the past decade. These trends are expected to continue. Some other video innovations on the horizon include teleconferencing, video compact discs, and home theater systems.

Teleconferencing

Since travel is so expensive and time consuming, it has become increasingly desirable for some meetings to be held without participants meeting face to face. A **teleconference** is a conference held simultaneously among participants who are in different locations. Participants are connected by telecommunication lines. In the simplest type of teleconference, participants are linked only by telephone lines. If telephones equipped with external speakers are used, a small group of people in each location can participate in the multi-party conference. Sometimes, televisions are also used to link participants. This is called *videoconferencing*. This allows participants to see as well as hear each other, making the conference more effective. However, this is very expensive.

Video Compact Discs

You have probably already heard compact discs of most of your favorite music recording stars. However, have you heard or seen another type of compact disc that features both video and audio? It's called a video compact disc (video CD). A video CD player can be hooked up to your television just like a video cassette recorder (VCR). However, there's a big difference. The video CD brings you interactive entertainment. You don't just watch it—you use it and interact with it! You can play video games with it or create your own cartoons, for example. Fig. 7-7.

In many ways, the video CD gives you some of the capabilities of using a computer. However, you don't need a separate computer. The computer program is part of

Fig. 7-7. Video compact discs make interactive entertainment easier for some people. They don't need a separate computer. The computer program is right on the disc.

the disc. The disc player looks like a regular CD player. However, it is specially designed to handle both video and audio CDs.

Video CDs are a convenient way to get interactive home entertainment, especially for people who find computer systems confusing and difficult. The video CD system is as easy to use as a VCR. With the growing trend toward convenience, this innovation could have a great impact on communication developments in the future.

Home Theater Systems

Imagine having a room in your home that was specially designed for watching television and movies. It would be like a home theater! You would view a wide, large screen and your *surround sound* audio would come from many directions for a more realistic and exciting effect. The room would be designed to promote good acoustics. **Acoustics** refers to how clearly sounds can be heard in a room.

This idea is already being put into use in some places, but it is not affordable for the average consumer. The expense of components (parts that make up the whole unit) alone—large projection screens, video disc players, special audio equipment—is enough to drain the savings account of the average consumer. Fig. 7-8. However, costs are expected to decline as the technology is

Fig. 7-8. The expense for home theater systems is still quite high, but these costs are expected to decrease as the technology is improved.

improved and the idea becomes more fashionable. Home video equipment has gained great popularity during the past decade. This factor, combined with the convenience of home entertainment, is evidence enough that home theaters cannot be that far behind.

Apply Your Thinking Skills

1. Consider the video innovations described here. Which one sounds more appealing to you? Which one do you think will become more popular during the next decade? Why?

Space Communications

You could hardly discuss trends in communication technology without mentioning the role of space communications. We have already learned that satellites —vehicles that are launched into space to reflect signals back to earth—play an important role in telecommunications. What other forms of space-related technology hold promise for more excitement in communications of the future? Digital image processing and space observatories are two examples.

Fig. 7-9. Digital image processing makes it possible to create images of distant planets from data sent back to earth from spacecraft. This is a picture of Jupiter taken from the *Voyager I* spacecraft.

Digital Image Processing

Digital image processing uses computers to change electronic data into pictures. The computers can manipulate these pictures in certain ways to make them more useful. For example, contrast can be enhanced, certain features can be emphasized, and colors can be changed.

In space communications, this advancing technology is used to interpret images sent to earth by spacecraft traveling through our solar system. It is this innovation which brings you colorful pictures of distant planets such as Jupiter, Neptune, and Uranus. It has enabled people to see things that would otherwise be impossible to see. Fig. 7-9.

Space Observatories

For many years, scientists and engineers have been exploring space. However, most of this exploration has taken place on earth! Spacecraft travel through space sending signals (data and pictures) back to earth. Scientists and engineers on earth study and interpret this information to learn more about many aspects of this vast universe. Fig. 7-10.

The spacecraft sending this information are called **space observatories**. Space observatories are launched into space by space shuttles above the earth's atmosphere. From this vantage point, they can more accurately record information and relay it back to earth. Fig. 7-11.

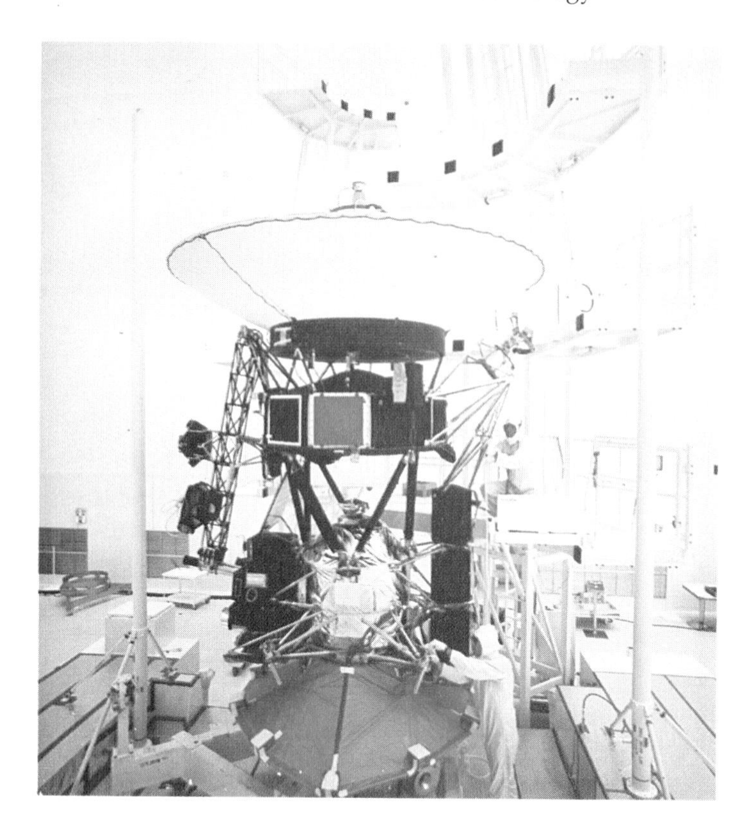

Fig. 7-10. The *Voyager* spacecraft are highly sophisticated vehicles that traveled through our solar system sending images back to earth. People on earth have reprogrammed these vehicles during their missions to make communications more efficient.

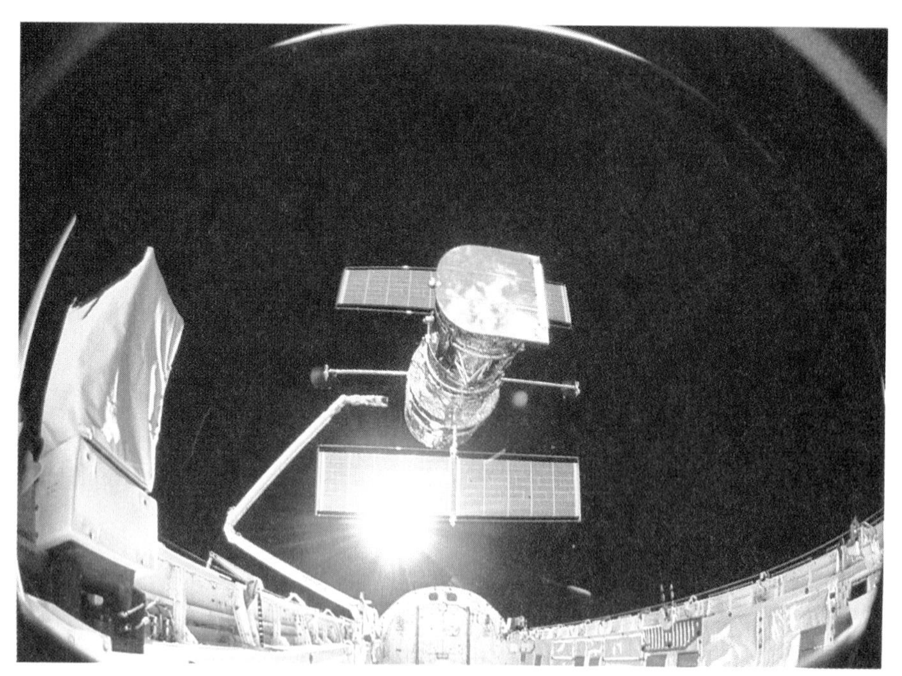

Fig. 7-11. The *Hubble Space Telescope* is one of the space observatories that will expand our view of the universe. This telescope is sending images back to earth of a variety of objects, from nearby planets to remote galaxies.

Fascinating Facts

Satellites have revolutionized our communication networks—and yet the original idea came right out of science fiction! Back in 1948, astronomer and science fiction writer Arthur Clarke first suggested the concept of using satellites to reflect radio waves. That vision became a reality in 1960 with the launching of the historic *Echo I*. *Echo I*, an aluminum-coated plastic balloon measuring 100 feet in diameter upon being inflated in space, paved the way for today's sophisticated communication satellites.

Today and in the future, space observatories will expand our vision of the things that make up the world far beyond our reach. Scientists will use these advanced communication tools to explore astronomical phenomena (events) such as black holes, pulsars, and quasars. They even plan to use these tools to study the origin of our universe and the birth of stars.

Apply Your Thinking Skills

1. Space exploration is an expensive endeavor. Some people think that it is necessary in order to increase human knowledge. Other people think the money should be spent on social or educational programs. What do you think?

Biotechnology

Our look at trends in communication technology is not complete without a view into what's going on in biotechnology. The innovations are many. You have already read how expert systems can be used to help diagnose illness and determine chemical dosage. Here we will examine two other exciting areas: prosthetics and computerized tomography.

Prosthetics

Due to accidents and other unfortunate circumstances, some people have had to deal with the loss of their limbs or other body parts. Many body parts can be replaced with artificial devices called *prostheses*. The name for this area of medicine is *prosthetics*.

Prosthetics have changed a great deal over the years. Advancing technology is improving prostheses all the time. Today, some prostheses are like communication devices. They use electrical signals to "communicate" with body parts such as muscles or nerves. This helps patients lead a more normal life by giving them more and better movement capabilities.

For example, a patient who loses a hand has at least two choices today. Using a less innovative form of technology, doctors can fit the patient with a prosthesis that has a hook at the end of the arm or wrist. This enables the wearer to hold and move things to some extent. The wearer operates the device by moving cables that are connected at the shoulder. Unfortunately, the device is noticeably different from a normal hand. Using a more innovative technology, doctors can fit a patient with a **myoelectric hand**. This device looks very similar to a human hand, but it is made of plastic and has a flesh-tone glove fitted over it. Fig. 7-12.

The myoelectric hand also works differently than the other prosthesis. It has electronic sensors that detect signals from the nerve endings in the remaining portion of the arm and relay these signals to activate the wrist, hand, and fingers. Microchips are used inside of the myoelectric hand to make it function more effectively.

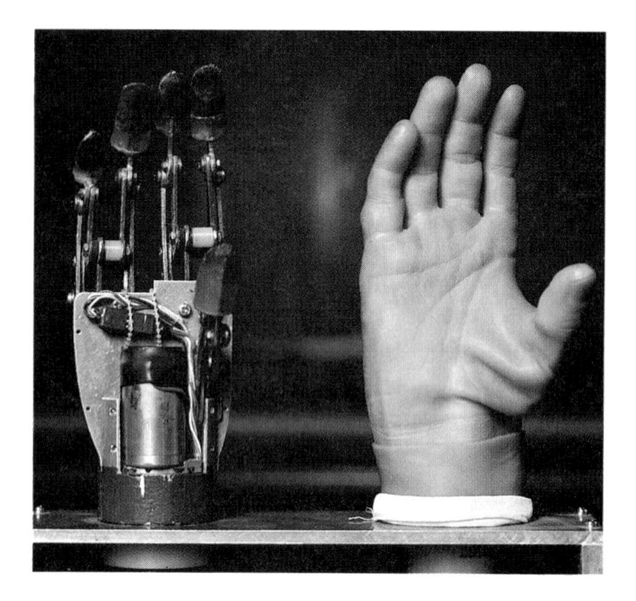

Fig. 7-12. This myoelectric hand looks and operates similar to a natural human hand. Electronic sensors in this prosthetic device detect signals from the nerve endings in the remaining arm and relay these signals to activate the wrist, hand, and fingers.

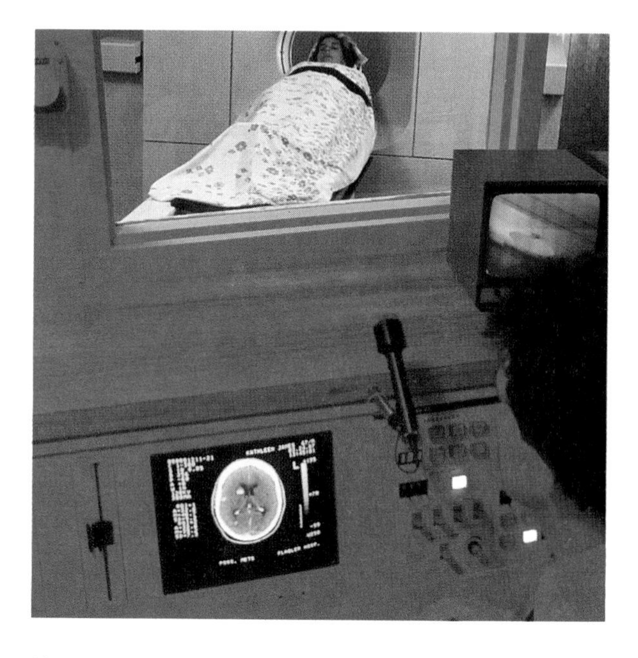

Fig. 7-13. Computerized tomography uses digital image processing technology to help doctors "see" inside the human body to help better diagnose many health problems.

Computerized Tomography

Using digital image processing, doctors have new ways to peer inside the human body. One way is commonly referred to as **computerized tomography (CT)**. Fig. 7-13. In CT, the patient's body is scanned by several X-ray machines. Data is collected and the image is graphically reconstructed on a computer screen. These computer graphics can help doctors diagnose many problems. Improper heart functions can be detected. Damage from heart attacks or strokes can be seen. Doctors can follow the flow of the blood and even locate brain tumors.

Many views of the inside of a body can be seen without taking many X-rays. The pictures can be changed to highlight certain areas. Doctors can view problem areas close up.

In computerized tomography, the computer is being used as a communication tool. It displays messages for doctors to analyze. It stores the information for doctors to review and to compare with later results. This technology is helping doctors effectively apply their skills and knowledge. Many people can live longer, healthier lives because of it.

Apply Your Thinking Skills

1. Innovations in medicine and medical treatment sometimes seem like miracles. Some situations that once seemed hopeless can now be remedied with new and innovative technologies. How would you feel about being a human "guinea pig" for an innovative treatment? In other words, if you were seriously ill or injured, how would you feel about having a brand new medicine or procedure tried on you?

Global Perspective

Mapping Our Progress

A map is a representation of an area of the earth or of the universe. Throughout history, people have created maps and changed them as their knowledge increased, as the amount of information available grew, and as technological advances were made. This trend continues today.

Early mapmakers included the Egyptians and the Babylonians (from the area now known as Iraq). One Babylonian map, a clay tablet, is the oldest map known to be in existence today. It was made about 2500 B.C.

The science of *cartography* (mapmaking) was established by the Greeks during the late centuries B.C. The Greeks were among the first to realize that the earth is spherical (ball-shaped).

The Chinese are credited with producing the first printed maps around 1150 A.D. This development of printing increased the use and usefulness of maps. Around this same time, the Arabs were producing quite accurate maps based on astronomical observations.

In the 15th and 16th centuries, European explorers made long voyages into uncharted territories. They brought back sketches of the areas they explored. Mapmakers would then add the new information to their maps.

When the use of airplanes became common, in the 1930s and '40s, large-scale use of photogrammetry began. *Photogrammetry* is the science of using aerial (taken from high overhead) photographs to make maps. Large areas could be mapped accurately and in great detail.

New technology is revolutionizing cartography. Cameras and other equipment are much improved. New devices such as sensors can be used in planes and satellites to gather a wide variety of useful facts about the earth.

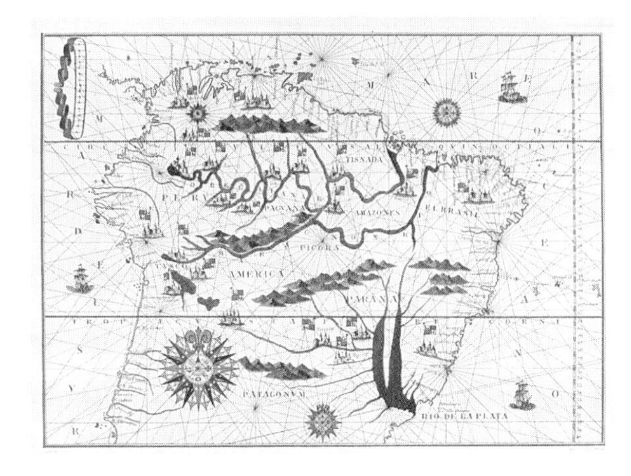

Fig. GP-7. This is a map of a portion of South America from the Spanish atlas that was done at Messina in 1582.

All details used on maps can be entered into computers and continuously updated as new information is gathered. Then maps of various types and sizes, showing details of locations around the world, can be produced by the computer.

Maps have changed greatly through the years. They provide us with a fascinating historical record. Maps are a reflection of people's knowledge of the world at any given time. Now, we map the universe.

Extend Your Knowledge

1. Accurately representing the features of the round earth on a flat surface is very difficult. Such a drawing is called a projection. Find out about various types of map projections, such as the Mercator Projection, and the techniques that were used to try to ensure accuracy.

Chapter 7 Review

Looking Back

Most trends in communication technology can be categorized into one of three areas: improved quality of communications, convenience and portability of devices, and speed and efficiency of devices and systems.

Among the new uses of computers are videotex, vehicle navigation systems, and artificial intelligence.

Video compact discs and home theaters are two innovative video technologies that may lead to new communication trends.

Digital image processing can be used to send colorful pictures of distant planets from spacecraft back to earth. Space observatories relay information about our universe back to earth.

Microchips and electronic sensors are communication devices that can be used in some prostheses to help them operate like a natural human limb.

Review Questions

1. Define trend and innovation.
2. List the three major trends in communication technology today. Give at least one example for each trend.
3. Describe a videotex system. List at least five things a user can do with a videotex system.
4. Briefly describe how a vehicle navigation system works.
5. Define artificial intelligence.
6. What is a computer program?
7. Briefly describe how a program for an expert system is developed.
8. Name some advantages of video CDs.
9. Define digital image processing.
10. Describe computerized tomography.

Discussion Questions

1. Imagine that administrators in your school district decided that it is no longer necessary to hire a school nurse. They feel that it is inefficient and costly. Instead, they have purchased a computer and an artificial intelligence system that will diagnose students' health problems. Discuss the advantages and disadvantages of using this method to handle medical situations in schools.
2. Imagine ways that increased use of videotex systems will affect businesses in your community. Which businesses will prosper? Which businesses might not thrive? What new businesses might be developed?
3. Often changes in communication technology also mean that changes are in store for employers and employees. Choose one of the communication innovations discussed in this chapter and evaluate the influence that innovation might have on a selected group of employers and employees.
4. People are overwhelmed by the changes taking place in communications today. How do you feel about these changes? Why do you think there are differing views on this subject?
5. Discuss some of the impacts the widespread availability of VCRs has had on our personal lifestyles and on the entertainment industry. What effects would a widespread use of home theaters have on these things?

Chapter 7 Review

Cross-Curricular Activities

Language Arts 

1. Investigate the role of the computer in your school, library, or police or fire department. Interview the people who are involved. Write an essay that details the use of this technology.

Social Studies 

1. To show how improved technology in communication can affect the outcome of a war or military conflict, research the communication tools used in Operation Desert Storm versus the lack of good communication tools in the early 1800s at the Battle of New Orleans during the War of 1812.

Science

1. The human nervous system is a biological communication network with message transmitting and receiving systems that work together to achieve a common goal. Think about how your brain, nerves, and sensory organs (eyes, ears, taste buds, etc.) work together to achieve the goals of keeping you alive and healthy. Draw a similar diagram for the communication network of the body.
2. NASA's TAU mission spacecraft will continue moving through space long after all its fuel is exhausted because of inertia. (*Inertia* means an object at rest tends to remain at rest and an object moving in a straight line tends to continue moving in a straight line unless affected by an outside force.) To test this principle: Place a ball at the center of the bottom of a child's wagon. Pull the wagon at a steady speed. Now stop. How does the concept of inertia explain what happened to the ball?

Math

For this activity, you will need access to a computer workstation that has had a BASIC Compiler booted. Enter the following:

```
10 READ X1,X2,X3
20 LET M=(X1+X2+X3)/3
30 PRINT "MEAN="M
40 DATA 16,27,33
50 END
```

Then type RUN

You will see the following: MEAN = 25.3333

You have entered a simple BASIC program into the computer. This program will calculate the mean (average) of three numbers.

The numbers 10, 20, 30, 40, and 50 are line numbers. Line 10 tells the computer to accept three numbers that you will enter as data. The numbers are represented by variables X1, X2, and X3. Line 20 instructs the computer to add the three variables and divide the sum by three. This generates the mean, which is assigned the symbol "M." Line 30 is a print statement that tells the computer to take the value of M and print it as the mean. Line 40 is where you enter the numbers you wish to average (data). Line 50 is needed to end the program.

You can experiment with this program by entering new numbers to average. Type your new numbers in line 40, and then type RUN.

Chapter 7 Technology Activity

Using Technical Communication Systems

Overview

We are developing more and more ways to communicate. Are some ways more effective than others? How does *what* we want to communicate influence our decision about *how* to communicate? In this activity, you will try three ways to communicate technical information. Then you will decide which way was most effective. These three ways to communicate technical information include:

- Auditory communication
- Visual communication
- Audiovisual communication

The class will be divided into small groups. Each group will develop a message to be communicated to a larger audience using one of the three technical communication systems listed. You will also evaluate the advantages and disadvantages of each system. The message you develop may be as simple as describing how to sharpen a pencil or as complex as describing how to use a videocassette recorder.

Goal

To try three different ways to communicate technical information, and then to decide which way was most effective.

Equipment and Materials

cassette tape recorder or reel-to-reel tape recorder and blank tape
overhead projector and transparency film (8 1/2" x 11" or 9" x 12")
video recorder, blank tape, camera, and monitor
paper, 8 1/2" x 11"

Procedure

1. Your teacher will assign you to one of three groups: an "audio group," a "visual group," and an "audiovisual group."
2. Choose one of the following messages to communicate:
   - How to tie a square knot.
   - How to tie a necktie.
   - How to use a videocassette recorder.
   - How to travel from school to a group member's home.
   - How to make a paper airplane.
   - How to sharpen a pencil.
   - Message of your choice, approved by the teacher.
3. The goal of each group is:
   Audio group—To use the tape recorder to communicate your message to the class.
   ASSIGNMENT: Write a script and produce a recording.
   Visual group—To develop drawings, diagrams, and/or charts that will communicate your message to the class. (No text—words or numbers—should appear on the visuals. Use only illustrations.)
   ASSIGNMENT: Develop visuals and convert these into overhead transparencies.
   Audiovisual group—To use a videocassette recorder to produce an audiovisual recording that will communicate your message to the class.

Chapter 7 Technology Activity

ASSIGNMENT: Write a script, plan the actions of the characters to be used, and produce a video tape.

An example message for each type of group is given in Fig. A7-1.

After completing the assignment:

4. Present your message to the class. Your group should provide the materials needed for each member of the class to follow the directions as you give them your message. Your teacher will help you locate materials.
5. After each group's presentation, evaluate the advantages, disadvantages, and limitations of that communication system. See Fig. A7-2 (p. 165) for an example of a chart that you can make up for this evaluation.
6. If time allows, you may be reassigned to a new group and repeat the activity using a different technical communication system.

A. Audio group—Script.

Directions

1. Get an $8\frac{1}{2}" \times 11"$ piece of paper and place it on a table.
2. Position the paper so that the long edge is parallel to you.
3. Take the left edge of the paper and fold it so that the edge is even with the right edge.
4. Once the edges are squared up, crease the fold.
5. Make sure the crease is on the left-hand side.
6. Take the bottom edge of the paper and fold it so that the edge is even with top edge.
7. Once the edges are squared up, crease the fold.
8. Turn the paper so that the creased edge is to your left.
9. Place a staple in the upper left-hand corner parallel to the creased edge about $\frac{1}{8}"$ from the edge.
10. Place a staple in the lower left-hand corner parallel to the creased edge about $\frac{1}{8}"$ from the edge.
11. Cut the creases that hold the paper (pages) together.
12. Number the pages 1-8 using both sides of the paper.
13. Congratulations! You now have a four-page, eight-sided booklet in front of you. Use it wisely.

Fig. A7-1. The message is the same from all three groups: How to create an eight-page booklet from an 8 1/2" x 11" sheet of paper.

Chapter 7 Technology Activity

B. Video group—Illustrations.

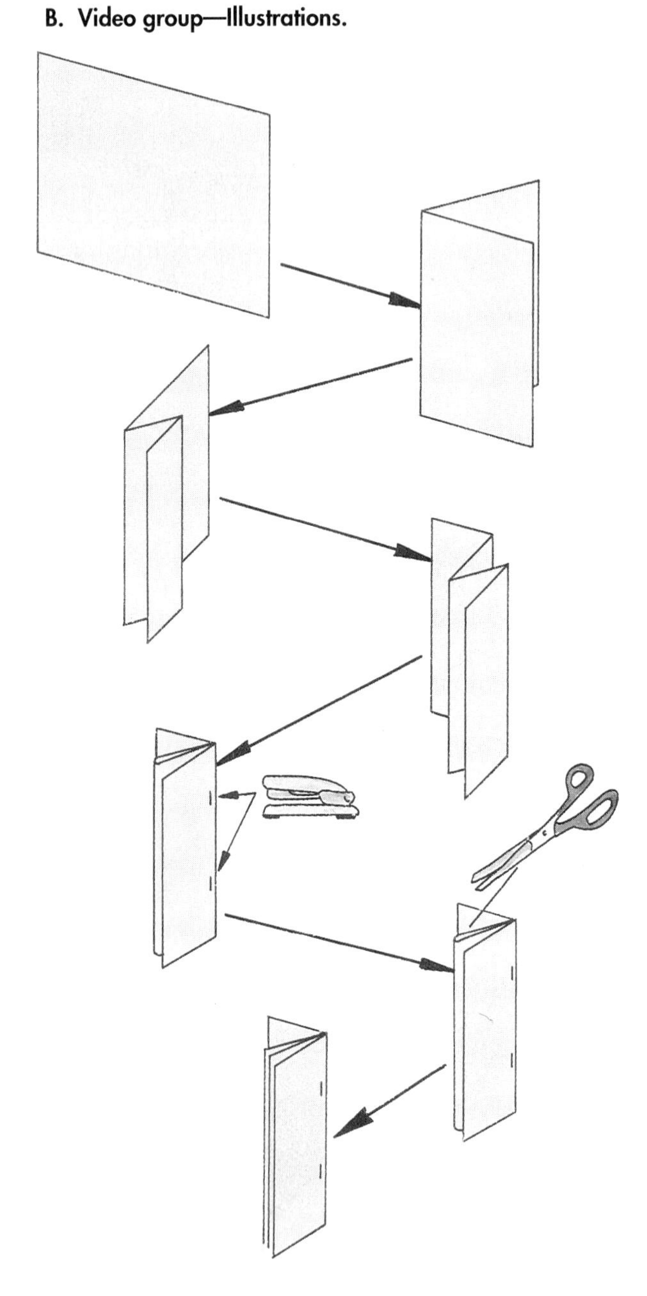

C. Audiovisual group—Illustrations and audio instructions.

AUDIO
1. GET A SHEET OF PAPER.
2. FOLD IT IN HALF.
3. FOLD PAPER IN HALF, THE OPPOSITE WAY THIS TIME.
4. STAPLE.
5. CUT CREASES OF PAPER.
6. NUMBER THE PAGES 1-8.

Fig. A7-1 (continued)

Chapter 7 Technology Activity

Evaluation

1. How well did your group's method of communicating a technical message work?
2. If you could choose another way to communicate the same technical message, which method would you choose and why?

Credit:
Developed by Ron Kovac for the Center for Implementing Technology Education
Ball State University
Muncie, Indiana 47306

Technical Communication Systems

NAME ________________________ DATE ________________

TASK COMMUNICATED ________________________________

TYPE OF SYSTEM & TASK	ADVANTAGES	DISADVANTAGES	When could you use this system?
AUDIO TASK:			
VISUAL TASK:			
AUDIO VISUAL TASK:			

Fig. A7-2. Make up a chart similar to this one to evaluate each group's presentation.

Section II Activity

Radio Scheduling

Overview

A local radio station has decided to change from its "talk radio" format to a "rock" format. It has decided to air a program every day called "Top 10," a countdown of the top ten songs for the week.

In this problem-solving activity, you will develop a one hour schedule that is needed for the "Top Ten" program.

STEP 1: State the Problem (Design Brief)

Typical radio broadcasts often involve many hours of pre-production efforts for each hour of air time. Announcers, sportscasters, and DJs usually spend considerable time preparing for their individual shows. Commercials must be arranged for broadcast in the proper time slot during the program. Music, guests, audio special effects, etc. must all be organized for their part of the show. The scheduling and organization of the material are critical to the success of the broadcast.

On a clean sheet of paper, type or write neatly a design brief (a short statement of your problem or task), including some of the major points to be considered in solving the problem. (In this case, those might include allowing enough time for commercials, scheduling guests, or writing any necessary material for your show.)

STEP 2: Collect Information

Here are possible sources of ideas for developing your program:

- Listen to your favorite radio and TV stations, and pay special attention to the commercials.
- Check the daily weather reports in your newspaper.
- Listen to the weather reports on the radio to learn the terminology and phrases that are used.
- Listen to the radio to determine your list of your favorite top ten songs.

STEP 3: Develop Alternative Solutions

On the same sheet of paper as your design brief, write down several ideas. For example, there may be formats other than the traditional one that we usually hear on the local stations. You might want to schedule your format differently, such as:

- More music with fewer commercials.
- More talk with fewer songs being played.
- More "soft rock" or more "hard rock."
- A phone-in request show.

STEP 4: Select the Best Solution

Consider the pros and cons of each solution. Questions that you might consider are:

- Would the station make as much money if there were fewer commercials?
- Would your target audience listen if there were fewer songs played in the hour? How would you fill that extra time?
- Could you get people to call in if it were a request show?
- Do you have the personnel to host a talk show?
- Do you think that your target audience would like more hard rock?

Song	Title	Artist	Time	Lead-In*
NAME: Lori Ramirez	SCHEDULING WORKSHEET		GROUP: #4	
#10	"This Is Fun"	Foxfires	3:15	:07
#9	"Why Do I Care?"	Fred Young	5:00	:15
#8	"What You Mean To Me"	Eli Jones	4:35	:12
#7	"Not Today"	Spectrum	6:05	:22
#6	"Life In The South"	Down Here	5:25	:15
#5	"This Day Forever"	Brenda James	5:00	:09
#4	"Hooked On Fun"	Michael Rea	4:20	:17
#3	"Today....and Always"	Third Edition	5:10	:10
#2	"Loving You"	John Kaufmann	5:00	0
#1	"No Place To Run"	High Five	6:00	0

*Until vocals begin

Commercials

	Firm	Time
#1	ABC Cable Co.	1:00
#2	Johnson Builders	:30
#3	Red Bird Restaurant	:30
#4	Noble Real Estate	1:00
#5	Stanley Computer Services	1:00

Other Features

	Estimated Time
Introductions for Songs	1:00
Weather Reports	1:15
Other Add 3:55 filler hit song	3:55
TOTAL TIME	60:00

Fig. SCII-1. Scheduling Worksheet.

STEP 5: Implement the Solution

Prepare the scheduling worksheet (Fig. SCII-1) and the programming planning form (Fig. SCII-2) and then test out your program to see that it will fit into the one hour time frame. Record your one hour show on tape so that it can be re-broadcast.

STEP 6: Evaluate the Solution

Play your preliminary tape for your friends and family. Have them make suggestions on how to make the program better. Evaluate their suggestions and include those that you think would make the program more enjoyable.

Credit:
Developed by Richard Seymour for the Center for Implementing Technology Education
Ball State University
Muncie, Indiana 47306

PROGRAMMING PLANNING FORM

Name: Lori Ramirez Group #: 4

Time Required	Running Time	Feature
:30	:30	Lead-in theme Music
:15	:45	Introduction to show
3:15	4:00	#10 hit song
:15	4:15	Weather and promo. spot
5:00	9:15	#9 hit song
:10	9:25	Introduction to next hit
4:35	14:00	#8 hit song
6:05	20:05	#7 hit song
:30	20:35	Weather and station call letters
5:25	26:00	#6 hit song
5:00	31:00	#5 hit song
4:00	35:00	Commercials
4:20	39:20	#4 hit song
5:10	44:30	#3 hit song
:30	45:00	Weather and promo. spot
5:00	50:00	#2 hit song
:30	50:30	Intro. last week's #1 hit
3:00	53:30	Play 3 min. of last week's #1
6:00	59:30	#1 hit song
:30	60:00	Exit remarks and theme

Fig. SCII-2. Programming Planning Form.

Section III

Manufacturing Technology

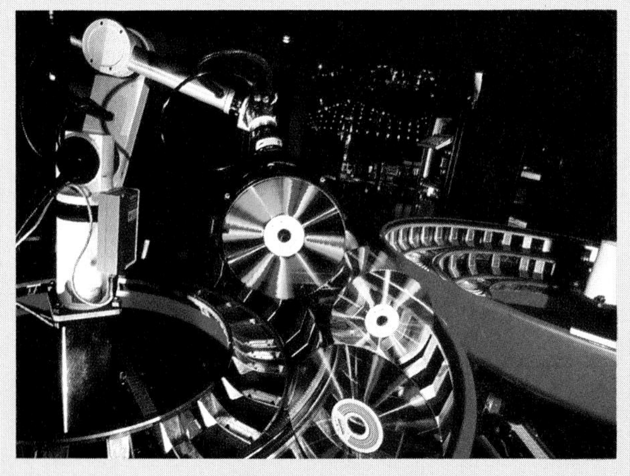

The introduction of compact discs revolutionized the field of audio recording. CDs are an example of recent advances made possible by lasers in manufacturing.

After being manufactured, most products must pass a final quality inspection. Some are examined for tears and flaws, while others are tested in laboratories to make sure that they work properly.

Technology Time Line

8000 B.C. | 6000 B.C. | 4000 B.C. | 2000 B.C. | 1 | 500 | 1000 | 1500

8000 B.C. Agriculture begins.

6000 B.C. First woolen textiles.

3600 B.C. Important tools such as the wheel and plow are perfected.

3500 B.C. People learn that by melting copper and tin they can make bronze.

1400 B.C. Metal smelting and casting is done in Egypt.

1000 B.C. Industrial use of iron in Egypt and Mesopotamia.

A.D. 400 Romans make cement; when their empire falls, cement making stops until 1,300 years later.

550 The secret of making silk is brought to Istanbul. The silk industry becomes a state run monopoly in Byzantine Empire.

1100 Use of stained glass in Europe becomes widespread due to improved construction techniques.

1242 Directions for making gunpowder become known to Western world.

Through such methods as manufacturing cells and flexible manufacturing systems, computers can be used to coordinate several machines. Even entire factory operations can be monitored from one central location.

Computers aid the process of quality assurance in capacities such as zone of acceptance testing and random sampling.

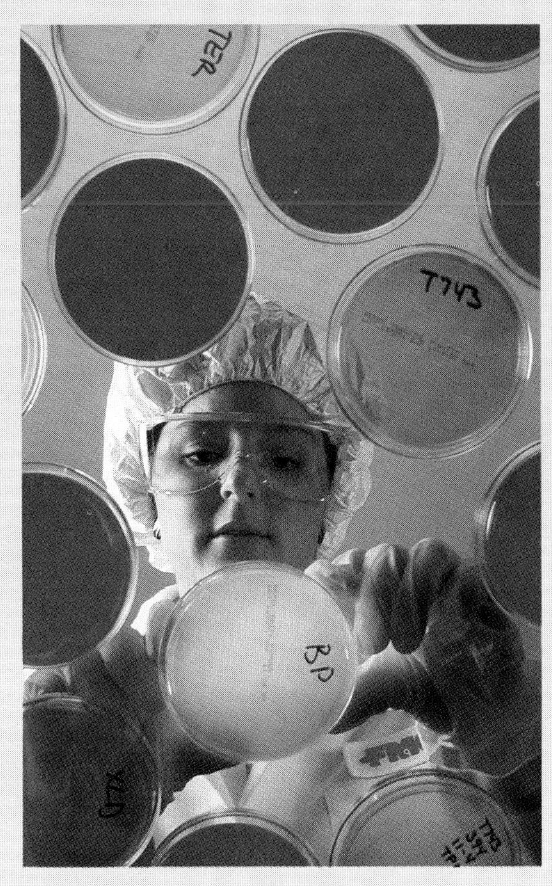

Quality assurance allows workers to correct problems at the earliest possible stage of manufacturing to save money and improve customer satisfaction.

1600 | 1650 | 1700 | 1750 | 1800 | 1850

1756 John Smeaton re-invents cement in England.

1770 The "spinning jenny," the first machine to spin many threads at a time, is patented by James Hargreaves.

1785 Edmund Cartwright patents the power loom.

1793 Eli Whitney is credited with the invention of the cotton gin.

1800 Industrial Revolution spreads to North America.

1817 Construction of the Erie Canal begins. It is completed in 1825.

1820 Hans Christian Oersted discovers electric current.

1846 Elias Howe patents a practical sewing machine.

Today, computers are used in many phases of manufacturing. Computers can be linked together to plan, design, and manufacture products at a lower cost and with less waste.

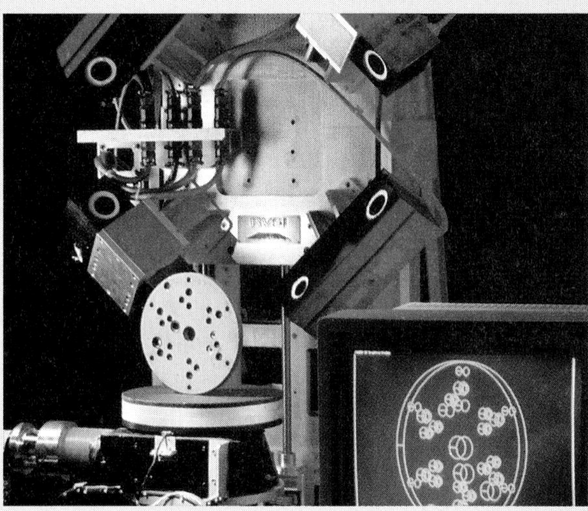

Laser scanners have uses that range from manufacturing and design engineering to translating bar-coded prices at a grocery store.

Electronic circuit boards allow equipment to be reprogrammed so that different parts of a product can be made on the same machine. They can also be used to coordinate various manufacturing processes from a central location.

1856 Henry Bessemer develops the Bessemer process of converting pig iron to steel.
1856 First synthetic dye is accidentally discovered by Sir William H. Perkin.

1873 James C. Maxwell stated that heat, light, and electricity are all electromagnetic waves.

1884 Hilaire Chardonnet manufactures artificial silk, which later becomes rayon.

1900 The vacuum cleaner is invented.

1918 An electric food mixer with two blades and a stand is produced.

1850 | 1860 | 1870 | 1880 | 1890 | 1900 | 1910

1858 The first successful use of an electric light at South Foreland Lighthouse in England.
1859 First successful oil well is drilled in Pennsylvania.

1872 Westinghouse perfects automatic railroad air brake.

1879 Edison makes the first practical electric light.

1913 The first domestic electric refrigerator, called the "Domelre," goes on the market in Chicago.

Computer-aided manufacturing (CAM) puts advanced technologies to work to produce products faster and less expensively. When combined with computer-aided design (CAD), design changes can be communicated to manufacturing without having to write new software.

Lasers concentrate the power of a great amount of light into a small area. The resulting beam can cut, drill, or bore—a capability put to good use on today's production lines.

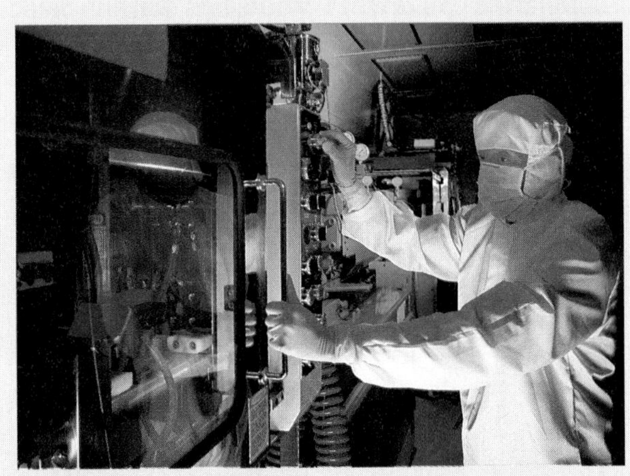

Manufacturing that takes place in sterile, dust-free rooms is vital to the production of many high-tech products such as computers, medical products, and compact discs.

1920 | 1930 | 1940 | 1950 | 1960 | 1970 | 1980

1924 Fax machines are first used.

1930 First analog computer is invented by Vannevar Bush.

1942 The first controlled nuclear reaction is achieved at the University of Chicago.

1945 The U.S. explodes the first atomic bomb in New Mexico.

1965 IBM produces the first floppy disk.

1978 First test-tube baby is born in England.

1980 Fax machines come into wide use because of size and because they use telephone lines.

1982 First artificial heart is transplanted in a human being.

1985 Canon markets a color photocopier.

1988 Frictionless bearings are the first practical use for superconductors.

Chapter 8

Manufacturing Systems

Looking Ahead

In this chapter, you will discover:

- what manufacturing is.
- how manufacturing developed.
- the three basic types of production systems.
- three ways to change the shape of materials.

New Terms

assembly line
assets
combining
continuous production
custom production
economy
efficiency
entrepreneur
forming
Industrial Revolution
job-lot production
raw materials
separating
standard stock
subsidiaries
value added

Technology Focus

High Tech Entrepreneurs

High technology creates new ideas that imaginative entrepreneurs can market. An **entrepreneur** is a person who starts his or her own business. To encourage high technology growth, KPMG Peat Marwick, one of the world's largest accounting and consulting firms in information technology, presents annual awards to high technology entrepreneurs in Illinois. Since 1984, the program has honored thirty-six very successful high tech entrepreneurs.

In 1969, Phillip Rollhaus, the winner of the 1986 High Tech Entrepreneur award, used his management skills to save a high tech company from bankruptcy. The company now dominates the market for highway safety cushions. He then formed Quixote Corporation. Quixote develops or acquires businesses that earn money from new technologies. Quixote's member companies include LaserVideo, which manufactures compact discs, and Amtel Systems Corporation, which turns existing structures into smart buildings. (Smart buildings are structures in which the electrical, heating, cooling, and security systems have computerized controls.) Quixote is now a $70 million public company that employs more than 1,000 people.

Bar codes are used in marketing (selling), in production, and in inventory control (keeping track of parts, materials, and products on hand). Zebra Technologies Corporation president Ed Kaplan, the winner of the 1988 award, foresaw the future of his numerical control printing business changing from punch tape to bar coding. He designed and developed bar code label printers that combine state-of-the-art high resolution thermal printing technologies in innovative (new and different) ways.

The winner of the High Tech Entrepreneur award in 1990 was Stephen Wolfram. In the late 1970s, Wolfram, a very young physicist at Caltech, needed to make some complex mathematical calculations. He developed a complete computer system for performing them. By 1986, microcomputers had become advanced enough to accommodate his system. This enabled Wolfram to develop the product, *Mathematica*, which was released in 1988. In just two years, *Mathematica* proved useful in almost every area of scientific and technical computing. *Mathematica* is used in space engineering, robotics, medical imaging, and other high tech applications. More than 50 major universities, colleges, and high schools are equipped with *Mathematica* labs. Wolfram Research, Inc., increased its employees from five to more than 100 people in the first two years of operation. It tripled its revenue from 1988 to 1989 and again in 1990.

Fig. TF-8. Each year, KPMG Peat Marwick presents this award to high technology entrepreneurs in Illinois who have contributed to the growth of technology.

What Is Manufacturing?

Imagine your life without manufacturing. You would have no bicycle to ride, no television to watch, and no sneakers to wear. There would be no cloth, no furniture, no airplanes. Lifesaving devices such as artificial hearts could not even be imagined. We are all very dependent on manufacturing, which is the making of parts and putting the parts together to make a product. The products can be large or small, simple or complex.

If you make parts and put them together to make a product, you are manufacturing. However, today, when most people think of manufacturing, they think of the manufacturing industry.

The manufacturing industry is important to our society. It's important to our economy. An **economy** is a system for producing and distributing products and services. Many people work in manufacturing. They help produce products. They also buy products with the money they earn. The more products people buy, the more products are manufactured. This allows more people to work.

Manufacturing is also important to the economy in another way. A piece of material is worth more after it's been changed into a useful product. That's **value added**. The value is increased by the manufacturing process.

Apply Your Thinking Skills

1. Look around the room and think about all the manufactured products you see. How would you have to adjust to life without these products?
2. How would the construction of a new factory in your area affect your community?

The Development of Manufacturing

Manufacturing began long ago when a person first changed a material into a useful item. Through the years, it has developed into a complex system of production.

In earlier times, each person or family made the products that were needed. This type of production was known as the domestic system. Any surplus (extra) products were bartered (traded) for other items. Cloth, brooms, and utensils were common products. All items were handmade. There were no power tools.

Cottage Industry

Later, families began to specialize. One family specialized in baking, another in weaving, and so forth. These families were not just making their own things and selling the rest. Now they were actually making goods to sell. Commercial manufacturing had begun. Everything was still handmade. Manufacturing took place in the home. Fig. 8-1. Workers were members of the family. This was called *cottage industry*.

Fascinating Facts

Not all advances in manufacturing have been welcomed by everyone—as a tailor named Barthelemy Thimonnier discovered in Paris, France, back in the 1840s. Thimonnier had invented a machine that could stitch seams at 200 stitches per minute. He was using 80 of these machines to make uniforms for the French Army. Other tailors, afraid that Thimonnier's invention might mean an end to their jobs, broke into his factory and smashed the machines. In spite of this, Thimonnier kept working on his invention, built an improved model, and started up again. The mob struck again. Thimonnier fled, but his idea for a "sewing machine" lived on.

Fig. 8-1. Manufacturing grew as families began to specialize in producing certain products.

Fig. 8-2. Large buildings called factories were built to house manufacturing machines. Workers now went to the factory to work instead of staying in their homes.

The Factory System

In the 1800s, the factory system came into being. Machines were being developed. A steam engine or waterwheel provided power. People went to work in a special building that housed these machines. This building was called a factory. Fig. 8-2. A company owned the factory and paid the workers. These changes were the beginning of the Industrial Revolution. The **Industrial Revolution** refers to the great changes in society and the economy caused by the switch from products being made by hand at home to products being made by machines in factories.

At first, the factory system had many faults. Workers were paid very little. Many worked 16 hours a day. Working conditions were usually poor. Often, there was no fresh air, no concern for safety, and no breaks in the work routine. Even young children were made to work under these conditions. Gradually, concerned people caused changes to be made. Laws were passed to protect workers.

Modern Manufacturing

Today manufacturing is still done in factories. However, modern factories are much safer and more efficient than old-time factories. Fig. 8-3. Production systems have become very technical and specialized. Much knowledge has been gained over the years.

Fig. 8-3. Today's factories are clean and neat and emphasize safety.

There are three basic types of modern production systems:

- Custom
- Job-lot
- Continuous

The type of production system chosen depends on the kind of product to be made and the number or amount to be produced. For example, consumer products that are widely used, such as candy bars and compact discs, are produced *for stock*. The company makes large amounts of the product and ships them to stores. When a product is not widely used, the manufacturer may not start producing the product until the customers have placed orders. The product is produced *on-demand*. For example, aircraft manufacturing companies make aircraft only when they have a firm order. The order may be for a single plane or for 100 planes, but the companies don't start production until they have the order.

As you read about the three basic types of production systems, you will be able to see why each has its own advantages.

Custom Production

In **custom production**, products are made one at a time according to the customer's specifications. Each one is different. This type of production is usually the most expensive per number of parts made. Custom-made products may be large or small, simple or complex. Many ships and pieces of jewelry are custom-made. Fig. 8-4.

Job-Lot Production

In **job-lot production**, a certain quantity of a product, called a "lot," is made. A "job" is producing one lot. (See why it's called job-lot production?) Then any necessary retooling or other changeovers can be made so another job of a different part or product

Fig. 8-4. Products such as race cars are produced one at a time. Each one is unique. Is this an example of custom, job-lot, or continuous production?

can be produced. Many seasonal items, such as lawn mowers and snow blowers, are manufactured this way. Per part, this type of production is less expensive than custom production. The cost can be spread over more products.

Continuous Production

Continuous production is the system used for mass producing products. This means a large quantity of the same product is made in one steady process using an assembly line. In an **assembly line**, the product moves from one work station to the next while parts are added. This type of production is also called *line production* or *mass production*. A production line is set up and the products are continuously produced. Thousands, maybe millions, of the same product are made. Continuous production is the most economical type of manufacturing system. Cars and electronic products, such as radios and computers, are made this way. People commonly think of continuous production as "manufacturing."

Apply Your Thinking Skills

1. Consider the example of candy bars produced for stock. Which of the three production methods is likely to be used?
2. Why do you think custom production is so expensive?

Who Does Manufacturing?

Anyone can manufacture a product, even you. If you built a bookcase for your room, you would be custom-producing a product. Usually, however, manufacturing is done by companies that specialize in making products.

Companies

A company is an organization formed by a group of people for the purpose of doing business. You probably know many manufacturing companies by name, like Apple Computers, IBM, Ford Motor Company, and General Electric. Other companies you might not recognize by name but you would know their products. Parker Brothers (games), Levi Strauss & Co. (jeans), and Sony Corporation of America (electronic products) are examples.

Large companies may manufacture many different products. The General Motors Corporation manufactures over 50 different types of cars and trucks. The Procter & Gamble Company makes many different types of products. Soaps, shampoos, cake mixes, and peanut butter are examples, and there are many more. Fig. 8-5.

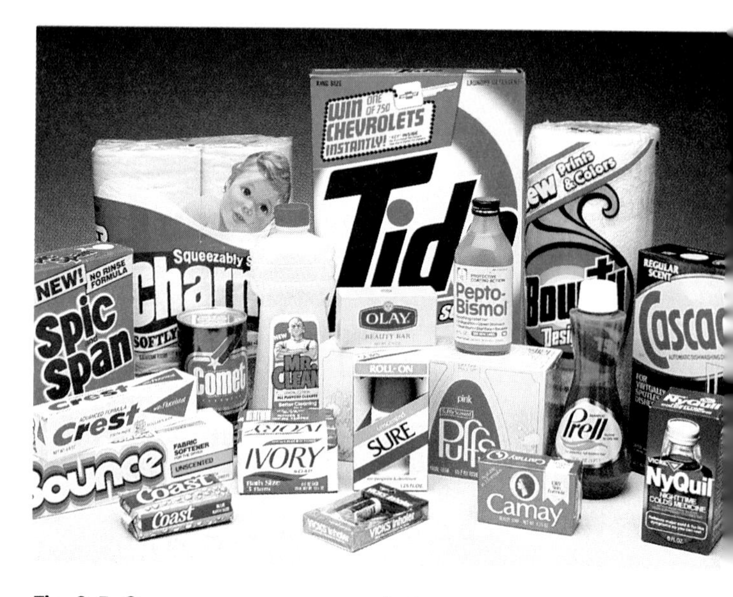

Fig. 8-5. One company may manufacture many different products. All of these products were made by the same company.

Sometimes large companies own smaller companies that manufacture different products. The smaller companies are called **subsidiaries**. They are like separate companies, but are all controlled by the same "parent" company.

Sometimes a corporation may be *diversified*. That means the subsidiaries don't necessarily make similar parts or products. United Technologies has more than 100 subsidiaries. Otis makes elevators and escalators. Carrier makes air conditioners. Essex makes wire and cable. Pratt & Whitney makes jet engines. All of these companies, however, are owned by United Technologies Corporation.

Sometimes a corporation has subsidiaries that are suppliers for the main product they manufacture. A company that makes telephones may also own a company that does plastics molding, another company that makes electronic chips, and still another company that manufactures wire.

Apply Your Thinking Skills

1. If a nation-wide fast-food chain has its own subsidiaries as suppliers, what might these different subsidiaries produce? What do you think the advantages are of a parent company being supplied by its own subsidiaries?

A Managed Production System

As you learned in Chapter 2, a system is a group of parts that work together to achieve a goal. A system is needed to efficiently produce manufactured goods. The manufacturing system works efficiently because it is managed. It is a managed production system. This system has inputs, processes, outputs, and feedback. Fig. 8-6.

Inputs

Input includes anything that is put into the system. The seven inputs in manufacturing are:

- People
- Materials
- Tools and machines
- Energy
- Capital
- Information
- Time

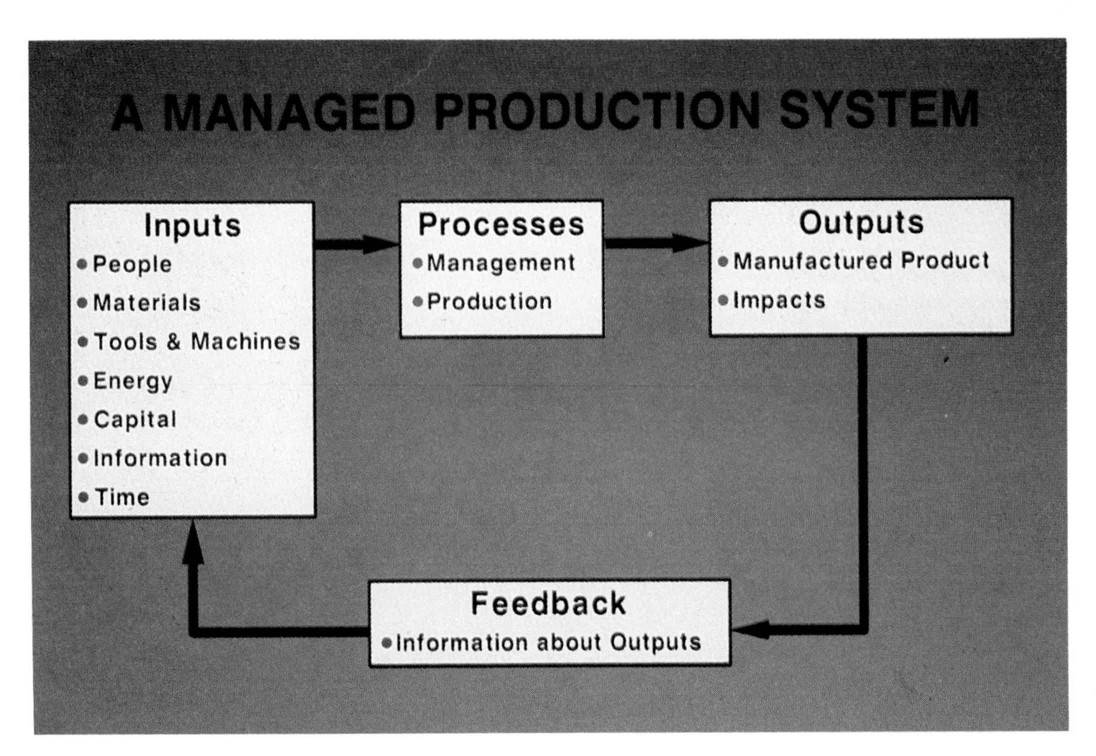

Fig. 8-6. A managed production system is the result of inputs, processes, outputs, and feedback.

People

People make the system work. They are the most important input. People in manufacturing may design products, purchase materials, run machines, assemble parts, inspect products, or sweep floors. No matter what their job is, they all contribute to a team effort. Fig. 8-7.

Fig. 8-7. In manufacturing, people are important. This worker is inspecting tires for defects.

Materials

Products are made from one or more materials. Metals and plastics are two materials commonly used in manufactured products. There are two categories of materials: raw materials and industrial materials.

All materials are first raw materials. **Raw materials** are materials found in nature. Iron ore, plant fibers, and petroleum are all examples of raw materials. Fig. 8-8.

Most raw materials need some refining or processing to convert them into *industrial materials*. These are materials that are in a form that can be used to make products. For example, metal ore must be heated to a very high temperature to remove its impurities before it can be used to make a product.

Fig. 8-8. Some raw materials are mined from the earth. Here coal is being mined.

Industrial materials are usually made as **standard stock**. This means that the material is formed or packaged in a widely used (standard) size, shape, or amount that is easy to ship and to use. Fig. 8-9. Standard stock includes sheets of plywood, steel, and aluminum. Bolts of cloth and barrels of liquid chemicals are also standard stock.

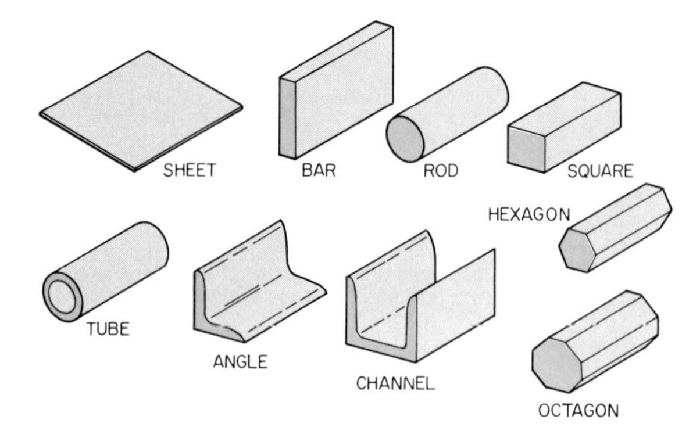

Fig. 8-9. Metal is available in a variety of standard shapes and sizes.

Tools and Machines

Tools and machines are used to help change materials into a finished product. The tools and machines used in manufacturing can be classified into three main categories:

- Hand tools
- Portable power tools
- Machines and equipment

Simple tools powered by humans are called *hand tools.* You have probably used some. Hammers, screwdrivers, and wrenches are all hand tools. Special hand tools are often used in manufacturing. Fig. 8-10.

Fig. 8-10. These hand tools are used to tighten nuts on bolts.

If a tool is powered by electricity or air and is small enough to carry, then it is a *portable power tool.* Portable electric- or air-powered tools, like drills, grinders, and wrenches, are used. Fig. 8-11.

The *machines and equipment* category includes the large, powerful machines used in manufacturing. They are usually installed permanently. This means they are not movable. Saws, milling machines, and drill presses are examples of manufacturing machines. Equipment includes ovens, paint booths, and welding outfits. Fig. 8-12.

Fig. 8-11. This electric drill is a commonly used portable power tool.

Energy

Energy is used to provide power and light for manufacturing. For example, electricity powers motors that run machines, and gas is burned to provide heat for furnaces.

Fig. 8-12. These saws are stationary. They are bolted to the floor.

Capital

Capital is another name for money or financial resources. The two basic types of capital are fixed capital and working capital.

Fixed capital refers to buildings and equipment owned by the company. It's called fixed capital because it represents money spent to buy things that belong to the company and will be there permanently. Tools and machines, though often listed separately as an input, are also fixed capital. Another term used to describe anything the company owns that has value is **assets**.

Working capital refers to cash. It's the money the company uses to buy materials and supplies, pay workers, buy advertising, and pay taxes.

Information

Information is an important input to manufacturing. A toy manufacturer, for example, must know the characteristics of different materials so that no toys will be made from a toxic plastic. A manufacturing company wants to know about consumers who may buy the product, information that can help as products are being designed. Information is often available in a computer data bank.

Time

Time is needed to order materials, produce parts, and assemble products. Manufacturing companies try to figure out the best ways to use production time in order to produce as many quality products as possible in the least amount of time. This is called **efficiency**. Companies that are efficient make more money than those that are inefficient.

Processes

Processes are all the activities that need to take place to make the product. There are two basic kinds of processes: management and production.

Management Processes

In order to work properly, every part of the manufacturing system must be managed. Basically, managers make decisions in three areas:

- Planning
- Organizing
- Controlling

Planning means deciding how something should be done before you do it. You make a plan of action so that things will work smoothly. *Organizing* is gathering and arranging everything needed to do a job. *Controlling* means keeping track of things. If something goes wrong, a correction is needed. Correcting an error is an example of controlling. Fig. 8-13.

Production Processes

All the processes used to actually produce the product are production processes. These processes can be classified as:

- Preprocessing
- Materials processing
- Postprocessing

"Pre" means before. *Preprocessing* happens before any work is done on the material. When the raw material or standard stock is received at the factory, it must be unloaded, stored, and protected until used. These activities are examples of preprocessing. Note that preprocessing activities don't actually change the material to make it worth more. (Remember the discussion of "value added" earlier in this chapter.)

Fig. 8-13. Manufacturing must be managed. Today, managers use computers to help them make decisions about planning, organizing, and controlling the manufacturing system.

In *materials processing*, the shape or size of the material is changed to increase its usability and value. This can be done in three basic ways (Fig. 8-14):

- Forming
- Separating
- Combining

Changing the shape of a material without adding or taking away any part of the materials is called **forming**. Casting molten aluminum in a mold is one example of forming. If you melted one pound of aluminum and poured it into a mold, the finished part would still weigh one pound.

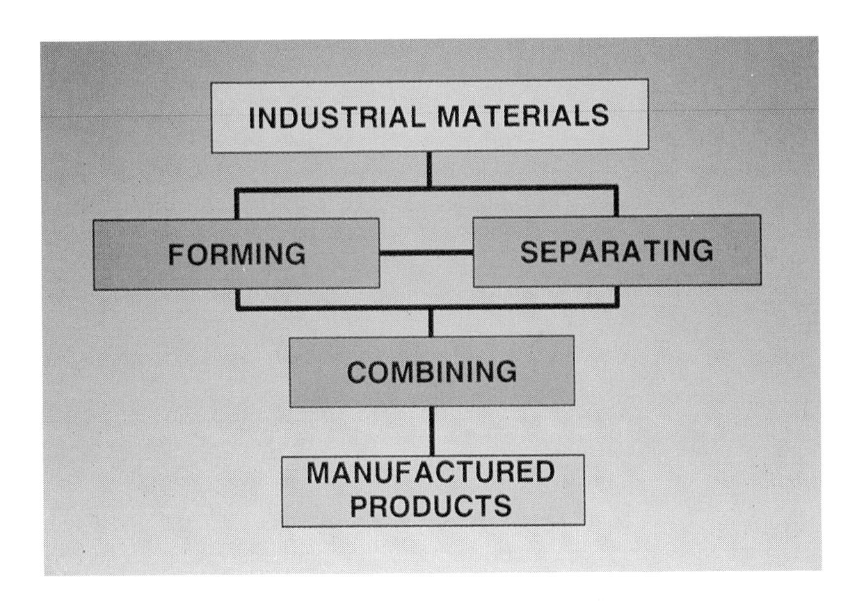

Fig. 8-14. Industrial materials are converted into manufactured products by materials processing.

However, it would be in a different shape than before. Likewise, if you hammered a round piece of steel into a flat shape, you would still have the same amount of steel; it would just be flat instead of round.

One special type of forming is called **conditioning**. In conditioning, the inside structure of the material is changed. Two pieces of steel may look identical, but one may be harder than the other, or one piece may be magnetized. Hardening and magnetizing are two kinds of conditioning.

Changing the shape of material by removing some of it is called **separating**. If you saw a 2 x 4 in half, you have changed its shape by separating. Cutting a piece of cloth from a large roll in order to make a shirt sleeve is another example of separating. The key point here is that the amount of material has been changed. One special category of separating is called *nontraditional*. Included in this category are some newer methods of cutting materials, like abrasive water jet cutting and electrical discharge machining.

Putting two or more materials together is called **combining**. When you bolt or weld two pieces of metal together, you are combining. Mixing chemicals to make a product is another example of combining. Putting a coat of paint on a part is also combining.

Each of the main categories of materials processing has several subcategories. The chart in Fig. 8-15 shows the categories and gives some examples of each.

MATERIALS PROCESSING

CATEGORIES	EXAMPLE
FORMING	
Casting or Molding	Injection Molding Plastic Cups
Compressing and Stretching	Forging Tractor Axles
Conditioning	Heat Treating
SEPARATING	
Shearing	Cutting Sheet Metal
Chip Removing	Drilling Holes in Plastic
Nontraditional	Laser Cutting Saw Blades
COMBINING	
Mixing	Alloying Steel
Coating	Painting Car Bodies
Bonding	Welding Bicycle Frames
Mechanical Fastening	Bolting Parts Together

Fig. 8-15. The main categories of materials processing and their subcategories.

"Post" means after. *Postprocessing* activities include things done to the product after the materials have been changed in form or shape. Handling, protecting, and storing products are all examples of postprocessing activities. Once again, there is no value added at this stage.

Postprocessing may include recycling activities. Collecting glass or plastic bottles, sorting them, and grinding them into chips is an example of recycling. The chips can then be used as raw material or standard stock for manufacturing other products. Perhaps your community has an organized recycling program. Fig. 8-16.

Output

The result of inputs and processes is called output. In manufacturing, as in any system, there are many outputs. One output, of course, is the manufactured product. This output is expected and desirable. Other outputs, such as waste and pollution, may not be expected, and they certainly are not wanted.

Fig. 8-16. Some products can be recycled after their functional life. This plastic will be ground up and remelted to be made into other products.

Impacts

The outputs of a system affect us and our world. In other words, they have an impact. For example, cars are an output of a manufacturing system. They have become so numerous and widely used that they have an impact on nearly every aspect of our lives.

- They affect our economy. Think of all the people whose jobs depend on the automobile industry.
- They affect our society. Cars are more than just a way to travel. They are part of our culture. We use cars as a way to express who we are, or who we would like to be.
- They affect our politics. Who should pay for road-building? Should the government protect the American car industry from foreign competition? Should the speed limit be raised or lowered? All these are political issues that result from our use of automobiles.
- They affect our environment. To build and drive cars, we use nonrenewable resources, such as metals and oil. Car exhaust pollutes the air. Automobile junkyards are an eyesore, and they can pollute the land and water. Fig. 8-17.

Feedback

The feedback loop of a system links the output back to the input. It is the information about the outputs that is sent back to the system to determine whether the desired results are being achieved. Consider the car example. We know that car exhaust pollutes the air. This is an undesirable output. People let the government and the car manufacturers know that they wanted cars to pollute less. That's an example of feedback. That feedback has led to changes in the way cars are designed. Can you think of other examples of feedback that have affected the way cars are produced?

Apply Your Thinking Skills

1. You have read about the seven inputs to a manufacturing system. How might the quality, amount, or cost of one input affect the other inputs? For example, if materials become more expensive, what might happen with the other inputs?
2. If a manufacturing system produces undesirable outputs such as waste products and pollution, should we stop making the product?

Fig, 8-17. Automobile junkyards are unattractive and can cause land and water pollution.

Investigating Your Environment
Environmental Manufacturers

Manufacturing systems are different today than they were in the 19th century. In the past, little consideration was given to the environmental hazards caused by manufacturing. As long as profits were good, little thought was given to how much material and energy was used or wasted in manufacturing processes. With the growing concern for the environment today, companies are actually setting environmental goals. More money is being spent to address environmental issues. Current research and new technology will enable manufacturers to make products without harming the environment.

Changing manufacturing processes and modifying raw materials are two ways to achieve environmental goals. Many utilities are cutting their carbon dioxide emissions by using conservation programs and solar and geothermal technologies. The Volvo company is cutting solvent emissions from car plants by switching to water-based paints. Researchers at Southern Illinois University are developing a strain of bacteria that will eat the sulfur contained in some coals before the coal is burned. If coal containing sulfur is burned, the sulfur that is given off combines with the water and nitrogen oxide in the atmosphere to form sulfuric acid. This results in acid rain.

Fig. IYE-8. Nowadays, many manufacturers are trying to find ways to reduce or eliminate environmental hazards that are caused by their manufacturing processes.

New methods of packaging are one way to help reduce waste problems. Procter & Gamble now sells small packages of concentrated liquid fabric softeners. The consumer then adds water to this concentration instead of using a larger package that already contains the water.

The paper industry has proved that better use of raw materials is possible. In the past, only half of a tree was used and the other half discarded. Now, 90% of the tree is used. The other 10% is burned for energy or put on land as a fertilizer to encourage new tree growth.

Some industries have found that their "garbage" may be another industry's product material. For example, one of the main products produced by DuPont is nylon. Every year, DuPont gets rid of 3,600 tons of hexamethyleneimine (HMI), a chemical byproduct from making nylon. Now, HMI is sold to pharmaceutical industries, who use it for coatings on pills.

As you can see, manufacturers have come a long way from the days of soot-belching smokestacks, the widespread stripping of natural resources, and the "use-the-part-we-need-throw-the-rest-of-it-away" attitude. They are aware we must all strive to protect our most valuable resource—our environment.

Take Action!

1. Contact a local manufacturing company representative to find out what his or her company is doing to reduce pollution while its product is being manufactured. Report your findings to the class.

Chapter 8 Review

Looking Back

Manufacturing affects every part of our lives. It is not only very important to us as individuals, but also to our entire society and our economy. It also affects our politics and our environment.

Manufacturing started out with families making what they needed. Then they began to specialize; family members made products in their home and sold them. Later, with the gradual development of machinery, people began to work in factories instead of their homes. They worked very long hours for little pay, often in unsafe conditions. Today's factory system is much safer and more efficient than early factories. They are also very technical and specialized.

The three basic types of modern production systems are custom, job-lot, and continuous. Each type has its own advantages.

Manufacturing is a system of inputs, processes, outputs, and feedback. Inputs include people, materials, tools and machines, energy, capital, information, and time. Processes include management processes and production processes. Outputs are the finished manufactured products as well as all the impacts of the system. Feedback helps the system check itself.

Review Questions

1. Define economy.
2. What does the term "value added" mean?
3. Briefly describe the Industrial Revolution.
4. Describe custom production.
5. Describe job-lot production. What are some advantages of this type of production?
6. What is a subsidiary?
7. What is the difference between fixed capital and working capital?
8. Briefly describe the three types of management processes.
9. Define each of the three types of materials processing and give two examples of each.
10. What are the outputs of a manufacturing system? Discuss some possible undesirable outputs of manufacturing.

Discussion Questions

1. How does the manufacturing industry influence our economy?
2. How can on-demand production save money?
3. Why do you think a person or a company would want a custom-made product? Do you think having a custom-made product is worth the great additional expense? Explain your answer.
4. Why is it important for manufacturers to have information about consumers who want to buy their product?
5. Why is efficiency so important to a manufacturing system?

Chapter 8 Review

Cross-Curricular Activities

Language Arts 

1. Research the cottage industries that exist in the United States today. What types of products are still being produced this way? Use the *Reader's Guide* for current facts. Write a short essay on these products.
2. The early days of clothing manufacturing produced work places known as "sweat shops." In a descriptive essay, describe these shops.

Social Studies 

1. Write a brief report about a manufacturing industry in your community or a nearby community. Tell about the impacts it has on society in the community, the local economy, and the environment.
2. Research the following labor unions: National Labor Union, the American Federation of Labor, the Congress of Industrial Organizations, and the Knights of Labor. Find out what worker gains each of these unions was responsible for. Then find a currently existing labor union and tell what gains it is currently working for on behalf of the American work force.

Science 

1. The choice of materials used in manufacturing depends on their chemical and physical characteristics. For example, glass is silicon dioxide, a substance that is transparent, easily shaped, and does not react chemically with many substances. List examples of new products manufactured by (a) forming, (b) separating, and (c) combining glass with other materials. How did silicon dioxide's characteristics of transparency, ease of shaping, and low chemical reactivity make it a good choice for use in making these products?

Math

Below is a copy of William's time card for one week. Use this time card to answer the questions that follow it.

EMPLOYEE TIME CARD
IFCO WELDING

Name: WILLIAM CLARK
Part-Time

DAY	DATE	IN	OUT	IN	OUT	HOURS
MON	2/9	8:00	12:00	1:00	5:00	
TUE	2/10	7:30	11:30	12:30	4:30	
WED	2/11	8:30	11:15	11:45	4:00	
THUR	2/12	7:45	12:15	--	--	
FRI	2/13	7:00	12:00	12:30	5:00	
SAT	2/14	8:00	1:00	--	--	
					TOTAL HOURS	

1. How many hours did William work on each of the days?
2. What were his total hours for the week?
3. If William earns $8.40/hr., what was his total pay for the week?
4. If William was paid time-and-a-half for Saturday work, what would be his pay for the week?

Chapter 8 Technology Activity

Material Testing

Overview

This activity is designed to help you learn about material-testing devices and some of the physical properties of materials. The property to be tested is adhesion. *Adhesion* is the bonding of objects or materials using a substance such as glue. This physical property will be tested by laboratory experiments. A suggested testing setup is shown at the end of this activity.

Goal

To learn about material-testing devices and the adhesive properties of different types of glue.

Equipment and Materials

testing equipment (see drawings)
vise
timer
buckets
sand
scales
wood strips (hardboard, particle board, plywood, hardwood, and softwood)
glue (polyvinyl resin, aliphetic resins, contact cement, epoxy, hot melt, hide glue, others)

Procedure

1. *Safety Note:* For this activity, make sure that the room you are working in has adequate ventilation. Also, avoid contact with the different types of glue. Contact with some types of glue can irritate your skin.
2. Prepare the wood strips and gather the various adhesives that will be tested. Remember that "constants" are very important in an experiment. These are factors that are not changed. Use the same size and type of wood for all of the adhesion tests. Fig. A8-1.
3. Use the same size area of contact surface for each test. Fig. A8-2.
4. Test a minimum of three different types of adhesives. Use drying times of 15 minutes, 45 minutes, and 24 hours and clamping/not clamping as variables. Variables are factors that are changed.
5. On a sheet of paper, list the type of glue and the combination of variables you use for each of your tests. Set up and perform the tests as shown in Fig. A8-3. For each test, record the weight of the sand that causes the glued joint to fail or the wood to break.
6. Make a chart that reflects the results of your tests.

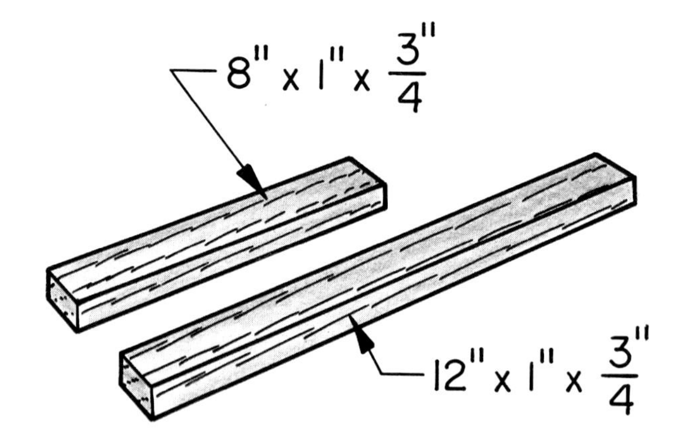

Fig. A8-1. Cut wood samples to size.

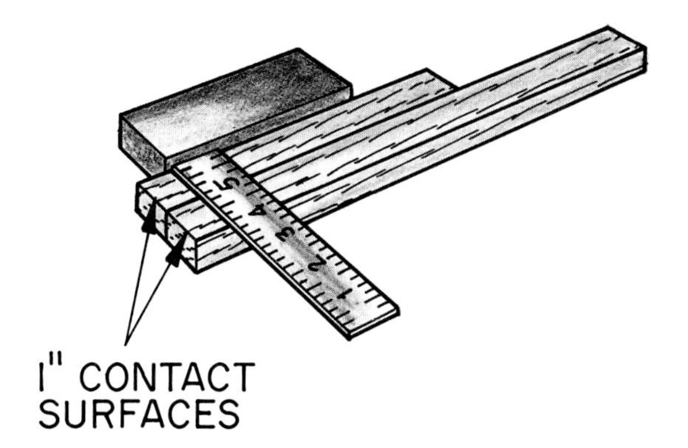

Fig. A8-2. Measure and mark one inch square contact surfaces.

Evaluation

1. Overall, which type of adhesive has the strongest bonding power? Which has the weakest bonding power? How much did the variables affect the results for the different glues you tested?
2. How well did your testing device work? If you could do this activity again, would you change the design of this device? If so, how?

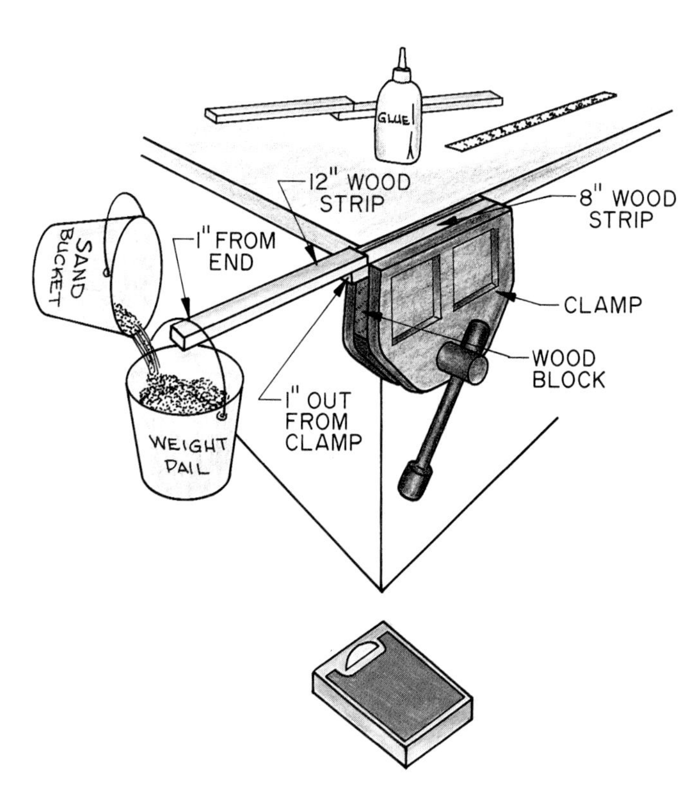

Fig. A8-3. Adhesion test. Load the weight pail slowly until the glue joint breaks. Weigh the load, calculate the "real load" (weight of the sand), and chart the data. Compare results of all tests.

Chapter 9

Product Development

Looking Ahead

In this chapter, you will discover:

- where new product ideas come from.
- how ideas are developed into products.
- what types of product drawings are needed.
- the major steps involved in engineering a new product.
- how computers are used in designing and engineering products.
- the importance of product testing.

New Terms

assembly drawings
computer-assisted design (CAD)
computer-assisted engineering (CAE)
detail drawings
functional design
interchangeability of parts
mock-up
modular design
product engineering
prototypes
schematic drawing
standardization
tolerance
value analysis
working drawings

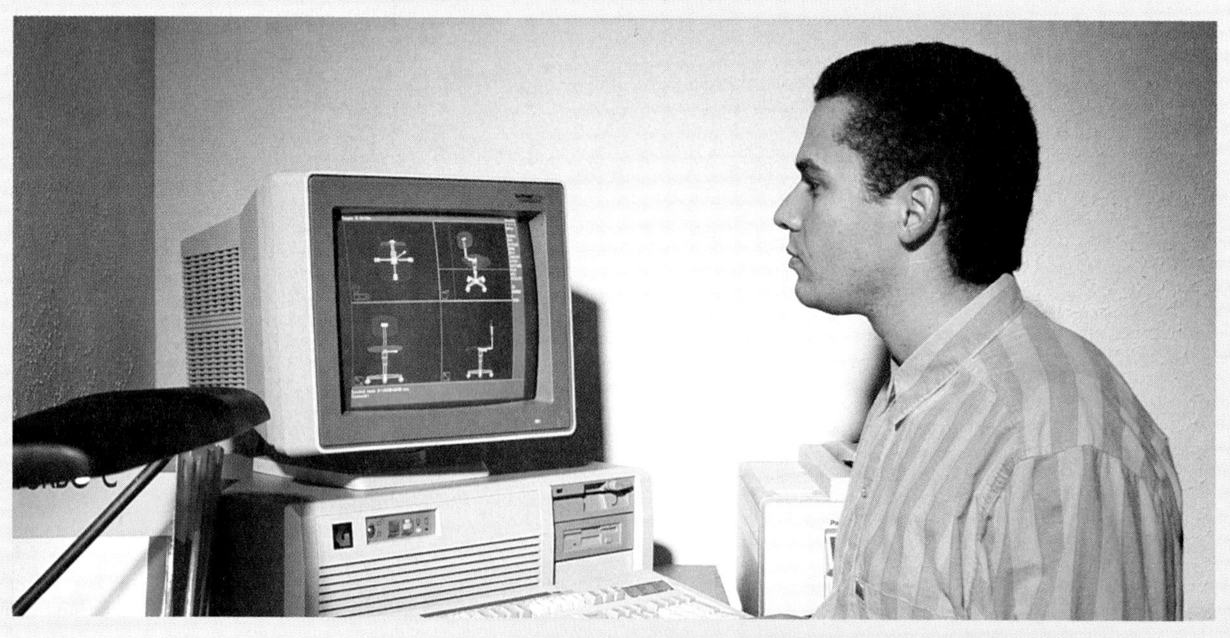

Technology Focus

Optical Glass Sorter

Like many other people, Michael Lewis has an interest in our environment. Unlike many, however, Mike has used his skills and talents to actually do something about improving the environment.

As a student at the University of Illinois, Mike was asked to design and build an engineering project. Mike knew that the average American discards three and a half pounds of trash each day. Eight percent of that trash is glass. He decided that his project would be to develop a process for recycling used glass containers.

Glass factories desire clear recycled glass because it can be used in new glass containers of any color. Clear recycled glass must be at least 99% color pure. Mike's goal was to find a way to mechanically sort clear, brown, and green glass containers.

He found that other mechanical devices have been developed to sort broken glass pieces, but they couldn't sort carefully enough to produce 99% clear glass. "Common sense told me it would be easier to sort whole, rather than broken, containers," says Mike.

His first problem was to develop a machine with vision that could "look" at glass containers and recognize which ones were the clear containers. His next problem was to mechanically separate the clear from the colored containers.

In the laboratory, Mike built a small working model of his optical sorter that worked well. He then built a full-sized, low-cost optical sorter and put it to work. Mike's classmates became excited about his work, and some of them assisted in his research. The Community Recycling Center of Champaign, Illinois, tested Mike's optical sorter.

Mike designed the sorter to work "upstream" of a manual worker in the recycling center. The sorter wasn't perfect. It was about 75% accurate. The worker "downstream" visually sorted the glass that the optical sorter missed. The optical sorter replaced at least one worker on the sorting line. It could process more than six tons of glass per hour. The final output from the glass sorting process was nearly 100% pure.

Mike Lewis has made a difference in our environment with his optical sorter. He provided a practical solution to an environmental problem.

Fig. TF-9. Michael Lewis's optical glass sorter has provided a practical solution to an environmental problem.

New Products, New Ideas

Hardly a day goes by without a new product being introduced. Sometimes the product isn't really new. It's just "new and improved." Why do we see such a constant parade of new or improved products? Where do manufacturers get all these ideas for new products or improved products?

The Product Life Cycle

All products are new when they are first introduced. For example, digital watches were new at one time, but now they're quite common. There are some you can buy for just a few dollars. When they were first introduced, however, they cost over a hundred dollars! What has happened? The watches are going through the product life cycle. Fig. 9-1.

The product life cycle starts as a product is *introduced*. Because it's new, the product is expensive and often not very reliable. However, as more people begin to buy the product, it enters the *growth period*.

During the growth period, product improvements are made. The "bugs are worked out." More of the products are made, and the price usually gets lower. More people can afford to buy the product.

Because of advertising and general acceptance by the public, the product enters the *maturity stage*. During this period, the price has stabilized (stopped changing). Also, the product is usually very reliable.

After a while, sales start to taper off. Nearly everyone who wanted the product now has one. This period in the product life cycle is called *saturation*. The demand has been satisfied fully. More products are available to sell than can be sold. The market is saturated. Finally, sales start declining, and the product may be taken off the market.

The length of a product's life cycle varies depending on the product. It may last a few months, a few years, or many, many years.

To be successful, manufacturing companies must continue to produce and sell products. Because of the product life cycle, each company tries to keep creating new product ideas. That way, as one product is declining, a new product can be entering the growth stage. This is what keeps a successful manu-

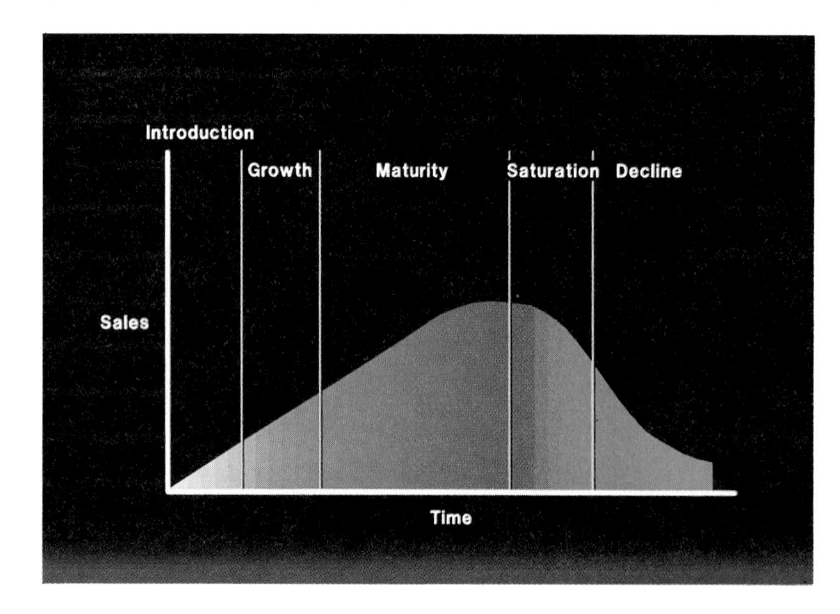

Fig. 9-1. Most products go through the product life cycle. The length of time for each stage may vary.

facturing company going and growing. It's a continual process.

New Ideas and Directions

Every new product starts as an idea. Generally, all product ideas are the result of one of two occurrences. Sometimes product ideas come from someone who thinks of a new product to solve a particular problem. The problem creates the demand for the product. Other times, someone suggests an idea that would make a useful or desirable product. The company may then, through advertising, create the demand for the product. Fig. 9-2.

Sometimes the idea for a new product comes from an individual, who may then sell that idea to a manufacturer. However, most product ideas come from research and development (R&D) departments. The research and development department may be a part of the company or may be an independent R&D company hired by the manufacturer to create workable ideas for new products. People working in R&D departments use brainstorming, market research, and problem-solving techniques to come up with new product ideas. Fig. 9-3. It is the responsibility of the R&D group to develop ideas and then refine them so they can be developed into workable products.

Actually, not every idea turns out to be a good product. Of about every sixty new product ideas, only one is good enough to finally make it as a new product. That's another reason a company is always looking for new ideas.

Research and Development

Research and development are processes used to find and develop new ideas into products. *Research* is done to gather information. How can we make a car that gets better gas mileage? How can we make lighter bathtubs? What is the best way to arrange the keys on a computer keyboard? These are just a few of the questions that could be answered by research. *Development* is using research information to solve a problem.

Fig. 9-2. Every product begins as an idea.

Fig. 9-3. Brainstorming is one way engineers can come up with new product ideas.

Fascinating Facts

Ideas for new inventions can come from anywhere, as Swiss engineer George de Mestral discovered back in 1948 during a hunting trip in the Alps. As he was walking along, burrs stuck to his clothes. He looked closely and found that the burrs had tiny hooks that caught the loops of thread on fabrics. He set out to develop a fastener that worked along the same principle. His discovery—Velcro®—was patented worldwide in 1957.

There are two types of research: basic research and applied research.

Basic research is done to learn more about the world around us. This kind of investigation (careful search) helps us understand things. Often new information can be put to use immediately. Sometimes, however, it can't. Scientists discovered the transistor as they investigated electronics. It wasn't until later that they found uses for it, such as making smaller radios and TVs.

Most research done in manufacturing is *applied research*. This type of research is done to solve a problem. The new information can be applied immediately. After transistors were in use for a while, researchers looked for a way to reduce the size of a radio still further. They tried to find ways to make electronic components even smaller. The result was the development of the microchip. Fig. 9-4.

The same basic methods are used for both basic and applied research: retrieving, describing, and experimenting.

Retrieving is getting information that is already known. This method identifies what already exists. That way we don't "reinvent the wheel."

Describing is finding present conditions. If you wanted to manufacture stereo headphones, you'd need to know the sizes of human ears. If you measured some ears to find out, you would be describing.

Finally, research is done by *experimenting*. When experimenting, researchers try things out and see what happens. This method provides answers about how things will work. If necessary, they can make corrections and try again. They do this until they find the answers they need.

A. Early radios used vacuum tubes.

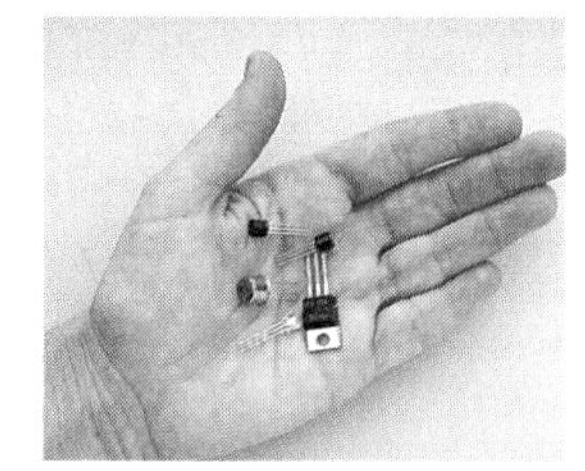

B. Transistors are small devices that transfer electrical signals.

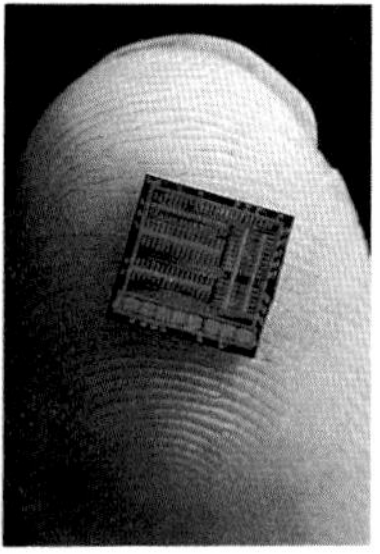

C. A microchip is a tiny chip of silicon (usually) that is chemically treated to both carry and control the flow of electricity.

Fig. 9-4. Discoveries made during research have led to the miniaturization of electronic products.

Apply Your Thinking Skills

1. What are some products that you have seen go quickly through the life cycle? Why do you think their life cycle was so short?

Product Design

After the product idea is established, the product itself must be designed.

Designing Product Appearance

Designing the product's appearance is an important part of the development process. The design should be eye-catching and appealing.

Drawings

Product designers start by thinking about the product idea. They try to visualize (picture in their minds) what the product should look like. Then they begin making simple drawings. First, they make lots of *thumbnail sketches*. Fig. 9-5. As you learned in Chapter 6, these are small drawings, done quickly. They are called "thumbnail" because of their small size. They don't include much detail. The designer tries to show the general idea. The purpose of making thumbnails is to capture many different ideas.

The next step is to make some *rough sketches*. The product designer reviews the thumbnails. Then he or she picks the best parts of each and combines them into a larger, more detailed drawing. Features like color and surface texture may be shown. As each new drawing is made, the product's appearance becomes more and more apparent.

Renderings are made next. These realistic drawings are done very carefully. A rendering shows the designer's idea of the final appearance of the product. It can be shown to potential (possible) consumers to see if they like the design.

Fig. 9-5. The designer begins by making thumbnail sketches.

Computer-Assisted Design

In **computer-assisted design (CAD)**, the designer creates designs right on the computer instead of using pencil and paper. The designer "draws" complex shapes right on the screen. This is called *interactive computer graphics*. It's interactive because the designer interacts with the computer. Fig. 9-6. The computer reacts to what the designer does, and the designer reacts to what happens on the computer.

The designer can see the design on the video screen. Drawings on the screen can be rotated, enlarged, or reduced. If the designer wants to make changes in the design's appearance, it can be done instantly. When satisfied, the designer can push a button and get a hard copy (drawing on paper or film) of the design.

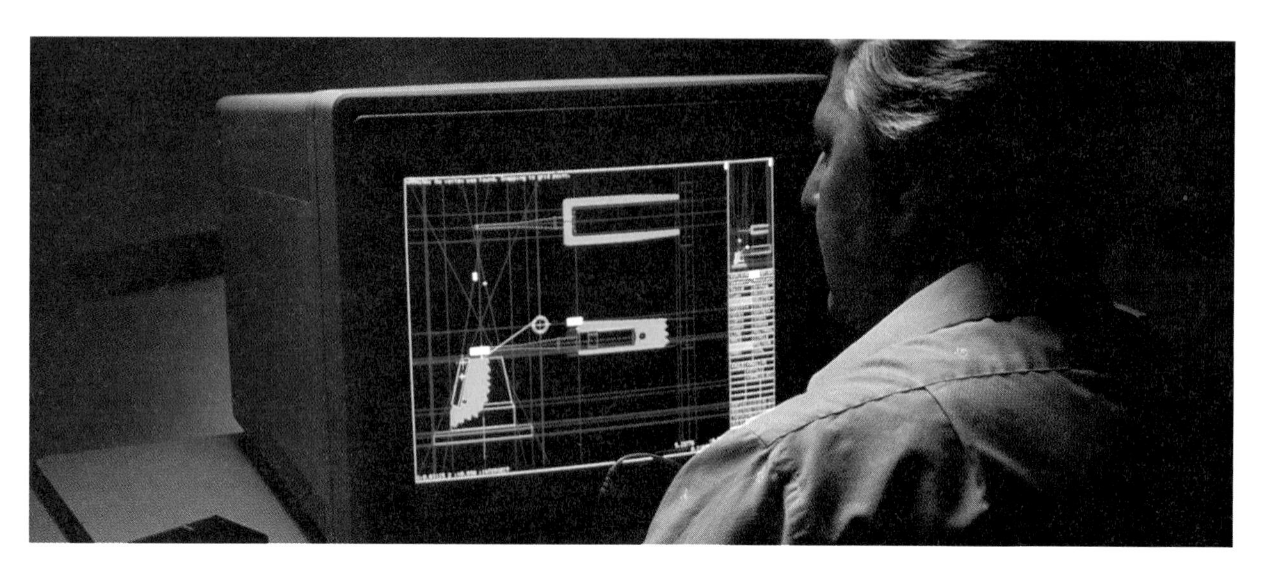

Fig. 9-6. Interactive computer graphics allow the designer to make instant changes to the product design.

Mock-ups

Designers also make mock-ups of the finished drawing. A **mock-up** is a three-dimensional model of the proposed product. It looks real, but has no working parts. People who see it can understand what the final product will look like.

Most of the time a mock-up is built full size. However, sometimes a scale model mock-up is made. Fig. 9-7. Mock-ups may be made of wood, plaster, plastic, or even cardboard. The type of material used does not matter as long as the mock-up looks like the real product.

Fig. 9-7. This is a scale model mock-up of a new aircraft. In a scale model, one size is used to represent another size. For example, one inch of the scale model might be equal to one foot of the actual product. This model is much smaller than the actual product, but size relationships are constant.

Management Approval

During the product design process, various people can make comments or suggest improvements. For example, managers from different departments in the company are asked to give their opinions about the design. The marketing manager is concerned with how the product will sell. The manufacturing manager is concerned with how easy or difficult it will be to manufacture the product. The financial manager is concerned with costs. All of these managers and more must agree that the product design is a good one. Fig. 9-8.

When everyone finally agrees on the product design, an official approval is given. Much developmental work remains to be done, but this approval is the signal that development should proceed.

Fig. 9-8. Managers approve the new product design if they are satisfied it is a good idea.

Apply Your Thinking Skills

1. Think of a product that has a poorly designed appearance. How would you improve the product's appearance without harming its usability?
2. What are some advantages of using CAD over drawing designs by hand? Are there any disadvantages to using CAD?

Product Engineering

Engineering involves predicting the future behavior of a material or system. Engineers base predictions on known facts, mathematics, and scientific principles. **Product engineering** means planning and designing to make sure the product will work properly, will withstand extensive use, and can be manufactured with a minimum of problems. There are many things to consider when engineering a product. Every detail has to be planned carefully.

Every part, including each nut and bolt, must be planned or chosen. Parts must be just the right material, size, and strength. Two major engineering design activities are done: functional design and design for manufacturability.

Functional Design

Functional design activities ensure that a new product will work properly or serve its planned purpose. Proper function is determined in part by the materials used and the structure of the product. For some products, the power system is important. For example, the success of a new aircraft design depends upon the materials specified, the arrangement of the aircraft's structural framework, and the output of the engines.

Materials

A very important part of engineering any product is specifying the exact materials to be used. Should the product be made of plastic or metal? Does the material need to be fireproof? Maybe it has to be able to withstand physical impacts. Perhaps the material must be able to endure sub-zero temperatures. The product engineer has to know exactly what is required. Then he or she can select just the right material. Fig. 9-9.

The composition of a material has an effect on its strength. Steel, for example, is available in degrees of strength referred to as grades. The engineer must determine which grade of steel is best for a particular product. For example, a tractor axle and a door hinge are both steel products. However, since the strength requirements are different, different steels would probably be specified.

Fascinating Facts

The search for a cheaper billiard ball led to the invention of one of the most common materials in the world today—plastic. A popular game in the 1800s, billiards used balls that were carved of ivory, which was very expensive. In 1863, a New York firm offered a $10,000 prize to anyone who could develop a practical substitute for ivory.

John Wesley Hyatt, who had a factory that made checkers and dominoes, set out to win the prize. For several years, he tried molding billiard balls of a variety of different materials, including sawdust and glue. He finally tried a new experiment, adding camphor and a dash of alcohol to treated cotton. He put the soft solid into a mold, heated it, and let it cool. The result? The first molded plastic!

Although the billiard balls didn't always follow "predictable" paths and also caught fire easily, the invention was the beginning of a whole new industry.

Structural and Mechanical Elements

Structural elements are frameworks and supporting members. *Mechanical elements* are the working parts. What holds a car body together? Why does the cap on a pen snap on and stay there? Why does the door open when you turn a doorknob? It's because the structural and mechanical elements of these products were engineered properly.

Fasteners, like rivets or screws, are chosen for particular reasons. The shape of a piece of aluminum tubing—round or square—can affect product performance. Mechanisms like gears and levers have to be planned so that everything works properly. Fig. 9-10. Once again, an engineer carefully plans exactly how each part will fit and work together with the other parts.

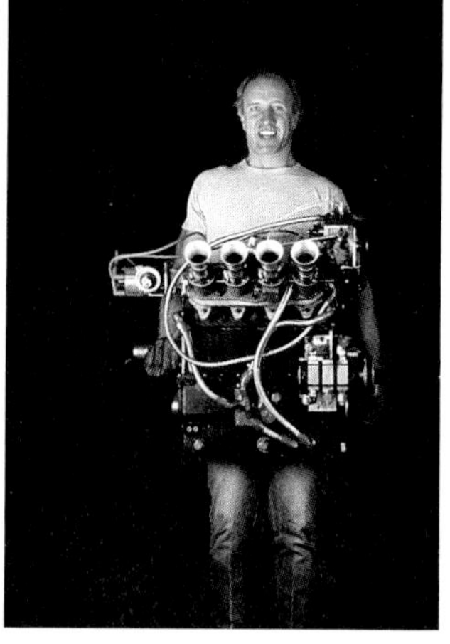

Fig. 9-9. This automobile engine is made of *plastic!* It weighs about half what an ordinary metal engine weighs. The engine performed efficiently in this race car.

Fig. 9-10. The U.S. cycling team rode this type of bicycle in the Olympics. All of the structural and mechanical elements were carefully planned and chosen.

Power Elements

Not all products are powered. Dishes, hats, and baseballs aren't powered. However, toasters, computers, mopeds, and lamps are. If a product requires some kind of power to make it work, then the power elements must be engineered.

Many products are electrically powered. Some, like a car lift, are powered by the force of a moving liquid. This is *hydraulic* power. Still others may be mechanically powered, such as by one or more springs. The size of each wire, the horsepower of each motor, and the stroke of a cylinder are all determined by the engineer. Every detail of the power elements must be planned. Fig. 9-11.

Fig. 9-11. Products may be powered in different ways.

A. This driver-reamer is powered by rechargeable batteries. It is used in hip replacement surgery.

B. This farm tractor uses a combination of electrical, mechanical, and hydraulic power.

C. This pneumatic jackhammer is powered by compressed air.

Design for Manufacturability

What effect will the design of the product have on the ability to actually manufacture the product? How will the design of a part affect the production of the part? (For example, a part with square corners may be harder to make than the same part with rounded corners.) These are the major considerations of *design for manufacturability*, which is also known as *production design*.

Engineers in production design try to design parts for easier manufacture and assembly. They accomplish this by including the following in their design for manufacturability:

- Interchangeability of parts
- Standardization
- Simplification
- Modular design
- Design for assembly

Fascinating Facts

Eli Whitney, best known as the inventor of the cotton gin, also introduced mass production with interchangeable parts to the United States. In 1798, he entered a two-year contract to make 10,000 muskets for the government. At that time, each musket—and actually everything else that had more than one part—had to be made by hand. Each separate part had to be adjusted to fit into the part next to it.

By 1801, Whitney had delivered only 500 of the 10,000 guns that had been ordered. He went to the War Department with a large box that—he said—contained 10 muskets. However, when he opened the box, the "muskets" were simply a collection of barrels, stocks, triggers, locks, and other parts. He then proceeded to pick a part of each type and put together a working musket. He had developed a way of machining the parts so carefully that they no longer required the time-consuming process of adjusting or specially making each part to fit into an adjoining part. The War Department forgave Whitney for the delay in delivering the muskets, and Whitney went on to make a fortune with his newest "invention," interchangeable parts, which led to the mass production of many different products.

Interchangeability of Parts

Mass production involves producing large quantities of a particular product. It is not possible unless a company is able to make large quantities of identical parts (parts that are alike). That way, any one of thousands of a specific kind of part may be used when assembling the product. Each part does not have to be specially made. This idea is called **interchangeability of parts**.

In order for parts to be interchangeable, they must be the same size. However, because of human error, or differences in machines, or tool wear, there will be slight differences. If you examined the parts closely, you would find they are not exactly the same. To allow for this, engineers plan for a certain amount of variation in each part. This variation is called tolerance. **Tolerance** is the amount that a part can be larger or smaller than the specified design size and still be used. It is tolerance that allows mass production of parts and rapid assembly of parts into a product.

Standardization

Agreement on common sizes of parts is **standardization**. When users of particular products agree to use the same sizes, then the sizes are standardized. Standard sizes make manufacturing easier. For example, bolt sizes are standardized. Instead of every manufacturing company making their own sizes of bolts, they can buy standard-size bolts. It's cheaper and faster.

Lots of things are standardized. Threads on a light bulb, tire sizes for cars, and electrical plugs and outlets are examples. What others can you name? Standard parts in standard sizes reduce the amount of new engineering work that must be done on a new product.

Simplification

Reducing the number of different size parts in a product makes the product easier to manufacture. Suppose that a design for a typewriter specified sixteen screws of different sizes. Then, sixteen different screws would have to be ordered and assembled correctly. If the same size screw fit in each place, however, you could order just one size. Assembly would be simpler, too. Any one of the screws could be placed in any one of the sixteen locations.

Modular Design

Still another important idea is modular design. A *module* is a basic unit. In manufacturing, parts are often made as modules. Designing a product that will use these modules or modular parts is **modular design**. For example, an automobile has a basic chassis and body. However, parts such as engines, wheels, and radiators are modules. It's possible to have a variety of products by using basic parts and adding different modules. The same basic car could be made with a four-cylinder engine, a six-cylinder engine, or even an eight-cylinder engine. Fig. 9-12.

Design for Assembly

Design for assembly means planning to make it easier to put product parts together. For example, if parts can be held in place with clips that are molded into the product instead of with nuts and bolts, fewer parts will be required since the nuts and bolts can be eliminated. The assembly can also be done much faster, since using clips is quicker than threading nuts onto bolts with wrenches. Many products, such as computer printers, automobile dashboards, and refrigerators use clips instead of screws or nuts and bolts to hold parts together. Fig. 9-13.

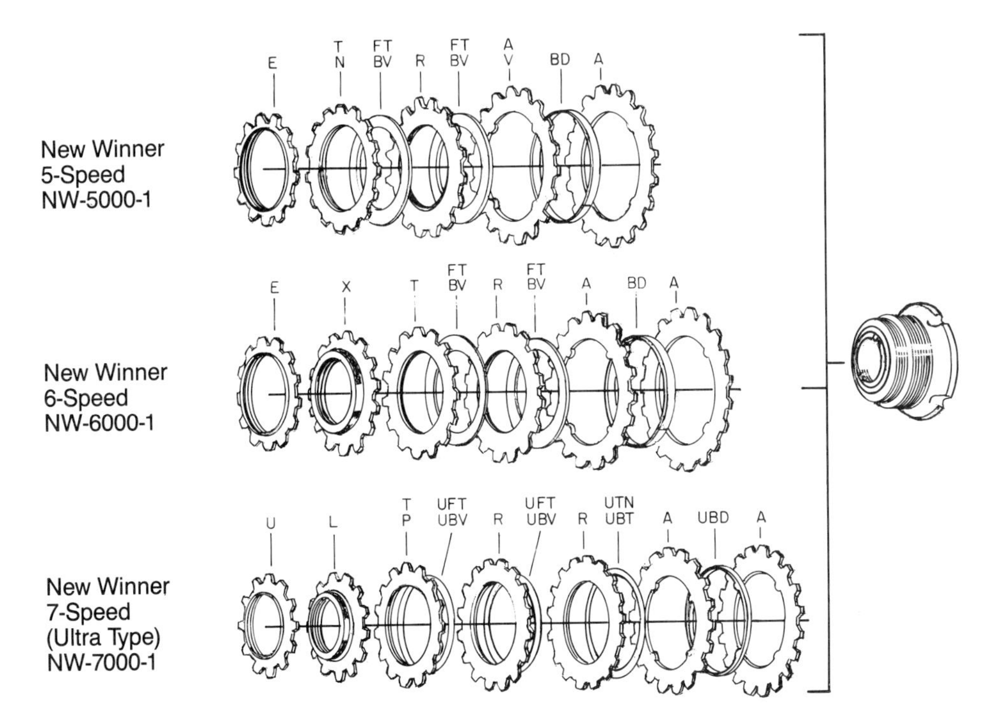

Fig. 9-12. Because of modular design, this bicycle hub can be used as a 5-speed, 6-speed, or 7-speed rear freewheel.

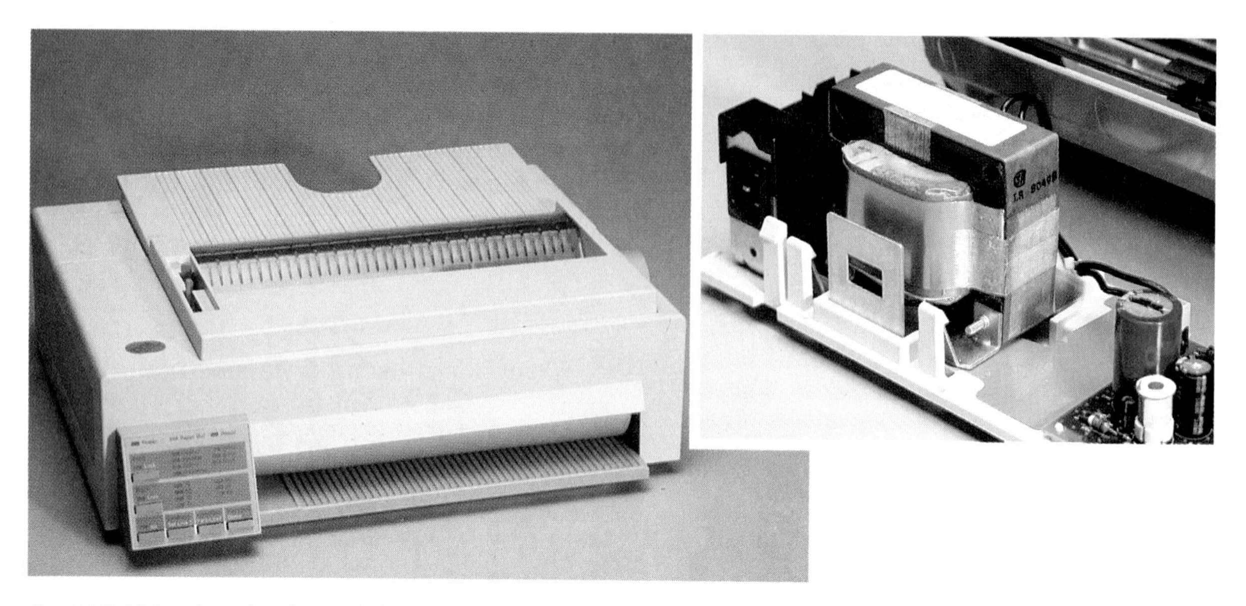

Fig. 9-13. This printer has been designed so that parts can be clipped in place. Robots are used to assemble this product.

Working Drawings

As the details are determined, they must be recorded. This is done by preparing a set of **working drawings**. These drawings show exact sizes, shapes, and other details. Products are manufactured according to the information these provide, so all working drawings must be accurate and readable. As you learned in Chapter 6, developing these drawings or plans is called *drafting*.

There are three main types of drawings in a set of working drawings: detail drawings, assembly drawings, and schematic drawings.

A **detail drawing** specifies the details of a particular part. All the necessary information is given so that the part can be manufactured. The detail drawing shows the shape of the part. Dimensions (sizes) are given. Also shown is the location of features like holes and bends. Tolerances are given for the shapes and sizes. Special information about materials and surface finish may also be provided. Fig. 9-14.

Generally, a detail drawing is needed for every part of a product. However, detail drawings are usually not needed for standard parts. Just imagine how many detail drawings are needed for a large or complex product like an airplane!

Assembly drawings show parts in their proper places and how they fit together. Fig. 9-15. There are several different types of assembly drawings, but they all serve the same purpose. They are used to show the relationship of parts.

Also given on an assembly drawing is a parts list. This is a listing of all parts needed to assemble the product. Usually, the quantity needed of each part is also shown.

Sometimes a special drawing called a **schematic drawing** is needed. This is a kind of diagram that is used to show the position of parts in a system—that is, the scheme of things. Again, see Fig. 9-15. Parts are not shown as they actually look. Instead *symbols* are used to represent the parts. Electrical and hydraulic (fluid-powered) systems are shown in schematic form.

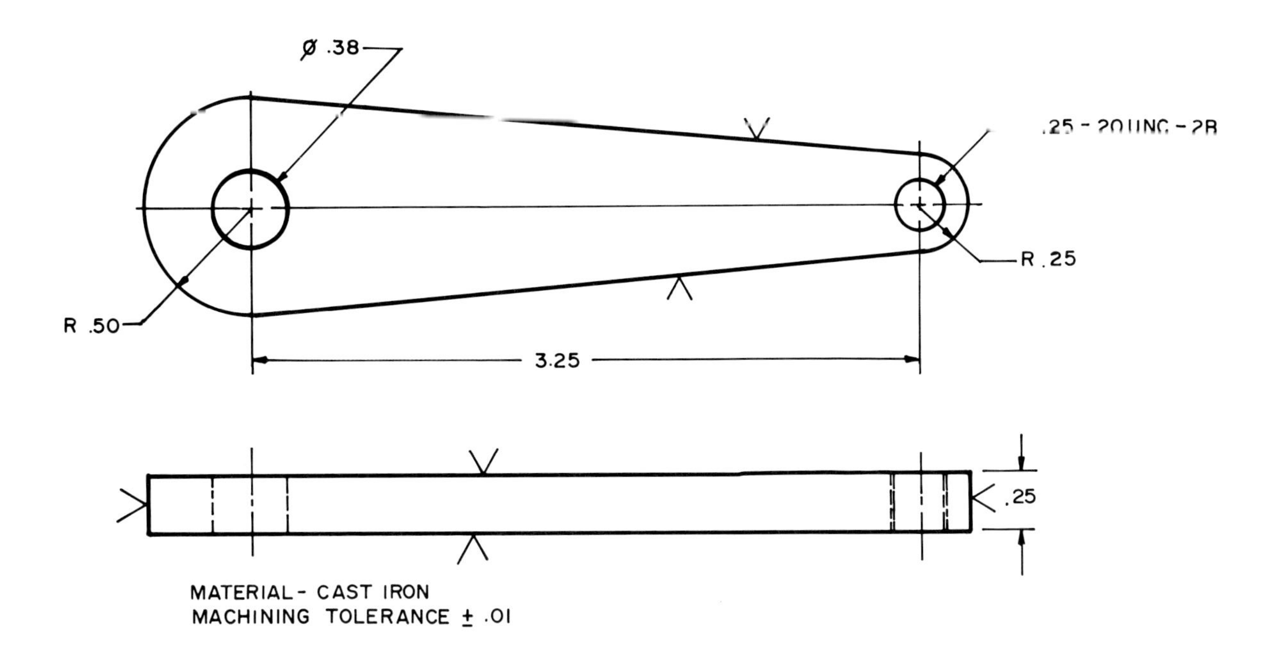

Fig. 9-14. This is a machining detail drawing for the control handle of an electric motor limit switch. It gives all the information needed to machine the part.

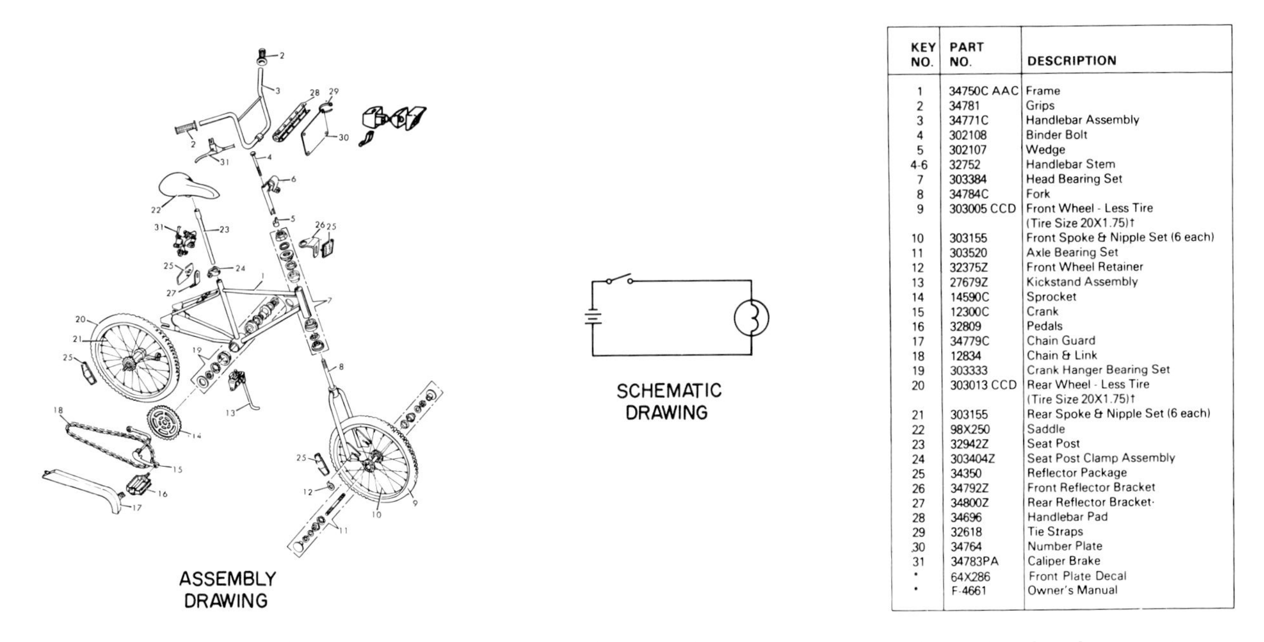

KEY NO.	PART NO.	DESCRIPTION
1	34750C AAC	Frame
2	34781	Grips
3	34771C	Handlebar Assembly
4	302108	Binder Bolt
5	302107	Wedge
4-6	32752	Handlebar Stem
7	303384	Head Bearing Set
8	34784C	Fork
9	303005 CCD	Front Wheel - Less Tire (Tire Size 20X1.75)†
10	303155	Front Spoke & Nipple Set (6 each)
11	303520	Axle Bearing Set
12	32375Z	Front Wheel Retainer
13	27679Z	Kickstand Assembly
14	14590C	Sprocket
15	12300C	Crank
16	32809	Pedals
17	34779C	Chain Guard
18	12834	Chain & Link
19	303333	Crank Hanger Bearing Set
20	303013 CCD	Rear Wheel - Less Tire (Tire Size 20X1.75)†
21	303155	Rear Spoke & Nipple Set (6 each)
22	98X250	Saddle
23	32942Z	Seat Post
24	303404Z	Seat Post Clamp Assembly
25	34350	Reflector Package
26	34792Z	Front Reflector Bracket
27	34800Z	Rear Reflector Bracket·
28	34696	Handlebar Pad
29	32618	Tie Straps
30	34764	Number Plate
31	34783PA	Caliper Brake
•	64X286	Front Plate Decal
•	F-4661	Owner's Manual

Fig. 9-15. Assembly drawings show what parts are needed to make a product and how to put together these parts.

Analyzing the Design

Product engineering includes making sure that everything about the new product is right. Each part, as well as the completed product, is carefully analyzed during the design process. Various methods are used to assure that the product will serve its intended purpose. They include computer simulation and testing and analyzing prototypes.

Computer Simulation

A computer is a very useful tool for analyzing a new product design. Information about the product is entered into a computer. The computer runs the information through a program (set of instructions). The program electronically simulates (imitates) the conditions under which the product will be used. Fig. 9-16. This type of analysis can be done before any parts are made. Needed changes can be made before producing real parts.

Prototypes and Product Testing

A **prototype** is a full-size working model of the actual product. It is the first of its kind, and is usually built by hand, part by part. Only one may be built or maybe a few hundred. The number depends on the size and complexity of the product. Because prototypes are hand-built, they are usually very expensive. A prototype for a new automobile can cost over half a million dollars! Fig. 9-17.

Prototypes serve as experimental models for analyzing product design and engineering. Each one is subjected to a variety of tests. The tests are designed to try out the product under extreme and difficult conditions. Fig. 9-18. Some tests actually destroy the prototype. This is called *destructive testing*. Crashing a prototype car into a brick wall in order to test its "crashworthiness" is an example. Most tests, however, do not actually destroy the prototype. *Nondestructive tests* are used to analyze the prototype product and its performance.

Analyzing the prototype to see if it works as predicted is called *functional analysis*. During the early stages of product engineering, engineers predict (mathematically) how the product will work. After the prototype is built, it can be tested to see if the predictions were correct. For example, a prototype of a new flashlight might be subjected to a temperature test. The prototype would be tested in freezing cold and boiling hot temperatures for various amounts of time. The purpose is to make sure the product works in a variety of temperatures.

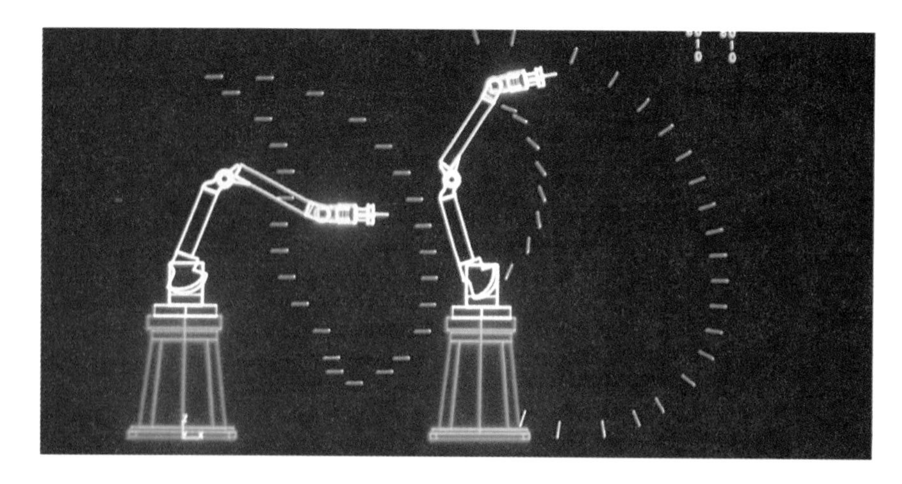

Fig. 9-16. Shown here is a computer simulation of the movement of an industrial robot.

Fig. 9-17. Each part for this prototype car was individually made just for this car. Then each part was assembled by hand.

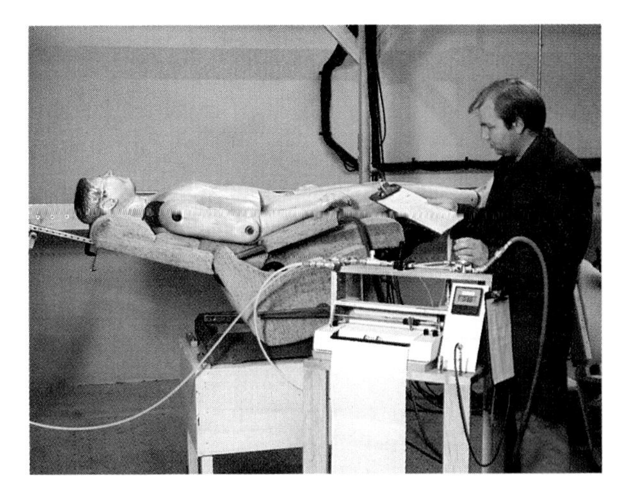

Fig. 9-18. If the prototype can perform well in extensive tests, then the manufacturer knows the product will be a good one. Here, a life-sized dummy is used to test a reclining chair.

Suppose the prototype didn't work as planned. Suppose it failed. Analyzing the prototype to see why it didn't work as it should is called *failure analysis*.

Usually a prototype is tested under certain conditions until it breaks down or fails. Then engineers tear it apart and analyze the parts very carefully. They look for clues that would show why the product failed. For the flashlight example, they might use a drop test and a switch test. In the drop test, a prototype would be dropped repeatedly from different heights. The engineers want to see how well it can stand shock. The switch test would show how many cycles (times switched on and off) the switch could stand before it failed. Depending on test results, the product may be redesigned to make it better.

Prototypes serve another purpose beside product testing. They can be used to check the manufacturing procedure used for the design. As the prototype is being built, manufacturing engineers keep track of the progress. They note any difficulties. This helps in planning the manufacturing processes that will be used to mass produce the product.

Reliability

One very important aspect of product engineering is *reliability*. It's one thing for a product to work properly, but will it continue to work? How long before a critical part quits or wears out? Product reliability is something that engineers spend a lot of time working on. They use the results of testing to make statistical predictions about the working life of the product. For example, based on their design data and on test results, engineers can calculate and predict the life of a car battery, a motorcycle tire, or a computer keyboard.

Value Analysis

A special kind of product engineering is called **value analysis**. Value analysis is an attempt to reduce the cost of materials and purchased parts without sacrificing appearance or function. Every part of the new product design is studied carefully to see if the right material has been selected. A certain part, for example, might be made of steel in the original design. A value analysis might reveal that the part could be made of plastic. Plastic would cost less, and yet the plastic part would still work as well as the steel part. Then the plastic part would have to undergo the same kind of product testing as the steel part. The engineers specify the most functional, yet lowest cost, material for every part. Fig. 9-19.

Computer-Assisted Engineering

A computer can be an effective tool in product engineering. Engineering work involves lots of mathematics. A computer can solve math problems much faster than a human can. In **computer-assisted engineering (CAE)**, computers are not only used to perform needed math calculations, but also to generate (produce) working drawings and to analyze parts.

Computers can be used when drafting needed detail, assembly, and schematic drawings. The computer works with numbers to make the drawings. The locations of certain points on the screen are given a numeric address. Just like your home has an address, a point in space can have an address. This address is called its *X,Y,Z coordinates*. Two points are identified. Then a line can be drawn between the points. This process is repeated until the drawing is complete.

After the drawing is created on the computer screen, it can be turned into a hard copy using a plotter. Fig. 9-20. A plotter is a special kind of drawing machine. The computer causes the plotter to move a pen point across a piece of paper or film to make the printed hard copy drawing.

You already learned that a computer simulation can help analyze a product's design even before any parts are made. Because it works so fast and can handle large amounts of information, the computer is also valuable for functional, failure, and value analysis.

Apply Your Thinking Skills

1. What might happen if tolerance levels for parts are given too wide a margin?
2. How is standardization beneficial to consumers?

Fig. 9-19. Because of value analysis, many heavy metal parts of cars have been replaced by lighter plastic parts that serve the purpose just as well.

Fig. 9-20. A plotter produces a hard copy of the computer drawing.

Global Perspective
The Superbulb

One of the brightest lights in the world is produced by a lamp about four inches long. The lamp is also capable of reaching temperatures greater than that of the surface of the sun. Sounds impossible, doesn't it? However, Vortek Industries in Vancouver, Canada, produces just such a lamp.

The lamp is an arc lamp. Electrically charged argon gas spirals through a quartz tube, producing the extremely bright light and high temperatures. A cooling stream of water is also spiraled through the tube to prevent the lamp from bursting.

The lamp was first developed in 1980 at the University of British Columbia by Gary Albach (now president of Vortek) and his associates. Since it is capable of lighting a 20-acre area (about the size of 18 football fields), they planned to sell it for use as a floodlight. However, few people needed such a lamp, and it was costly to operate.

Eventually, the lamp's developers realized that it has many industrial uses as a source of heat. In manufacturing, it can be used in quality testing and materials research. Some companies use it to test the effects of heating and cooling on products such as automobiles. Others use it to test materials such as those being considered for use in spacecraft.

Manufacturers of metal products use the lamp in thermal-conditioning processes. It can be used for surface hardening of parts or for deeper conditioning. In addition, it can be used to meld (combination of melt and weld) a material onto the surface of another material to give the surface a special quality, such as resistance to wear or rust.

Manufacturers of computer chips are also making use of the lamp. Hundreds of chips are produced on a single silicon wafer, which must be heated to remove defects. The lamp can heat all the chips to the same temperature and maintain the temperature evenly for a period of time.

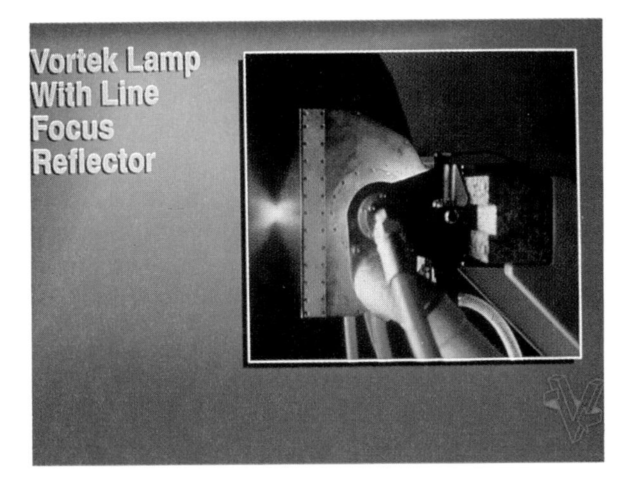

Fig. GP-9. The arc lamp is encased in a reflector that can be adjusted to various beam widths, depending on the size of the area to be heat-treated.

Word is spreading about the lamp. Someday the "superbulb," developed for light but used for heat, may shine around the world.

Extend Your Knowledge

1. Thomas Edison's incandescent lamp was the first electric light bulb. Since then, many people have made changes and improvements to his invention. One person was Lewis Howard Latimer, who actually worked with Edison for a time. Who was Lewis Howard Latimer? What improvement did he make to the electric light bulb? How was he able to save Edison millions of dollars?

Chapter 9 Review

Looking Back

New products are being introduced every day. Manufacturing companies are constantly planning new products. All new products start as ideas. Each new idea is developed by designing and engineering.

Product appearance is very important. A great deal of effort is put into product design. Designers use sketches and models to show their designs to other people. Managers choose the designs most likely to succeed.

When engineering a product, details are carefully planned. The best materials, structural elements, mechanical elements, and power elements must be chosen to make sure the product will work properly and serve its planned purpose. Parts are designed to make manufacture and assembly as easy as possible. After all the details are determined, they are recorded on working drawings.

The design must be tested to make sure everything about the product is right. Computer simulation is an excellent method of analyzing the design before any parts are actually made. Prototypes are made for product testing. Engineers can use some of the results of product testing to predict product reliability.

The computer is a useful tool in product design and engineering. Computers are used for creating designs, drafting working drawings, simulating product performance, and making math calculations.

Review Questions

1. What is the responsibility of the research and development department?
2. Briefly describe CAD.
3. Define product engineering.
4. What is functional design? What four types of elements determine functional design?
5. What is meant by "interchangeability of parts"?
6. What is tolerance?
7. What is standardization? What are some benefits of standardization?
8. Briefly describe what is shown on each of the three types of working drawings.
9. What is a prototype? What are the two main functions of prototypes?
10. Discuss several ways that computers can be used in product engineering.

Discussion Questions

1. Discuss why companies try to keep creating new product ideas.
2. Discuss some ways consumers benefit from mass production of goods.
3. Why are interchangeability of parts and tolerance so important to mass production?
4. How is product testing beneficial to manufacturers? How is it beneficial to consumers?
5. Explain the importance of product reliability.

Chapter 9 Review

Cross-Curricular Activities

Language Arts 

1. Research the development in design of a household appliance. Write a research paper outlining the advances made since the product's inception (beginning). (An outline will help you to organize facts.)
2. Research the assembly line techniques used by at least three manufacturers. Then write a brief paper discussing the similarities and differences in the techniques used by the manufacturers.

Social Studies 

1. Henry Ford revolutionized the car industry when he introduced assembly line production in the manufacture of his Model Ts. Research the manufacturing techniques used by Henry Ford at the turn of the century. Then compare these with the manufacturing techniques used today by the Big Three car manufacturers (GM, Ford, and Chrysler).

Science 

1. Scientists do much of the *basic research* that results in new technology. The ancient Greeks discovered that a type of iron ore called lodestone always pointed north-south when hung from a string. This was the basis for the development of the compass. You can make a compass similar to those used by early navigators. Rub one end of a magnet along a needle. (Rub only in one direction, not back and forth.) Lay the needle on a slice of cork floating on water. The needle will move to align itself in a north-south direction.
2. *Applied research* is done to solve a problem. Can you see some problems with the compass described above? List them and tell how retrieving, describing, and experimenting might have led to the development of the modern-day compass, pictured below.

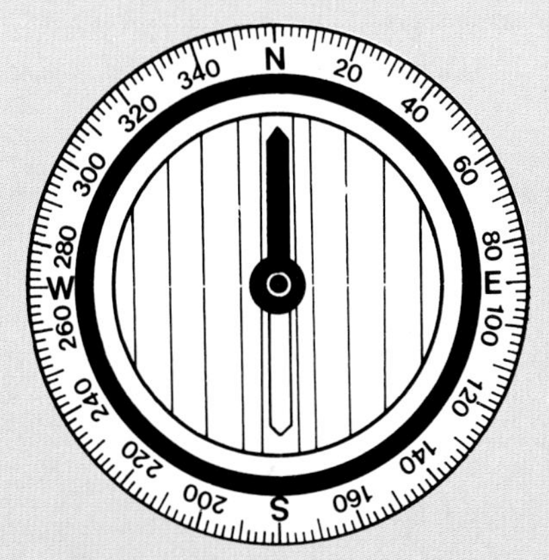

Math

Tolerance is the amount a piece can be oversize or undersize and still be considered within acceptable limits.

1. The dimension given for the length of a flange is 13.12 (+ or - .06). What is the maximum acceptable length of the flange? What is the minimum acceptable length?
2. If a dowel rod for the back of a chair is dimensioned 18 (+ 1/16, - 1/8), what would be the range of acceptable lengths?

Chapter 9 Technology Activity

Product Development

Overview

The purpose of this activity is to develop problem-solving skills in the product development process. You will be given a design problem to solve and a nine-step design process to follow. Some of the design steps you will use to solve the problem include brainstorming, sketching, and model making. Skills that you should develop in this activity include creative thinking, group dynamics (teamwork), verbal (spoken) communication, and graphic (drawing) communication.

In this activity, all of the design specifications are provided. You must design a product that will meet these specifications.

Goal

To design a product by following the nine steps of the design process.

Equipment and Materials

drafting equipment
1/4" graph paper
pencils
drawing paper
cardboard
poster board
rigid-foam insulation
masking tape
white polyvinyl resin glue
utility and modeling knives
straight pins

The Problem

You work for a research and development firm called Bright Ideas Company. Your company has been contracted by Martin Manufacturing to develop a new product for them. They make floppy disk organizers, cassette tape carrying cases, and record racks.

The marketing department at Martin Manufacturing has reported an all-time high in video recorder sales. Videotape cassette sales have also increased. Market research indicates that even after the videotape recorder sales taper off, prerecorded and blank videotape cassette sales will continue to climb. The marketing department believes that a product that organizes, stores, and displays videocassettes would be profitable for the company.

Market research shows that most of the videotape recorders are being used in rooms with wood furniture. Videotape recorder cabinets come in black, black with a metal face (front), or wood grain with a metal face.

Martin Manufacturing would like a sketch, working drawings, and mock-up of your product design idea to be submitted for their approval. Your product design will be among many considered.

Design Specifications

Martin Manufacturing has provided the following design considerations:

- The videotape organizer should be able to hold at least eight videotapes 1 3/8" wide, 8" high, and 5" deep.

Chapter 9 Technology Activity

- The organizer should be designed so it can be located on a bookshelf or on a tabletop.
- The labels located on the 1 3/8" x 8" edge of the videotape cassette should be exposed and in plain view.
- The organizer can be manufactured out of a variety of materials.
- The design should have aesthetic appeal (beauty) and complement (go well with) electronic equipment, furniture, and books.

Design Procedure

1. *Safety Note:* Before doing this activity, make sure you understand how to use the tools and materials safely. Have your teacher demonstrate their proper use. Follow all safety rules.
2. You will be assigned a partner for this activity.
3. Look at the nine steps of the design process listed in Fig. A9-1. Follow these steps as you solve this product design problem.
4. Create as many ideas as possible. Sketch each idea on paper. Do not include a lot of detail in your sketches. Keep them small and simple. Do not evaluate your ideas at this time. Refer to Fig. A9-2 for possible design ideas.

Identifying the Problem
Define the problem by identifying cause-and-effect relationships.

Researching and Gathering Data
Find background information about the problem. Do this by retrieving recorded facts and data about the problem.

Developing Preliminary Ideas
Brainstorm for possible solutions. This step also includes preparing rough sketches of your ideas.

Selecting a Possible Solution
Select one or two of your best ideas.

Refining the Design
This design step includes preparing refined sketches and scale drawings as well as "engineering" your chosen design.

Preparing a Model
During this step of the design process, you will construct a model (mock-up) of your design solution, using the required materials.

Analyzing the Design
Using standard formulas and information, the strength and predicted performance of your design can be calculated (determined).

Experimenting
This design step will help you make sure your design is sound and workable. During this stage, your design may be tested, changed, and tested again.

Implementing the Final Solution
The final product or solution is ready for presentation.

Fig. A9-1. The nine steps of the design process.

Chapter 9 Technology Activity

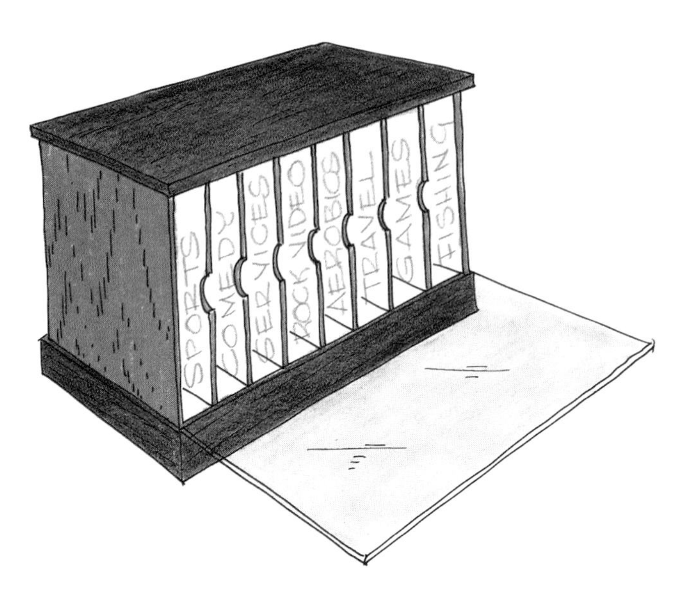

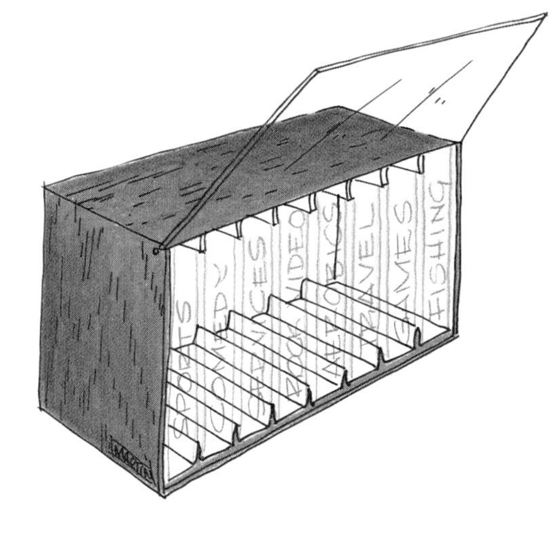

Fig. A9-2. Possible solutions to the design problem.

Chapter 9 Technology Activity

5. After you and your partner have eight designs, evaluate each design together. Try to find the best features in each design. You may find one design that is clearly the best. However, you may need to combine good features from several designs to arrive at the best design. Prepare a rough sketch of your design and make sure it meets the design considerations.
6. Make a final rendering sketch of your best design idea. Your final sketch should be larger and include more detail than your thumbnail sketches. Try to improve your idea as you draw.
7. Using your final rendering as a guide, prepare a set of working drawings of your product design.
8. Next, use the working drawings to build a mock-up of your product design. Construct the mock-up from cardboard, poster board, foam insulation, glue, and tape.
9. Prepare a product design presentation. Your presentation should include a review of the problem, how your product design meets the design considerations, and the unique features of your design. Decide which parts of the presentation you will do and which your partner will do. Be prepared to answer questions about your product design. Your presentation will be evaluated on organization, content, creativity, and clarity.

Evaluation

1. What were the advantages of having a partner for this activity? Were there any disadvantages to having a partner?
2. Which part of the design process did you enjoy the most? Which part did you enjoy the least?
3. How well did your product meet the design specifications? If you could design your product again, what would you do differently?

Chapter 10

Production Planning

Looking Ahead

In this chapter, you will discover:

- how production processes are chosen.
- various techniques used to achieve production efficiency.
- how the layout of a manufacturing plant is planned.
- how various types of materials-handling equipment can be used in production.
- how all parts of the production process are organized and tested.

New Terms

automated storage and retrieval system (AS/RS)
automatic guided vehicle system (AGVS)
bill of materials (BOM)
die
fixture
group technology
jig
part print analysis
pilot run
plant layout
process chart
tooling-up

Technology Focus

MacFactory

The Macintosh factory in Fremont, California, manufactures computers. It is one of today's most respected manufacturing operations. What makes it so? Careful planning.

Engineers planned all parts of *plant layout* (arrangement of machinery, equipment, materials, and traffic flow) and production, down to the finest detail. Engineers designed most of the parts of the computers to be snap-in and surface-mounted rather than inserted through holes in the board. The surface-mounted parts are easier to handle by people and machines. In the assembly area, there are several straight, carefully-paced (timed) assembly lines. On each of these lines, a product is put together from start to finish.

Each assembly line is connected to a special automated materials-delivery system. Using automated equipment minimizes handling of a large number of parts. As they are needed, parts are automatically taken from storage by *automatically guided vehicles*. These driverless vehicles know where to go by following a wire path that's been installed in the floor. Small parts are delivered in trays to each workstation on the line just as they are needed. Large parts are delivered, also just in time, by an overhead rail system.

Just like automobile customers, Macintosh customers often want their products with different "extras." The assembly lines have been designed to build more than one kind of computer at a time. During assembly, each Macintosh computer sits on a special pallet or platform. Microprocessors direct the pallets along a special conveyor according to pre-programmed instructions. (A *conveyor* is a special device that moves materials, parts, or products over a fixed path.)

After assembly, the computers are temporarily stored in special testing racks. From there, they are moved to the inspection area, where they are checked for defects, and then to packaging.

Much of the Macintosh manufacturing operation is automated. However, many tasks must be performed by people. Planners carefully choose the best method for each operation.

The Macintosh manufacturing operation is efficient and productive. Careful planning made it so.

Fig. TF-10. This Macintosh factory in Fremont, California, is very efficient due to careful plant layout planning.

Process Planning

Changing the form or shape of an industrial material requires processes. The processes are chosen carefully. The company wants to make high-quality products in the least costly way.

Production Approval

As you read in Chapter 9, management gives approval to go ahead with the design of the product. Management also decides whether the company should proceed with production. Managers review information about the product design. They also consider future sales. If these factors look promising, then the managers allow production planning to begin.

Product Analysis

Manufacturing engineers plan production. To do so, they must know all the details about the product. They study the bill of materials and the working drawings.

A complete list of the materials or parts needed to make one product is called a **bill of materials (BOM)**. The quantity or amount of each part or material is given. Items are listed in the order in which they are used. Figure 10-1 shows a sample BOM.

Working drawings provide valuable information for the process planner. By carefully studying the drawing (print) for a part, a process planner can begin to get ideas about how the part could be made. This is **part print analysis**. Fig. 10-2. Specified features and dimensions are identified on the drawing. Information about the finished shape and size of the part, the materials needed, and the tolerances is also collected during part print analysis.

BILL OF MATERIALS

Part Number	Part Name	Quantity
256100	Wheel	1
256110	Rim	1
256120	Hub	1
256121	Axle	1
256122	Cone	2
256123	Bearing	16
256124	Locknut	2
256125	Nut	2
256126	Washer	2
256130	Spoke	36
256131	Nipple	36

Fig. 10-1. A bill of materials lists all the parts needed to make one product.

Fig. 10-2. A process planner begins by analyzing the working drawings. This is called part print analysis.

Make-or-Buy Decisions

Should the company make or buy the parts needed? Three questions must be answered for every part in the product.

- Can the item be purchased? (Availability)
- Can the company make the part? (Manufacturability)
- Is it cheaper to make the part or to buy it? (Cost)

Information is gathered about parts that are available for purchase. Quality is important. Also, suppliers must be able to provide the quantities of parts needed at the time they are needed.

Engineers also consider whether the company is able to manufacture the part. Can the tools and equipment now owned by the company be used? Do workers have the knowledge and skills needed to make the part?

Parts that cannot be purchased must be made by the company. However, "make-or-buy" decisions are usually based on cost. Which way will be least expensive?

Process Selection

Manufacturing engineers make production plans for each part that will be made in the plant. The first step is to identify the processes needed to make the part. Then equipment is selected for each process.

The next step is to arrange all the processes into a logical sequence (order). One common way of doing this is to make a **process chart**. Fig. 10-3. There are several types of these charts. The amount of detail varies, but all process charts show the sequence of manufacturing steps.

Fig. 10-3. A process chart shows the sequence of operations used to manufacture a product.

Group Technology

An interesting technique used in production planning is called **group technology**. This is a method of keeping track of similar parts that a company manufactures. Using a classification scheme, all parts can be coded according to their shape and features. When the company has to manufacture a new part, the engineers first check the file of existing parts to find similar parts. Suppose the company manufactures a certain cylindrical part with a hole through the center. For a different product they may need to make a similar but different size cylindrical part. Since they already have the production plans for one cylindrical part, all they have to do is to retrieve the information about the first part and change the dimensions to make the size part needed for the new product. Production planning can be done more quickly since the manufacturing engineers don't have to start all over each time. The process plans for a general cylindrical part with a hole through it can be used for any similar part; only the dimensions have to be changed. Fig. 10-4.

Apply Your Thinking Skills

1. What might happen if a supplier cannot supply enough parts to a manufacturer by the time they are needed for production?

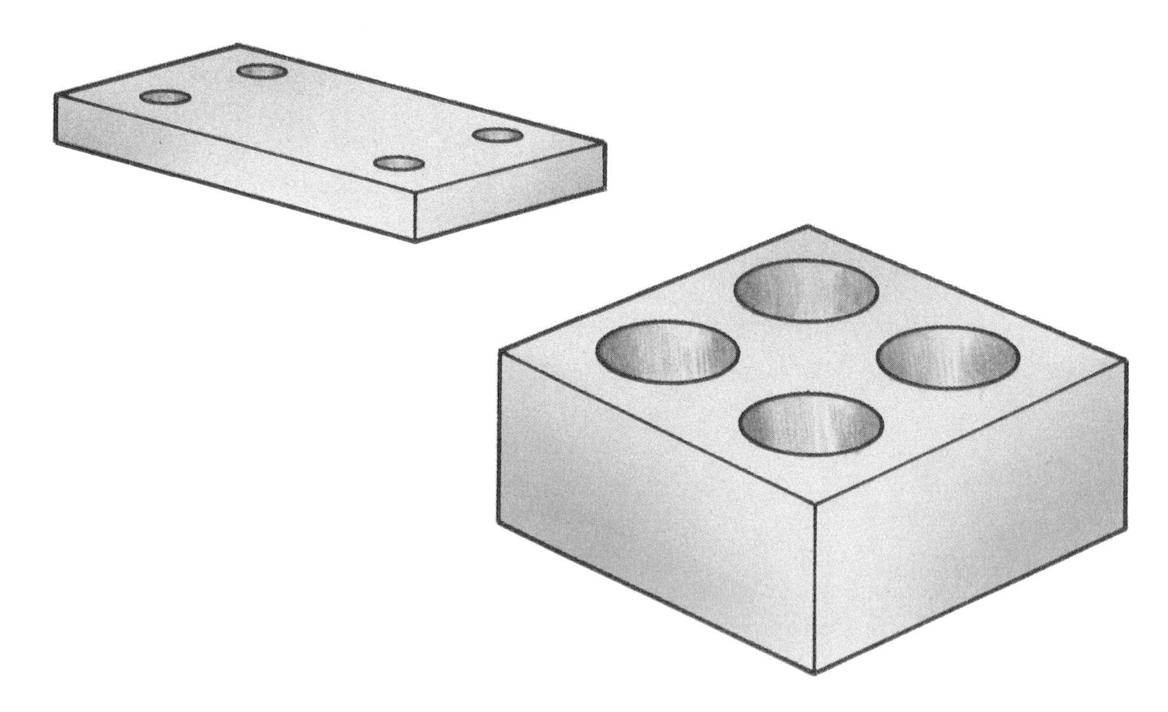

Fig. 10-4. In what ways are these two parts the same? In what ways are they different? In group technology, one set of production plans can be used for both, if appropriate changes in dimensions are made.

Facilities Planning

Well-planned facilities are important in manufacturing. Buildings must be planned. Everything that goes into the buildings must be placed according to the plan. Then products can be produced efficiently.

Plant Location

Where should a factory be located? Which part of the world? Which part of the country? Which state? Which city? Which part of the city?

Manufacturing facilities are located in various places for various reasons. A plant may be located near a large city because many workers are available there. Sometimes the best location is near a source of raw materials. Other times, a location near the marketplace is better. Each company must decide which is best for producing and selling its product. Fig. 10-5.

Fig. 10-5. The location of a manufacturing plant is important. Transportation services must be available to bring in materials and to ship out finished goods.

Plant Layout

A plan is developed for the interior (inside) of the plant. This plan shows the arrangement of machinery, equipment, materials, and traffic flow. This is called the **plant layout**. This layout is determined by production flow, related activities, and space requirements. Fig. 10-6.

Fig. 10-6. Some layout planners build a model of the plant from sketches to help them visualize production flow and space requirements.

Using the process charts, a layout planner develops a plan for production flow. The plan shows how materials and parts are moved into, through, and out of the plant. The production flow can be a kind of "road map." Sometimes lines are drawn on the factory floor plan showing where things go.

Related activities are usually located near each other for efficient production. For example, a drying room would be located near the painting room. Some activities that occur in the factory are not directly involved with production. Every factory has a shipping and receiving area. There are also offices, rest rooms, and maybe even a cafeteria. These areas support the manufacturing process and are planned in relation to the production area.

The amount of space for each activity has to be determined. Usually this is done mathematically. Consider cafeteria space, for example. Suppose 50 people will be using the space and each person needs 12 square feet. Then 600 square feet of space will be needed (50 x 12 = 600). In the same way, if you know how much space is needed for each machine and you know how many machines will be used, then by multiplication you can determine how much space will be needed. Fig. 10-7.

Fascinating Facts

When Henry Ford first began completely assembling his cars on conveyor belts back in 1914, he revolutionized American manufacturing. He found that instead of building a car in 12 1/2 hours, he could now build one in just 1 1/2 hours and could sell it at a profit for $440!

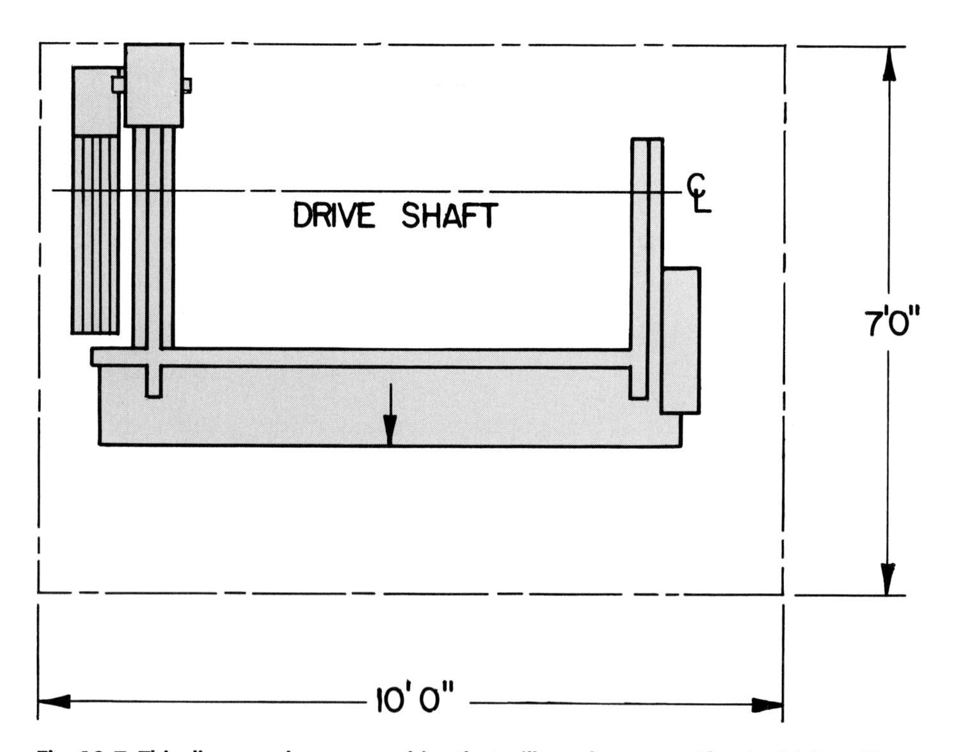

Fig. 10-7. This diagram shows a machine that will need a space 7 feet by 10 feet. How much space would ten of these machines occupy?

Materials-Handling Systems

Materials handling is moving and storing parts and materials. A variety of special types of equipment is used. All of the equipment and automated systems together are known as the *materials-handling system.*

Equipment

Conveyors are used to move materials, parts, and products from one place to another along a fixed path. Fig. 10-8. There are many different types of conveyors, each used for a specific purpose. You have probably seen conveyors. Many supermarkets have belt conveyors at checkout counters. Belt conveyors are also often used in manufacturing plants. Roller conveyors are good for carrying heavy loads. Skate-wheel conveyors are good for transporting boxes.

A *hoist* has one or more pulleys and is used to lift heavy loads. A *crane* is actually a hoist that can be moved about in a limited area. One kind of crane travels on overhead rails. Fig. 10-9. The parts move along the track suspended from the hoist.

The trucks used for materials handling are different from the trucks you see on the road. The proper name for them is industrial trucks, but most people just call them *forklifts.* They have forks on the front that can be raised and lowered. Forklifts are used for lifting and carrying loads from one place in the plant to another. Fig. 10-10. Forklifts are often used to move *pallets,* which are special platforms on which large parts or materials are often stored. Loaded pallets can be stored on pallet racks to save floor space.

Fig. 10-8. Roller conveyors are used to move boxes from one place to another.

Fig. 10-9. An overhead crane is used to lift large heavy items and move them to another spot nearby. The crane shown here can also transfer its load to another crane.

Fig. 10-10. Industrial trucks, like this forklift, are used to lift and move materials from one place to another.

Automatic Guided Vehicle System

One of the newer ideas in material handling is the use of an **automatic guided vehicle system (AGVS)**. Specially built driverless carts (AGVs) follow a wire "path" installed in the floor. Fig. 10-11. The movement of these vehicles is controlled by a central computer. The computer keeps track of each vehicle's location. It directs

Fig. 10-11. These automatic guided vehicles (AGVs) are part of a system that moves materials and products around a factory.

each vehicle's starting, stopping, and speed and causes the vehicle to switch from one path to another. Some AGVs can carry heavy loads like car engines. Others are made to carry lighter loads, such as computer chips and circuit boards.

Automated Storage and Retrieval System

Another type of system used in manufacturing today is an **automated storage and retrieval system (AS/RS)**. This is a special set of tall racks with a computer-controlled crane that travels between the racks. Fig. 10-12. The racks have "cubbyholes," or sections for storage. Loads are usually on pallets. The computer causes the crane to pick up a load. Then it selects an empty slot and directs the crane to travel to that spot and store the load. The crane can also be directed to retrieve or pick up loads from storage.

One kind of AS/RS is called a *miniload*. It is designed to handle items small enough to fit in drawers or tote pans. The miniload is like an automatic set of cabinet or dresser drawers. Imagine if your dresser drawers were unloaded and loaded automatically.

Apply Your Thinking Skills

1. What factors other than availability of workers might make a manufacturer decide to locate near a large city?
2. Why is plant layout so important to production planning?

Fig. 10-12. Computer-controlled stacker cranes store and retrieve parts.

Putting the Plan into Action

After planning is complete, further preparations are made. Workers are chosen and organized. Machines and other equipment are obtained and prepared for production. The system is tested and refined. Production can begin!

Organizing People

People are needed to work in production. The *personnel department* employs and trains workers. However, specifying the number and types of workers needed is part of production planning. Fig. 10-13.

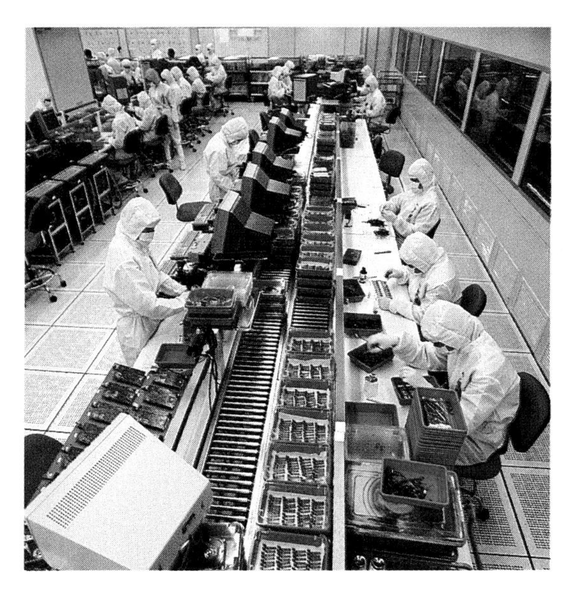

Fig. 10-13. Production planners determine how many workers are needed and what kind. What skills might these workers need in order to assemble hard disk drives for computers?

Production planners have carefully studied all the possible methods for manufacturing the product. They decide which methods and machines to use. Then they know how many workers will be needed, and exactly what each worker will be required to do.

Tooling-Up

Getting the tools and equipment ready for production is called **tooling-up**. All the machines and tools are prepared for certain jobs. If necessary, the company buys new tools, machines, and equipment. Tooling-up often includes adapting (changing) existing machines. Special tools may be needed. These are made or purchased. Special tools include jigs, fixtures, molds, and dies. Gages are instruments used as an aid to tooling-up. They will be discussed in Chapter 11.

Jigs and fixtures are used to adapt general purpose machines to do certain operations. They are frequently used during the drilling and cutting of metal parts. A **fixture** is a special type of holder. It clamps and *holds* the part in place during processing. Fig. 10-14. A

Fig. 10-14. Most fixtures are custom-made to hold a specific type of part.

jig is like a fixture in that it *holds* the workpiece (part being processed), but it also *guides* the tool. A jig, for example, might hold a piece of wood in place while it also serves as a guide for the saw being used to cut the wood.

Some machines require molds or dies to shape materials. A *mold* is a hollow form. Liquid material is poured, squirted, or forced into the mold cavity (space inside the mold). As the material hardens, it takes the shape of the cavity. Fig. 10-15. A **die** is a piece of metal with a cut-out or raised area of the

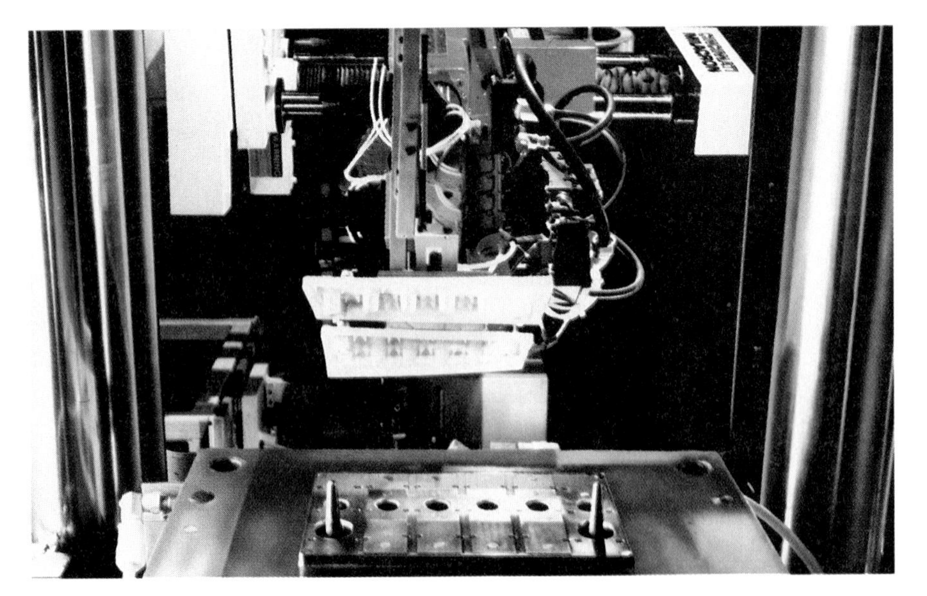

Fig. 10-15. Molds are used to shape parts or products. This picture shows a robot removing a finished part from an injection molding machine. In injection molding, a material such as plastic or metal, in heated liquid form, is forced (injected) into a mold where it is allowed to cool and harden.

desired finished shape. The material is then forced through or against the die to take on the shape of the die. Fig. 10-16.

A punch and die set is used to punch or cut shapes from materials. This is called *die punching*. It works like an ordinary paper punch. Coins are made this way. Fig. 10-16. Sheet metal and plastics are frequently cut to shape using a punch and die.

Another kind of die is used for forming sheet material. The material is stretched over the die or the die is pushed into the material. This is *die stamping*. The body of a toy car may be shaped this way. Many times die stamping and die punching are done at the same time on the same machine.

One machine can be used to make different parts by simply changing the mold or die. For example, an injection molding machine might be used to make plastic hairbrushes. By changing the mold, the same machine could be used to make toothbrushes.

Fascinating Facts

Sometimes new tools have to be invented to solve a specific problem in producing an item. Back in 1763, when James Watt began his experiments with the steam engine, he found that he needed a tool that could create a perfect hole or steam would leak from his engines, causing a loss of power. In 1774, John Wilkinson invented the boring machine, which made the perfect hole needed to prevent steam from leaking from the engine.

The Pilot Run

Will the plan work? One way to know for sure is to conduct a **pilot run**. A pilot run is like a practice where all parts of the system are operated together before production really starts. The main purpose is to find and correct production problems, but it also gives workers a chance to learn new tasks.

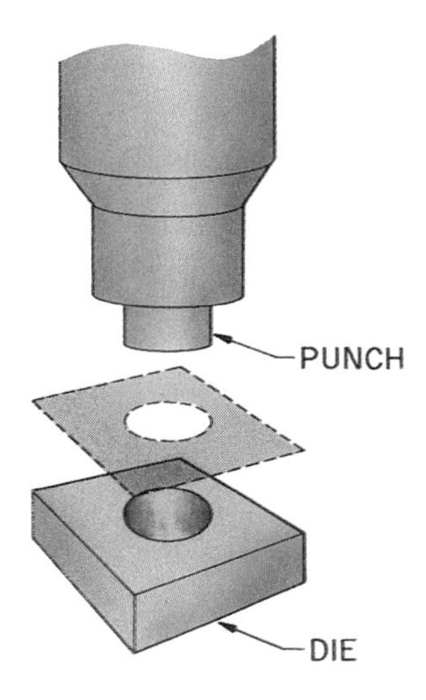

Fig. 10-16. Coins are made by using the die-punching process.

During the pilot run, engineers watch closely and keep records. As problems are identified, corrections are suggested and tried. This process of finding and correcting problems in the operation of a system is called *debugging*. Because this is a pilot run, the system can be stopped while corrections are made. Actual production is not delayed.

The pilot run also allows engineers to check and adjust the timing along the production and assembly lines. They may speed up or slow down equipment to see the effects on production. Fig. 10-17. Speeding up may increase the production or assembly rate, but the quality of the product may be affected. Workers must have enough time to do tasks properly.

The pilot run also serves as a training time for workers. They can learn new tasks or see how their jobs have been changed. Workers need to know exactly how to make the new product. They also need to understand how their work fits into the system.

Apply Your Thinking Skills

1. How does having a machine that can use different molds or dies help reduce production costs?
2. Discuss several consequences of not making a pilot run.

Fig. 10-17. An engineer times the assembly process during a pilot run.

Investigating Your Environment

Planning for a Cleaner Environment

Planning every detail of production ensures producing a product of quality in the least expensive way. It is also at the planning stage that environmental considerations can best be addressed. One company leading the way in this area is 3M, a diverse manufacturer that makes more than 60,000 products. You probably have used some of them, such as Scotch® brand adhesive tape.

3M has a program called "3P"—"Pollution Prevention Pays." The program, started in 1973, has had more than 2,500 3P projects. For an idea to qualify it must eliminate or reduce a pollutant, use less energy or make better use of resources, be technically innovative (new and different), and save money. 3M employees at all levels are encouraged to participate in the 3P program, and awards are given for the best projects. In the first 15 years, the 3P program prevented more than 500,000 tons of pollutants.

For example, at a manufacturing plant in Texas, the battery-powered forklifts were modified. New gages and low-charge warning lights were installed. Now, workers know when it is time to recharge the industrial trucks. In the past, they were charged at the end of each day whether they needed it or not. The cost to modify the forklifts was $4,000. However, it saved 3M $25,390 a year in battery and energy costs.

Since 1973, 3M has employed full-time experts to help 3M use less electricity and fuel. This is accomplished by using the "3Rs": "Reduce, Reuse, Recover." Using these three rules, *reducing* the energy used, *reusing* energy exhaust, and *recovering* energy exhaust for reprocessing, 3M was able to reduce its amount of energy consumption by one-half.

Fig. IYE-10. This 3M researcher is working to improve coating processes for the company's products.

Not all of the methods are complex or high tech. We are all familiar with the idea of turning down the heat to save energy. An energy survey in an office building revealed that the temperature and humidity controls ran during hours when no one was there. Special thermostats were installed to maintain a nighttime temperature of 55° F. Though it cost almost $12,000 to install, 3M saved $56,000 a year in energy costs. With measures like these, 3M has been a leader in proving that what's good for the environment can also be good for business.

Take Action!

1. Investigate how your school uses energy. Prepare a report on ways 3M's "Reduce, Reuse, Recover" policy could be applied to your school.

Chapter 10 Review

Looking Back

Every detail of production is planned. Proper planning ensures that the product will be produced efficiently. The bill of materials and working drawings are studied to determine what is needed to make the product. Decisions must be made about which parts will be manufactured by the company and which will be purchased.

Processes needed to make each part are identified. The processes are arranged in a logical sequence and listed on a process chart.

Plant locations and layouts are important factors in producing products. When preparing a plan for the inside of the plant, the flow of materials, activity relationships, and space requirements are considered.

A materials-handling system includes equipment such as conveyors, cranes, and hoists. These are used to move parts and products around inside the plant.

Organizing workers and getting tools and equipment ready are also important steps in preparing for production. A pilot run is conducted to test the system before actual production starts. The system is debugged, and the timing of the assembly line is adjusted for efficient production.

Review Questions

1. What information is given on a bill of materials?
2. List at least four types of information that are collected during part print analysis. What is the purpose of part print analysis?
3. When deciding whether to make or buy parts, what three basic questions need to be answered?
4. Briefly describe group technology and its benefits.
5. Define plant layout. What three things determine plant layout?
6. Explain how an AGVS works.
7. What is tooling-up?
8. What is the difference between a jig and a fixture?
9. What is a die? How is a die used in production?
10. What is a pilot run? What are the advantages of making a pilot run?

Discussion Questions

1. What effects does plant layout have on the efficiency of production from start to finish?
2. How can a computer be used in plant layout?
3. What are some safety considerations that might be involved in materials handling?
4. Discuss why there are few changes in the design of mass-produced products from year to year.
5. Discuss some of the different ways that jigs and fixtures can be used during the manufacture of a product.

Chapter 10 Review

Cross-Curricular Activities

Language Arts

1. Careful planning is essential to every job. One way to plan is to outline what needs to be done. Make an outline for the production of a product that you could present to a company.
2. A process chart shows the logical sequence of manufacturing steps. Make a process chart for writing a research paper.

Social Studies

1. Choose a product that was invented in the twentieth century. Research its development. Discuss the pilot runs and debugging operations needed before the final version of the new item could be completed.

Science

1. Sheet metal is used to make coins through the process of die punching. The metals used for coins are chosen because of their properties (characteristics). These properties include such things as hardness, melting point, ductility (ease of forming), and corrosion resistance. Use a reference book to find out which properties would make these metals poor choices for coins: mercury, magnesium, aluminum, and iron.
2. Substances used in molds are those that are solids at normal environmental temperatures, but can be melted to form pourable liquids. Use a reference work to find the melting points of the following substances: silicon dioxide (glass), tin, lead, and copper. Which has the lowest melting point? The highest?

Math

On a trip to the builders supply store, Carlo made the following purchases:

One 20′ extension ladder at $112.00/each
Two boxes of 1″ x 8″ FH wood screws at $3.19/box
Five sheets of 100c abrasive paper at $.49/sheet

1. What was the total cost of Carlo's purchases?
2. If the tax on the items was 6 1/2 %, what was Carlo's total bill?

Chapter 10 Technology Activity

Selecting and Buying Standard Stock

Overview

After a product has been researched and developed, the production stage of manufacturing is put into action. Before a product can be manufactured, the tools, equipment, and consumable supplies needed for production must be purchased. People working in the purchasing department must buy the needed supplies from vendors. A *vendor* is a wholesaler or a retailer that sells tools, materials, and equipment.

The supplies needed during the production of a product are usually purchased in the form of standard stock. *Standard stock* is material that has been formed or separated into widely accepted sizes, shapes, or amounts. See Figs. A10-1 and A10-2 for examples of standard types and sizes of fasteners and wood.

In this activity, you will play the role of a buyer. You will research catalogs from various companies for standard stock wood, metal, and fasteners.

Goal

To find out the costs of standard stock wood, metal, and fasteners from at least two sources.

Equipment and Materials

notebook paper
pencils and ballpoint pens
catalogs from companies that sell standard stock materials (for example, structural steel companies and lumberyards)

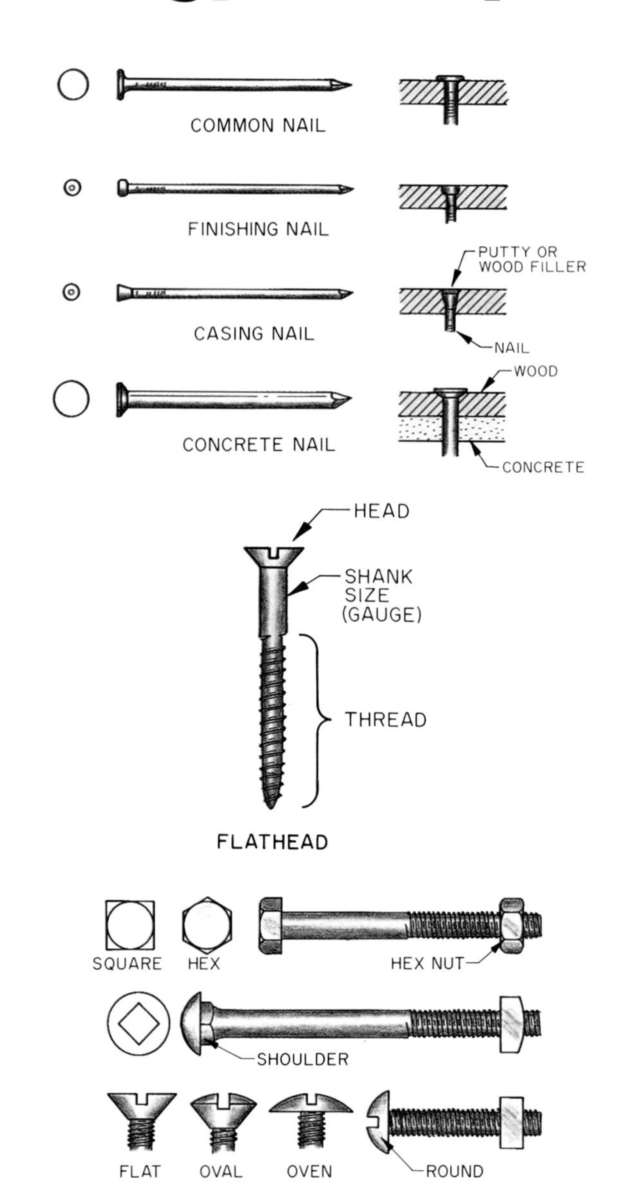

Fig. A10-1. Examples of standard fasteners.

Procedure

1. Obtain at least two catalogs for standard stock materials.
2. Compute the cost of each of the following from two different sources:

Chapter 10 Technology Activity

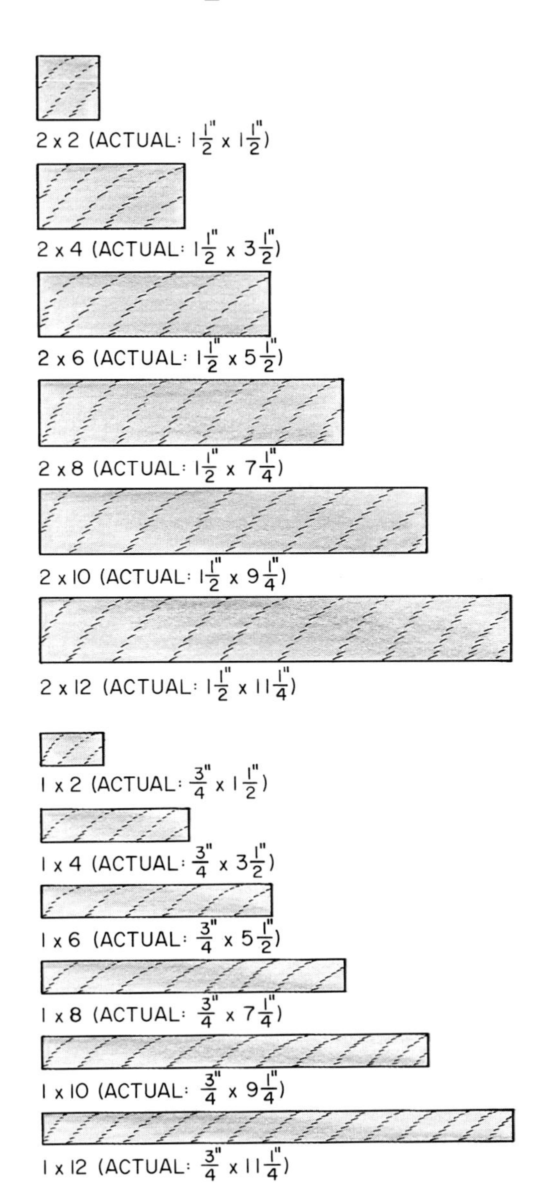

Fig. A10-2. Wood is cut into standard sizes, such as 2" x 4". After the wood is surfaced (smoothed), it is smaller. The actual size of a 2 x 4 that you buy is 1 1/2" x 3 1/2".

- Dimensional lumber
  - A. Eight 2 x 4 boards, each 16 ft. long
  - B. One 1 x 10 board, 12 ft. long
- Plywood, 3/4", one 4' x 8' sheet, exterior grade
- Galvanized sheet metal, 22 ga., one sheet, 3' x 8'

3. What is the price per pound (from each source) for each of the following types of fasteners:
   - Nails, 16d, common
   - Nails, 10d, common
   - Nails, 6d, finishing
4. Hardwood is purchased in board feet. See Fig. A10-3.
   - Compute the amount of board feet in four pieces of oak, 2" x 8", 3 ft. long.
   - Figure the costs (two sources) of this wood if it is surfaced on two sides.

Evaluation

1. For each type of standard stock that you researched, how much did the different sources' costs differ? Share your findings with the class.
2. Why do you think these cost differences exist between different companies' standard stock?

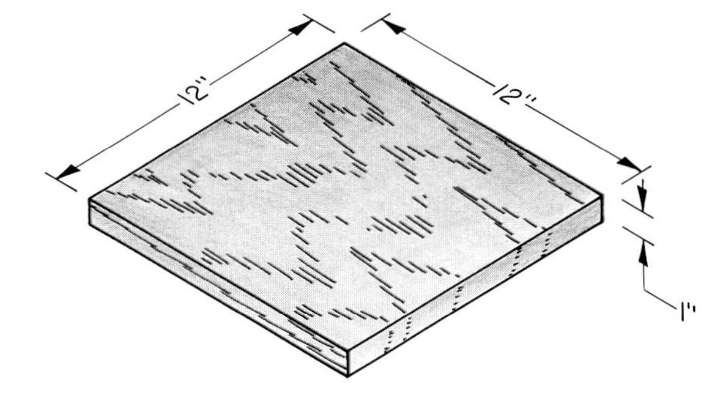

Fig. A10-3. Use this formula to figure board feet:

$$\frac{\text{length (ft.)} \times \text{width (in.)} \times \text{thickness (in.)}}{12}$$

Chapter 11

Production

Looking Ahead

In this chapter, you will discover:

- the difference between components and assemblies.
- some innovative processing methods.
- how production is controlled.
- what is meant by inventory.
- how product quality is checked.

New Terms

acceptance sampling
assemblies
automatic processing
component
computer numerical control (CNC)
gages
inventory
inventory control
laser cutting
production control
quality assurance
robots
specifications
subassembly
waterjet cutting

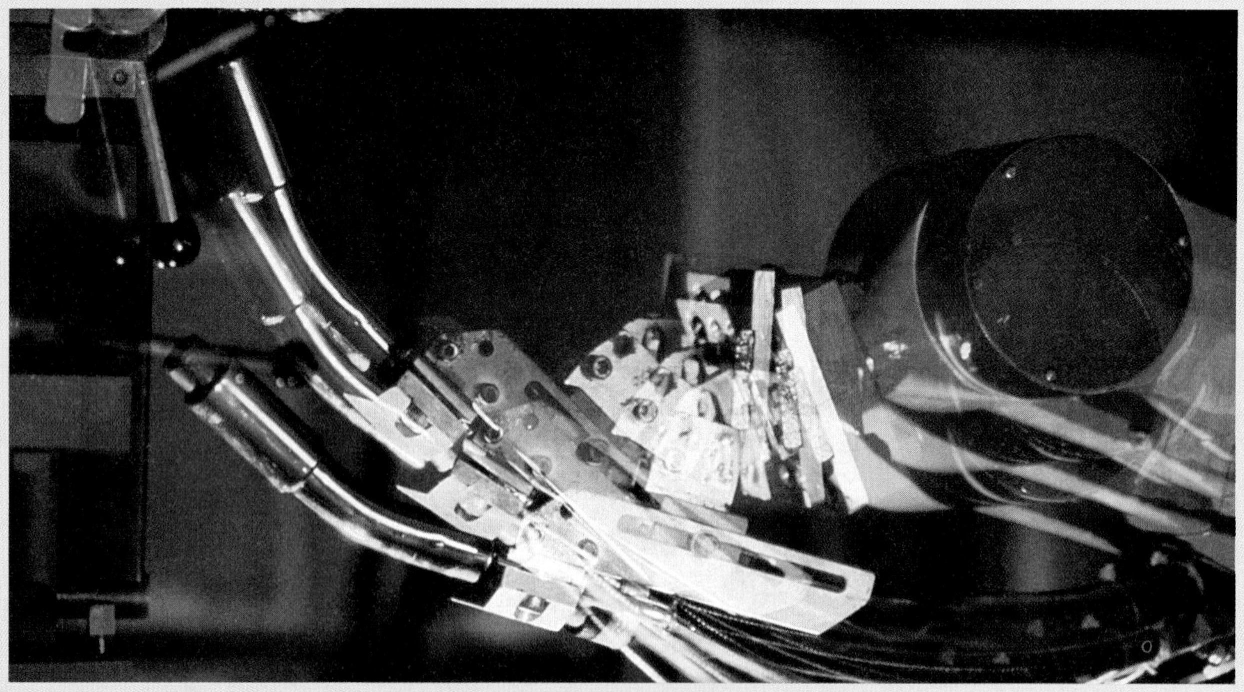

Technology Focus

Applying New Production Ideas

How do you plan for more than 4,000 separate parts from over 600 different suppliers to arrive just as they are needed to assemble them into an automobile? It isn't easy, but it's being done at the General Motors automobile plant in Flint, Michigan, known as "Buick City." The plant itself was built in 1907, but it has been turned into a modern automobile assembly plant that produces today's Buick LeSabres and Oldsmobile 88s.

A new style of production is used at this factory. It is called *synchronized production*. This type of production is very complex. Timing is the critical factor.

Only a minimum amount of parts is kept on hand. Most parts are scheduled to arrive just in time for use. Each order must be delivered to the factory during a set twenty-minute time period. There are eighty-six unloading docks around the complex. Each part order is delivered to a certain dock. The assigned dock is near the place on the assembly line at which the particular parts will be installed. Robots unload some items. Robots and automated guided vehicles (AGVs) are used extensively. Parts are carried by on-site transportation devices to the spot on the line at which they're needed. They are used practically immediately.

These new production procedures are part of a team concept (idea) in car manufacturing. To be successful, production workers, managers, suppliers, and transportation companies must all work closely together and meet their commitments. Computer links between Buick City and their suppliers help keep scheduling precise.

Fig. TF-11. Robotic welders are used at the General Motors plant known as "Buick City."

Producing Products

Producing products is what manufacturing is all about. *Production* is the multi-step process of making parts and assembling parts into products.

Components

Each individual part of a product is called a **component**. Some components are simple, like a wire. Other components, such as a casting for an automobile engine, may be very complicated. Fig. 11-1.

As you know, a manufacturer decides whether to make or buy each part. A computer manufacturing company, for example, may buy many electronic components. However, the company would probably make some of the special parts.

Assemblies

Components are *assembled* with other components. This means they are put together in a planned way. Assembled components are called **assemblies**. If an assembly will be used as a component in another product, it is called a **subassembly**. The handlebars and brakes of a bicycle are examples of subassemblies. Fig. 11-2.

When assembly operations begin, all the necessary parts must be available in the right quantities. Sometimes assembly work is done by hand. That is, workers pick up parts and put them together. They may glue, tighten, or perform other tasks. Sometimes automatic assembly machines are used.

Components may be assembled into subassemblies, and subassemblies may be assembled with additional components or

Fig. 11-1. A casting is a very complicated component of an automobile engine.

Fig. 11-2. Even a fairly small product like a portable electric saw can contain many components and subassemblies.

other subassemblies. The point at which all the parts are combined to form the product is called *final assembly*.

Packaging

After final assembly, many products are packaged. There are many reasons for packaging products. Fig. 11-3. Some products require more than one package. Think of chewing gum. Many companies place each stick of gum in its own package. Then several sticks are put in another package. These packages may in turn be placed in a bag to be sold as a multi-pack. For shipping, the packages are placed in a large fiberboard or cardboard carton.

Most manufacturing companies buy packages from a package manufacturer. The packages may already be printed. If the packages are not preprinted, they can be printed or labeled after packaging. A printed code is also often applied after packaging.

Apply Your Thinking Skills

1. Think of a product you or your family has recently bought. How was it packaged? What purposes did the packaging serve? Can you think of any ways the packaging might have been improved?

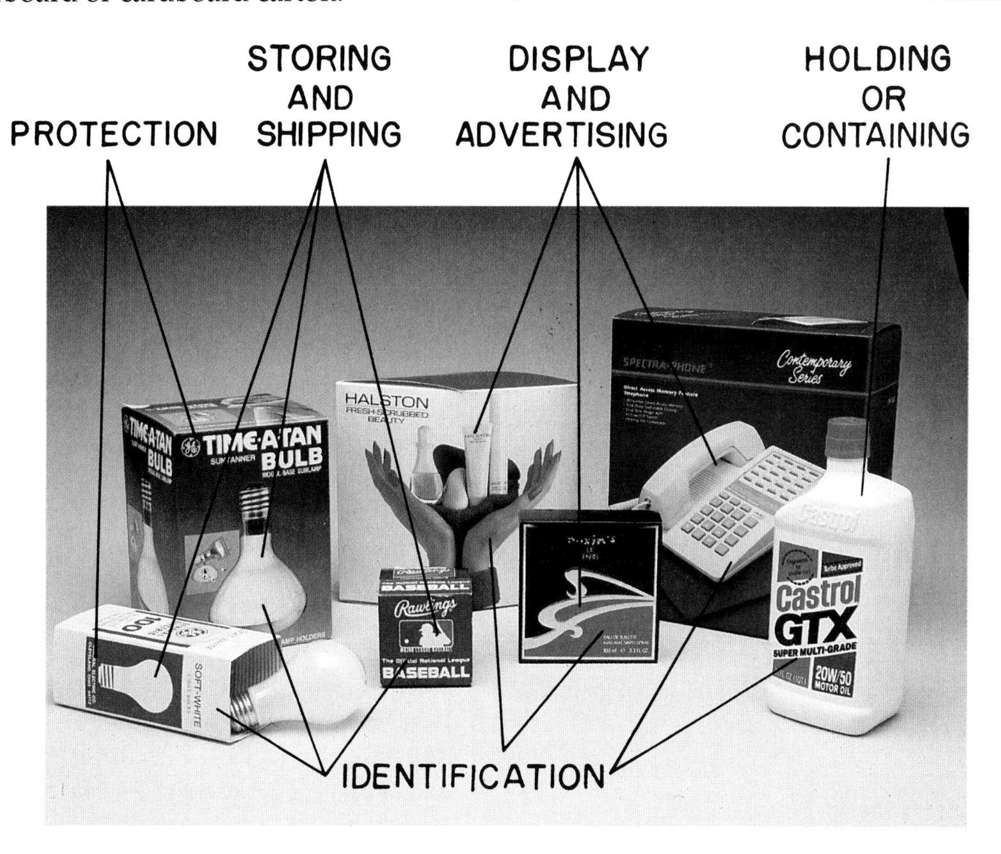

Fig. 11-3. Products are packaged for many different reasons.

Automatic Processing

The manufacturing industry is very dynamic. Something is always new. Something is always changing. Some new and interesting ways are now being used to shape and form materials and to assemble components. Computers are used to control processing. These computer-controlled methods for processing materials are referred to as **automatic processing**.

Computer Numerical-Control Machining

In the past all machines were controlled by operators. They turned handles and pressed levers to make the machine work. Then *numerical control (NC)* was developed. Many machines were controlled according to a numerical code. An electrical controller was connected to a machine by wires. The special numerical code was entered into the controller by a reader that "read" a punched paper tape. The holes in the tape represented a series of numbers. These were the directions to the machine.

Now **computer numerical control (CNC)** is commonly used. Fig. 11-4. This is a form of NC, but it doesn't need paper tape or a special reader. The numerical directions are contained in a computer program. Not only can the computer give directions to machines, but it can also receive feedback. This means it can detect what is happening. For example, if a cutting tool is broken, the computer is sent this information and it stops the machine.

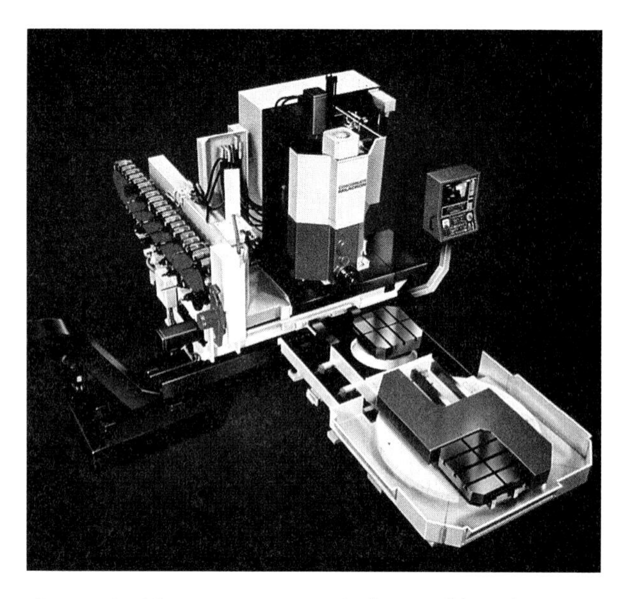

Fig. 11-4. This computer numerical control (CNC) machine performs many operations quickly and automatically.

New Technology for Cutting

For many years cutting had to be done using a saw, knife blade, or cutter. Much cutting is still done that way, but thanks to technology, new processes are being developed. Materials can now be cut to shape or size by unique methods.

Waterjet cutting is the process of using a highly pressurized jet of water to cut a material. The water pressure at your house is about 50 pounds per square inch (PSI). At that pressure, water won't cut anything. Squirting water through a very tiny hole at 50,000 PSI, however, turns the water into a "knife blade!" Fig. 11-5. Sheet materials like cloth, plywood, rubber, and plastic can be cut easily and quickly this way. The waterjet follows a path guided by a computer program.

Lasers strengthen and direct light to produce a narrow, high-energy beam. Whatever the beam strikes becomes so hot that the material vaporizes (turns into a gas). In **laser cutting**, this concentrated, high-energy beam of light is used to cut materials. Another use of laser cutting is to engrave the molds used to make compact discs (CDs). Like waterjet cutting, laser cutting is computer-controlled.

Fig. 11-5. This waterjet cutter can cut metals up to six inches thick, leaving a smooth finished edge.

Fig. 11-6. This automatic assembly machine installs and tightens nuts on the connecting rods in this engine. How many people do you see in the picture?

Fascinating Facts

Although Albert Einstein worked out the "theory" of the laser back in 1917, the first laser wasn't actually developed until 43 years later.

American physicist Theodore Maiman designed a special ruby cylinder, which he used to produce the first laser beam in 1960. The narrow beam of intense light, when focused, could drill a hole through a diamond. The laser soon became valuable in a variety of areas, including communications and medicine, as well as industry.

Innovative Assembly Machines

Most assembly machines are specially built to assemble a certain product. This may be done by **robots,** which are special machines that are programmed to move things or do certain tasks automatically. Robots are discussed in detail in Chapter 13.

Manufacturers of electronic products often use special inserting machines. These automatically place components into products such as computers, video recorders, and stereos. Fig. 11-6.

Apply Your Thinking Skills

1. What are some advantages of automatic processing? Can you think of some possible disadvantages?

Production and Inventory Control

While production is being done, someone must make sure there's enough material on hand. Someone also needs to see that the right number of parts is being made. These tasks are part of production and inventory control.

Production Control

Production control is controlling what is made and when it is made. How do workers know when to start on a certain product? How many parts should be made at one time? When will products be ready to ship to the customer? A plan for controlling production provides answers to these and other questions.

The master *production schedule* is very important to production control. This is a time chart that lists parts and shows how many of each the company plans to make in a certain period of time. Fig. 11-7. Usually a schedule is prepared several months in advance. It gives start and stop dates as well as the number of machines to be used. The schedule is a projection (prediction) of what will happen. The actual time spent may be longer than planned. If so, the schedule is revised (changed).

Controlling production involves keeping track of what work has been done, when it was done, and who did it. After raw material is released to the production department, production control must know what is happening to it at all times. All material that is being worked on is called *work in process* (WIP). A system called *shop floor control* is often used to keep track of work that has been done. The needed information is collected "on the shop floor." Workers record information about the work they've done on computer cards or special note pads. They

Fig. 11-7. A production schedule shows what work is to be done and the dates for starting and ending production.

may enter data into the computer at a terminal near their work station. The information goes into the computer and updates the computer records.

Inventory Control

Inventory is the quantity of items on hand. In a manufacturing plant, **inventory control** means keeping track of:

- Raw materials
- Purchased parts
- Supplies
- Finished goods

For purposes of inventory control, a *raw material* is any material before it enters processing. Potatoes, corn, and cheese are examples of raw materials for a snack food factory.

Purchased parts are ready-made parts that the company buys. A lawn mower factory might buy engines and wheels.

Supplies are different kinds of items needed to keep the plant running smoothly. These items do not become part of the product, but they are needed to support the production process. Supplies include typewriter ribbons, computer paper, oil, and light bulbs.

Products that are completed but not yet sold are called *finished goods*. In a furniture factory, tables and chairs might be the finished goods.

Keeping good records of all inventories is important. For example, when the inventory for materials gets low, more must be ordered. Without records the company would not know how much was on hand. Inventory records are usually kept in computer files.

The purchasing department helps make sure proper inventory levels are maintained. Buyers or purchasing agents buy things a company needs, such as materials, parts, equipment, and supplies. They make sure the items ordered are the proper quality and are a reasonable price. They also make sure the items are delivered at the right time. Timing is very important.

Material Requirements Planning

A system used to help in both production and inventory control is *material requirements planning* (MRP). A computer analyzes information from the bill of materials, the master production schedule, and inventory records. Then it accurately predicts when the various materials will be needed.

Apply Your Thinking Skills

1. Imagine your company needs to buy a certain part for the product you produce so you can fill an order promised for delivery in two weeks. Company A can provide the needed parts tomorrow, but the parts will cost 30% more than if you ordered them from Company B, who cannot deliver the parts for 10-14 days. What would you do? Explain your answer.

Quality Assurance

The quality of a product is how well it is made. Manufacturing companies want to produce high-quality products. They want each product to fulfill its purpose in the best way possible. They also want to satisfy consumers. **Quality assurance** means making sure the product is produced according to plans and meets all specifications. Sometimes this is called *quality control.*

The level of quality must be set in advance. This level of quality is called a *quality standard.* The problem is that it is impossible to do something perfectly over and over again. There will always be variation (slight change from one part or piece to another). Variation occurs because of differences in workers, materials, machines, and processes. Controlling this variation is the goal of quality assurance.

There are two basic ways to approach quality assurance: prevention and detection. Prevention involves doing everything possible to prevent variation in materials or processes before parts are made. Detection means inspecting to find variations in parts or products after they have been produced. Preventing mistakes is better than finding mistakes; it costs less to correct errors before or during production instead of after. Fig. 11-8.

Process Improvement

Process improvement involves continually working to improve the processes by which things are made. This could involve changes in machines, for example, or in the ways workers do their jobs. Some variation is normal in any process. With careful checking and analysis, quality assurance workers can discover the type and amount of variation that is normal for a process. Then they can monitor (watch to check) the process and determine when the process is either *in control* or *out of control.*

Statistical process control (SPC) is one technique used in process improvement. It is based on a special type of mathematics called statistics, which involves collecting and arranging facts in the form of numbers

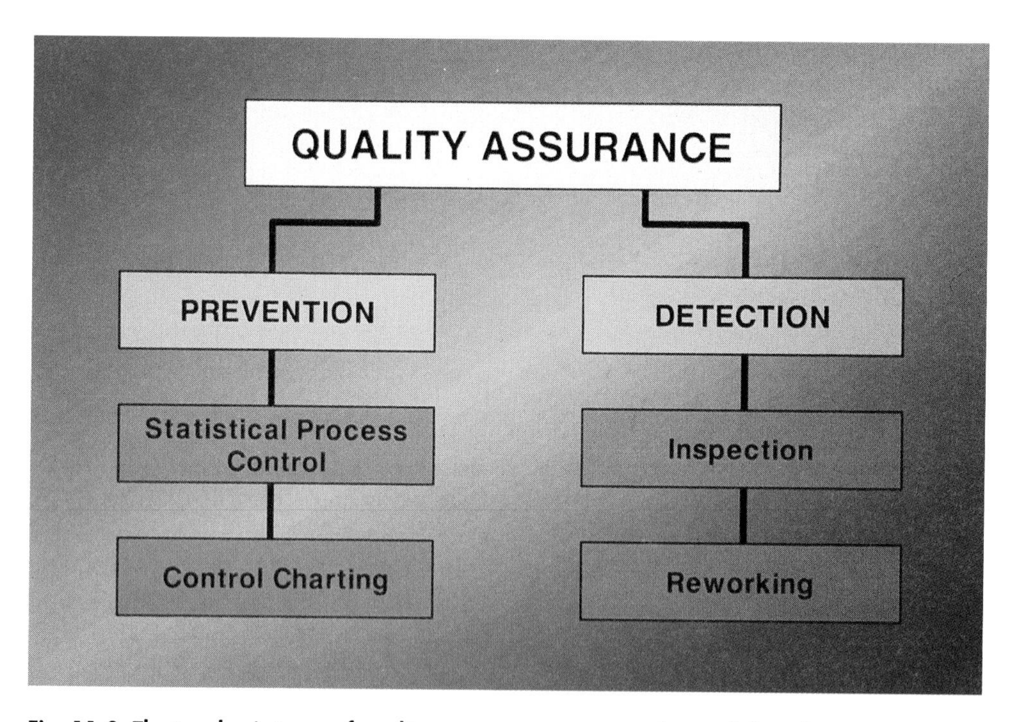

Fig. 11-8. The two basic types of quality assurance are prevention and detection. Prevention is preferred over detection because it occurs before or during production instead of after.

to show certain information. Computers and other recording devices keep a record of what a particular machine is doing. This information is recorded on a *control chart*. Fig. 11-9. The chart has a mean (average) and shows the upper and lower acceptable limits of variation from that mean. (As you learned in Chapter 9, this is called *tolerance*.) Quality assurance workers use the control charts to analyze the process. Let's look at an example.

Suppose a certain machine automatically fills empty cereal boxes with cornflakes. The label on the box says that there are 16 ounces of cereal in the box, but the boxes are not all filled exactly the same. There is a tolerance of plus or minus one-half ounce. That means the box can actually contain any amount between 15.5 and 16.5 ounces. By checking the weight of every 100th box and plotting that information on a control chart, it's possible to keep an eye on the machine's correctness. As long as the machine is running within the upper and lower limits, the product is OK. If the control chart shows that the boxes are being over- or underfilled, workers can stop the machine and make the necessary adjustments.

Inspection

To *inspect* something means to look at it and compare it to some standard. Inspectors examine a part or product to see if it meets the specifications. **Specifications** are the detailed descriptions of the design standards for a part or product. These standards include rules about the type and amount of materials, size, shape, function, and performance.

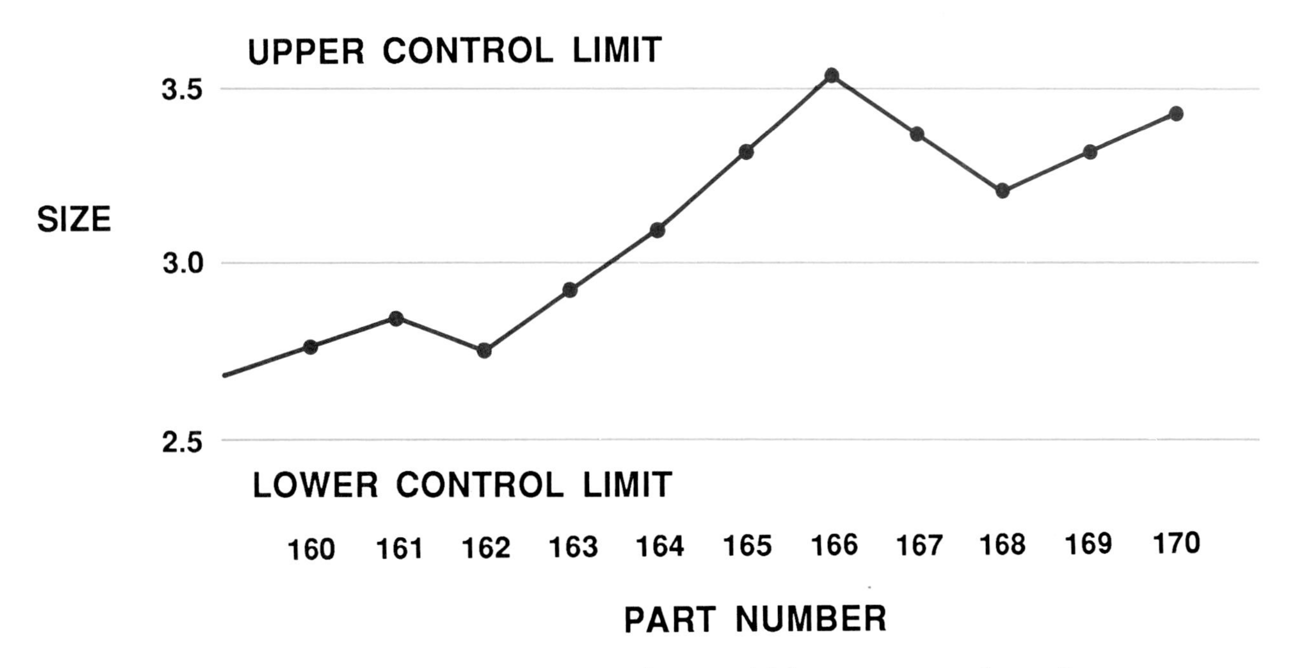

Fig. 11-9. A control chart is used to track the process. At what point did the process go out of control?

Inspectors check on materials, parts, and processes. Sometimes they visually inspect the part (look at it). Most often they use some kind of measuring device.

Inspections are made at key times in the production cycle:

- Delivery of materials
- Work in process
- Finished goods

Materials are inspected as they are delivered to the plant. Materials that don't meet the standards are rejected and returned to the supplier. Work in process (WIP) is inspected to make sure the work is being done properly and that the parts are correct. After the product is made, it is given a final inspection. Everything is checked to make sure it works and looks right. Fig. 11-10.

Inspection Tools

Various inspection tools and devices are used to check materials, parts, and products. Some are used for measuring and others for comparing.

Various **gages** are used as inspection tools to compare or measure sizes of parts and depths of holes. One simple gage is a go/no go gage. Fig. 11-11. By slipping a part into this gage, the inspector can tell at a glance whether or not the part is the right size.

Not all inspection tools are as simple as gages. Computer-controlled devices can make very precise measurements. Fig. 11-12. Optical comparators are used to magnify small parts. Special scanning microscopes are needed to inspect very tiny parts. X-ray

Fig. 11-10. Final inspections are done on products before they leave the factory to make sure that everything works properly. In this photo, the inspector is checking the backup lights of a new automobile before it leaves the factory.

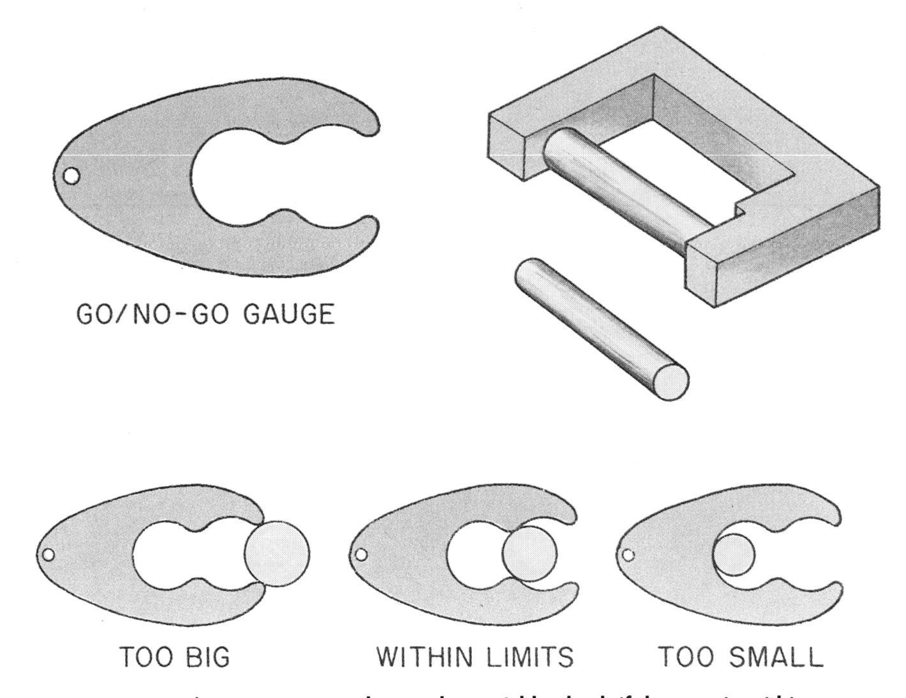

Fig. 11-11. A go/no go gage can be used to quickly check if the part is within tolerance levels.

machines are used to "see inside" welded metal parts. Some devices emit sound waves to check product characteristics.

Laser beams can also be used to detect variations in a part. To check variations in distance, the laser beam is directed at various areas of the part. Measuring the time it takes the light to return reveals the distance it traveled. To check whether a part's surface is flat, a laser beam can be projected across the surface. The beam will hit anything that sticks up.

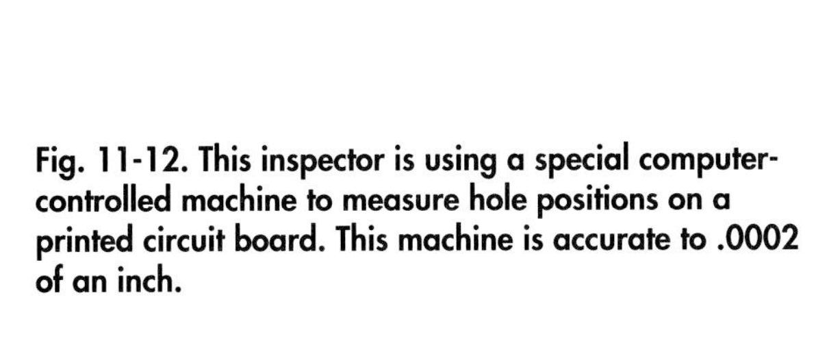

Fig. 11-12. This inspector is using a special computer-controlled machine to measure hole positions on a printed circuit board. This machine is accurate to .0002 of an inch.

Acceptance Sampling

Many products are made in large quantities, so it is not possible or practical to inspect each product. An inspection procedure called acceptance sampling is used in these cases. **Acceptance sampling** is randomly selecting a few typical products from a production run, or lot, and inspecting them to see whether they meet the standards, then the whole lot is approved. If the samples are rejected, the other products from that lot are also rejected.

The size of the sample depends on the lot size. Generally speaking, the smaller the lot size, the greater the percentage that should be inspected. The table in Fig. 11-13 shows one company's sampling plan. You can see that for a lot size of two, all products are tested. However, for a lot size of 10,000 the sample size is only 125. The *acceptance level* is how many of the sample must pass if the lot is to be accepted. For example, in a lot size of 280, the sample size is 20. Of the 20 pieces inspected, you would reject the entire lot if more than four of the sample pieces did not meet acceptable standards.

Burn In

One special quality assurance measure is called *burn in*. It is done to electronics products like computers. Electronic products that fail tend to do so in the first few hours of operation. Because of this, a computer manufacturer actually runs every computer for the first few hours. Those that fail are repaired, if possible. If the computer passes the burn-in test, then it will probably last a long time.

Apply Your Thinking Skills

1. Your textbook lists three key times for inspection. What might be the result if any one of these three inspections were overlooked?
2. What might happen if the acceptance level in acceptance sampling were set too low?

XYZ COMPANY SAMPLING PLAN

Lot Size	Sample Size	Sample Percentage	Acceptance Level	Acceptance Percentage
2-8	2	100%-25%	1 out of 2	50%
9-15	2	22%-13%	1 out of 2	50%
16-25	3	19%-12%	2 out of 3	67%
26-50	5	19%-10%	3 out of 5	60%
51-90	8	16%-9%	6 out of 8	75%
91-150	13	14%-9%	10 out of 13	77%
151-280	20	13%-7%	16 out of 20	80%
281-500	32	11%-6%	27 out of 32	84%
501-1,200	50	10%-4%	43 out of 50	86%
1,201-3,200	80	7%-3%	71 out of 80	89%
3,201-10,000	125	4%-1%	114 out of 125	91%

Fig. 11-13. Notice that the smaller the sample percentage, the greater the percentage of sample products that must be acceptable. Why is this?

Global Perspective

Old Is New

In 1913, Henry Ford revolutionized the manufacturing industry by introducing the assembly line system of producing products. Today, the Volvo automobile factory in Uddevalla, Sweden, may be starting another revolution by reverting to older manufacturing methods.

The plant is a final assembly operation. Instead of working on assembly lines, workers are grouped into teams of eight to ten persons. Each team works in its own special area, referred to as an "assembly cell." The entire final assembly operation is completed there. Materials are sent to the cells in kits made up in a separate storage area. The kits are carried on AGVs.

Workers in the cells can work on assembling three or four cars at a time. Usually, three team members at a time work on each car. Workers rotate from car to car performing their assigned tasks. Movement patterns are carefully planned and organized so that work is completed efficiently.

The layout in the plant allows for up to 48 assembly cells. When the plant eventually reaches full production, its goal is to produce 40,000 cars on each work shift in a year.

Fig. GP-11. Volvo's Uddevalla plant uses no assembly lines. Teams of 8 to 10 persons build an entire car at one location.

Why did management at the Volvo company decide not to use the assembly line system? The basic idea they adopted is that workers should control machines instead of having to adjust their work activities to what the machines are doing. The plant and the system at Volvo were planned and designed taking ergonomics into consideration. (*Ergonomics* involves designing machines and adapting working conditions to suit the workers' needs.) All equipment within the plant was chosen with the workers in mind. For example, each cell contains a special assembly stand. Cars are hung on a tilt device on the stand as final assembly is done. Workers can tilt the cars to positions comfortable for working.

In the days before the assembly line system was developed, workers completed products in one place just as these workers at the Volvo plant are. However, the process used in those times was very slow. Workers had to gather parts and then put them together. Today, ergonomics and the great advancements that have been made in the technology used for materials handling make final assembly in one area a good alternative to the assembly line system. Technology can often help make old ideas new again.

Extend Your Knowledge

1. Relative to its population, Sweden is one of the world's top users of robots. Do research to find information and statistics related to the use of robots and to the general increase in automation in manufacturing in various countries around the world.

Chapter 11 Review

Looking Back

Products are made up of components or subassemblies which are assembled into the final product. Once the product is completed, it is packaged.

Computers can be programmed to control machines in processing operations. Waterjet cutting and laser cutting are two new processing operations controlled by computers.

There must be people to oversee that there is enough material on hand to make needed parts, that the right number of parts is being made, and that the product will be finished to ship to the customer on time.

Manufacturers want to produce high-quality products. They want the product to fulfill its purpose. They want the consumer to be pleased with the product. To accomplish all of this, they work to prevent any defects during processing and they inspect for defects after processing.

Review Questions

1. What is production?
2. Give five reasons for packaging.
3. Briefly describe computer numerical control.
4. Describe one new computer-controlled cutting operation.
5. Describe production control.
6. What is inventory? What four things must be kept track of in inventory control?
7. Describe the responsibilities of the purchasing department.
8. What are specifications?
9. What is the role of inspectors? At what key times are inspections made?
10. Briefly describe acceptance sampling.

Discussion Questions

1. Now that you have read about the burn- in process used to test electronic products, how do you feel about buying a VCR that has been "used?"
2. You've probably heard your grandparents and possibly your parents complain that things "aren't made the way they used to be." However, manufacturers today are using many quality assurance programs. How do you explain this?
3. Describe how you would inspect the following: a watch with alarm and stopwatch, grease remover, an automatic timer, break-resistant dishes, and a board game.
4. Imagine you had to design the packaging for a new snack product. Describe the construction of the packaging as well as its appearance and the information it would give.
5. Discuss some things you use every day that are ergonomically designed. Then discuss some things you use every day that should be ergonomically designed and suggest some possible design changes.

Chapter 11 Review

Cross-Curricular Activities

Language Arts 

1. Production is a multi-step process that should run smoothly. Write a sequence chain to analyze a production process.

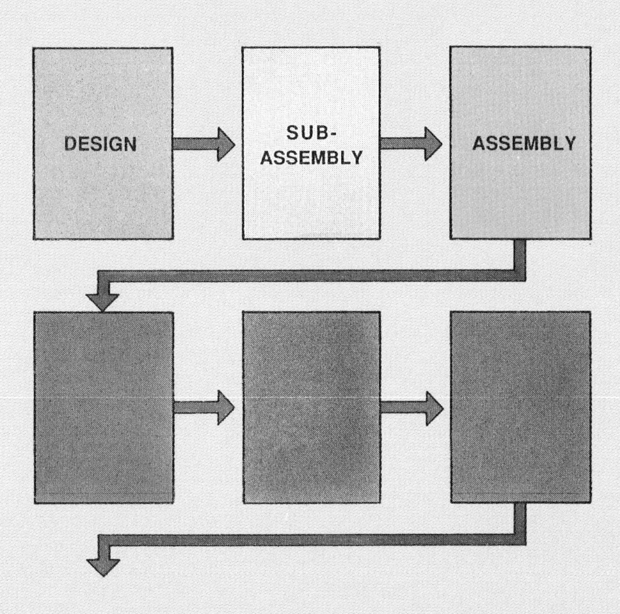

2. Manufacturers make a production schedule or time chart. Make a production schedule for writing a term paper. Consider lengths of time needed for research, note-taking, rough draft, and final copy. Use a calendar to set up a production schedule from start to finish.

Social Studies 

1. Research to find ten current or potential future uses of lasers in one of the following fields: industry, medicine, science. Briefly describe the types of lasers used and how they are used.

Science 

1. The cutting action of a blade, such as a saw or knife, depends on a simple machine called a wedge. This is what happens: The very thin, sharp leading edge of the wedge (blade) pushes bits of the material apart a little, so that the thicker part behind can move deeper into the material and force the two sides apart. The wedge works by concentrating a lot of force along a very thin line. What examples of the wedge can you find at home? At school?

Math X÷

Carol secured a summer job as quality assurance checker on a production line that produced assembly spacers. She was told that in order to be considered acceptable, each spacer must measure 1/4" in thickness, 1/2" in width, and 4" in length. She was also told that each dimension had a certain tolerance assigned that would allow the spacers to vary slightly and still be considered acceptable. The tolerance for thickness and width was 1/16" and the tolerance for length was 5/16".

1. If Carol found a piece that measured 5/16" x 7/16" x 3 5/8", should she have accepted it?
2. What about a piece that measured 3/16" x 9/16" x 3 3/4" ?

Chapter 11 Technology Activity

Manufacturing a Product in Your Lab

Overview

In this section, you are learning about manufacturing and the manufacturing industry. You are also applying many of the manufacturing principles that you have learned. This activity provides the opportunity for you to actually manufacture a product—a desk lamp.

Goal

To manufacture a desk lamp using the working drawings, parts and materials chart, flow chart, and procedure chart supplied within this activity.

Equipment and Materials

(See Parts and Materials Chart in Fig. A11-1.)

Procedure

Safety Note: Before doing this activity, make sure you understand how to use the tools and materials safely. Have your teacher demonstrate their proper use. Follow all safety rules.

You will be given working drawings, a parts and materials chart, a flow chart (Fig. A11-2), and a procedure chart (Fig. A11-3) much like those used in the manufacturing industry.

PARTS AND MATERIALS CHART			
Qty.	Part	Size	Material
1	Base	2"×4"×A	Wood
2	Ends	E×F	Sheet Metal (26 gage)
1	Shade	A×B	Sheet Metal (26 gage)
1	Bracket	M×U	Sheet Metal (26 gage)
2	Stems	$\frac{1}{2}$"o.d.×T	Metal Tube
1	Standard Light Socket		
1	Switch, Single Pole	6A, 125V	
1	Pipe Nipple with Locknuts	$\frac{1}{8}$" dia. ×I	
1	Electrical Lamp Cord #18-2, Brown	7'	
1	Plug		
3	Wire Nuts	73B	
1	Felt	A×W	
1	Self-stick Vinyl	18"×14"	Plastic
4	Sheet Metal Screws PH	#6×$\frac{3}{8}$"	
1	Washer (Flat)	$\frac{3}{16}$"	
2	Machine Screws RH	#6-32×$\frac{3}{4}$"	
2	Hex Nuts	#6-32	
			White Paint
			Contact Cement

Fig. A11-1. The parts and materials chart for the desk lamp.

Chapter 11 Technology Activity

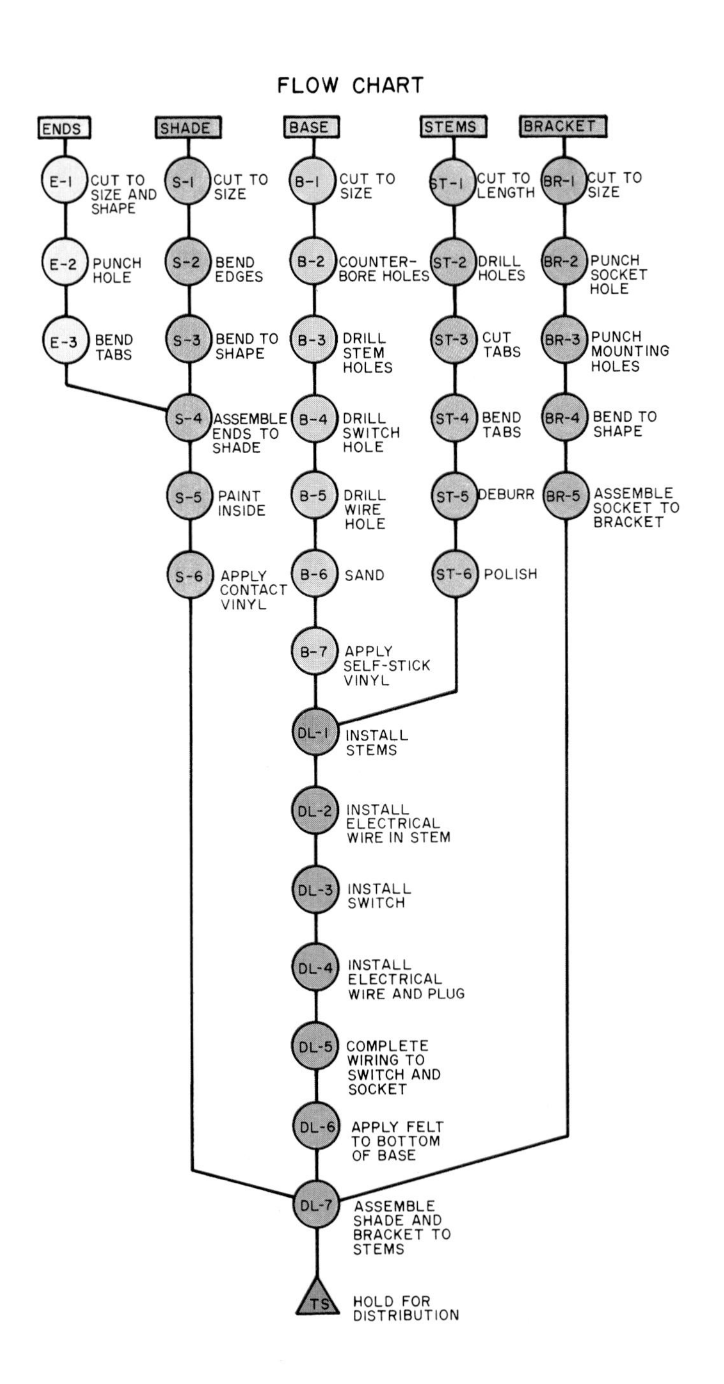

DIMENSIONS CHART		
Dimension Symbol	Metric mm	Cust. in.
A	254	10
B	267	10 1/2
C	5	3/16
D	94	3 11/16
E	143	5 5/8
F	86	3 3/8
G	38	1 1/2
H	62	2 7/16
I	12	1/2
J	75	3
K	6	1/4
L	124	4 7/8
M	25	1
N	215	8 1/2
O	203	8
P	32	1 1/4
Q	44	1 3/4
R	20	3/4
S	10	3/8
T	264	10 3/8
U	112	4 1/2
V	56	2 1/4
W	88	3 1/2
X	56	2 1/4

Fig. A11-2. The desk lamp's flow chart, dimensions chart, and working drawings.

Chapter 11 Technology Activity

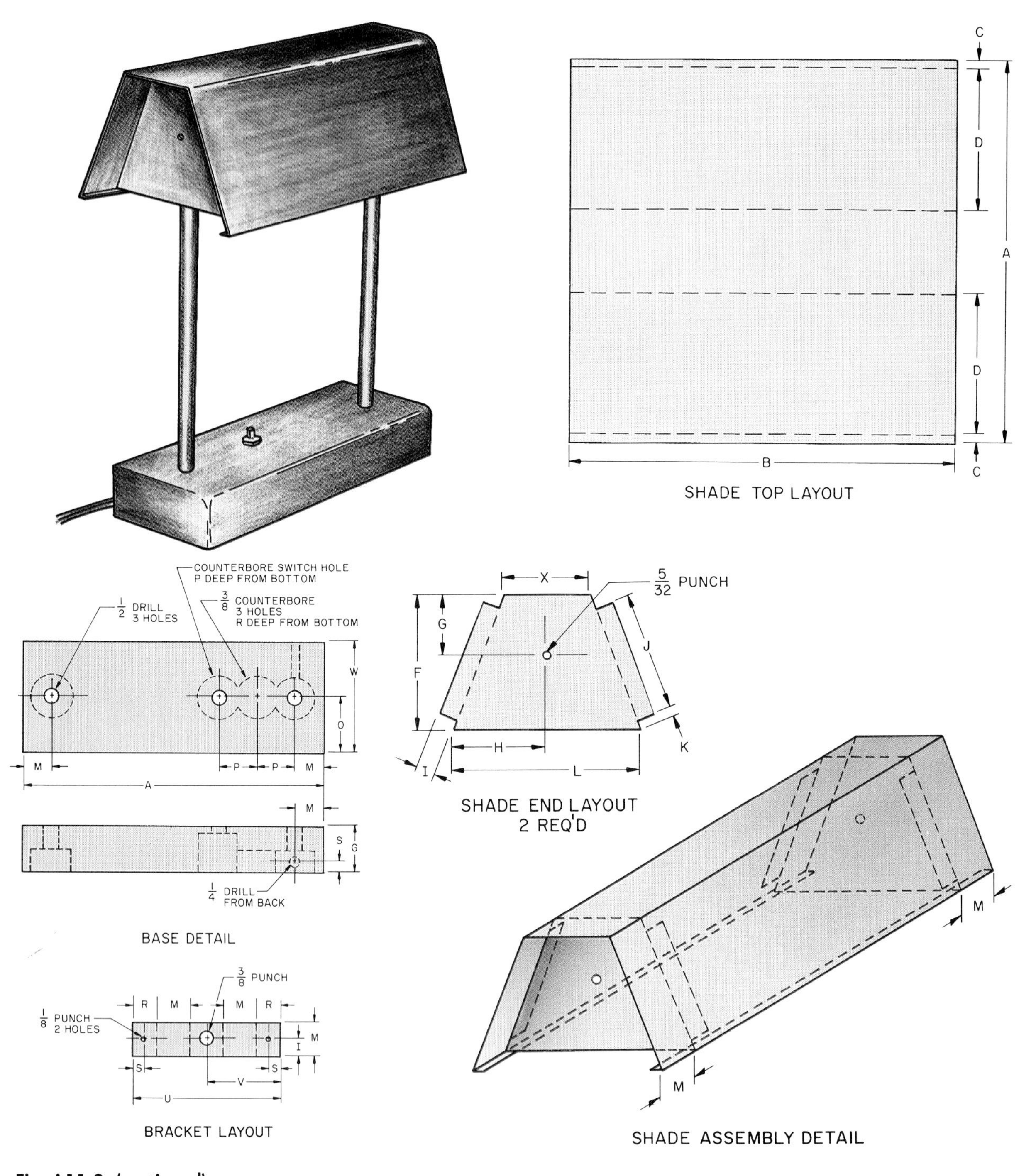

Fig. A11-2. (continued)

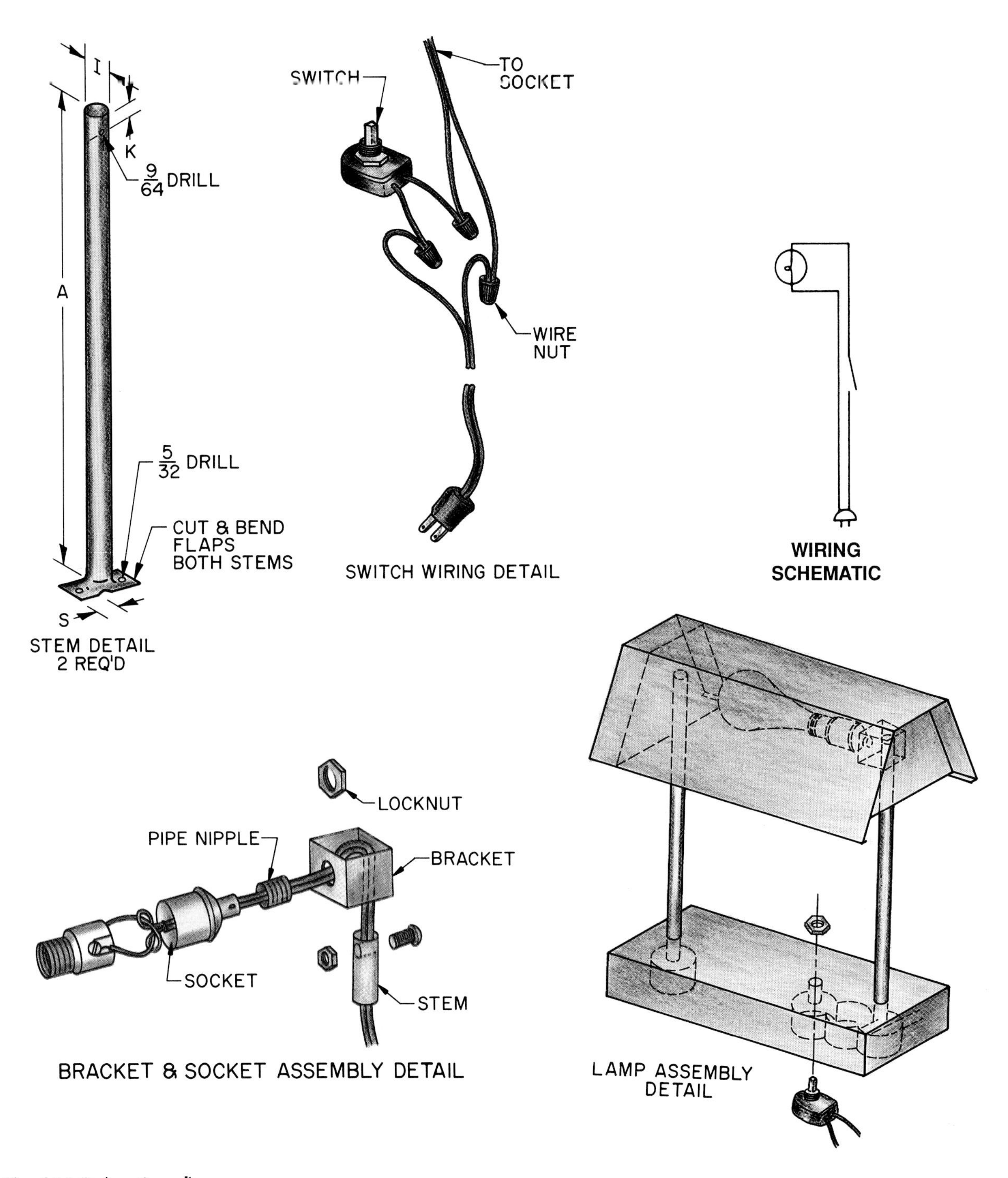

Fig. A11-2. (continued)

Chapter 11 Technology Activity

PROCEDURE CHART			
Operation Number	Operation	Tools & Equipment	Notes
		ENDS	
E-1	Cut to size and shape	Squaring shears, tin snips	
E-2	Punch hole	5/32" hand punch	Could be drilled.
E-3	Bend tabs	Bar folder or box and pan brake	Bend 90°.
		SHADE	
S-1	Cut to size	Squaring shears, tin snips	
S-2	Bend edges	Bar folder or box and pan brake	Bend 90°.
S-3	Bend to shape	Box and pan brake	
S-4	Assemble ends to shade	Spot welder	Could be riveted.
S-5	Paint inside	Spray or brush	Paint entire underside (including ends) with white paint.
S-6	Apply self-stick vinyl	Scissors, knife	Apply self-stick vinyl carefully to outside of shade.
		BASE	
B-1	Cut to size	Handsaw or power saw	
B-2	Counterbore hole	Drill press or portable electric drill, 1 3/8" spade bit	A drilling fixture may be used.
B-3	Drill stem holes	Portable electric drill or drill press, 33/64" twist drill	A drilling fixture may be used.
B-4	Drill switch hole	Hand drill or power drill, 7/16" twist drill	
B-5	Drill wire hole	Hand drill or power drill, 1/4" twist drill	
B-6	Sand	Belt sander or finishing sander	
B-7	Apply self-stick vinyl	Scissors, knife	Be sure to cut holes completely through vinyl.
		STEMS	
ST-1	Cut to length	Tubing cutter or hacksaw	
ST-2	Drill holes	Drill press or portable electric drill, 9/64" twist drill	A drilling fixture may be used.
ST-3	Cut tabs	Hacksaw, tin snips	Make the saw cut first. Then trim to size.
ST-4	Bend tabs	Pliers	
ST-5	Deburr	File	
ST-6	Polish		Use steel wool to give a satin finish

Fig. A11-3. The procedure chart for the desk lamp.

Chapter 11 Technology Activity

	BRACKET		
BR-1	Cut to size	Squaring shears, tin snips	
BR-2	Punch socket hole	3/8" hand punch	Could be drilled.
BR-3	Punch mounting holes	1/8" hand punch	Could be drilled.
BR-4	Bend to shape	Hand seamer	
BR-5	Assemble socket to bracket	Wrench, screwdriver	
	ASSEMBLY		
DL-1	Install stems	Screwdriver	Use two #6×3/8" sheet metal screws PH for each stem.
DL-2	Install electrical wire in stem		Thread 12" piece of wire through the stem that will support light socket.
DL-3	Install switch	Pliers	
DL-4	Install electrical wire and plug		
DL-5	Complete wiring to switch and socket	Wire strippers	Strip wire ends. Do wiring. Use wire nuts. No bare wire should show.
DL-6	Apply felt to bottom of base		Use glue. Four small felt circles may be used.
DL-7	Assemble shade and bracket to stems	Screwdriver	Use one #6-32 × 3/4" machine screw RH and nut for each stem.

Fig. A11-3. (continued)

Look at the working drawings (Fig. A11-2) and the parts and materials chart. You will note that most dimensions on the working drawings are indicated by letters. This has been done to allow you to use either metric or customary measurements. When you are making the lamp, locate the letters in the dimensions chart and choose *either* the metric or customary measurements. Do not try to mix measurements. The customary and metric measurements may not be exactly the same.

Now look at the flow chart. The letters and numbers on this chart refer to the manufacturing operations listed on the procedure chart. Follow the steps (operations) listed in the procedure chart to make your desk lamp. Refer often to the drawings and the flow chart.

Your teacher will provide additional instructions as needed for working in your lab.

Evaluation

1. What problems, if any, did you encounter while you were manufacturing your desk lamp?
2. If you could manufacture your desk lamp again, what would you do differently?

Chapter 12

Marketing

Looking Ahead

In this chapter, you will discover:

- who is responsible for selling products.
- what a market is.
- how market research is done.
- why advertising is important.
- the three major ways of selling products.

New Terms

advertising
chain of distribution
commission
consumers
direct sales
distribution
market
marketing
market research
retailers
sales forecast
test marketing
warehouse
wholesalers

Technology Focus

3M Post-It Brand Notes

There's hardly a refrigerator or a computer monitor that hasn't worn a sticky yellow note at one time or another. The same is true for telephone receivers, catalogs, office memos, and homework papers. You name it and somebody surely has already stuck a 3M Post-it brand note on it somewhere.

In 1974, Arthur Fry, a chemical engineer at 3M, was at choir practice when he got frustrated trying to keep track of his music in the hymnal. If only he could find a way to put in temporary markers that wouldn't fall out. Almost like the light bulb that blinks on above a cartoon character's head, Art suddenly remembered the glue that didn't stick right. A co-worker at 3M, Spencer Silver, had been trying to make a super-stick glue, but one of his experimental batches wasn't very sticky at all. Fry tried Silver's not-so-sticky glue.

The glue gave him problems. Sometimes the sticky stuff would leave a residue. He continued to iron out the problems with the sticky bookmarks by "moonlighting" time on his job. 3M allows employees to spend 15% of company time on other projects. They believe that if a person is interested in a project, he or she will work harder.

Confident that he had a good idea, Fry convinced a 3M production department to produce his product.

When 3M *test marketed* (sold in a limited area to test appeal and demand) its new Post-it note pads, however, sales were disappointing. Two 3M executives, Geoffrey Nicholson and Joseph Ramey, decided to give the idea another chance. They flew to a test city and went from office to office, giving away the note pads. They felt that people had to use Post-it notes to appreciate them. This unique marketing approach proved they were right. As soon as people saw one use for the notes, they found new uses. Post-it notes now rank as one of the most successful new product introductions in the history of 3M.

Fig. TF-12. Post-it™ notes are one of the most successful new product introductions in the history of 3M.

Markets

The goal of manufacturing is to produce and *sell* the product. Thus the job isn't complete until the product is sold. **Marketing** includes all the activities involved in selling the product.

Manufacturers often try to make products for a certain market. A **market** is a specific group of people who might buy a product. A market might be teenagers, senior citizens, young marrieds, mechanics, the military, or even a school district. For example, a watch manufacturer may make colorful plastic watches aimed at the teenage market. Fig. 12-1.

Markets change as population, incomes, and lifestyles change. For example, in more and more households all adults have jobs. This has increased the demand for prepared foods. People in charge of marketing must be aware of changes in the market.

There are two major types of markets: industrial and consumer.

Fig. 12-1. Marketing techniques are often aimed at particular parts of the market. Who is the market for each of these products?

The Industrial Market

Businesses and industries make up the industrial market. They buy products to use in their own companies. They also buy products to use as parts in the products they make.

Car dealerships, hospitals, and magazine publishers all buy paper goods, light bulbs, and pencils. They are part of the industrial market because they use the purchased products in their businesses.

As you remember from Chapter 10, manufacturers decide whether to make or buy parts for products they manufacture. If they buy parts to use in their products, they are part of the industrial market. One manufacturer buys another manufacturer's product. Automobile manufacturers buy tires from one company, windshield wipers from another, and so on.

The Consumer Market

Consumers are everyday people (like you and your family) who buy products for their own personal use—things like toothpaste, videotapes, and sneakers.

The consumer market is a big one. There are millions and millions of people "out there" to buy products. Manufacturers count on it. They spend a lot of time and money making sure their product is one consumers will want to buy and then persuading consumers to buy it.

Apply Your Thinking Skills

1. Every time you shop, whether it's for clothes, food, a CD, or a video tape, you are making consumer choices. What kind of consumer choices have you made in the past month? Why did you choose the items you bought over other similar items?

The Marketing Plan

Before a company begins to sell a product, the marketing department develops a marketing plan. This plan includes a **sales forecast**, which is a prediction of how many products the company will sell. Marketing arrives at this estimated figure by looking at the market potential (amount of possible sales). A percentage of this number is figured as the company's expected *market share*. (Each manufacturer competes with other manufacturers of similar products for its portion, or share, of the total sales of that type of product.) The plan also includes ideas for advertising and sales.

Market Research

Market research includes all activities used to determine what people want to buy and how much they will pay for it. The results indicate how well the company can expect the product to sell. If the market research shows that lots of people will buy the product, then the company can plan a large-scale production. Suppose, however, the research is inaccurate. The company can lose a lot of money. Imagine making thousands of a product and selling only a few hundred! Market research must be done carefully and accurately.

One way of researching the market is to *interview* people. Potential (possible) consumers are asked their opinions about a product. If it's a food product, people may be asked whether they like its appearance or taste. (You've probably seen "taste tests" for soft drinks on a TV commercial.) The most important question asked is "Would you buy this product?"

Another way of doing market research is **test marketing**. A company produces a small number of products and sells them in a very limited area, such as in one city or region of the country. If sales are good in the test market, then the company can usually expect that the product will also sell elsewhere. Fig. 12-2.

Fig. 12-2. A company often test markets an item, such as spreadable cheese, to see whether consumers will buy it.

Advertising

Advertising tells people about a product. It is the method or methods the company uses to persuade, inform, or influence consumers to buy a certain product. The whole goal of advertising is to convince consumers that they need the product. Market research is used to determine the best type of advertising to reach the target market. The target market is the group of consumers the marketing department decides is most likely to use and want the product.

Fascinating Facts

The giant industry of television advertising was born July 1, 1941, when Bulova sponsored the first TV commercial. In that commercial, airing on New York's WBNT, a camera focused on a Bulova wristwatch, and an announcer read the time: "10 minutes after 10." The cost of that commercial? $9!

Most of us are familiar with TV commercials. They cost a lot of money, but they reach a lot of people. Other places we see or hear advertising are magazines, newspapers, radio, and billboards. Fig. 12-3.

One interesting method of advertising is to give away free samples. Sometimes a company will send you a sample of their product in the mail. If you like the free sample, you will probably buy the product.

Apply Your Thinking Skills

1. What effect does advertising have on your shopping choices?

Fig. 12-3. Advertising is everywhere. Whenever you wear a T-shirt picturing your favorite rock group or movie, you are advertising.

Fascinating Facts

A young door-to-door book salesman launched Avon cosmetics back in 1886 when he discovered that customers liked his "marketing gimmick" better than his books! David McConnell, a resident of upstate New York, began selling books at the age of 16. In order to increase his sales, he decided to offer a free introductory gift to those who would listen to his sales pitch.

He created an original perfume with the help of a local pharmacist. He soon found that women loved his perfume, but weren't that interested in his books! He switched careers and set up a company that later became known as Avon.

Sales and Distribution

The processes of actually selling and getting the products to the purchasers are sales and distribution.

Sales

If a company can't sell the product, all is lost. Selling involves convincing consumers to buy the product. Sales personnel are trained to know about the company's products and services. They are also taught selling techniques. Fig. 12-4. Many people in sales work on a **commission**. That is, the money they make is a certain percentage of the amount received for the products they sell. The more they sell, the more money they make. This is a good incentive (encouragement) to sell more products. There are three main ways of selling products:

- Direct sales
- Wholesale sales
- Retail sales

Direct Sales

Sometimes a manufacturing company sells its product directly to the customer. This method of sales is called **direct sales**.

Fig. 12-4. Knowledgeable salespeople help customers find the products they need.

Manufacturers commonly sell directly to purchasers in the industrial market. An aluminum producer sells aluminum tubing directly to an air conditioner manufacturer. A manufacturer who sells products directly to another manufacturer is called an *original equipment manufacturer* (OEM). Fig. 12-5.

Fig. 12-5. Products such as engines for grass trimmers may be made by one manufacturer and sold to others for use in their products.

Some direct sales are also made to consumers, but this is less common. Perhaps you've gotten some advertising and an order blank for a product in the mail. Maybe a TV commercial invited you to send for a record or cassette or jewelry. These products are usually offered direct from the manufacturer.

Wholesale Sales

All sales made to any person other than the consumers who will finally use the product are wholesale sales. **Wholesalers** are people or companies who act as intermediaries (go betweens). They buy large quantities of products from manufacturers. Then they sell the products to commercial, professional, retail, or other types of institutions that also purchase in quantity. Wholesalers may also be involved in such activities as financing, storing, and transporting products.

Retail Sales

Retailers buy products from manufacturers or wholesalers. Then they sell the products directly to consumers who will actually use the product.

Fascinating Facts

In early days, people shopped in "general" stores. The store owner knew the townspeople and they bought items on credit. Often the buyer and seller would haggle over the price before the deal was made. The first department store, the Belle Jardiniere, was opened in 1824 in Paris by Pierre Parissot. The store revolutionized the business practices of the day. Items were sold for cash rather than on credit. They were also sold at a "set" price rather than a price that was simply agreed upon by the buyer and the seller.

There are millions of retail stores in the United States. Retail stores include department stores, chain stores, and discount stores. The variety of retail stores is almost unimaginable, but they all have the same function—selling products to consumers. Fig. 12-6.

Distribution

Distribution refers to methods used to get goods to the purchaser. As the result of distribution activities, we have many well-stocked stores from which to choose.

The "path" that goods take in moving from the manufacturer to the consumer is called the **chain of distribution**. The chain of distribution can be long. It may involve both a wholesaler and a retailer as well as the manufacturer and the consumer. A short distribution chain includes only the manufacturer and the consumer, as in direct sales.

Managing distribution is important. The product must be available at the right time and in the right place. A distribution manager has to make certain that products move smoothly through the entire chain of distribution.

Fig. 12-6. The stores in this mall are all retail stores. Locating stores near one another makes shopping convenient for consumers.

Warehousing

A **warehouse** is a building where products are temporarily stored until the next part of the chain of distribution is ready for them. Fig. 12-7. Having products already made and stored ahead ensures that there will be enough products on hand for the purchasers when they want them.

The number and location of storage facilities are very important. Just one company may have many warehouses located throughout the country. A large manufacturer of snack foods, for example, may have warehouses located in 20 states. This makes distributing the product easier.

Fig. 12-7. Products may be temporarily stored in a warehouse until they are ready to be distributed to wholesalers and retailers.

Transportation

Regardless of the chain of distribution used, some form of transportation is involved. All products need to be shipped from the factory and must eventually reach the industrial or consumer market. Trucks and trains are commonly used. Depending on the product, air freight may be used. For products that are exported (sent out) to other countries, ships are the most common method of transportation.

Apply Your Thinking Skills

1. What school courses do you think might be helpful to prepare you for a career in sales?
2. What might be some disadvantages of working for a commission?
3. What effects do you think the chain of distribution might have on the final price of the product?

Investigating Your Environment

Green Schemes

Manufacturers and retailers know that there is a growing market for products that do not harm the environment. Consumers are informed about environmentally friendly products through advertising. Or are they? Phrases like "safe for the environment," "biodegradable," and "recyclable'' are popping up everywhere. What do these phrases really mean, however?

Some plastic garbage bags carry the label "safe for the environment," but this doesn't necessarily mean that they are biodegradable (will break down and decompose). It may just mean that they won't poison the environment once they are in the landfill. They can't poison anything if they never break down! What about the products that really are "biodegradable"? Will it take one month, one year, or one hundred years for them to break down?

Many products that are "recyclable" can't even be recycled in your community yet. Many communities are able to recycle paper, glass, and metals. However, the ability to recycle plastics is still limited. Just because a product is marked "recyclable" doesn't mean *you* can recycle it.

Deceptive (deliberately misleading) advertising is only part of the problem. For example, one scientist claims that polystyrene cups have gotten a bad name from environmentalists. She says manufacturing a paper cup uses 6 times more steam, 24 times more electricity, and produces 200 times more wastewater than manufacturing a polystyrene cup. In addition, although the paper cup will break down faster in a landfill, the conditions must be just right. Paper buried in landfills that are in dry areas does not break down any better than polystyrene; and paper in moist landfills produces gases that may contribute to the greenhouse effect.

The next time you see a package or advertisement claiming a product does not harm the environment, you may want to check those claims for yourself.

Fig. IYE-12. Many products and/or packages are said to be environmentally friendly, but are they?

Take Action!

1. Take a survey of products at your local stores that claim they are safe for the environment. Research these products to find what effects they really do have on the environment.
2. Brainstorm to think of ways to avoid using products that are harmful to the environment.
3. Several kinds of plastic are recyclable, and these are marked with symbols. Prepare a display to educate people about these symbols. On your display, also list some of the ways that plastics are recycled.

Chapter 12 Review

Looking Back

The goal of manufacturing is to produce and sell products. The part of manufacturing that is responsible for selling the product is called marketing. The two major types of markets are industrial and consumer.

Companies develop marketing plans to determine how well their product will sell and how best to advertise it.

Products may be sold directly to the final user or to wholesalers or retailers. Part of the sales process includes distribution. This means actually getting the product to the buyers. Often this involves storing products in warehouses. Distribution also includes the transportation used to move the products from place to place.

Review Questions

1. What is a market? Give three examples of a market.
2. Define consumers.
3. What is market research?
4. Describe two market research methods.
5. What is the primary goal of advertising? Name at least five different places where you see or hear advertising.
6. What is a target market?
7. What is a commission?
8. What is direct sales?
9. What is the difference between wholesale and retail sales?
10. What is the chain of distribution? List all the different things and groups of people that could make up this chain.

Discussion Questions

1. Why is a marketing plan necessary?
2. How can advertising benefit the consumer?
3. As a consumer, what do you think are some negative aspects of advertising?
4. What kind of traits do you think a salesperson needs?
5. Choose three different target markets. Then discuss the type of products aimed at each market. Is there any difference in the type or style of advertising aimed at each of these markets?

Chapter 12 Review

Cross-Curricular Activities

Language Arts 

1. Adjectives and adverbs play a major role in the marketing of a product. Bring several ads to class and list the adjectives and adverbs used in the promotion.
2. Select a product (candy, soft drinks, cars, etc.) and write a marketing campaign that would appeal to different age groups.

Social Studies 

1. Go back to early colonization and find out why the monarch of England was anxious to establish workable colonies in the "New World." Then explain how this attitude eventually led to English laws such as the Sugar Act and the Tea Act. How did this change the course of history for the colonies?
2. Try to think of at least three new products that failed and were taken off the market. If you have trouble coming up with that many on your own, then ask your parents and/or your grandparents. What consumer needs were not fulfilled by these failed products? Can you think of any improvements that could have been made in these products that would have made them more appealing?

Science 

1. Market research shows that people want to recycle to save the environment. Plastic containers, for example, may be collected and then reused to make other plastic products. The symbol pictured below is printed on recyclable plastic containers. The number inside the triangle varies, representing plastics of different chemical structures. Each kind of plastic can only be mixed with others of its kind in recycling. Collect samples of as many different kinds of recyclable plastics as you can. How are those of one kind alike? How are the different kinds different from one another?

Math X÷

1. Daniel sells used cars for Pinewood Autos. He is paid a straight commission of 5% of the sale price of each car he sells. What is his commission if he sells an auto for $6,400?
2. Frances has taken on a part-time job selling water filtration systems. She receives $90 for each $600 system she sells. What is her rate of commission?

Chapter 12 Technology Activity

Planning an Advertising Campaign

In this chapter, you read that the goal of advertising is to convince consumers they need a product. Suppose you work for an advertising agency. Several large corporations have hired your agency to prepare advertising campaigns for new products that will soon be put on the market. Your boss would like you to choose one of those products and plan the campaign for it.

Goal

To plan an advertising campaign that will convince consumers to buy the new product.

Equipment and Materials

drawing paper
pencils and colored markers
typewriter or word processor

Procedure

1. Select the product for which you will plan an advertising campaign. The products' names and descriptions are shown in Fig. A12-1.
2. Determine the target market for the product. Who is most likely to buy it? Does the product appeal to a certain age group or fit with a particular life style?
3. Once you have determined the target market, consider how you will reach that market. For example, you might place ads in magazines, but which magazines are consumers of the product likely to read? What about newspapers, radio, television, and other media? Select the medium (or media) you will use to advertise the product. Give reasons for your selections.
4. Determine the product's major selling point. What benefit could a user of the product expect? Is it good for one's health? Will it improve appearance? Will it save money?
5. Create an advertising slogan for the product.
6. Prepare at least one type of advertisement for the product. For example, if you have decided to run ads in magazines, make a thumbnail sketch of one ad. (See Chapter 6 for a description of thumbnail sketches.) If you want to advertise on television, write a script for a TV commercial. In any case, be sure to include the advertising slogan.
7. Present your campaign plans, including the sample ad, to the class.

Evaluation

1. Did the class understand your advertising message?
2. What did they like most about it?
3. If you were going to run the ad in a real magazine (or on television, etc.), what might you do differently?

Chapter 12 Technology Activity

Healthy Bites
A breakfast cereal made from whole grains. It contains fruit and nuts.

Brite Toothpaste
Designed to whiten teeth, it also contains fluoride and breath fresheners. It comes in both paste and gel form.

Tempus Wristwatch
An expensive new timepiece with a jeweled face and gold band. Comes in men's and women's styles.

Olympic Gold Shoes
A new line of athletic shoes. Well made but priced lower than other popular brands. Comes in different styles for running, walking, and various sports.

Fig. A12-1. Plan an advertising campaign for one of these new products.

Chapter 13

Computer-Integrated Manufacturing

Looking Ahead

In this chapter, you will discover:

- what computer-integrated manufacturing means.
- three main ways in which computers are used in CIM.
- the differences between an automat*ed* factory and an automat*ic* factory.

New Terms

automated factory
automatic factory
computer-assisted production planning (CAPP)
computer-integrated manufacturing (CIM)
coordinate measuring machine (CMM)
flexible machining center (FMC)
just-in-time (JIT)
laser curtain
manufacturing resource planning (MRP II)
programmable controller
robotics
statistical quality control (SQC)

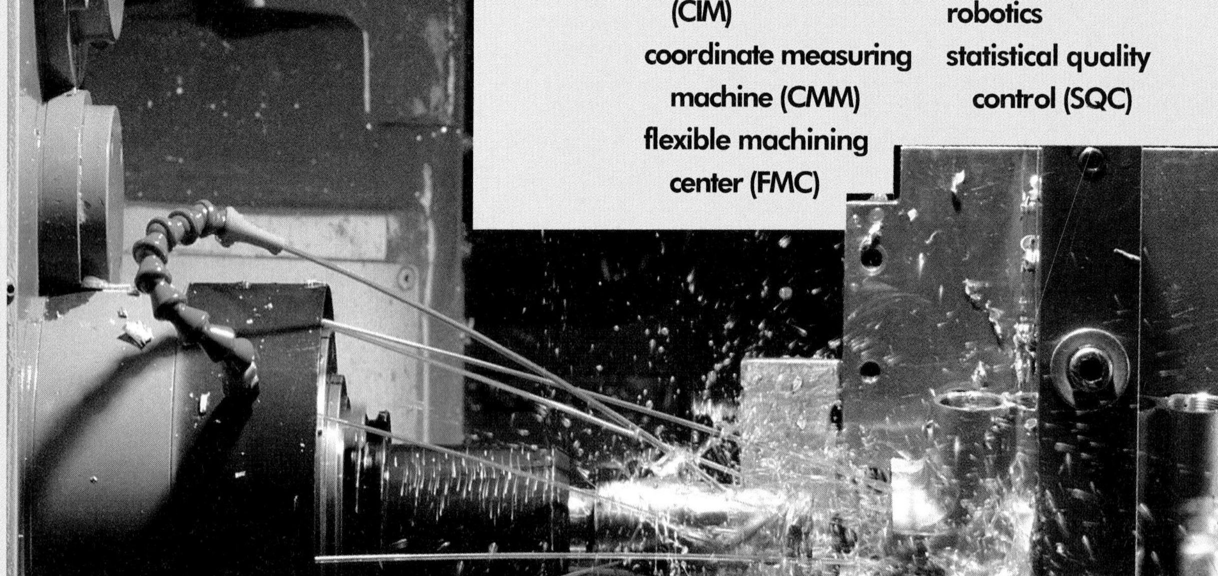

Stereo Lithography

Ideas are often difficult to visualize. Computers can help us see ideas in graphs, tables, and drawings. In manufacturing, computer-assisted design (CAD) gives us the ability to develop and change designs right on our desktops. Terry Kreplin has taken this one step further.

Terry Kreplin works in the Advanced Engineering Design Center for Baxter Health Care Corporation. Like other engineers, Kreplin has developed great skill in "seeing" his ideas before they are constructed. Kreplin says, "However, I have to be able to communicate the design of a product to other people." Sometimes a drawing is not enough. A *prototype* (model) must be made.

Kreplin is manager of Baxter's "rapid prototyping"organization. His group uses a technology called stereo lithography to make full-size, three-dimensional plastic prototypes of his CAD drawings. Kreplin's machine produces intricate models of small valves and other medical apparatuses. Prototypes are built in just a few hours using stereo lithography.

Kreplin uses a 3D Systems Stereo Lithography Apparatus (SLA), which produces plastic models directly from CAD drawings. The technique is called photopolymerization. ("Photo" means light; a polymer is a plastic with many molecules in a chain.) It is photography and chemistry combined. Some liquid polymers, when exposed to light, become solid. The SLA system uses an ultraviolet laser beam to "draw" the first layer of the object in liquid polymer. The ultraviolet light causes that polymer layer to harden. As soon as it hardens, the laser beam draws another thin layer. As layers rapidly build up, complex objects with precise details take form.

Kreplin recently designed a full-sized helmet. A full-size model of the shell, bill, and all the internal supporting mechanisms was constructed in one operation right from his CAD drawing. It took nearly twenty-four hours for the SLA to "draw" and build the product in polymer. "That's a long time," says Kreplin, "until you realize that if someone were to cut it out of solid plastic or build a mold, it would take weeks to accomplish the same thing." Stereo lithography quickly turns ideas into reality.

Fig. TF-13. CAD drawings such as these are used at Baxter Health Care Corporation to make prototypes using stereo lithography.

What Is Computer-Integrated Manufacturing?

In **computer-integrated manufacturing (CIM)**, computers are used to help tie all the phases of manufacturing together to make a unified whole. Computer-integrated manufacturing helps make all areas of manufacturing as efficient as possible. In CIM, computers are an essential part of the following areas:

- Planning
- Production
- Control

Fig. 13-1. Products can be designed and engineered faster and more accurately using CAD/CAE techniques.

Computerized Planning

Computers can be used for planning products and production. They help engineers and planners do their work faster and more accurately.

Product Planning

In Chapter 9, you learned about computer-assisted design (CAD) and computer-assisted engineering (CAE). Figure 13-1 shows a computer being used for product planning.

Production Planning

Process planners use computer software designed for **computer-assisted production planning (CAPP)**. Using it, they can quickly determine the best processes for production flow and manufacturing times.

Part programming for numerical control and computer numerical control (NC and CNC) can be done much faster using a computer. Many decisions can be made by the computer. Less human input is required.

In Chapter 11, you learned about material requirements planning (MRP). These programs help plan material usage. Another type of program, called **manufacturing resource planning (MRP II)**, involves planning not only for material requirements, but also for people, time, and money requirements.

As you read in Chapter 9, *computer simulations* are used to analyze product designs. They are also helpful in production planning. (Remember, to simulate something means to imitate.) In a computer simulation, estimates of production factors are entered into the computer. Then the computer simulates what will happen. Fig. 13-2. Problems or "bugs" can be discovered and corrected before actual production begins. This saves time and money later on.

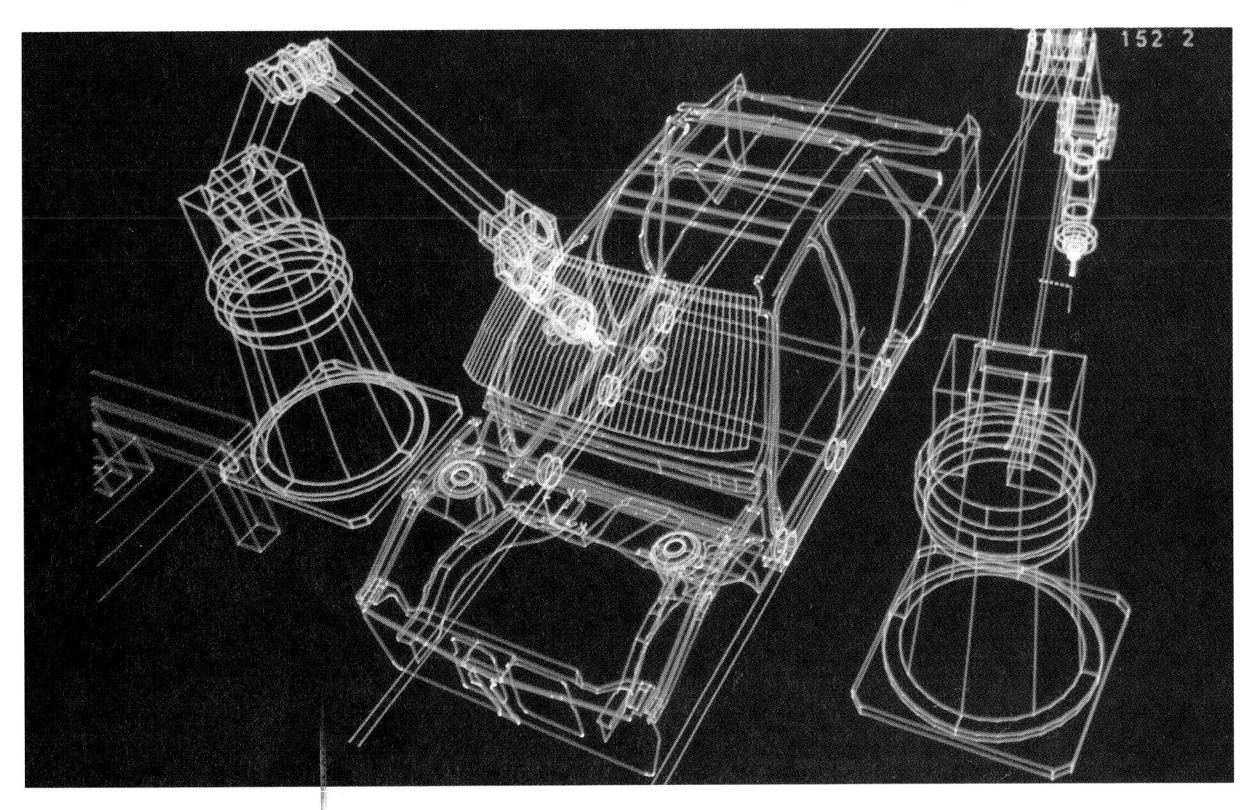

Fig. 13-2. Computer simulations help planners to anticipate the amount of time required to complete each process. Shown here is a simulation of the process of robots placing and gluing a windshield on a car.

Computerized Production

Computers are very helpful in production, too. Parts and products can be made more efficiently using computerized production methods.

Flexible Machining Centers

A **flexible machining center (FMC)** is a combination machine tool. It's capable of drilling, turning, milling, and doing other processing. Fig. 13-3. A computer controls the various parts of the machine.

One place a flexible machining center might be used is in an engine factory. The FMC could be used to work on different kinds of engines. It might work first on an eight-cylinder engine. Next, it might work

Fascinating Facts

Computer systems, designed to streamline production and increase productivity, have revolutionized manufacturing.

Bakers in a cookie factory can use a computer to oversee the mixing of ingredients and the production of the cookies. Robots can paint cars in an automated automobile assembly and finishing plant.

A conventional factory that manufactures cellular telephones may need 300 to 400 workers to produce 1,000 units a week. However, in an automated plant, 70 workers can complete 3,700 units a week. In a sheet metal plant, a computerized system can cut the time required to make parts from 40 days to 4 days.

Fig. 13-3. A flexible machining center is capable of performing several processing operations.

on a four-cylinder engine, followed by a six-cylinder engine. Each engine is different and would require a different tool setup, but an FMC can handle the differences easily.

Automated Assembly

Assembly work is often automated. For example, automated assembly is common in the electronics industry. Small components, such as resistors or diodes, are packaged on a role of tape and loaded into an automatic insertion machine. As circuit boards travel past the insertion machines, the parts are rapidly and accurately inserted into the right places. Fig. 13-4.

Automated Materials Handling

Computer-controlled handling equipment can make sure that the right material is in the right place at the right time. Remember the automatic guided vehicles and automated storage and retrieval systems (AGVs and AS/RS) discussed in Chapter 10. Those are examples of automated materials-handling equipment.

Robotics

Robotics means using robots to perform tasks. As you learned in Chapter 11, robots are special machines that are programmed to

Fig. 13-4. An automatic assembly machine can assemble small parts much faster than a human worker.

Fascinating Facts

A milestone in factory automation occurred in 1983 in Japan with the opening of a plant especially designed to use robots to manufacture robots. In three days, the 68 industrial robots can do the work that 220 workers would take a month to complete.

automatically do tasks that people usually do, such as moving objects from one place to another, assembling parts, welding, or spray painting. There are several types of robots, but they all have common features. They all have a "hand," usually called an *end effector*. The end effector may be a gripper for holding things or it may be a built-in tool like a drill or a welder. Robots also have joints just as a human has joints. Robots typically have a wrist, an elbow, a shoulder, and a waist. These joints allow a robot to stretch, turn, raise, and lower within a limited work area. This work area is called its "working envelope." Fig. 13-5. A robot can be programmed to manipulate its end effector anywhere inside the envelope.

One main advantage of a robot is that it can be programmed to repeat a very complicated task over and over. It doesn't need a coffee break or rest period. Also, it's very accurate. Robots are very good for doing work that is hazardous to humans, such as in paint shops filled with fumes, in high or low temperature conditions, or where very heavy loads must be lifted repeatedly. Fig. 13-6.

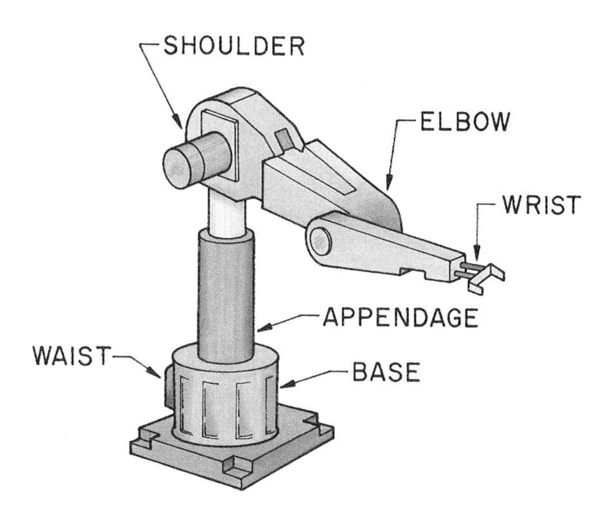

Fig. 13-5. Robots have movable joints, much like a human. This allows them to reach any point inside their "working envelope."

Fig. 13-6. Robots are used to quickly perform manufacturing processes such as welding. They can do many tasks that would be dangerous for humans to perform.

Quality Control Devices

One kind of quality control device is a **coordinate measuring machine (CMM)**. Fig. 13-7. A CMM, as it is usually called, is a very accurate computer-controlled measuring device. The main advantage of a CMM is its high degree of accuracy and its consistency. A CMM is usually used to measure "hard-to-measure" parts, like rounded or spherical parts. It can be programmed to measure a part and compare the measurements with the specifications for the part's dimensions, which have been stored in its memory.

Another quality-control machine is a **laser curtain**. This machine uses a moving laser beam to record the measurements of a part. The measurements are automatically entered into the computer for comparison.

Fig. 13-7. Parts can be measured very accurately using this coordinate measuring machine (CMM). This CMM is inspecting 70 different features of an engine block.

A computer is used to perform **statistical quality control (SQC)**. It uses a sampling system to determine how well parts are being made. The computer gathers information from a small number of parts. Then, by using statistics, it can predict the percentage of parts not meeting specifications.

Computerized Control

Another very important part of CIM is *control*. As you've just read, computerized production uses computers to control the various machines that handle materials, make products, and inspect products. Computers also monitor and control the flow of work. For example, if one work station develops a production problem, the other work stations adjust their schedules.

Many methods and devices are used for control. These include just-in-time manufacturing, automatic identification, voice recognition, and programmable controllers.

JIT Manufacturing

In CIM, the supply of materials is controlled by computer. Computer control helps make **just-in-time (JIT)** manufacturing practical. In JIT manufacturing, materials are delivered as they are needed. This reduces the need for warehouse space. It also means fewer workers are needed to organize and keep track of the materials. However, careful planning and control are needed to assure that the right materials, in the right amount, will arrive at the right time. Computers are therefore used to keep track of inventory, to order materials, and to schedule deliveries.

Automatic Identification

Controlling requires current information. One way that current information is entered into a computer is by a process called *automatic identification*. A special tag is attached to a part or product. The tag contains an identification code. A machine reader can read the code to identify the part. The best example of "auto ID" is *bar coding*. You've seen bar codes on products you buy in the store. The code is a series of black and white lines. The computer can read the code and thus identify the product. Fig. 13-8. In manufacturing, the information in a bar code can be used to keep track of inventory, to direct a part to the right work station, and to otherwise control what happens to it.

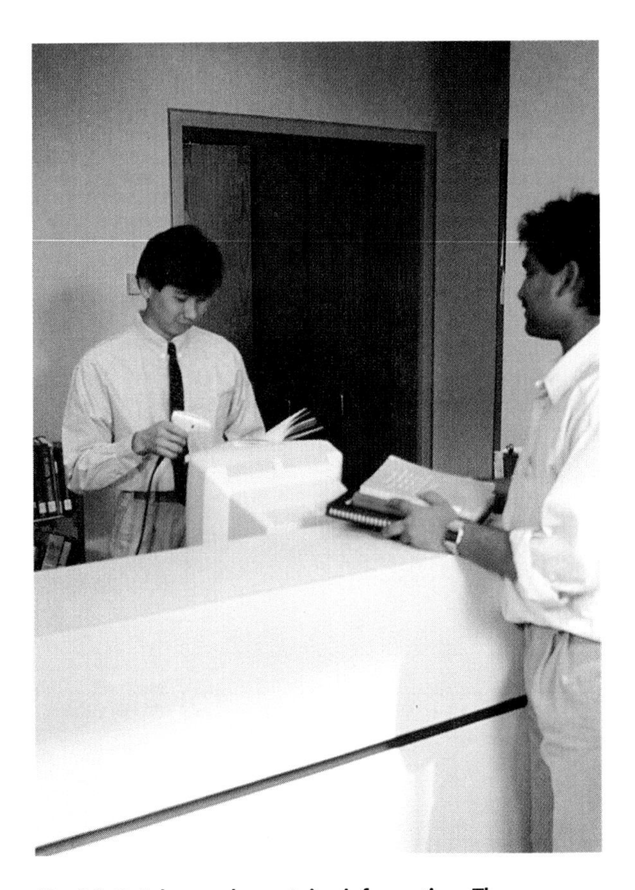

Fig. 13-8. A bar code contains information. The information can be entered into the computer by scanning the code with a special reader device.

Voice Recognition

Another method of control is voice recognition. A worker speaks commands into a microphone. The computer recognizes the voice and the words, and it carries out the instructions. That way, the worker's hands are free to do other work. Voice recognition is not yet widely used, but the time may come when we can easily communicate with computers just by talking to them.

Programmable Controllers

A **programmable controller** is a small self-contained computer used to run machines and equipment. It's housed in a heavy-duty case. The case protects it from the "wear and tear" of the factory environment. Fig. 13-9. The fact that it is programmable means that workers can change the way it functions. This makes it more useful than controllers that are built to do only one thing.

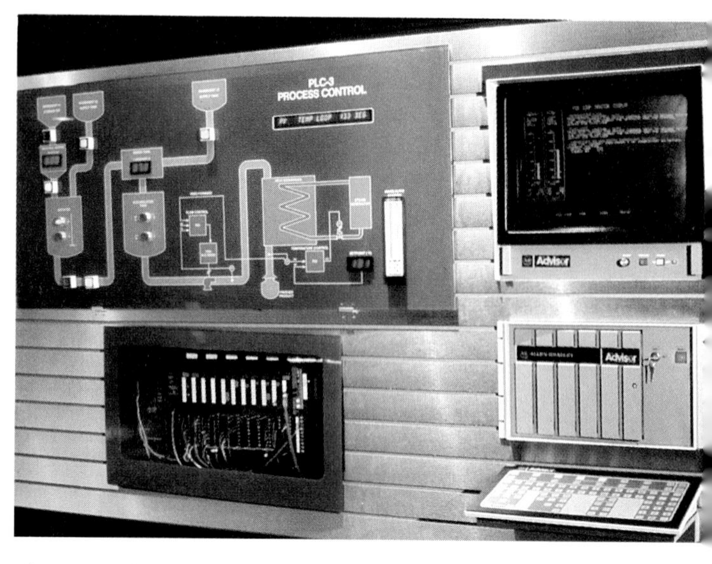

Fig. 13-9. This programmable controller is running a process control.

Computer Networks

As you can see, there are many computer-controlled machines and devices being used in manufacturing. Some machines have their own computer built in. Others require a direct link to a larger computer. A company's main computer may be in one central location. Each of its factories or plants would have its own computer. In addition, there would be many small computers located throughout the various departments of each plant. Fig. 13-10. For CIM to work, all the computers must be connected so that

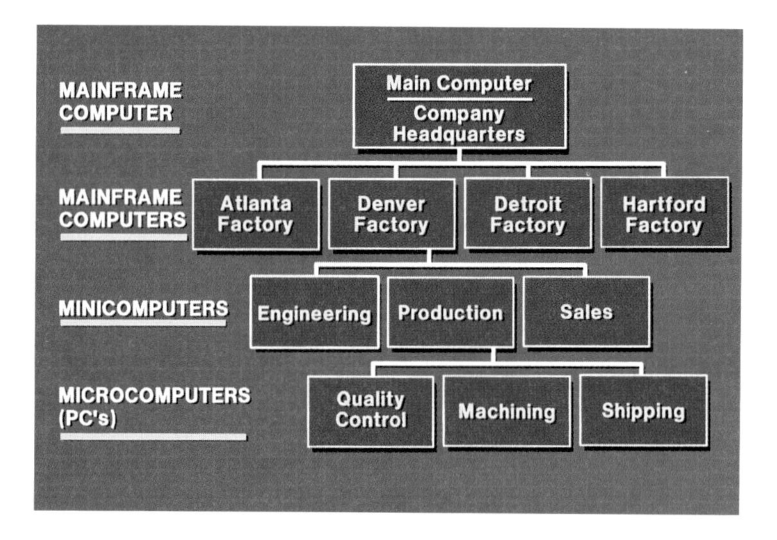

Fig. 13-10. This diagram shows how various types of computers are linked together into one large system.

they can all communicate ("talk" to each other). This is done using computer networks.

A *network* is a way of hooking together several computers. The computers are linked by wires or fiber optic cables. Instructions and information are transmitted from one device to another over the network. Inside a plant there is a *local area network (LAN)*, while between plant locations there is a *wide area network (WAN)*. Because of networking, all of a company's computers (and the machines they control) can be linked together for greater efficiency.

Apply Your Thinking Skills

1. Your text discusses the advantages of JIT. What might be some possible disadvantages in using JIT?
2. What might be some problems in using voice recognition as a control method?

Computer-Integrated Manufacturing in Action

The combination of computerized planning, production, and control (CIM) is an efficient way to manufacture parts and products. However, buying all the hardware and software is expensive.

There are two stages of development in CIM: automated factory and automatic factory.

The Automated Factory

An **automated factory** is one in which many of the processes are self-operating. That is, the processes are directed and controlled by computers. The automated factory includes:

- Manufacturing cells
- Islands of automation
- Flexible manufacturing systems

Manufacturing cells are groups of machines working together as directed by a computer. Fig. 13-11.

Fig. 13-11. In a manufacturing cell, several machines are grouped and operated together. A computer controls the operation.

Manufacturing cells that aren't connected in any way are called *islands of automation*. They're like islands in an ocean. Each is independent of the others.

Flexible manufacturing systems (FMS) are groups of manufacturing cells and flexible machining centers. The cells and FMCs are tied together by an automated materials-handling system and by computer control. Fig. 13-12.

The Automatic Factory

Imagine a factory in which there are no people working. All the parts are made by automatic machines. All materials are moved by automatic materials-handling equipment. All assembly work is done by automatic assembly machines. The quality control checking is also all done automatically. The factory may be dark inside. Most machines do not need lights to operate. A factory in which everything is done by machines—automatically—is an **automatic factory**.

In an automatic factory, all the various systems work together. This integration of the various systems into one giant system makes the factory "automatic." All the information collected from all the subsystems is kept in a main computer. The main computer directs the other computers.

Since all records are kept in computer files there is no need for file cabinets, typewriters, and similar types of equipment. A copy of a report or drawing can easily be printed at any time.

Will people be needed at all in such a factory? Yes; they will be needed to control, service, and repair the computers and other equipment. However, far fewer workers will be needed in this factory than in today's factories.

Apply Your Thinking Skills

1. What effects do you think automatic factories will have on the economy and the work force in the future?

Fig. 13-12. A flexible manufacturing system is a computer-controlled system that can produce a variety of parts or products in any order, without the time-consuming task of changing machine setups.

Global Perspective

A Global Challenge

On construction sites around the world, giant pieces of equipment painted a distinctive yellow rumble and roar—lifting, digging, hauling—performing the heaviest of tasks. The manufacturer of this heavy equipment is Caterpillar® Inc.

Caterpillar® is a large, multinational company headquartered in the United States. It operates factories in eleven different countries outside the U.S. and works with independent manufacturers in about a dozen others. Caterpillar® products are sold around the world.

In years past, Caterpillar® had its markets practically to itself. However, in the 1980s, world economic conditions worsened and foreign competition increased. Caterpillar® was forced to find ways in which it could remain competitive.

To reduce costs, company leaders began streamlining production operations, refining layouts and material flow patterns, reducing inventories by producing products on order, and downsizing production plants. Obviously, CIM plays a very important role in their push to remain competitive in the world market. Caterpillar® leaders also looked for innovative ways in which they could work with countries affected by poor economic conditions. In many countries, goods were available but cash was lacking. For this reason, the Caterpillar World Trading Corporation (CWTC) was established in 1984. Through this company, Caterpillar® is able to arrange for the trading or selling of goods from one country in another country in order to obtain money to pay for Caterpillar® products. This process is called *countertrading*.

Global competition presents a global challenge. The use of efficient production techniques and innovative marketing ideas are ways in which Caterpillar® and other companies are meeting that challenge.

Extend Your Knowledge

1. Manufacturers today are becoming increasingly aware of the advantages in simplifying product designs. Compare the design of two similar products—one old and one fairly recent—and identify the changes that have been made to make the product less complicated.

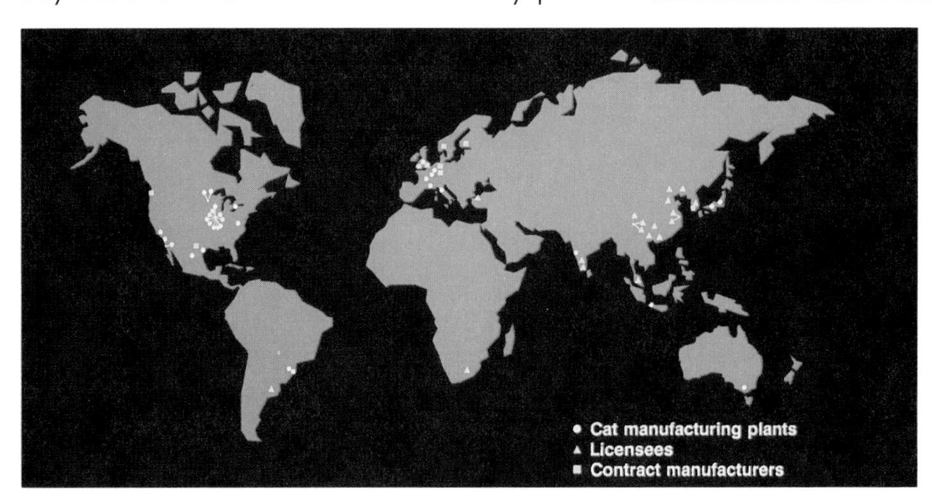

Fig. GP-13. This map shows Caterpillar's® manufacturing plants, licenses, and contract manufacturers around the world.

Chapter 13 Review

Looking Back

In computer-integrated manufacturing, computers are used to tie all the phases of manufacturing together so products can be manufactured as efficiently as possible.

Computers can help in planning products and production. In computerized production, parts and products can be made more efficiently. Computers are also used for controlling the manufacturing processes.

CIM has two stages of development. These are the automated factory and the automatic factory. The main difference between the two is the amount of automatic equipment and the way it is controlled.

Review Questions

1. In what ways are computers used in product planning?
2. Describe three ways computers can be used for production planning.
3. Describe a flexible machining center.
4. Define robots. List at least three types of jobs robots might do in manufacturing.
5. What is the robot's end effector? What is its working envelope?
6. Discuss three advantages of robots.
7. Describe just-in-time manufacturing.
8. What are manufacturing cells?
9. Describe flexible manufacturing systems.
10. Discuss the differences between automated and automatic factories.

Discussion Questions

1. Describe computer networks and discuss their importance in CIM.
2. Many factories today operate around the clock all year round. Automation allows many more parts to be made in a much shorter period of time. How do you think the increasing use of automation will affect factories' operational times in the future? Will they still operate year round? Will they still continually produce the same products?
3. Discuss the advantages and disadvantages of using computerized quality control devices.
4. Discuss Caterpillar's® countertrading. What might be some risks involved in countertrading? What effects might it have on the economy of the countries involved in the countertrading, especially that of the country receiving the Caterpillar® product?
5. Discuss the advantages and disadvantages of automatic factories.

Chapter 13 Review

Cross-Curricular Activities

Language Arts 

1. Computer-integrated manufacturing (CIM) has two stages of development. In an essay describe the differences between the automated factory and the automatic factory.
2. An "acronym" is a word that is made from the first letters in a series of words or from groups of letters from each word in a series. CIM is an acronym used for computer-integrated manufacturing. Many acronyms are used in manufacturing. How many can you name?

Social Studies 

1. Do some research to find out what products that are manufactured in the United States are most frequently exported (sold to other countries). Also find out to which countries each of these products is sold.

Science 

1. Robots are an important part of many computer-integrated manufacturing systems, especially in jobs that would be dangerous or unpleasant for workers. For the same reasons, robots are useful in scientific applications. List at least five environments where robots could be used to advantage in place of humans for scientific work.
2. Computer simulations are used to plan the processes of manufacturing. They are also being used by scientists. One subject being explored through computer simulations is earth's future climate. Computer simulations of climate involve gathering data on many different weather factors and examining how they might interact to change climate. What factors can you think of that might be central to such a computer simulation of climate?

Math X÷

On the graph below, points A, B, and C represent points where holes are to be drilled. Each position is associated with an address represented by its position relative to the X and Y axes (X,Y). For example, the address of point A is (4,2) because it is vertically aligned with 4 on the X axis and horizontally aligned with the 2 on the Y axis. The address of point C is (6,8).

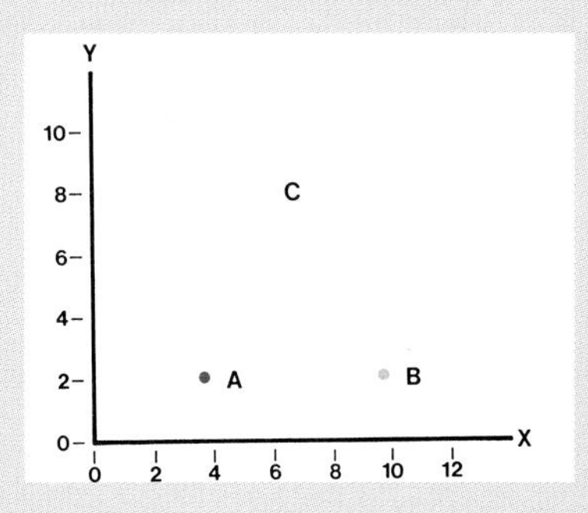

1. What is the address of point B?
2. What would be the address of a point halfway between points A and B?
3. What is the distance from point C to the origin (0,0)?

Chapter 13 Technology Activity

Bar Coding

Overview

Bar coding is a common form of automatic identification. It is a system in which wide and narrow lines and spaces are used to represent alphabetical and numerical characters. A laser is used to "read" the patterns of light and dark. It converts this data into digital information that is usable by a computer.

One type of bar code is Code 39. This system includes "quiet areas" at the beginning and ending of a code as well as start and stop characters. These help ensure that the reader functions properly and that complete codes are read accurately. Fig. A13-1.

In this activity you will use Code 39 to create a bar code for your initials. The code for each character is given in Fig. A13-2.

Goal

To use Code 39 to create a bar code for your initials.

Equipment and Materials

pencil and eraser

Procedure

1. Make a grid similar to the one shown in Fig. A13-4. (Do not use the one that's in this book!) Remember, however, that no grids are visible in actual bar codes. Use the grid lines as guides but do *not* count these lines as part of the code. One space on the grid is a thin line or space. Two spaces on the grid make up a thick line or space. Study the example given in Fig. A13-3.
2. Print your initials on the lines below the grid.

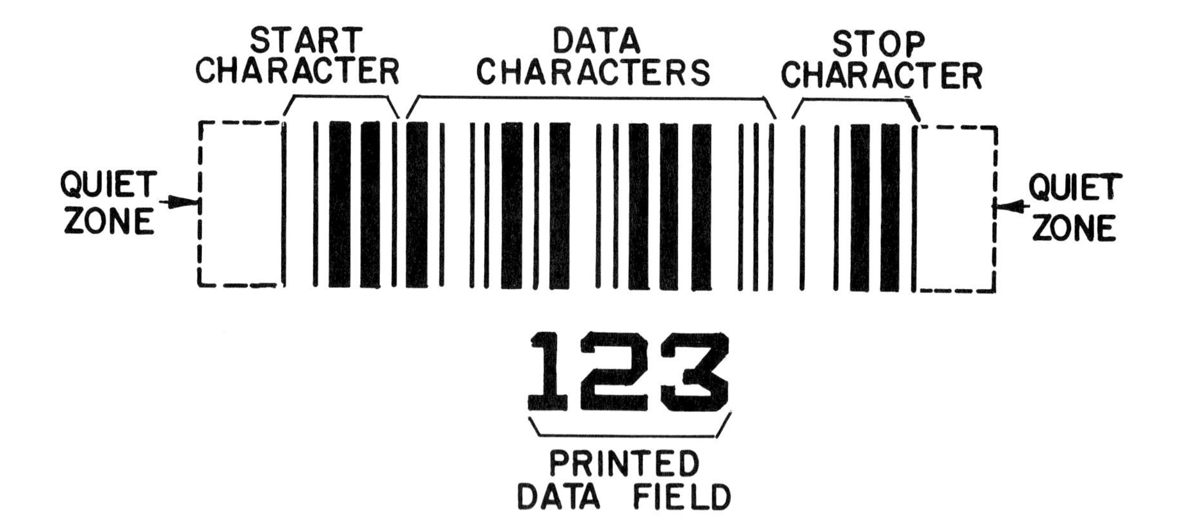

Fig. A13-1. Basic parts of a typical Code 39 bar code.

Chapter 13 Technology Activity

CHAR	PATTERN	BARS	SPACES	CHAR	PATTERN	BARS	SPACES
1		10001	0100	M		11000	0001
2		01001	0100	N		00101	0001
3		11000	0100	O		10100	0001
4		00101	0100	P		01100	0001
5		10100	0100	Q		00011	0001
6		01100	0100	R		10010	0001
7		00011	0100	S		01010	0001
8		10010	0100	T		00110	0001
9		01010	0100	U		10001	1000
0		00110	0100	V		01001	1000
A		10001	0010	W		11000	1000
B		01001	0010	X		00101	1000
C		11000	0010	Y		10100	1000
D		00101	0010	Z		01100	1000
E		10100	0010	–		00011	1000
F		01100	0010	•		10010	1000
G		00011	0010	SPACE		01010	1000
H		10010	0010	START/STOP		00110	1000
I		01010	0010	$		00000	1110
J		00110	0010	/		00000	1101
K		10001	0001	+		00000	1011
L		01001	0001	%		00000	0111

Fig. A13-2. Character coding system.

Chapter 13 Technology Activity

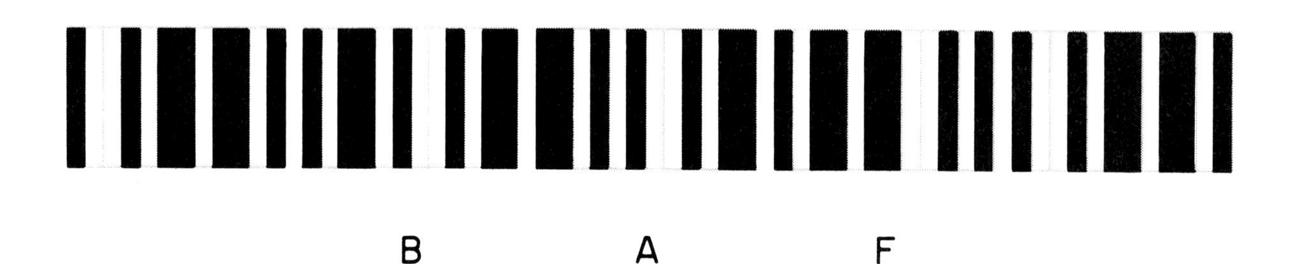

Fig. A13-3. Example bar code.

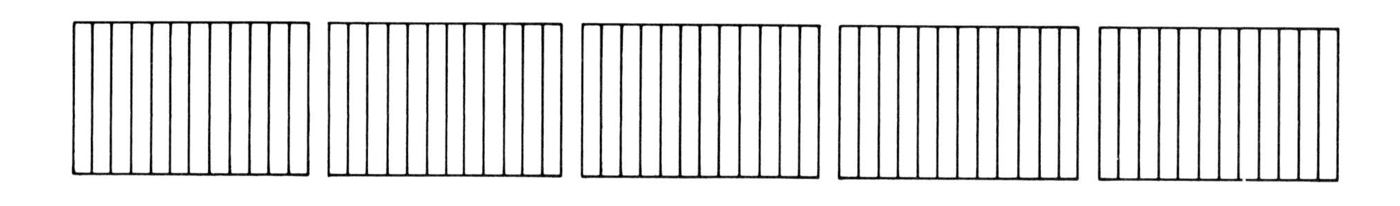

Fig. A13-4. Make a blank grid like this one so that you can do a bar code of your initials.

3. Fill in the start character in the first section of your grid as shown in Fig. A13-3.
4. Fill in the stop character in the last section of your grid.
5. Locate the letter of your first initial in the chart in Fig. A13-2. Study the pattern of the bar code used to represent this letter. Fill in the code for your first initial on the second section of your grid.
6. Follow the same procedure to fill in the code for your middle initial on the third section of your grid.
7. Finally, fill in the code for your last initial on the fourth section of your grid.
8. See if you can decode the mystery message given in Fig. A13-5. Write the letters of the message on a separate piece of paper.

Evaluation

1. How long did it take you to create a bar code for your initials?
2. What is the mystery message given in Fig. A13-5? How long did it take you to decode this message?

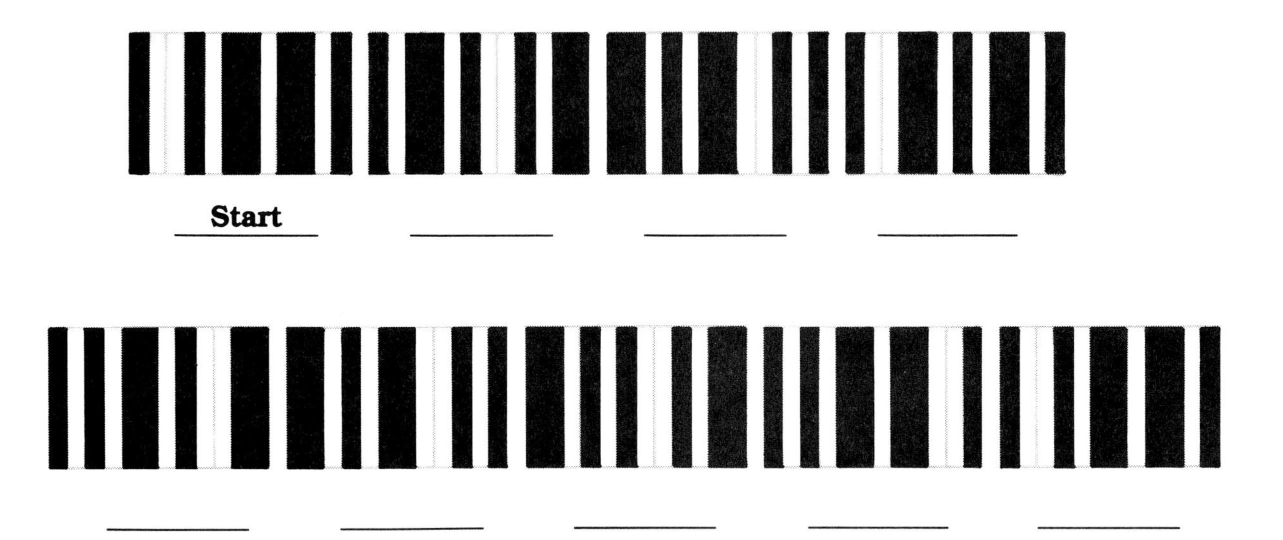

Fig. A13-5. Can you decode the mystery message?

Chapter 14

Trends in Manufacturing Technology

Looking Ahead

In this chapter, you will discover:

- future ways products will be manufactured, including manufacturing in space.
- changes taking place in the need for new and different products.
- how the marketplace for manufactured products is changing.
- how manufacturers will compete for their share of the world marketplace.

New Terms

alloy
biomaterials
microgravity
productivity
profit
quality circles
superconductor

Technology Focus

Spencer Plastics

Industrial processes are changing rapidly. Workers with new skills are needed to produce more modern products. However, in many occupations, there aren't enough skilled workers to satisfy the needs of industry. A shortage of skilled workers has caused businesses to find ways to retrain their seasoned employees so they can perform new jobs. Needless to say, retraining workers is more considerate than laying off older employees and hiring others.

Spencer Plastics Products Company, located in the tiny midwestern town of Dale, Indiana, is a relatively small manufacturer of custom-made plastic products. Although small, it is typical of industries that don't want to lose their valued employees. "There is a retraining process around here all the time," says Kenneth Sparrow, personnel manager for Spencer. "We are a custom plastic parts manufacturer—our product changes all the time." Spencer Plastics must constantly retrain its workers to perform new tasks in order to produce new products.

Sparrow says, "A lot of our success is due to our 'participative management system'." The workers supervise, train, and retrain each other. Teams of about ten people who work together on the job meet on a regular schedule. Everyone is on a team. Even the company president sits in on different meetings. Some teams consist of people who work on the same machines together. Other teams are made up of people from different areas who meet to work on special projects. In the meetings, the workers discuss problems on the job, attendance, and safety. Most importantly, training is always part of every meeting.

For each job in the plant—even the very newest jobs—there is a training manual. At Spencer, the employees write their own training manuals. Sparrow says, "Who else knows the job better?" By helping manage the plant and by training each other, Spencer Plastics employees constantly learn new skills. They know they are part of the company and that its success depends on their efforts.

The participatory management system "works very well," concludes Sparrow. That is how Spencer Plastics has become a successful small manufacturer.

Fig. TF-14. At Spencer Plastics, team meetings such as this one have helped the company become a successful small manufacturer.

The Major Trends in Manufacturing

Like all technologies, manufacturing is always changing. To help you understand the coming changes in manufacturing, we will look at three major trends:

- The way products are manufactured. The future will see an increased use of computers and robots. Some manufacturing will begin to take place in space.
- The products that are needed. The future will see an increased need for electronic products, for biotech and other health-related products, for improved transportation, for new energy sources, and for preserving and protecting the environment. These trends will all affect the products we manufacture.
- The people who use our manufactured products. The future will see the present trend towards a single world economy continue. Improved transportation and communications along with a steady reduction in trade barriers combine to provide manufacturers with a worldwide market for their products.

As you look at manufacturing trends, it is important to consider the three areas just presented. Fig. 14-1. The products that people want and need always determine what will be manufactured. The size of the market determines the number of products needed. These two principles plus available technology are considered when choosing the methods that will be used to manufacture the product. The combination of these three elements—product need, market size, and manufacturing methods—

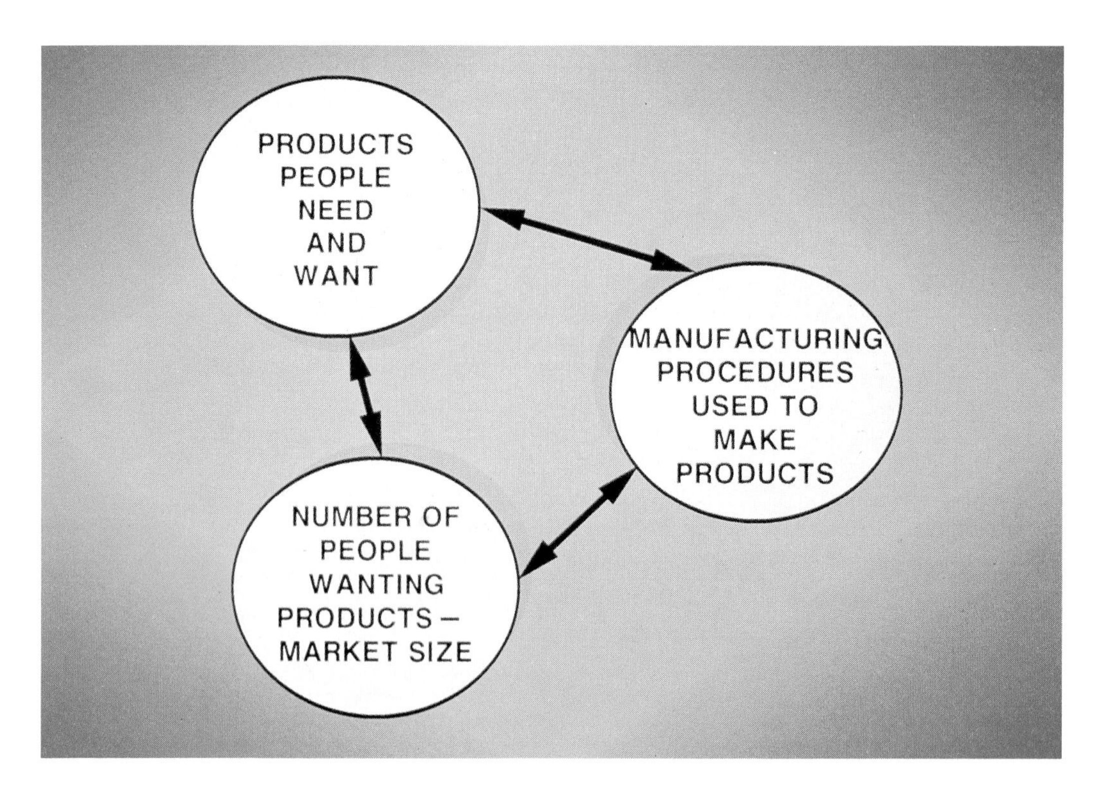

Fig. 14-1. The interaction of the three elements controlling trends in manufacturing.

will help you understand manufacturing trends. Also, when you apply these three principles, you will be able to predict manufacturing trends for the future.

Apply Your Thinking Skills

1. Since manufacturing is always changing due to rapidly changing technologies, how do you think you could best prepare for a possible career in manufacturing?

Trends in Manufacturing Processes

The increasing capabilities of computers, the increase in automation, the growing use of outer space, and the discovery of new materials are all affecting manufacturing processes. The following sections will help you understand changes that are being planned for the future.

Computer Design

The use of computers in manufacturing will continue to grow in the future. Each year, computer chip manufacturers are able to place greater memory capacity on smaller chips. The result is increased power and capability, and smaller size. Computer designers are identifying new ways that these improved computers can serve manufacturing.

Figure 14-2 shows a woman entering "a computer-generated world." A computer, special glasses, and other devices such as gloves or special keyboards are used to create a "virtual reality." Instead of viewing a flat computer screen, the user is surrounded by the image. As the user's head moves, the scene changes to adjust to the new position, just as the view changes when you move your head in real life. Users can interact with this computer-generated world. For example, wearing a special glove, they can move objects in the computer-generated world.

Virtual reality permits a designer to work in three dimensions when designing automobiles, airplanes, or other products.

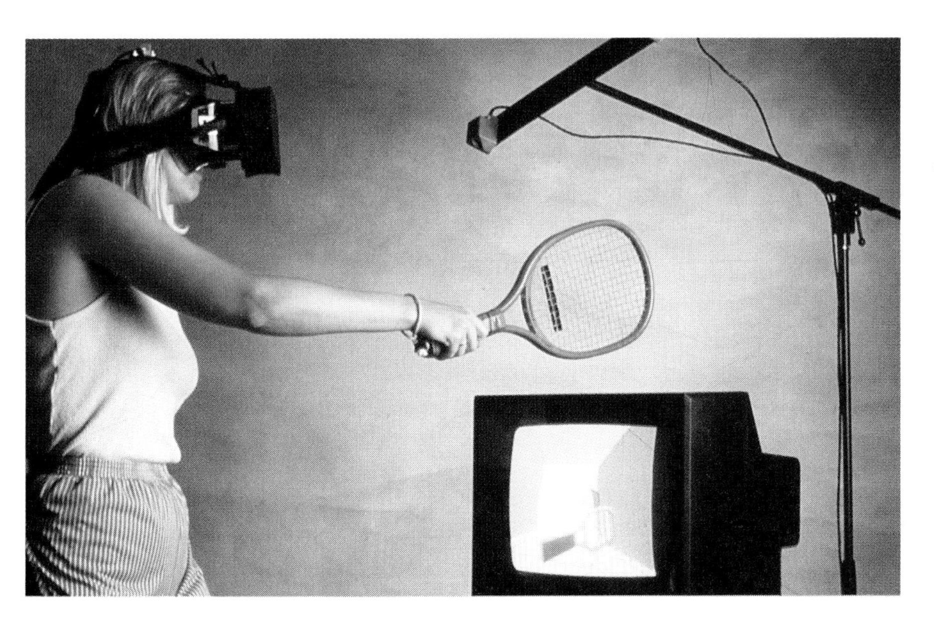

Fig. 14-2. This woman is playing racquetball in virtual reality. On the computer monitor, we see the room that she is seeing through her headgear.

This technology will also have numerous other applications, such as training people for space exploration.

Robots and Automation

In the future, more and more of the manufacturing processes will be done by robots and other automated machinery.

Robots

Early books and movies about robots depicted them as mechanical devices that looked like humans. They would walk and work in the same manner we do. As you learned in Chapter 13, however, the robots commonly used in manufacturing don't look like humans. They look like multi-jointed arms—robotic arms—attached to a base. These devices can pick up and move objects and perform operations such as welding or painting an automobile. They are often used to handle dangerous materials, such as radioactive products, or used in environments that can be hazardous to humans. These include paint shops filled with fumes or areas that have very high or low temperature conditions. The role of the robot will increase in future manufacturing operations. Fig. 14-3.

Automated and Automatic Factories

You read in Chapter 13 about increasingly automated factories and about automatic factories of the near future. There already is one manufacturing plant that is completely automatic, using automated systems and robots. It is a Japanese manufacturer of heavy machinery. Their machinery is used throughout the world. The plant operates in the dark, except when humans are needed for repair or modification of manufacturing procedures.

This type of automated plant will gain greater use in the future. Since the cost of such plants is very high, they can be used only to manufacture products that will require few design and tooling changes. The products will need to be in demand for many years.

Fig. 14-3. Robots are being used more and more in place of human workers to perform hazardous manufacturing tasks.

While automatic plants will probably increase in number very slowly, the use of automated manufacturing will increase. More machines that produce products or parts without human assistance will be used. Most of these machines can be adjusted and changed as product needs change. Their use results in high quality control and low production cost per item. Fig. 14-4.

Manufacturing in Space

Manufacturing in space will emphasize materials processing. Fig. 14-5. As you learned in Chapter 8, materials processing involves changing the size or shape of a material to increase its usability and value. The processed material may be the finished product itself or may be used in making a finished product.

Fig. 14-4. This machining cell operates without worker supervision or assistance. Machines of this type will increase in use. This will help decrease product cost, and often help improve product quality.

Manufacturing in space will involve the use of both manned and unmanned space flights in which materials will be processed automatically.

In space, objects are almost weightless because the forces of gravity acting upon them are very small. This condition is called **microgravity**. Microgravity permits droplets of metal to be formed into near perfect spheres, making superior ball bearings. This small amount of gravity also makes it possible to make mixtures from materials of different densities, which is not possible on earth. On earth, if you tried to make a mixture from materials with different densities, the materials would separate—just like when you try to mix oil and water. In space, new metal alloys can be made because heavier materials do not settle to the bottom. (An **alloy** is made by combining two or more metals or a metal and a nonmetal to form a new metal with improved properties.)

There are many other materials and products that can be produced better in space than they can on earth. These include crystals for electronic applications, ceramics, glass, and ingredients for new medicines or actual new medicines. Additional applications of materials processing in space are being developed as many nations of the world look to the commercial uses of space. Fig. 14-6.

New Materials and Processes

Scientists continue to study materials and develop new alloys, composites, and ceramics capable of providing specific properties not available in conventional

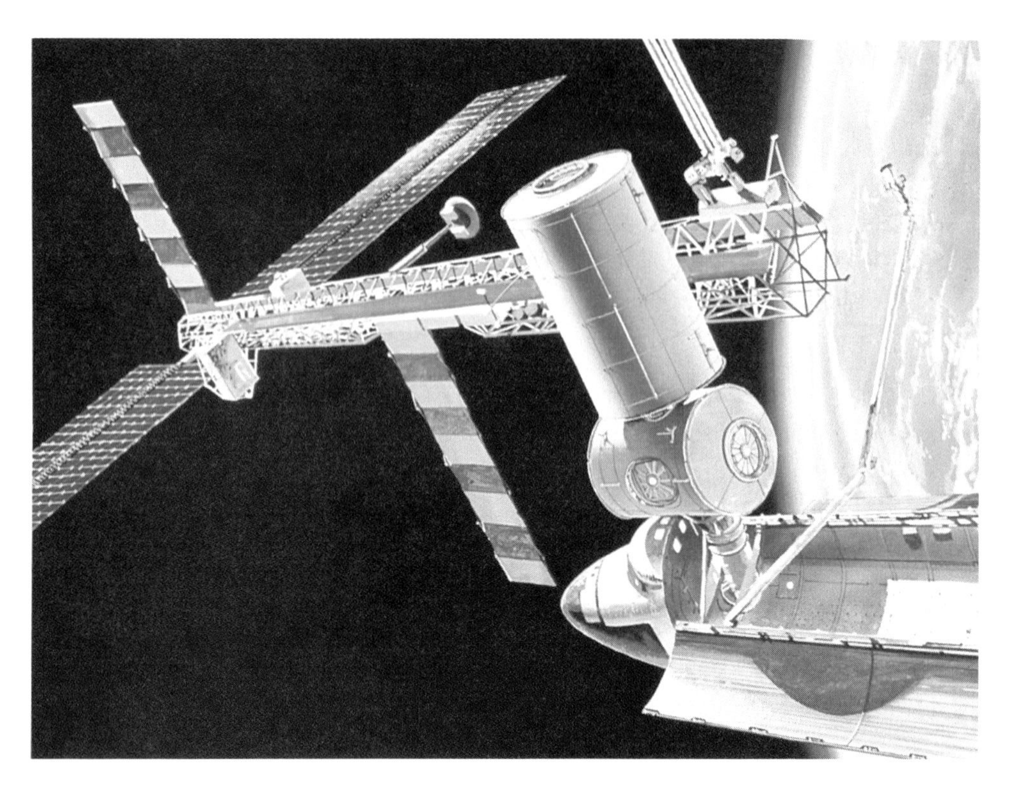

Fig. 14-5. Shown here is an artist's conception of a future space station. This space station will provide new materials and products for us to use here on earth.

materials. A *composite* is a new material made by combining two or more materials. A composite is usually much stronger and more durable than the materials from which it was made. *Ceramics* are materials made from nonmetallic minerals that have been heated to exceptionally high temperatures. Ceramic materials are strong, hard, and resistant to corrosion and heat. Because of their great heat resistance, ceramic tiles are used to cover the exterior of NASA's space shuttles. Ceramic materials are also used to make cookware that can go directly from the freezer to the oven.

Fig. 14-6. Crystals grown in space are used in electronics and other precision applications. The crystals grow free of defects since there is very little gravity acting upon them.

Research continues in the development of superconductor materials. A **superconductor** is a material that will carry electrical current with virtually no loss of energy. Using superconductors, small generators can be made to be very powerful and efficient. Fig. 14-7. Superconductors can also be used to develop powerful electromagnets. Electricity can be transmitted great distances with minimal energy loss when superconductors are used.

These new materials, with their special properties, will permit the manufacture of many new and different products. However, they will also create manufacturing problems. Some of the new materials are very hard and brittle. New methods to form and shape these materials, both in space and on earth, will need to be developed. New tools, improved quality control, and new machine designs will be needed to meet the continual development and use of new materials.

Biotechnology

As you learned in Chapter 1, *biotechnology* is all the technology connected with plant, animal, and human life. Often, it involves combining knowledge about biology and engineering to solve health problems. For example, knowledge about the joints, muscles, and nerve endings in our bodies can be combined with engineering to develop an artificial hand. Fig. 14-8. Biotechnology is also responsible for the development of such things as artificial joints and blood vessels. These human-made materials designed to be placed within the human body are called **biomaterials**.

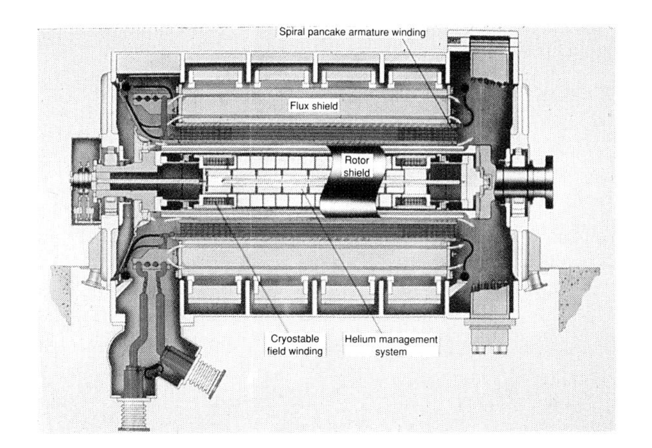

Fig. 14-7. Superconducting generators provide greater power and efficiency than conventional electric generators.

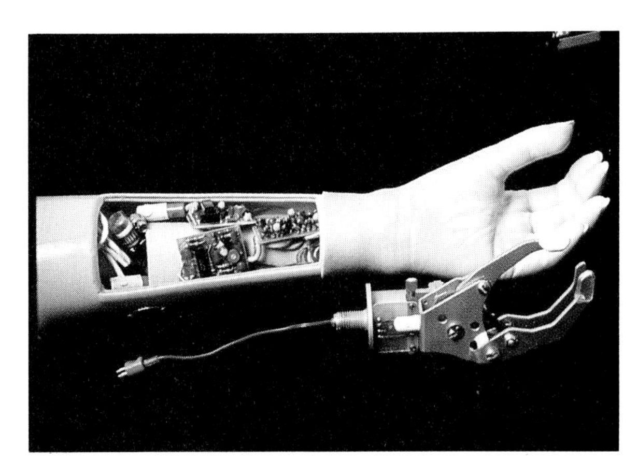

Fig. 14-8. Biotechnology was instrumental in manufacturing this artificial arm and hand. Constant research in biotechnology will lead to the development of many new products.

Another aspect of biotechnology involves using living organisms to develop commercial products, such as vaccines and medicines. Researchers have already been able to manipulate *DNA* (deoxyribonucleic acid), the source of genetic information in living organisms, to create some new medicines and vaccines. Once they are developed and proven safe to use, these medicines and vaccines will be manufactured and sold to the medical community for our use. Research is being done toward developing new medicines to help fight or prevent such diseases as cancer, AIDS, and multiple sclerosis. The future will see many more medicines developed by biotechnology.

Biotechnology is still in its early stages of development. However, the potential is great, and it is emerging as one of the world's most rapidly growing industries. In 1990-91, a year in which much of the world suffered from a recession, biotechnology sales increased by 39 percent. During the same period, the work force in biotechnology grew by 6 percent, while most other industries reduced their work force.

Manufacturing Originals and "One-of-a-Kind"

On one hand, technology will continue to grow and advance. This growth will bring new products and new materials, methods, and procedures to manufacture these products.

On the other hand, individuals will continue to hold onto traditional values and seek the unique and different. While furniture manufacturing is becoming more automated, the most valued pieces of furniture are handcrafted pieces, carefully designed and constructed. The creative designs and products of individual efforts will continue to be of special value to people. Fig. 14-9.

Fig. 14-9. Individual design and "one-of-a-kind" products will continue to grow as a counterbalance to mass production and repetitive designs.

Fascinating Facts

Biotechnology is not new. In fact, it is one of the world's oldest manufacturing procedures. For thousands of years, people have used the process of fermentation to make bread. In this process, the microscopic plant yeast reproduces rapidly because it is provided with the proper environment—food (sugar and flour), warmth, and moisture. As the yeast grows, it gives off carbon dioxide. The carbon dioxide forms bubbles in the bread dough. The bubbles cause the dough to rise. Once the dough has risen enough, it is baked to make bread.

At one time, manufacturers believed that making products one at a time by hand would become a thing of the past. Instead, handcrafting industries have survived and grown. There are many small businesses producing specialized and/or custom-made, handcrafted products. Their number should continue to grow in the future.

Apply Your Thinking Skills

1. What effects do you think the processing of materials in space will have on production costs? How will this affect the final cost of the finished product?
2. If you were buying new kitchen cabinets, would you rather have handcrafted cabinets or mass-produced ones? Explain your answer.

Trends in the Need for Manufactured Products

The demand for products determines what will be manufactured. There are two types of needs that should be considered. The first is the need of the individual. Individuals determine their needs for home and personal products such as appliances, clothes, and recreational devices. They also determine their needs for health care.

The needs of groups of people, nations, and the world also should be considered. We often refer to these needs as national or world priorities. The need for good health and health care is a worldwide, as well as an individual, concern. There is an ever-growing need for more and better methods of transportation—and especially fuel-efficient transportation. The whole world needs an inexpensive, environmentally clean, and readily available energy supply. Alternate sources must be developed to replace our dwindling supply of fossil fuels. The need for protection of the environment is a growing priority for many nations, and is becoming a worldwide need.

As both individual and group needs change and grow for manufactured products, industry will adjust to meet these needs. In this section, we will look at just five areas of changing and growing needs: electronics, health, transportation, energy, and the environment. You may be able to identify many others.

Electronics

You read earlier about the increasing use of computers and computerized equipment in both communication and manufacturing. You also read about the increased use of robots in manufacturing. The technology in these areas changes rapidly. This means these products soon need to be updated or even become so outdated that they are replaced. Manufacturers of electronic and computerized equipment, including manufacturers of robots, will have to keep up with the demand for the newest, most up-to-date computers and other electronic equipment. Fig. 14-10.

Health

As long as diseases, accidents, and natural disasters exist, there will be demand for improved medicines, health-related products, and medical care. The ever-growing population and ever-increasing life expectancy for individuals add to this demand. Research and product development in the health industry are never ending, even though there have been great advancements in this field.

You just read how biotechnology holds promise for producing more and better medicines and biomaterials in the future. There have also been many new surgical

Fig. 14-10. This "Ambler" robot was designed to walk on the rough surfaces of the moon and Mars. Here on earth and in space there will be a growing demand for robots and robotic equipment.

techniques developed that involve microsurgery or the use of lasers. These techniques require many special instruments, tools, and equipment that must be manufactured.

People are becoming increasingly health-conscious. They want products that will help them have a longer and healthier life. There is increasing demand for healthier foods. Exercise equipment and special shoes for different exercise activities are in increasing demand. Along with this increase in concern for health goes the increased attention to early diagnosis of illnesses. After all, the sooner an illness or disease is discovered, the better the chances of a cure and complete recovery. This means an increased demand for more and better diagnostic equipment. Fig. 14-11. There is also increasing demand for special devices that allow people to monitor their own blood pressure, heart rate, and insulin levels so they can keep a closer watch over their health.

Robots have even been introduced to the health care industry to help provide hospital care. A mobile "hospital helpmate" is already being used in at least one hospital. It is designed to bring patients food and medicine, carry medical records and blood samples, and dispose of contagious wastes. Fig. 14-12. If it proves successful, there will be increased demand for these service robots in other hospitals and health-care facilities, as well as in homes to care for the physically challenged.

Transportation

The United States already has a national commitment to diversify and increase public transportation. This commitment, coupled with the growing worldwide need to transport more people and more goods, will make manufacturing for transportation a growing area in the future. There is also growing demand for energy efficiency in transportation. You will learn more about manufacturing trends for transportation systems in Chapter 19.

Fascinating Facts

A robot has been developed to perform surgery. Robodoc, a 250-pound, three-foot arm with a precision cutting tool at its end, assists human surgeons during hip-replacement operations. First, the surgeon removes the diseased part of the bone. Next, Robodoc carves out an area of the healthy bone into which the artificial joint will fit. The surgeon then completes the operation.

Fig. 14-11. This diagnostic machine is used in hospitals to help doctors diagnose people's illnesses early enough to increase their chances of a cure and complete recovery.

Energy

The entire world depends on an inexpensive and easily available supply of usable energy. At the present time, oil is the principal source of energy, followed by coal and natural gas. As you learned in Chapter 2, these fuels are called fossil fuels. The supply of these fossil fuels is limited. They are nonrenewable sources of energy, and the world is rapidly using the remaining supply.

In the future, alternate sources of energy (alternatives to fossil fuels) will continue to be sought and developed. These efforts will require manufactured products such as solar cells, wind turbines, and other devices needed to capture and convert energy to usable power. Fig. 14-13.

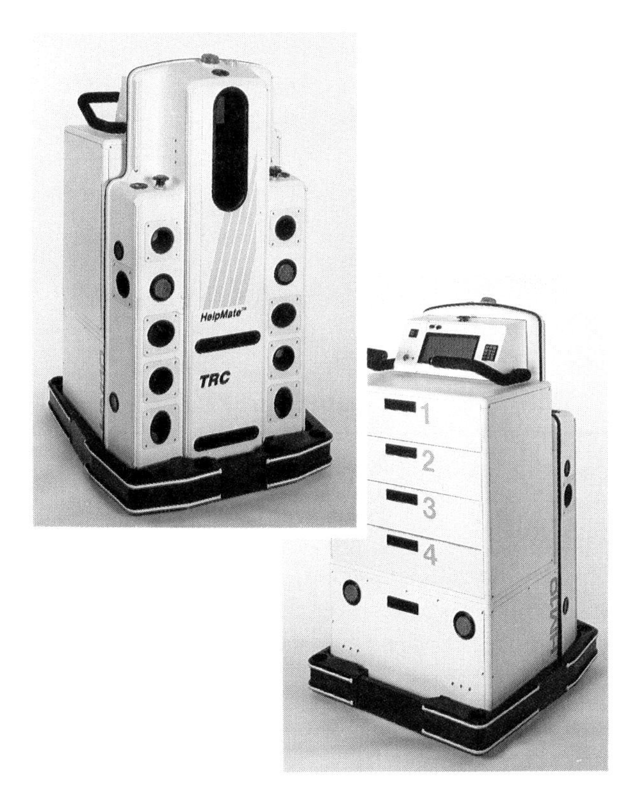

Fig. 14-12. This four-foot-tall robot is designed to perform many necessary hospital procedures. It uses 23 sensors, varying from bumpers to devices that measure distance using sound (sonar sensors).

Fig. 14-13. Solar energy can be collected using photovoltaic cells (cells designed to convert solar energy to electrical energy) arranged in large banks. These "solar farms" may become major sources of electricity in the future.

Environment

The concern for the environment that exists in the United States is shared by many other countries in the world. There is growing world concern about controlling pollution and protecting the environment.

Pollution prevention and the cleanup of past pollution will be important and growing concerns in the future. Manufacturing plants will have the responsibility of controlling their own pollutants during production. We also will see entire new industries develop in the area of pollution control and cleanup.

Apply Your Thinking Skills

1. Why do you think people are becoming increasingly health-conscious?
2. It is important that we all work to conserve energy. What are some ways your school could save energy? How could you help save energy at home?

The Developing Global Economy

Rapid transportation and efficient communications via satellite have helped people understand what is happening in all parts of the world. One outcome is the tendency for products to become popular throughout the world rather than in just one nation. Automobiles are manufactured and sold worldwide. Many clothes, food products, and other products have worldwide markets.

This growing need for similar types of products is being helped by a gradual and steady reduction of trade barriers. The European Common Market permits the countries of Europe to sell goods throughout Europe without import taxes or restrictions. Similar trade agreements are also being planned among the countries of North America. This tendency to decrease trade barriers will help manufacturers find larger, more global markets for their products.

Manufacturing companies have already taken advantage of these changes. Many U.S. companies have manufacturing plants in different parts of the world. Sometimes these locations are chosen to take advantage of lower labor costs and/or lower cost of materials. Other times, these locations are chosen so that products can be manufactured in the countries in which they are to be sold.

This helps eliminate the high cost of transporting finished products.

We already have companies from different countries entering into manufacturing agreements. In some cases, we have manufacturing plants managed and operated in a cooperative manner. Toyota Company of Japan and General Motors Corporation of the United States have a joint agreement to cooperatively manufacture automobiles in California. Fig. 14-14. The future may see increased cooperation and possibly even mergers across national boundaries. The result may be multi-national companies manufacturing products.

One definite effect of the development of the global economy on manufacturing is the increase in competition for a share of the world marketplace. With goods available from all over the world, consumers have more products from which to choose.

Global Competition

Throughout the world, everyone wants a good buy for his or her money. To make consumers want to buy their product, manufacturers must make a quality product at a good price. As the market and the competition continues to grow, manufacturers will have to work harder than ever to meet consumer demands for good quality at a fair price.

Quality

Worldwide competition has made quality an important issue. Often, the major difference between products produced by different manufacturers is the reliability and potential life of the product. You already read in Chapters 11 and 13 about many ways processes are checked before and during production and about products being checked during and after processing operations. It is

Fig. 14-14. This automotive plant is a joint venture between Toyota, the largest automobile manufacturer in Japan, and General Motors, the largest automobile manufacturer in the United States.

apparent computerization has a growing role in manufacturing quality products. Another method of helping to ensure quality is quality circles.

The theme of the concept of quality circles is "employee involvement." Employees participate in a variety of planning and production improvement procedures. In **quality circles**, workers performing similar tasks meet periodically with managers to discuss production problems and offer suggestions for improvement. Fig. 14-15.

There are many variations of this process. All are designed to involve workers in the company's operation. This helps make better use of the abilities of all employees. Companies have also found that workers involved in decision making are better motivated and help maintain product quality.

Productivity

To stay in business, a manufacturer must make a profit. A **profit** is the amount of money a business makes after all expenses have been paid. However, the selling price of a product must be competitive with similar products. As stated earlier, global markets make sales more competitive. Manufacturers will have to try harder than ever to keep production costs down without sacrificing quality.

Fig. 14-15. In quality circles, workers are encouraged to offer suggestions on how to improve the production of a product.

You read in Chapter 9 how products are designed for ease of manufacturing and assembly (design for manufacturability). You learned in Chapter 13 how CIM is used to make all areas of manufacturing more efficient. These things help ensure quality and increase productivity. **Productivity** is the measure of the amount of goods produced (output) and the amount of resources (input) that produced them. If you increase your output without increasing input, you increase your productivity. High productivity helps keep costs down. This not only helps keep the manufacturers more competitive, it also makes the market larger. When a product is more affordable, more people will be able to buy it.

Toward the Use of Metrics

As the world moves toward a more global economy, the pressure will continue to increase for a unified worldwide measurement system. This worldwide system will be metrics. Countries such as Canada have already converted from the customary to the metric system. The United States government began trying to convert the whole country to metrics more than a decade ago, but it abandoned the attempt due to public resistance. However, much of U.S. industry has already converted to the metric system in an effort to compete throughout the world. We can expect the U.S. to renew its plans for total metrication in the future.

Apply Your Thinking Skills

1. What evidence do you see of the growing global market when you go shopping?
2. In some industries, productivity is higher in the factories of countries like Japan and South Korea than it is in U.S. factories. How does this affect us in the world market?

Investigating Your Environment

Letting It Rot

To make their products, manufacturers should choose materials that will last a very long time, right? Actually, long-lasting materials are not always a good thing. Take the case of plastic.

About 30 percent of the trash in landfills is plastic, and it can take centuries to decay. Other kinds of waste, such as food or plain paper, are broken down chemically by the action of bacteria. However, bacteria cannot digest most plastics. This is because plastic is smooth and tightly constructed. Even under a microscope, plastic doesn't have the little hills and valleys that other materials do. Another problem is that plastic repels (pushes away) water, and bacteria need to live in a moist environment.

Scientists are trying to solve this problem in many ways. New plastic blends have been created. For example, you can now get garbage bags made of plastic and starch. Bacteria are able to "grab on" to the starch parts of plastic bags and break them down more quickly.

Substances like talc and glass fibers are commonly used to strengthen plastic. These "fillers" are not recyclable. Chemists in Canada have discovered that pecan shells ground into a flour can replace some of these nonrecyclable substances found in plastic. This flour is even stronger than the substances currently used to strengthen plastics, so using it would actually improve the quality of plastic.

It may be possible to make plastic out of lignin, a byproduct of papermaking. Paper is made from wood pulp, so bacteria that naturally eat wood would get rid of these lignin-based plastics.

Biodegradable plastic can't solve all of our plastic problems. For example, it would not be appropriate for long-term storage. Imagine what a grocery store or warehouse would look like if all the plastic containers started breaking down! However, we can help reduce the amount of plastic waste in our landfills by re-using old plastic to make new products.

Whether it involves making plastic in new ways or using old plastic in new ways, we must find ways to eliminate plastic waste. Otherwise, we may be buried *under* mountains of it.

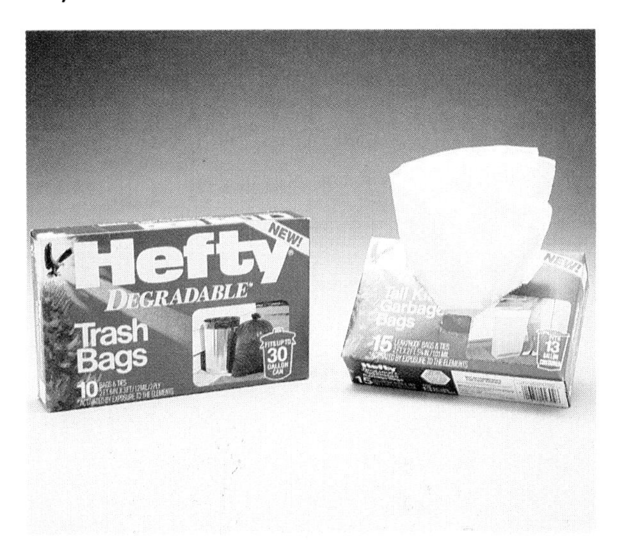

Fig. IYE-14. A new type of garbage bag is made of plastic that breaks down more easily when the bag is in a landfill and exposed to the different elements in the environment.

Take Action!

1. Make a chart listing at least 10 common products made with plastic. Next to each, list alternative materials that would be better for the environment, and explain why they would be better.

Chapter 14 Review

Looking Back

The main trends in manufacturing will be related to the way products are designed and made, the types of products that are in demand, and the growing market for manufactured products.

Manufacturing will see an increase in the use of computers and robots. Factories will continue to become more automated. Some manufacturing will take place in space. New materials are being developed both on earth and in space. Many new materials, medicines, and products will be developed using biotechnology. Handcrafting is still an important method of manufacturing.

There is an increasing demand for new products related to electronics, health, transportation, energy, and the environment.

The market for many manufactured products has become worldwide. There is increased cooperation among industrialized countries and some manufacturers. However, there is also increased competition among manufacturers for a share in the world marketplace. Manufacturers will strive to find new ways to stay competitive. Some U.S. industries have converted to metrics to help them stay competitive in the world market.

Review Questions

1. Briefly describe what "virtual reality" is and how it is created. Why is virtual reality important to manufacturing?
2. What is the main disadvantage of automatic manufacturing plants? What types of products will these plants produce?
3. Describe the benefits of manufacturing in space. Name at least three types of materials or products that will be manufactured in space.
4. Define alloy. Why do manufacturers want to make alloys?
5. Define composite. Why are composites desirable materials?
6. What is a superconductor? Describe one use of a superconductor.
7. Name five types of products that have been developed using biotechnology.
8. Why is there increased demand for health-related products and improved medical care?
9. What changes in technology and world relations have contributed to the growing world marketplace?
10. What is productivity? Why must manufacturers increase productivity?

Discussion Questions

1. What are some advantages and disadvantages of using automated equipment instead of using human workers in manufacturing?
2. How would our lives be changed if petroleum was no longer a source of energy? What jobs might be eliminated?
3. Discuss the positive effects of quality circles on both quality and productivity in manufacturing.
4. Productivity is higher in Japan and South Korea than it is in the U.S. Why do you think this is true? What can we do to help improve our productivity?
5. How do you feel about our nation converting to the metric system? How will converting to this system of measurement change your life?

Chapter 14 Review

Cross-Curricular Activities

Language Arts 

1. Trends play an important part in the clothing industry. Investigate the trends in clothing during the last ten years. Present your findings in a brief essay.
2. After World War II, Japan was not noted for the quality of its goods. Things have changed. Today, Japanese products are noted for their quality. Research and make a report about the changes that have taken place in Japanese manufacturing.

Social Studies 

1. Your text defines a trend. After the energy crisis in 1972, what trends related to energy conservation began to occur within the U.S.? Be specific.
2. Do some research to answer the following questions: Where are the primary markets for U.S. manufactured products? Who are America's best trade partners? What commodities do we trade with these nations? What products do we buy from them? When did the U.S. develop a trade deficit? What products does the U.S. need to make more cheaply and efficiently to eliminate this imbalance of trade?

Science 

1. Changing the composition of metals can make them more useful in manufacturing. Metal *alloys* are made by combining two or more metals. By careful alloying, new metals with desirable properties can be created. Aluminum is a lightweight, strong, nonmagnetic metal that can be easily shaped. It is used in more than 300 different alloys. Use a reference work to find some alloys made with aluminum. How does mixing aluminum with other metals make it more useful?
2. In January 1990, a research satellite called the Long Duration Exposure Facility (LDEF) returned to earth after six years in orbit around earth. The LDEF's 12 sides and 2 ends were covered with trays containing many different materials, from paints and plastics to fibers and alloys. After LDEF returned, the trays were sent to research facilities around the country for analysis. Find out how the different materials reacted to conditions in space. What implications does this project have for manufacturing in space?

Math

Marlin has designed a "widget" that requires the following materials to manufacture:

—four 1/2" dia. x 3' dowel rods at 75 cents each
—12' of binders twine at 6 cents per foot
—two No. 6 x 1 1/4" RH (round head) brass wood screws at 5 cents each

1. If the prices are as indicated, what is the total cost of each widget?
2. If Marlin were to sell his widgets, how much would he have to charge to make a 40 percent profit above his materials cost?

Chapter 14 Technology Activity

Designing and Producing an Industrial Robot

Overview

Just what is a robot, anyway? A *robot* is a machine that performs tasks usually done by people. It is a mechanical "worker."

Originally, people thought robots were machines that looked, walked, and talked much like people. Some people still do, but the main robot worker used in manufacturing today looks more like a person's arm. As a matter of fact, it is commonly called a *robot arm*. Another name for it is *manipulator*.

Sometimes a robot arm is referred to as an *articulated arm*. This is because it has a series of pivot points, or joints, around which it moves. The pivot points are called *axes*. The number of axes a robot arm has depends on the type of work it is designed to do. In manufacturing, robots commonly have six axes. However, the three basic axes can be identified as the joints in your arm are identified—shoulder, elbow, and wrist. Fig. A14-1.

The "hand" of the robot is called a *gripper*, or *end effector*. There is a wide variety of these devices because of the many different types of work that robots now do. A common type consists of two parts that open and close like pincers. These actions enable the robot to pick up, hold, and release parts or products being processed. Fig. A14-2.

The device or system that makes the robot move is called an *actuator*. It may be electrical, pneumatic (air-powered), or hydraulic (fluid-powered).

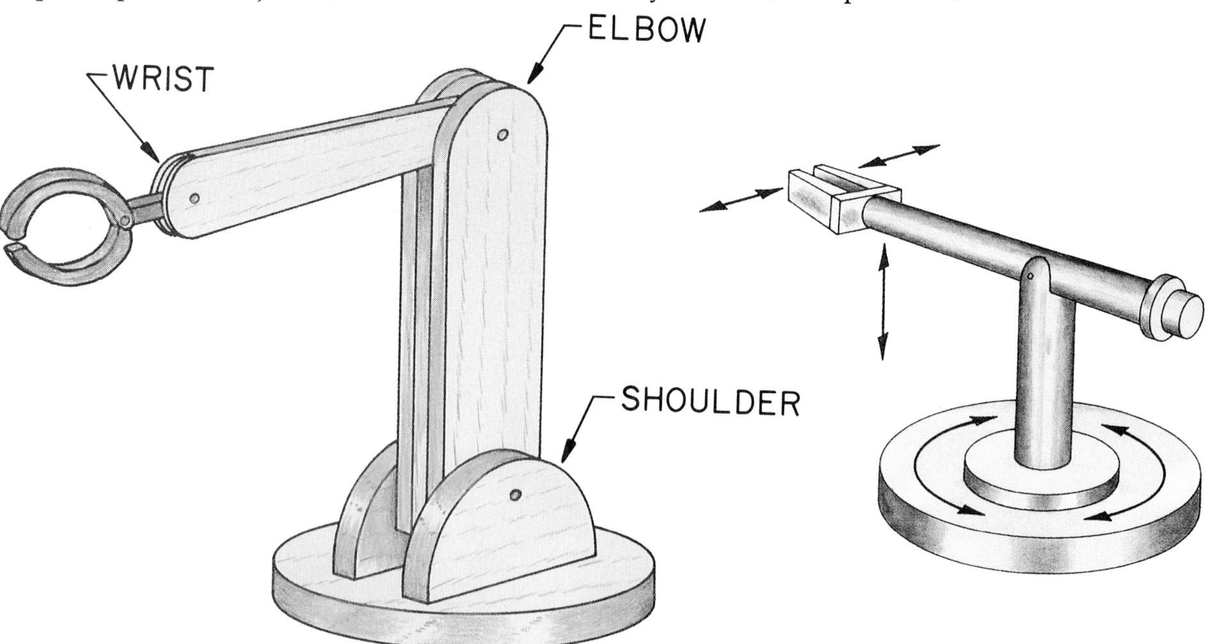

Fig. A14-1. Three basic axes of a robot arm.

Fig. A14-2. Basic movements of a robot arm.

Chapter 14 Technology Activity

Goal

To design and produce a hydraulic robot arm using common, everyday materials and creative thinking.

Equipment and Materials

paper, 8 1/2" x 11"
pencil
notebook paper
drafting paper
drafting equipment
any suitable materials such as:
- plywood, assorted sizes
- scrap lumber
- dowel rods, various sizes
- PVC plastic pipe
- sheet metal
- metal cans

clear plastic tubing, 1/8" diameter
syringes, 6-20 cc
ball bearings
springs
adhesives
fasteners
abrasive paper
hand tools
power tools, hand-held
band saw
belt sander
drill press
safety glasses and other safety devices as needed

Procedure

1. *Safety Note:* Before doing this activity, make sure you understand how to use the tools and materials safely. Have your teacher demonstrate their proper use. Follow all safety rules.
2. Your teacher will assign you into groups of two or three students. Each group will research and develop a prototype industrial robot. Your robot must be hydraulically controlled and must have two or more axes. It must be able to pick up, move, and put down an object. This means that your design must allow for both horizontal (sideways) and vertical (up-and-down) movement.
3. A key element in the design of the robot is the power/control system. In this model, you will use syringes and 1/8" plastic tubing to create a hydraulic system. Any particular movement will require a water-filled piece of plastic tubing with a syringe on each end. Basically, this system works as follows:

   One syringe will be part of the actuator. The other syringe will be mounted appropriately, usually on the part of the arm you wish to move. Pushing in the plunger will cause the other syringe to extend or "go out," moving the part on which it is mounted. Pulling out the plunger of the actuator syringe will cause the mounted plunger to retract or "go in," creating movement in the opposite direction. Fig. A14-3.
4. Examples of robots are shown in Fig. A14-4. These are presented just to start you thinking. Your robot is to be a *prototype*. That is, it should be a unique, one-of-a-kind item. As a group, discuss various possibilities.
5. Make thumbnail sketches of possible robot designs.
6. Combine your group's best ideas into a rough sketch and submit it to your teacher. Modify your design as needed.

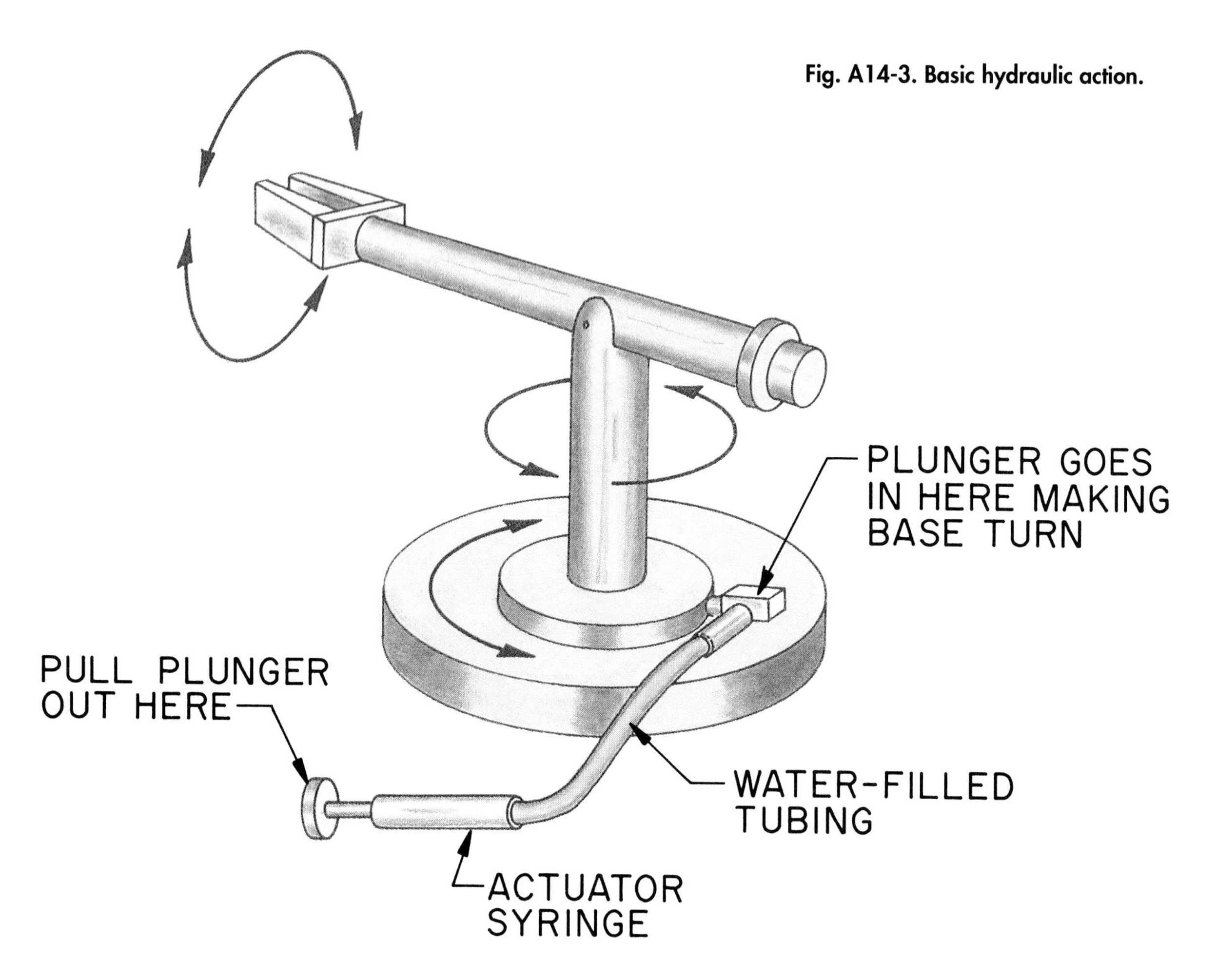

Fig. A14-3. Basic hydraulic action.

7. After your teacher approves your rough design, make working drawings. Prepare at least three views—front, top, and right side—and any detail drawings that are needed. Have these approved by your teacher.
8. Create a bill of materials for making your robot and submit it to your teacher.
9. As in the manufacturing industry, production must be well-planned and organized. Prepare a plan of procedure for making your robot. Following a plan will help you work more efficiently. Submit your plan to the teacher.
10. After your drawings, bill of materials, and plan of procedure have been approved, gather the materials and supplies that you will need.
11. Following your teacher's instructions for working in the lab, build your robot. Test and modify it as needed.
12. When your robot is completed, prepare and give a presentation for the class. Explain your design. Relate problems you encountered and describe how you solved them. Demonstrate what your robot can do. For example, can you get

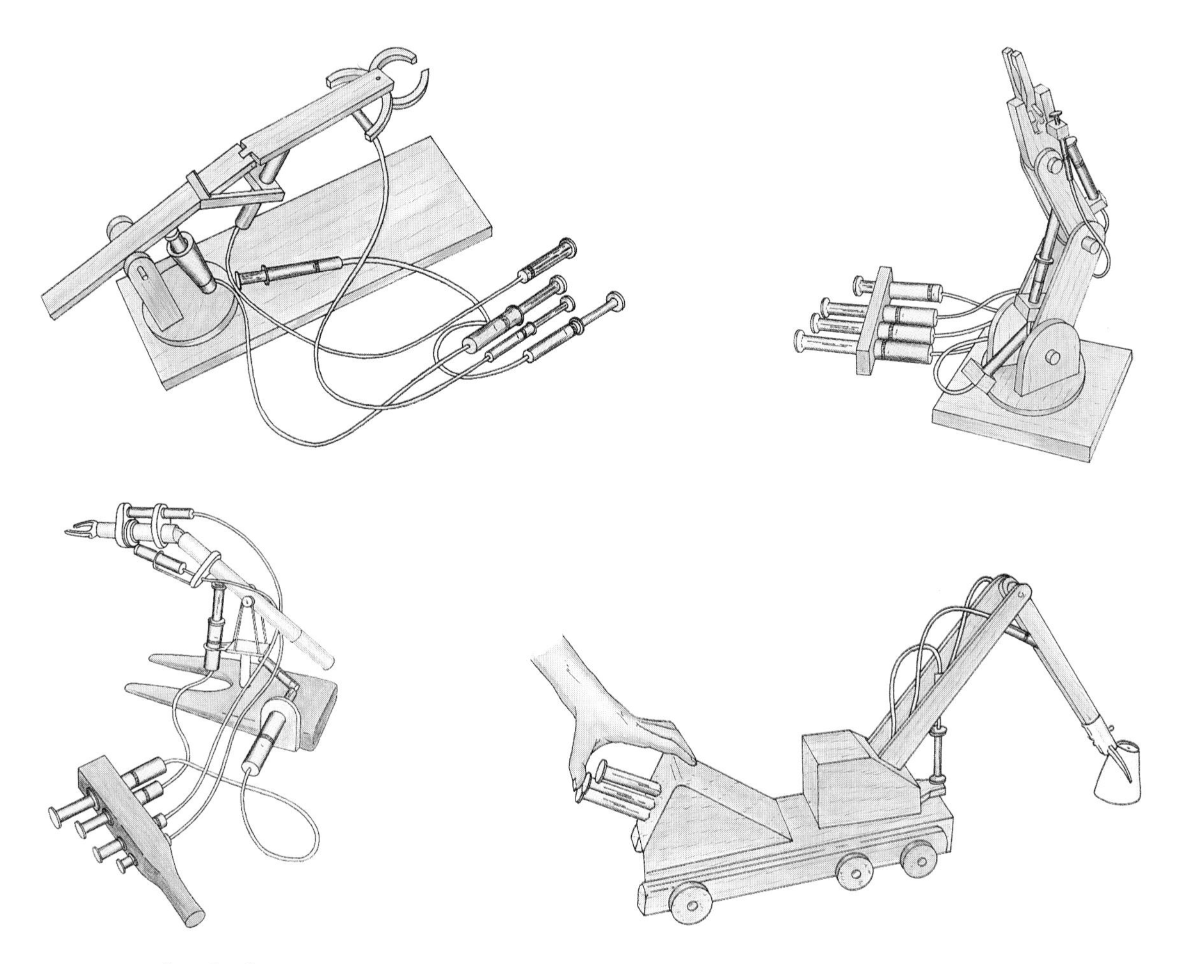

Fig. A14-4. Examples of robots.

it to pick up the telephone receiver and hand it to you?

Evaluation

1. On a separate sheet of paper, write a brief evaluation of your robot and the work you did on it. If you had it all to do over again, what would you do differently? Submit your evaluation to your teacher.
2. If you were to make another robot, what new or additional features would you try to include?
3. Think about these things:
   - How might your robot be used on a production line in your lab?
   - How could a motor be used to power the robot?
   - How could a computer be used to control robot operations?

Section III Activity

Designing a Product for Use in Microgravity

Overview

In the future, there will probably be a permanent space station orbiting the earth. Gravity will be far less than it is on earth. In order for people to live and work under microgravity conditions, many products will need to be redesigned.

In this problem-solving activity, you will choose a familiar object, such as a salt shaker or pen, and redesign it for use under microgravity conditions. To do this, you will follow the steps of the problem-solving process.

STEP 1: State the Problem (Design Brief)

In microgravity, materials behave differently than on earth. They do not fall down (there is no "down"). Fig. SCIII-1. Once in motion, they tend to keep moving because gravity is not slowing them down. For example, salt shaken out of a regular shaker will tend to float all over the spacecraft. The problem is to design a device that will accomplish its purpose under microgravity conditions.

STEP 2: Collect Information

Here are possible sources of information:

- This textbook (look in the index under "space" and "microgravity")
- The library (look in the *Reader's Guide to Periodical Literature* for magazine articles about space stations)
- NASA

STEP 3: Develop Alternative Solutions

Write down several ideas. For example, if you are designing a new type of salting device, you might write the following:

- A liquid salt solution in a squeeze bottle (Fig. SCIII-2)
- Spreadable salt
- A special chamber for adding salt and other seasonings to food

STEP 4: Select the Best Solution

Consider the pros and cons of each solution. Questions to consider might include:

- What materials could be used to make the device?
- Would the device be expensive in terms of material or energy used?
- How difficult would it be to manufacture the device?
- How large would the device be? Would it be easy to store on a spacecraft?
- Would the device be easy to use?
- What about storage? For example, would the spreadable salt need to be refrigerated?

Select what you feel is the best solution and write your reasons why.

STEP 5: Implement the Solution

Prepare drawings of your solution. Make a model or prototype based on your drawings.

STEP 6: Evaluate the Solution

Your teacher and classmates will help evaluate your solution. Changes may be suggested to make the product work better or make it easier to manufacture. Incorporate the changes into your drawings to prepare a final version.

Fig. SCIII-1. Living and working in microgravity presents special challenges.

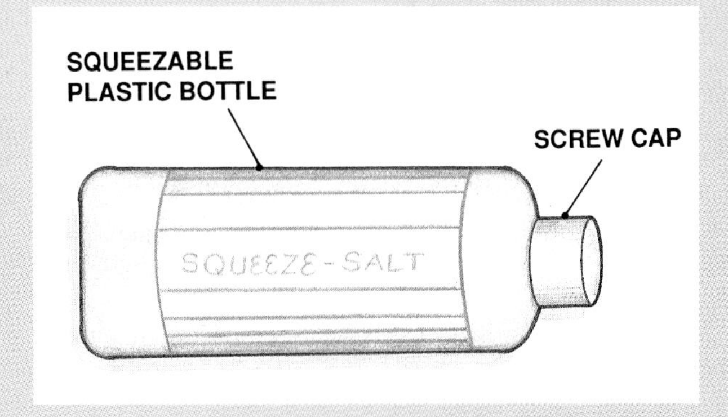

Fig. SCIII-2. Here is one possible solution. What might be its advantages and disadvantages?

Section IV
Transportation Technology

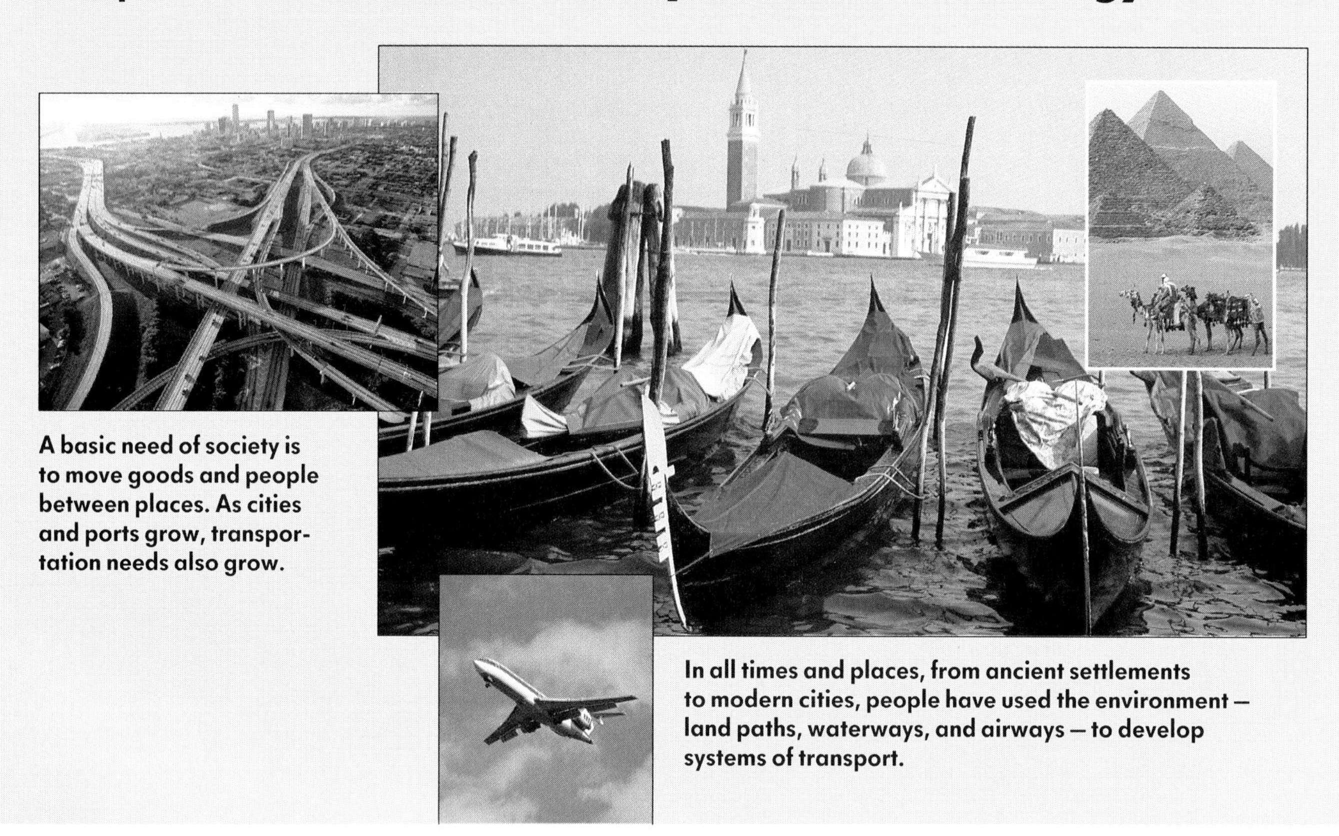

A basic need of society is to move goods and people between places. As cities and ports grow, transportation needs also grow.

In all times and places, from ancient settlements to modern cities, people have used the environment — land paths, waterways, and airways — to develop systems of transport.

Technology Time Line

3500 B.C. Wheeled carts are used in Sumer.

1000 B.C. Phoenicians dominate trade on Mediterranean with wooden ships that can be sailed or rowed by slaves.

550 B.C. Anaximander of Miletus makes the first map of the known world.

230 B.C. Archimedes, a Greek mathematician, makes first elevator with rope and pulleys. It can hoist one person.

1492 Columbus reaches the New World.

1769 First steam-powered land vehicle (car).

1783 First hot-air balloon flight.

3500 B.C. | 2500 B.C. | 1500 B.C. | 1 | 1400 | 1700

Transportation systems provide ways to deliver goods and people over both short and long distances.

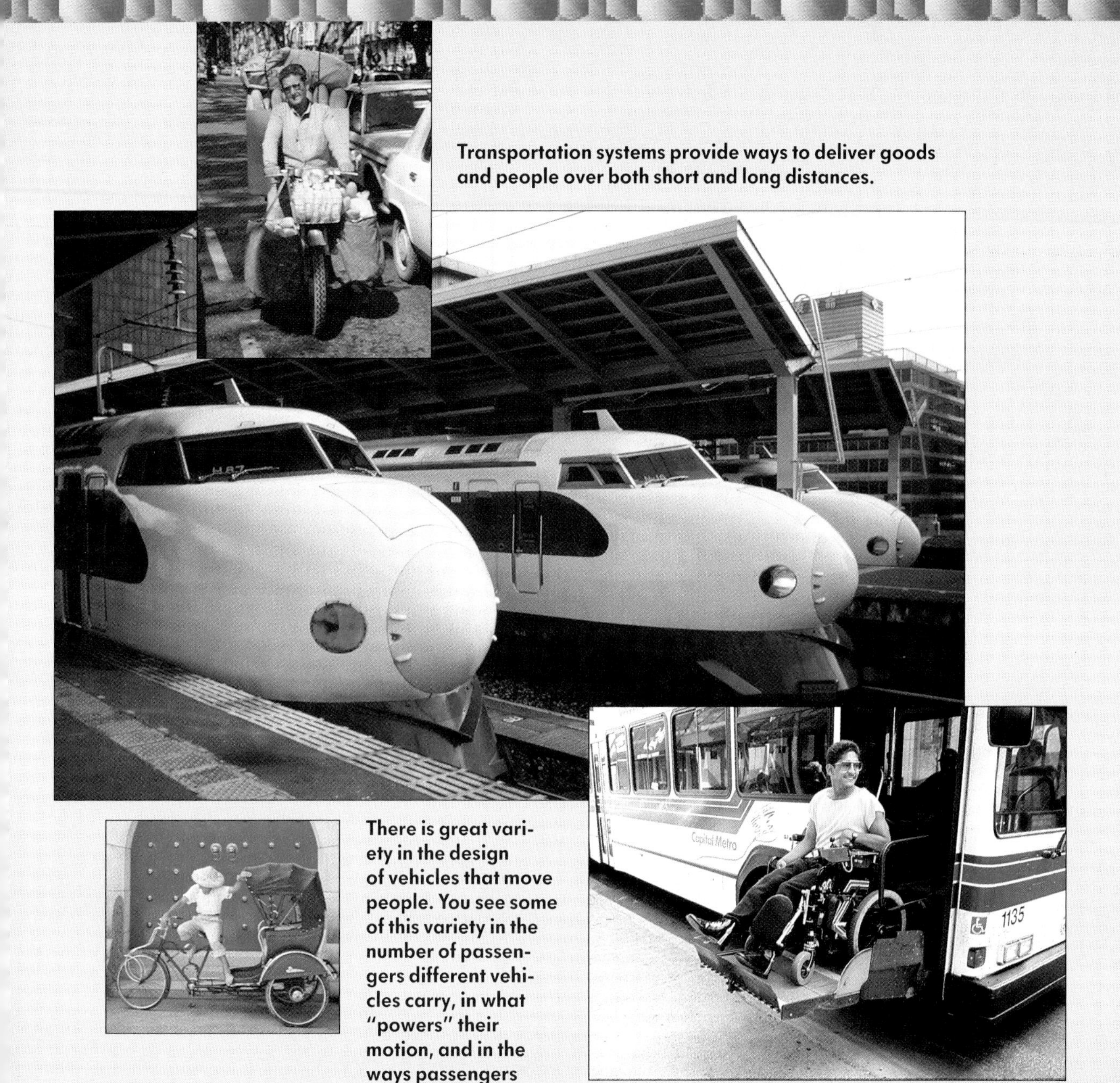

There is great variety in the design of vehicles that move people. You see some of this variety in the number of passengers different vehicles carry, in what "powers" their motion, and in the ways passengers enter and exit the vehicles.

1800 | 1825 | 1850 | 1875 | 1900

1807 Robert Fulton and his steamboat make their first successful voyage.

1828 First U.S. passenger railroad is begun.

1838 *Sirius*, a British ship, is the first ship to cross the Atlantic under steam power.

1857 First passenger elevator is installed in New York.

1869 First transcontinental railroad is completed.

1869 The Suez Canal is opened.

1880s First electric cars appear in Europe.

1893-94 Charles E. and J. Frank Duryea build the first successful American gasoline-powered car.

1897 The first practical subway system in the U.S. is opened in Boston.

1900 The U.S. Navy buys its first submarine and names it the U.S.S. *Holland*.

1901-03 Using assembly line, Oldsmobile car company increases production from 425 to 5,000 cars per year.

1903 Wright brothers are first to fly with a gasoline-powered plane. The first flight lasts 12 seconds and goes 120 feet.

1913 Henry Ford reduces production time of cars by using conveyor belts on his assembly line.

How do you transport the vehicles that carry goods and people? Some ways include moving trailers by huge cargo planes, shipping cars on trucks or ferries, and even carrying satellites into space.

Engines are still measured in horsepower, even though horses were replaced by engines driven by steam, internal combustion, and rocket fuel.

Technology Time Line

1925	1930	1935	1940	1945	1950	1955

1926 First successful liquid-propellant rocket is launched by Robert H. Goddard.

1927 Charles Lindbergh becomes the first to fly solo across the Atlantic Ocean.

1929 Lt. James Doolittle pilots airplane solely using instruments.

1932 Amelia Earhart is the first woman to fly across the Atlantic.

1933 The *Autobahn*, a superhighway, is opened in parts of Germany.

1947 Capt. Charles Yeager breaks the sound barrier in a jet airplane.

1954 The *Nautilus* is the first nuclear-powered submarine commissioned.

1957 The Soviet Union's *Sputnik I* becomes the first artificial satellite to orbit earth.

1958 The Federal Aviation Administration (FAA) is established.

1959 The St. Lawrence Seaway opens and allows seagoing vessels access to the Great Lakes.

Today, mass transportation vehicles make commuting and other types of travel by land, air, and water very convenient. Power for these vehicles must meet our need for top efficiency and capacity as well as our desire for increased speed and range.

The technology for transportation needs of tomorrow will probably use principles of aerodynamics and hydrodynamics that haven't been discovered yet, and they will probably use power from fuels, solar collectors, and batteries that haven't been invented yet.

1964 Japanese Tokaido "Bullet Train" is opened and travels in excess of 130 mph.

1969 Neil Armstrong becomes the first man to walk on the moon.

1976 *Viking I* lands on Mars to record data.

1978 Americans Maxie Anderson, Larry Newman, and Ben Abruzzo make the first balloon flight across the Atlantic Ocean.

1981 The U.S. launches the first reusable spacecraft, the space shuttle *Columbia*.

1986 *Voyager*, piloted by Dick Rutan and Jeana Yeager, makes the first nonstop flight around the world without refueling.

1987 Japan and West Germany test prototype trains with speeds up to 250 mph.

1990 More than 9 million cars are sold in the U.S.—25% are imported.
1990 World motor vehicle production exceeds 48 million; 20% are made in U.S.

1960 | 1965 | 1970 | 1975 | 1980 | 1985 | 1990

Chapter 15

Transportation Systems

Looking Ahead

In this chapter, you will discover:

- what transportation is.
- how transportation is used to provide services.
- how transportation affects the economic value of products.
- the parts of transportation systems.
- some of the positive and negative impacts of transportation.

New Terms

break bulk cargo
bulk cargo
cargo
freight
gridlock
on-site transportation
passengers
routes
time and place utility
transportation
vehicles
ways

Technology Focus

Wildcats in Search of Oil

Petroleum, a nonrenewable natural resource, is a treasure sought by many. Petroleum, also called crude oil, was formed from decayed plant and animal life that lived millions of years ago. With pressure from the buildup of layers of the earth's crust over it, this decayed matter was gradually converted to coal, natural gas, and petroleum. Vast underground stores of petroleum are found in many places throughout the world. People like Robert O. Anderson make it their business to try to find it.

In 1940, Anderson, then in his early 20s, became a wildcat oil driller. A wildcatter takes financial risks in drilling for oil in places not known before to have oil in the hopes of striking it rich. Anderson found the outdoor life and the risks to be exciting. Many of the wells he first drilled did not produce oil, and others produced only a little. By studying land formations and other characteristics of successful oil fields, he eventually located some very productive wells.

Anderson also purchased a small refinery, where the petroleum is heated and cooled in order to condense it into refined petroleum products such as aviation fuel, gasoline, diesel fuel, kerosene, and asphalt. Each refined product's characteristics depend on how high the petroleum is heated before condensing. Anderson constructed pipelines from his wells to his refinery. In this manner, he soon built a fully integrated (unified) business of producing, transporting, refining, and selling petroleum. Profits from his refinery were used to finance new wildcat ventures.

Anderson eventually became chief executive officer of a large oil company, Atlantic Richfield (ARCO). Anderson retired from ARCO, however, in 1986. At age 70, he returned to his first love, wildcatting. Today, however, although an expensive undertaking, wildcatting is not quite as financially risky a venture as it was in Anderson's earlier wildcatting days. Now, seismology technology can be used to help locate an oil field. Sound waves are transmitted into likely areas beneath the earth's crust. Seismographs read the echoes to help pinpoint areas likely to hold oil. This method has been used to locate oil as much as three miles below the earth's surface.

Anderson called his new company Hondo Oil and Gas. In 1990, Hondo had sales of $350 million. Not bad for an old wildcat!

Fig. TF-15. Robert Anderson has returned to his first love, wildcatting, with his company called Hondo Oil and Gas.

What Is Transportation?

To *transport* means to carry from one place to another. **Transportation** is the movement of people, animals, or things from one place to another using vehicles. A **vehicle** is any means or device used to transport people, animals, or things. Some examples of vehicles include buses, trucks, airplanes, ships, railroad cars, and bicycles. Fig. 15-1.

The distance traveled to transport someone or something need not be long. **On-site transportation** always happens within a building or a group of buildings. Moving from one story of a building to another in an elevator is an example of on-site transportation. Loading or unloading freight using an industrial truck is another example. Fig. 15-2. You are performing on-site transportation tasks when you carry your schoolbooks from your locker to the classroom.

Fig. 15-2. A variety of on-site transportation vehicles are available for efficient movement of people and cargo.

Fig. 15-1. A surge of power and a heavy load of cargo is on the way to its destination.

We see and use the services of transportation each day. For example, how do you and other students get to school? Look around you. Everything you see has been transported to the room you are in. Transportation affects almost all parts of our lives.

Apply Your Thinking Skills

1. What types of transportation do you use in a typical day?

Using Transportation

All vehicles need a place to operate and facilities to support their operation. Transportation is used to move people or cargo, and it can increase the value of what is moved.

Ways and Routes

Vehicles are usually operated on ways and routes. **Ways** are the actual spaces set aside especially for use by the transportation system. The way may be a specific strip of land or measured altitude above the earth. Pipeline right-of-ways, highways, and railways are examples of ways. **Routes** are particular courses traveled. Shipping lanes and air routes are examples of routes.

The air, water, and highway transportation industries are operated on ways and routes that are owned or controlled by the local, state, or federal government. Fig. 15-3. The railroad and pipeline industries own and maintain their own ways.

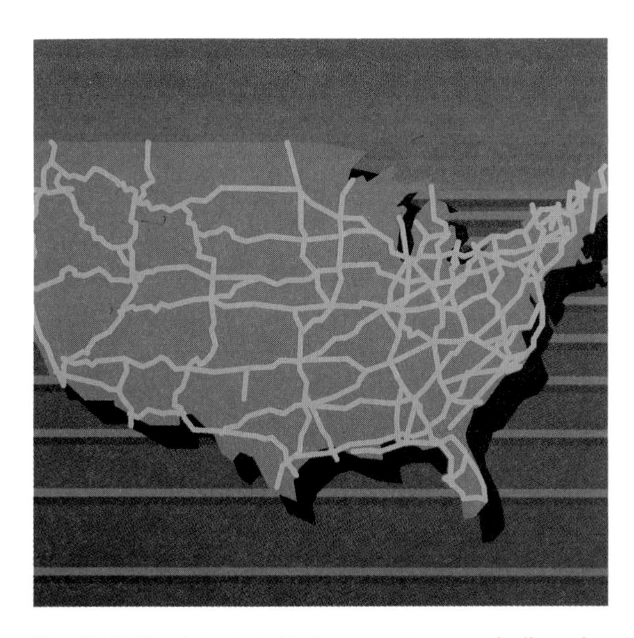

Fig. 15-3. The interstate highway system was built and is maintained by the federal and state governments working together.

Support Facilities

Transportation support facilities include a variety of terminals and warehouses, rail yards, landing strips, ports and docks, repair garages, and hangars. Your nearby gas station is a support facility for the transportation system.

In every method of transportation, the size, value, and cost of operating the support facilities exceeds (is greater than) the actual size, value, and operating costs of the transportation vehicles. Most of the people employed in the transportation industry work at the support facilities. Fig. 15-4.

Fig. 15-4. The airport is the main support facility for air transportation.

Passengers and Cargo

Everything that is transported in vehicles is classified as either passengers or cargo. **Passengers** are people who are moved from one place to another. Everything else transported, from automobiles to zoo animals, is **cargo**. Cargo is also called **freight**. There are two basic types of cargo: bulk cargo and break bulk cargo.

Bulk cargo is *loose* cargo. It may be a solid material like sand or a liquid such as oil. It may even be a gas. Bulk cargo is not packaged, and it is usually never mixed with another cargo in a transportation vehicle. Fig. 15-5.

Break bulk cargo consists of single units or cartons of freight. Fig. 15-6. Books, bicycles, and meat are packaged and shipped as break bulk cargo. Almost all the items in the stores where you shop were delivered as break bulk cargo.

It generally costs more to transport materials as break bulk than as bulk.

Fig. 15-5. Bulk cargo, such as coal, is loaded and unloaded quickly.

Fig. 15-6. Break bulk cargo must be packed with care to make the most efficient use of available space and to prevent damage.

Packaging costs are higher, and more handling is required. Cost is not the only factor that must be considered, however, when deciding how freight should be transported. For example, a school could charge less for milk if each student filled his or her own glass from a milk tank truck (bulk). However, imagine the problems that would cause! It is more convenient to provide milk in small cartons (break bulk).

Economic Value of Using Transportation

Transportation affects the economic value of services and products. ("Economic" refers to money.) The services of a person or the worth of an object may increase in value by being transported. Two things are required to make the value increase. First, transportation must satisfy a *need* for a person or item to be moved to another location. Second, the transportation must happen at the right *time*. For example, oranges cannot be grown in Canada. Orange growers in southern states transport their fruit to Canada. However, there would be no profit if it took too long to transport the oranges and many of them rotted while in shipment. Another example is seasonal items such as Christmas decorations. If they aren't delivered until December 30, they are of little value.

Change in value caused by transportation is called **time and place utility**. For the most value, items should be delivered to the place they are needed at the time they are needed. Fig. 15-7.

Apply Your Thinking Skills

1. Would it be practical to mix different bulk cargoes in one vehicle? Why or why not?

Parts of the Transportation System

Transportation is a system. In our world, there are few systems larger than the transportation system. However, every system, regardless of size, consists of four main parts. Fig. 15-8.

- Inputs
- Processes
- Outputs
- Feedback

These four parts were described in a general way in Chapter 2. On the following pages, you will learn how these parts relate specifically to transportation.

Fig. 15-7. Items should arrive *where* they are needed, *when* they are needed.

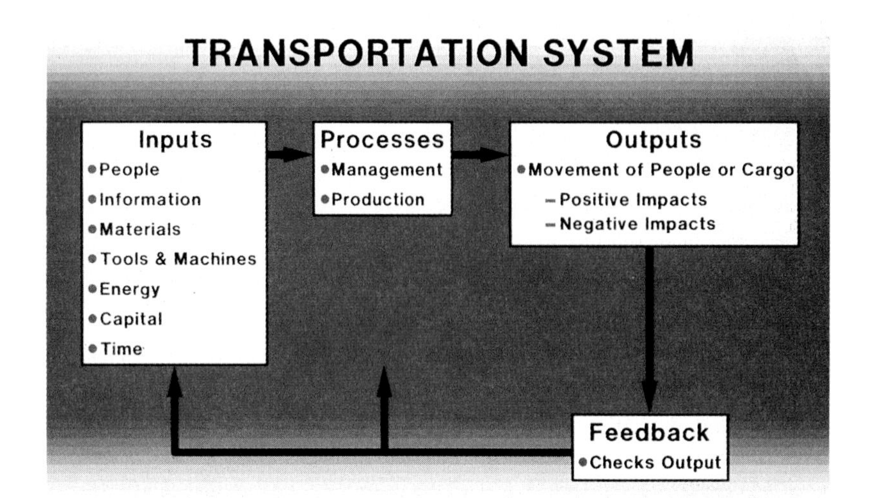

Fig. 15-8. A transportation system is made up of these parts.

Apply Your Thinking Skills

1. Can you think of any other system that could continue without a transportation system?

Inputs

Inputs are things that are "put into" a system. All systems of technology need seven basic types of inputs to operate:

- People
- Information
- Materials
- Tools and machines
- Energy
- Capital
- Time

People

In all systems of technology, people are the most important input. Without people to supply the knowledge and skills, transportation could not take place.

The people involved in transportation are either providers or users. Sometimes the same people can both provide and use the services of transportation. You are a user of transportation. How many times have you used the benefits of transportation today? If you've ever delivered newspapers or other items, you have provided transportation.

Do you know someone who works in the transportation industry? Truck drivers, airline mechanics, dock workers, and train engineers all work in transportation. Other examples include the travel agents who sell airline tickets, the railroad detectives who investigate freight theft, and the people who plan bus routes. You may have a job in the transportation industry right now. Do you work at a car wash? In a service station? For a company that delivers freight? There are many career opportunities in the transportation industry. Fig. 15-9.

Fig. 15-9. The transportation industry is one of the largest employers in the world.

Information

Information is necessary for every type of transportation system. People who control the systems need information to gain knowledge and skills. Workers need information about how to safely operate tools and vehicles or to load cargo. Managers need information to decide where to send freight. Without the information provided by road signs and maps, imagine trying to travel to or around an unfamiliar city. Fig. 15-10.

Materials

You have already learned that any material transported by the system is called cargo or freight. Materials are also needed as inputs for transportation systems. Water is used to cool engines, clean vehicles, and supply drinking water for people. Sand and salt are used to make icy roads safe for driving.

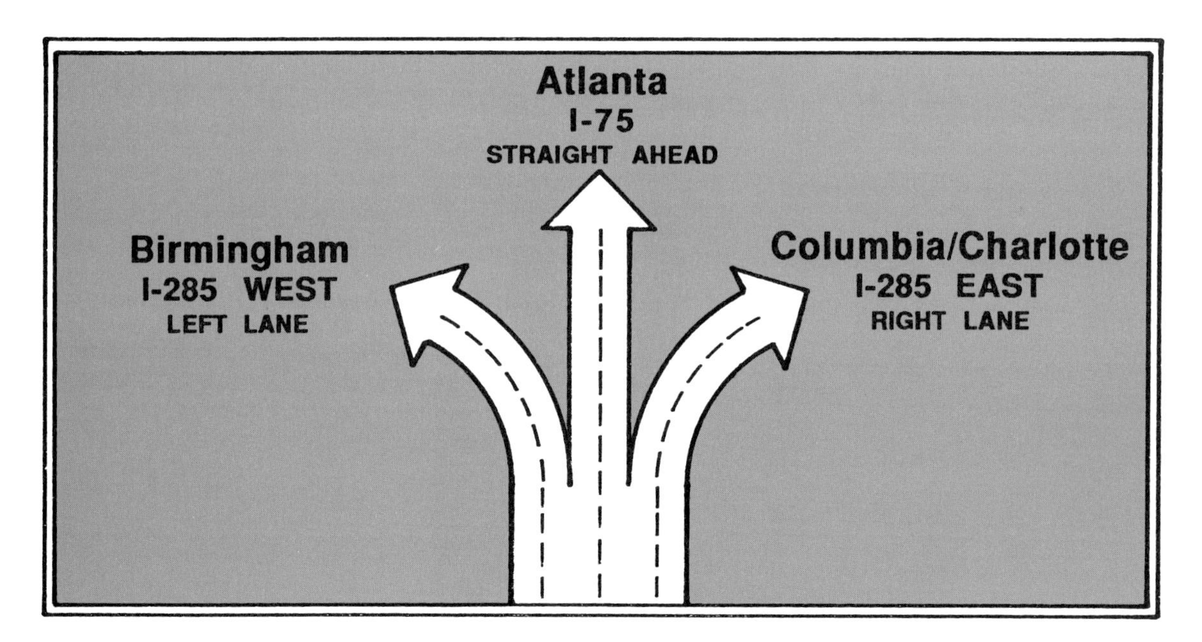

Fig. 15-10. Clearly presented information like this superhighway sign helps keep traffic flowing smoothly.

Tools and Machines

A transportation system needs tools and machines of all types. Most of the vehicles used in transportation are actually very complex machines. Without tools to maintain and repair these machines, no transportation system could function very long. Fig. 15-11.

Energy

Transportation systems consume large amounts of energy in many forms. Gasoline, diesel fuel, propane, and kerosene are the most common energy forms used to power vehicles. All of these fuels are made from crude oil, which is a limited natural resource. Electricity is used to power some vehicles, and it also supplies the energy for traffic lights, street lights, and computers.

Capital

Capital is another name for money or investments. A company's capital consists of working capital and fixed capital. Fig. 15-12. Working capital is the money that companies use to buy things and pay their employees. It includes more than cash. Other examples of working capital are shares of stock, money certificates, and property titles. All of these items can be exchanged on an open market.

Fixed capital includes items such as vehicles, facilities, and equipment owned by a company. The buses owned by your school are fixed capital. The computer that records your attendance and grades is another example of the school's fixed capital.

Time

Time used in transportation systems represents how long it takes to get from one place to another. It may be measured as the seconds it takes to go from one floor to another in an elevator or as the years it took the *Voyager* spacecraft to travel to another planet.

Saving time often means saving money to industry. Transportation systems that save time usually replace slower methods. Just as the truck replaced the horse-drawn wagon, something faster may someday replace the truck.

Fig. 15-11. This mechanic needs tools ranging from hand tools to computers to work on modern automobiles.

Fig. 15-12. There are two types of capital. Working capital is money. Fixed capital includes vehicles and facilities like these.

Apply Your Thinking Skills

1. Which, if any, of the inputs for the transportation system are in danger of being used up? How might we replace them in the future?
2. Could a transportation system function with any of the inputs missing?

Processes

Processes are the things that are done to actually cause the movement of passengers and cargo. In the system of transportation, there are two basic groups of processes: management processes and production processes.

Management Processes

Management processes include the activities the company does to keep the movement of passengers and cargo organized. Computers are important tools in the management processes. The three main management processes are:

- Planning
- Organizing
- Controlling

When *planning* transportation, people decide what must be done and how and when it should be done. They also consider costs.

Organizing processes include deciding who will do the tasks. Deciding which team of workers will do which loading operations is an example of an organizing process. Fig. 15-13.

During the *controlling* process, people perform a variety of tasks. For example, they may keep records of cars on freight trains, follow directional signals, or record how much money employees earn.

Production Processes

Production processes are the most visible parts of a transportation system. The flight of an airplane and the movement of trucks and cars on the highways are examples of transportation production processes.

Fig. 15-13. Deciding who does which job is part of management's organizing process.

Production processes are divided into three types of activities:

- Preparing to move
- Moving (operation)
- Completing the move

Preparing to move involves all the activities accomplished just prior to actually moving. Airline companies load the passengers' baggage into the plane. Shippers lock and seal cargo containers. The school bus driver may scrape the snow from the windshield before leaving the parking lot.

The actual operation (*moving*) of the vehicles is the most exciting process of a transportation system. Two kinds of activities are involved: those needed for vehicle operation and those providing en-route services.

Vehicle operation includes a wide range of activities. Fig. 15-14. One example is opening and closing valves on a pipeline. Another is driving an 18-wheel truck. Even slowing down or speeding up the elevators inside the Statue of Liberty is vehicle operation. How many different transportation vehicles have you ever operated or ridden in? What activities were involved in the vehicle's operation?

Providing en-route services is part of the moving process. These services include such things as serving food on an airline flight or changing a tire at a truck stop. Delivering mail to a ship while it is steaming across the ocean is also an en-route service.

The third process of providing transportation is *completing the move*. Cargo is unloaded. Passengers collect their luggage. Vehicles are shut down.

Fig. 15-14. When vehicles are actually moving passengers and cargo, they are contributing to the production process of transportation.

Fig. 15-15. After one move is completed, workers begin preparing for the next one.

Transportation systems are seldom idle, though. The production process may start all over again immediately. As soon as passengers leave an aircraft, workers begin to prepare for the next flight. Fig. 15-15.

Apply Your Thinking Skills

1. If you had a job loading freight onto trucks, in which type of production activity would you be involved?
2. Are highway maintenance crews part of the transportation process? Why or why not?

Outputs

The outputs of a system are the results achieved because the system completed the planned processes. In transportation, the basic output is the movement of passengers and cargo. Let's break this down further.

When the move is efficiently completed, changes occur. These are outputs. Perhaps the value of a cargo (time and place utility) increases. Maybe people make money for completing their part in the process. Maybe business people sell and buy merchandise. Perhaps it's just that family and friends get together.

Among the outputs of a system are the effects, or impacts, that system has on people and the environment.

Impacts of Transportation

Transportation has both negative and positive impacts. Let's first consider the positive.

Positive Impacts—The Rewards

Improvements in the technology of transportation have had a tremendous impact on our society. Think about the history of the United States. Developments in transportation supported the development of the country. Canals, railroads, and highways enabled people to explore and settle the far regions of the country. Today, we are still exploring. We are traveling down into the depths of the oceans and out into the far universe. Fig. 15-16. However, most of the benefits of transportation are not from exploration and discovery.

Our daily life is greatly affected by transportation. By transporting their products, producers expand their markets, which generally causes sales to increase. As sales increase, the demand for the product or one like it also increases. Increased demand encourages others to go into business. This is *business growth.*

As people and cargo are moved from one location to another, services and products become available to more people, allowing people to choose among products. Each provider or producer tries to offer the best price. This is *competitive pricing*. Fig. 15-17.

Because of transportation, the countries of the world are able to trade products.

Fig. 15-16. Remotely operated vehicles (ROVs), carrying lights, TV cameras, and other equipment, allow us to explore the oceans' depths from the safety of a surface vessel. ROVs can also be equipped to perform tasks such as repairing undersea pipelines.

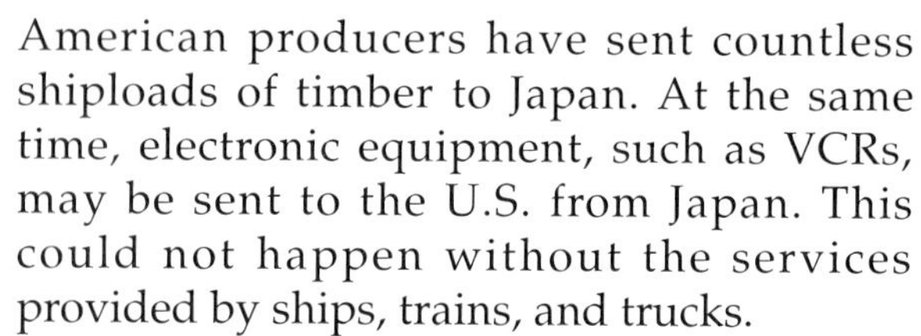

American producers have sent countless shiploads of timber to Japan. At the same time, electronic equipment, such as VCRs, may be sent to the U.S. from Japan. This could not happen without the services provided by ships, trains, and trucks.

Improvements in vehicles, ways, and facilities have enabled people to travel longer distances in less time. Business people meet more customers. Families can travel to different locations and see new things. Students from the South can snow ski in New York. People from one country can meet people from other countries. All of this and more happens because of an efficient transportation system.

Modern transportation has also improved and expanded public services. Many cities have public transportation, such as subways. In emergency situations, specialized vehicles bring help quickly. Fig. 15-18.

Negative Impacts—The Challenges

Just as transportation can increase the value of an item, it could also cause the value to decrease. If too much of the same product were available at a particular place, the value would go down.

Fig. 15-17. Transportation makes more products available to consumers at competitive prices.

Fig. 15-18. Modern transportation vehicles help us cope with emergencies.

Americans have come to rely upon the automobile as our most popular transportation system. A negative impact of this is that most major cities experience large traffic jams each day. Sometimes a traffic jam becomes a gridlock. A **gridlock** means that no vehicles can move in any direction.

The costs of transportation can be high. People must pay high taxes for the construction and maintenance of public ways and transportation facilities. Other costs cannot be measured as easily as dollars. Homes and land must sometimes be given up to make way for new roads and airports. Thousands of lives are lost each year due to transportation accidents.

Fascinating Facts

The first public mass transportation system was proposed back in the 1660s by the French philosopher and scientist Blaise Pascal. He suggested setting up a system of horse-drawn coaches in Paris that would travel along set routes at set times. The service began in 1662, with four coaches covering the route every eight minutes.

Transportation also creates pollution. Discarded or abandoned vehicles cause land and visual pollution. Vehicle exhaust smoke pollutes the air. However, new technology is being developed to help control this. Fig. 15-19. People must do their part in protecting the environment from these types of pollution.

Fig. 15-19. Locomotives used to produce a great deal of smoke. Today, thanks to advances in technology, modern locomotives contribute much less to air pollution.

As you can see, transportation has both positive and negative impacts. Most people feel there are more positive than negative impacts. Understanding how transportation systems work will help us make decisions about how to improve transportation and reduce the negative impacts.

Apply Your Thinking Skills

1. Have you ever been in a traffic jam or gridlock? Can you think of any ways our society could eliminate traffic jams?
2. Can you name any vehicles that do not pollute? What are they?

Fig. 15-20. This device can help tell where a package is located at all times.

Feedback

All systems are made to get a desired output. We need some way to see if the results of the system are what we wanted. The feedback loop is the part of the system that checks the output. If the output is what was desired, the system worked correctly. If the output is not what was desired, the inputs or the processes of the system must be changed.

In transportation systems, feedback can take many forms. If a package was to be shipped to a specific place on a specific day, a feedback loop would check if it arrived on time. The feedback might be as simple as a phone call to the destination or as complex as a computer system that tracks the package. Fig. 15-20. If the package was late, a part of the shipping process may need to be changed.

Feedback also comes from customers. Imagine that two airlines fly between the same two cities and one has a much lower ticket price. Most people would probably fly on the cheaper airline. Low sales would tell the more expensive airline to lower prices or go out of business. The number of tickets sold provides a feedback loop.

Apply Your Thinking Skills

1. Is the speedometer in an automobile part of a feedback loop? What does the driver do to control the output (speed) in this case? Does the driver change the input or the processes?
2. Are you ever part of a feedback loop when you buy things? How?

Global Perspective

"It's a Bird . . . It's a Plane . . ."

. . . It's a blimp! Actually, because of the enormous size and distinctive shape, few people would mistake a blimp for any other kind of flying object. These airships are self-powered by engines and propellers, as is an airplane. However, they have no rigid structure to help them hold their shape. The gas that makes the blimp lighter than air is contained by the strength of the fabric from which it is made, giving the blimp its unique shape.

Forerunners of today's blimps were dirigibles. Dirigibles looked like and operated very similarly to today's blimps, but they also had a strong internal skeletal structure. Dirigibles were used frequently from 1900 until about 1940 to transport people and cargo.

Many people considered dirigibles superior to airplanes as a means of transportation. One such person was Count Ferdinand von Zeppelin, a former German cavalry officer who worked determinedly to develop a dirigible that he hoped would rule the skies. Count von Zeppelin's goal seemed realized for a little while, but then disaster struck. In the early 1930s, the Zeppelin Company constructed the largest and most luxurious airship ever built, the *Hindenburg*. While on a trip to the United States in 1937, the *Hindenburg* burned and crashed, causing 36 deaths and ending the dreams of great airships.

Fig. GP-15. The *Hindenburg* during its trip to America in the 1930s.

Why did the *Hindenburg* burn and crash? As mentioned earlier, airships rise above the earth because they are filled with a gas that is lighter than air. Hydrogen, the lightest of all gases, was used to lift the *Hindenburg*. In addition to being light, however, hydrogen also burns very easily and quickly. No one knows whether sabotage (deliberate destruction) or static electricity started the fire, but hydrogen was blamed as a major cause of the disaster. From then on, up to the present day, blimps and dirigibles have been filled with helium, which does not explode or burn.

Because of their slowness, blimps and dirigibles may never again become a popular means of transportation. However, lighter-than-air enthusiasts see possibilities of using blimps or dirigibles not only in advertising and sportscasting but also as surveillance platforms to watch over suspects in the war against drugs or as in-the-air radar stations.

Extend Your Knowledge

1. Hot-air ballooning is becoming increasingly popular as a pastime. Find out how a hot-air balloon is flown and the extent to which its flight can be controlled. Compare flying a hot-air balloon with operating an airship.

Chapter 15 Review

Looking Back

Transportation is the movement of people, animals, and things from one place to another using vehicles. Ways, routes, and support facilities are needed in order for vehicles to operate.

People who are transported are passengers. Any other transported item is called cargo or freight. The two basic types of cargo are bulk, or loose cargo, and break bulk cargo.

A change in an item's value caused by transportation is called time and place utility.

Transportation is a system. A transportation system consists of input, processes, output, and feedback. People, information, materials, tools and machines, energy, capital, and time are necessary inputs. The two basic groups of processes are management and production. The basic output of transportation is the movement of people or cargo. Feedback checks the output. Transportation systems produce both negative and positive impacts on people and the environment.

Review Questions

1. What is transportation?
2. What is the difference between a way and a route?
3. What is the difference between bulk and break bulk cargo? Give an example of each.
4. Explain how time and place utility can cause the value of an item to increase.
5. List and give examples of the seven basic types of inputs needed in transportation systems.
6. Which input is so important that no system could exist without it? Explain why.
7. From what limited resource do most vehicle fuels come? List four of those fuels.
8. Name and explain the two types of capital.
9. List three positive and three negative impacts of transportation. Try to include one or more not given in the book.
10. What is gridlock? Which type of transportation industries might experience gridlock?

Discussion Questions

1. Highways are owned and maintained by federal, state, and local governments. The money for these roads comes mainly from taxes. Is it fair to someone who does not own an automobile to pay these taxes? Why or why not?
2. If you deliver newspapers on your bicycle, are you involved in transportation? Are you part of a larger system? If so, what part of the system would you be?
3. What types of fuels do you think will replace oil-based fuels like gasoline in the future?
4. Automobiles give us the freedom to go where we want to go whenever we want. Is this freedom worth the pollution that autos produce?
5. Should we use transportation systems to continue to explore our solar system and the universe? Is this exploration worth the billions of dollars it costs? Why or why not?

Chapter 15 Review

Cross-Curricular Activities

Language Arts 

1. There are many career opportunities in the field of transportation. List as many jobs as you can under the following headings: Input, Processes, and Output.
2. Using your lists of career opportunities, detail which jobs require special educational skills.

Social Studies 

1. Get a road map of the Midwest and Plains states. Use the map to chart a trip from Springfield, Illinois, to Denver, Colorado. As you plan the trip, write down all the interstates, county roads, and state highways you will take to get to your destination. Make an estimate of how many miles your trip will be. After everyone in the class has completed the same task, compare your routes and your mileage.
2. Look ahead toward future trends in the automotive industry. Find out information on projected trends for tomorrow's cars. How will the inputs and outputs change? Which of these two processes will change the most and why?

Science 

1. The invention of the wheel marked the true beginning of efficient transportation. The wheel is one of the six simple machines. It works by reducing the force needed to move a heavy object. Try this: Use a finger to push a food can (a 16-oz. can of tomatoes, for example) on its bottom across a tabletop. Now do the same thing with the can lying on its side. Which requires less force?
2. The first wheels on transportation vehicles were most likely made of wood. Tires, inflated with air, provide a smoother, safer ride because air is elastic. When tires hit a bump the air is suddenly squeezed, absorbing the force of the jolt. Fill a small plastic bag with air and hold it closed. Place a heavy object such as a book on it. Push down on the book. You will see that the air supports the book, but bounces back when you release the pressure.

Math

It is easy to figure your gas mileage (miles per gallon). Each time you get gas, record both the mileage on the odometer and the number of gallons needed to fill the tank. To calculate gas mileage, you divide the miles traveled since your last fill-up by the number of gallons needed to fill the tank.

1. Debbie stops in a service plaza to put fuel into her car. She needs 12.8 gallons of gas to fill her tank. If it has been 332.8 miles since Debbie put fuel in her car, how many miles did she drive for each gallon of gas consumed?
2. The odometer on Ernie's sports car read 39,406.3 the last time he put fuel into his car. After filling his gas tank with 14.2 gallons, he recorded his mileage as 39,860.7. What was Ernie's gas mileage?

Chapter 15 Technology Activity

Components of Transportation Systems

Overview

This activity is designed to introduce you to the five basic components of transportation systems:

- propulsion
- control
- guidance
- structure
- suspension

You will research and develop these components in designing a vehicle. It must be capable of transporting a chicken egg (passenger) along a line stretched horizontally between two points 50 feet apart. The line will be placed 4 feet above the surface of the floor. Fig. A15-1. The vehicle, with the egg on board, is to be transported along the line as rapidly as possible.

The vehicle will use a propulsion system of your choice. The loaded vehicle is to impact (hit) the finish point and have minimal (least possible) rebound without breaking the egg. All vehicles will be attached to the horizontal line with two support points (screw eyes). These will allow the vehicle to be suspended below the line. Fig. A15-2.

Goal

To design and build a vehicle that will transport a chicken egg and protect this egg when the vehicle hits a wall.

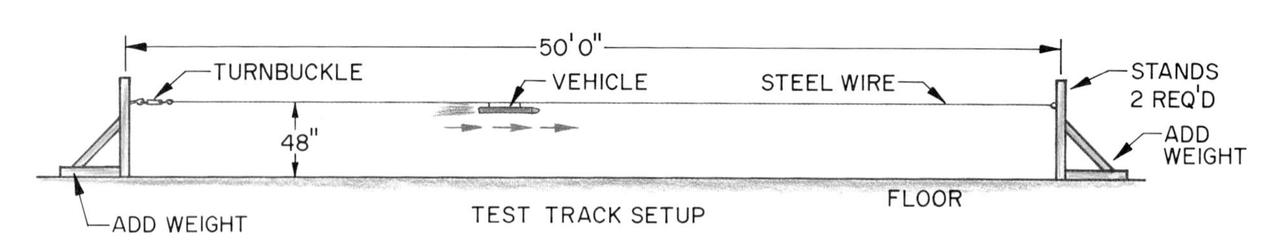

Fig. A15-1. This is what the test track should look like.

Chapter 15 Technology Activity

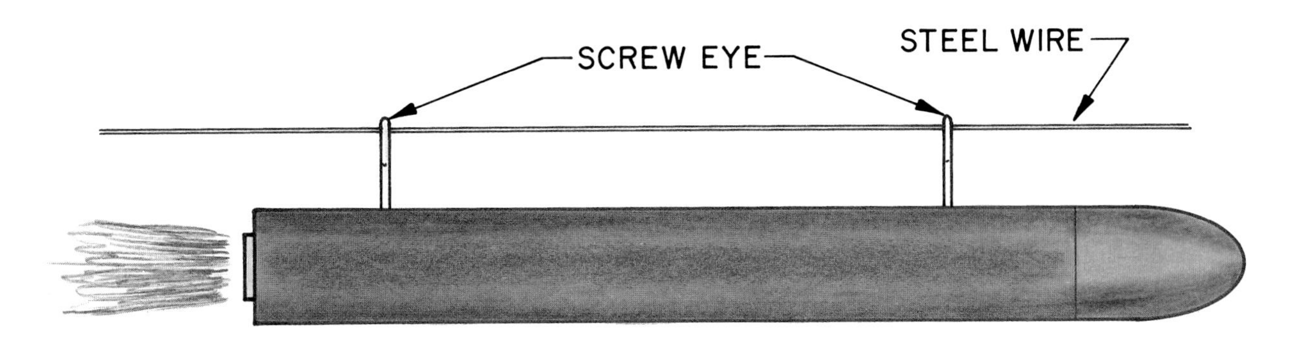

Fig. A15-2. Example of a vehicle design.

Equipment and Materials

cardboard
plywood
wood strips (assorted sizes)
urethane foam
rigid foam insulation
sheet metal (assorted sizes)
hot melt glue
wood glue
contact cement
masking tape (1/2" and 3/4")
duct tape (3")
wood screws
sheet metal screws
nails
acrylic plastic sheet
50 feet of #19 ga. galvanized steel wire for test track
turnbuckle to adjust tension of test track wire
eggs

Procedure

1. *Safety Note:* Before doing this activity, make sure you understand how to use the tools and materials safely. Have your teacher demonstrate their proper use. Follow all safety rules.
2. Follow the nine steps of the design process listed to complete this activity.
3. The first step of the design process is to state the problem. (Reread the overview.)

Material Limitations of the Vehicle

The fuselage or structure of the vehicle may be made of an industrial material such as wood, metal, plastic, or rubber. Do *not* use glass or ceramic materials. The interior of the vehicle (where the egg is to be placed) may be of any of the acceptable materials or any other method or device. The design of the vehicle cannot incorporate (use) nails, spikes, or any other sharp or pointed objects protruding from the front of the vehicle to make it stick into the finish-line board.

Design Limitations of the Vehicle

The completed vehicle must be able to fit inside a 5" x 5" x 12" box (excluding the two screw eyes used for suspending the vehicle from the track line). The maximum allowable weight of the vehicle is 24 ounces.

Method of Propulsion

Any safe form of propulsion such as heat engines, wind power, human power, or any other method is allowed. Get your teacher's approval for the propulsion system you will use.

4. Begin this activity by researching chicken eggs. Consider the shape. How would you describe the shape of a chicken egg? Note that one end is smaller than the other. Is one end of the chicken egg stronger than the other? What is a chicken egg made of? Do some parts of a chicken egg accept stress better than other parts? What is the average size of a chicken egg? What is the average weight of a chicken egg? Does the color of the eggshell indicate a difference in shell strength?

Your task is to find all the important information about the chicken egg necessary to proceed with the design of your vehicle.

5. Brainstorm ideas about how to protect the egg. Brainstorm about the type of propulsion system to be used and the structural design of your vehicle. Remember that the vehicle is to impact the finish-line board at a high speed. Will the vehicle and egg survive the impact? Make several thumbnail and rough sketches of your design ideas.

Name of Vehicle: ______________________________

Energy Source (Method of Propulsion): ______________________________

BILL OF MATERIALS

Quantity	Item	Description

Fig. A15-3. This is what your bill of materials should look like.

Chapter 15 Technology Activity

6. Select one or two of the possible solutions to the vehicle design problem. Are your solutions within the design and material limitations of the stated problem?
7. Prepare drawings of your final design, indicating its dimensions. These drawings should show how the propulsion system, structure, and egg-protecting device are to work and fit together. (Your teacher may require you to prepare a set of working drawings.)
8. Make a bill of materials based on your drawing. Fig. A15-3. Gather all of the required supplies as specified on the bill of materials.
9. Make a full-size model (prototype) of your vehicle.
10. After building your vehicle, analyze its design. Does it appear that it will be able to perform as you predicted? Will the egg survive the high-speed impact? Will the propulsion system work as planned? Is the vehicle's structure rigid enough to withstand the forces of the propulsion device and the impact?
11. The next step in the design process is experimentation. Test-run your vehicle several times with and without the egg on board. Record the time, rebound distance, and condition of the egg after each trial run. Use the following formulas to calculate actual and scale miles per hour. (*Note:* When testing your vehicle with an egg on board, place the egg in a sandwich bag in case it does not survive the impact.)

Actual MPH

$$\frac{\text{50 ft.}}{\text{time (sec.)}} \times .6818$$

Scale MPH

Actual MPH x 24

12. After all class members have completed their vehicles and test runs, have a contest. See which vehicle has the fastest time, rebounds the least, and protects the egg best.

Evaluation

1. How fast did your vehicle travel? Did your vehicle protect the chicken egg?
2. If you could design and build another vehicle, what would you do differently?

Chapter 16

Types and Modes of Transportation

Looking Ahead

In this chapter, you will discover:

- the five modes of transportation that make up the overall transportation system.
- the reasons for selecting one mode of transportation over another.
- what types of vehicles are used to move people and cargo.
- the facilities needed by each transportation mode.

New Terms

airspace
AMTRAK
barge
classification yards
commuter service
diesel-electric locomotives
Federal Aviation Administration (FAA)
fifth wheel
friction
navigable
rolling stock
sea-lanes
slurry
tractor-semi-trailer rig
unit train

Technology Focus

Cycle Entrepreneur

G. N. Gonzales is an entrepreneur. An entrepreneur is a person who starts a business. While a student at Louisiana State University, he opened a cycle sales and repair shop in Baton Rouge, Louisiana. Mr. Gonzales has been a motorcycle dealer for more than 50 years.

Since childhood, Mr. Gonzales has repaired, rebuilt, bought, sold, and traded about every kind of two-wheeled vehicle you can think of. He began working on motorcycles for his father while living in New Orleans. Today, he still repairs and rides the many different models of cycles he sells.

Mr. Gonzales knows a lot about motorcycles. In the time he has been in business, he has sold over 100,000 new cycles. If you don't count Sundays and holidays, that is nearly seven bikes a day for every day the store has been open. Some of the brands he once sold are not even produced any longer.

Today, Mr. Gonzales' store has the largest cycle showroom in the country. It has over 20,000 square feet of showroom space. That's about as big as twelve average-size houses.

Being an entrepreneur is not easy. The business idea must be a good one. The business owner must be willing to take a risk and work hard. Still, the life of an entrepreneur can also be interesting, challenging, and rewarding. It is for Mr. Gonzales.

Fig. TF-16. Even though being an entrepreneur is risky and requires hard work, for business owners such as Mr. Gonzales, it can also be challenging and rewarding.

Air Transportation

Air transportation is the fastest mode of transport in operation today. Airplanes can travel across the United States from coast to coast in five hours or less. To do this, the engines on the planes use large amounts of fuel. Fuel costs, plus the high price of building the aircraft, make air travel for short trips very expensive. However, long trips by air, usually over 500 miles, have become a good value for passenger and some cargo transportation. They save time, which, in turn, helps save money. Fig. 16-1.

Types of Air Transportation

Air transportation is divided into three basic types:

- Commercial aviation
- General aviation
- Military aviation

Companies involved in *commercial aviation* provide air transportation for a profit. Most commercial aviation planes are large. The more people traveling in a plane, the higher the profits.

General aviation includes all privately owned airplanes. These may be used for personal or business reasons. General aviation planes are usually smaller than commercial vehicles. Personal and business flights do not usually involve large groups of people.

Fig. 16-1. If time saved is money saved, air transportation may be the most economical to use.

Military air transportation includes many kinds of vehicles, large and small. Helicopters, fighter planes, bombers, and surveillance aircraft are some examples of military aircraft.

Aircraft

Vehicles that fly are called aircraft. Airplanes, helicopters, airships, and rocket vehicles are among the different types of aircraft that transport various types of cargo and passengers.

Airplanes are the most common aircraft, and come in a variety of sizes. They are flying vehicles that are heavier than air. If they are heavier than air, how can they fly? Air helps them. Let's see how it's done.

The plane's engines move the plane down a strip of pavement called a runway. Air flows over and under the wings. The shape of a wing increases the speed of the air flow on the upper surface. This reduces the pressure. At the same time, the pressure on the lower surface increases. The resulting upward movement is *lift*, and the plane flies. Fig. 16-2.

Fascinating Facts

Modern aviation dates back to December 17, 1903, with the Wright Brothers' first powered flights on a sandy beach at Kitty Hawk, North Carolina. Wilbur and Orville Wright, bicycle manufacturers from Dayton, Ohio, built that first successful airplane for less than $1,000!

In the first historic flight, with Orville as the pilot, the *Flyer I* traveled 120 feet, reached a speed of about 30 miles per hour, and remained in the air for 12 seconds. The brothers made three other flights that day. The fourth, with Wilbur as pilot, went a distance of 852 feet and lasted 59 seconds.

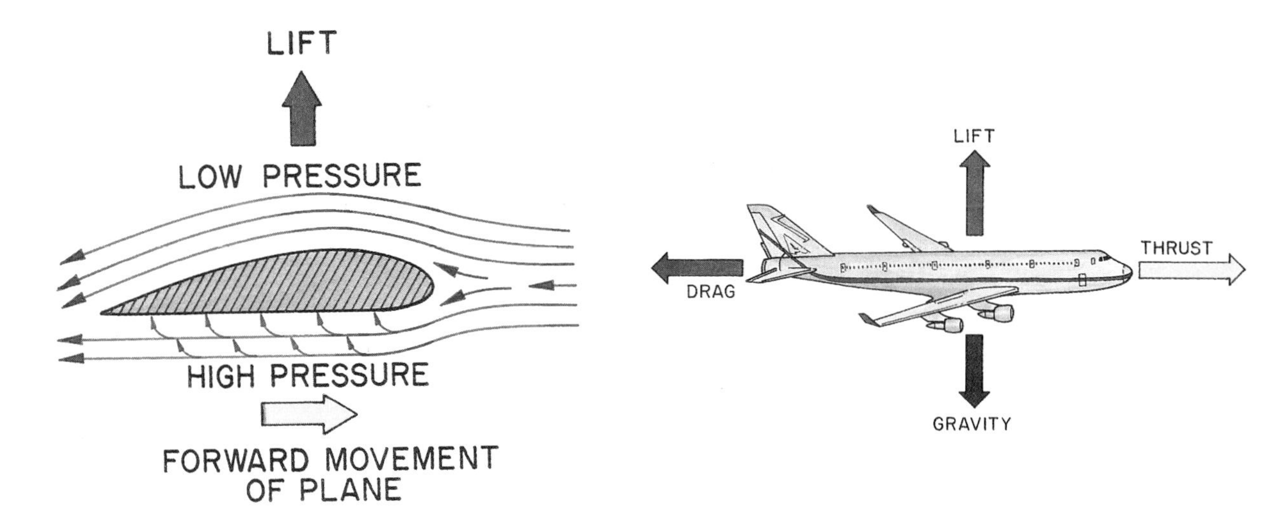

Fig. 16-2. The shape of an airplane wing actually helps the vehicle fly. It causes the air traveling over the wing to go faster than the air traveling under the wing. This creates lift. The drawing on the right shows the four forces that act on an airplane in flight.

V/STOL aircraft use the lifting shape of a wing somewhat differently. The most common V/STOL (vertical/short takeoff and landing aircraft) is the helicopter. The spinning blades of the helicopter are shaped like an airplane wing to provide lift. Some new designs give airplanes the ability to rotate their wings or engines so that they can take off like a helicopter and then fly like an airplane. Other airplanes use special wings that can provide great lifting power. They can take off and land on very short runways. Fig 16-3.

Another type of aircraft is lighter-than-air vehicles. These vehicles fly because they are filled with a gas that "floats" on air. You have probably had a helium-filled balloon before. What did it do if you let go of the string? (It floated away because helium is a gas that is lighter than air.) Hot air is also a lighter-than-air gas. A hot-air-filled balloon will rise just like one filled with helium. Fig. 16-4.

Lighter-than-air vehicles can be either rigid or nonrigid. Rigid vehicles, called dirigibles, have a strong internal skeleton. They were used frequently from 1900 until about 1940 to transport people and cargo. Nonrigid lighter-than-air vehicles are usually either hot-air balloons or blimps. These craft have no rigid structure to help them hold their shape. The strength of the fabric they are made of contains the gas and holds its shape.

Fig. 16-3. NASA's quiet short-haul research aircraft can take off and land on a very small runway due to its wing shape and engines.

Fig. 16-4. A gas-fired burner heats air to provide lift on this hot-air balloon.

The blimp, like its predecessor the dirigible, is usually self-powered by engines and propellers like an airplane. Blimps can transport people or cargo efficiently, but somewhat slowly. Today, blimps are most often used to carry television cameras and for advertising. (You may wish to reread the Global Perspective "It's a Bird...It's a Plane..." on page 331 to refresh your memory about blimps and dirigibles.)

Because they are moved only by the wind, hot air balloons are not very effective for moving people or cargo to a specific destination. They are, however, popular for recreation around the world.

Lighter-than-air vehicles are being developed that will be used for many other tasks in the near future.

Airways

The area above the earth is called **airspace**. For safety, aircraft moving through airspace follow specific routes. There are different routes for commercial, general, and military aircraft. The **Federal Aviation Administration (FAA)** is a government agency that controls all air traffic above the United States.

Facilities and Support Services

Transportation facilities are buildings or areas in which activities take place to keep the transportation system working. In air transportation, many ground vehicles as well as facilities are needed. Ground vehicles provide a variety of services. Fig. 16-5. Some bring supplies. Some clean the craft inside and out. Others move passengers and their baggage. Cargo and the planes themselves are moved by still other ground vehicles.

The airport is the main ground facility in air transportation. Aircraft move on taxiways and runways. Terminals, control towers, automobile parking lots, hotels, fire stations, and airplane hangars make up different parts of an airport. Fuel storage areas are usually located nearby. However, they must be far enough away for safety.

Fig. 16-5. Different types of ground vehicles are used to service an airplane at an air terminal.

Apply Your Thinking Skills

1. Why is it more economical for airplanes to fly long rather than short distances?
2. What type of tasks do you think lighter-than-air vehicles might perform in the future?

Rail Transportation

Some advantages of rail transportation can be seen as you look at a train. Trains are strong and tough! They have steel wheels that move on steel rails. Fig. 16-6. These are strong enough to efficiently transport large, heavy loads.

Trains offer other advantages, too. A loaded railroad car is a "rolling warehouse." It protects the shipper's cargo during transport. In addition, a single train engine hauling several cars of freight uses a lot less fuel than the several trucks it would take to haul the same amount of freight. Also, trains do not get into traffic jams in crowded cities, which can mean a big savings of time.

Passenger Service

Rail passenger service is a good example of how government and private industry work together to provide transportation. Today, all long-distance rail passenger service in the United States is provided by **AMTRAK**. The AMerican TRavel trAcK system is owned by the federal government. The trains are operated by the railroad lines that own the track the trains are using.

In the United States, most trains carry cargo rather than passengers. Most other industrial countries make much more use of railroads for passenger transportation than we do. Some European countries and Japan have high-speed passenger trains that can go more than 200 miles per hour. (See Fig. 19-9 for an example of a high-speed passenger train.)

High-speed rail systems that will travel over 300 miles per hour are currently being developed. Some of these systems are now being planned to link various American cities with high-speed passenger service. These modern trains will probably be cheaper than air transportation and almost as fast.

Fascinating Facts

Rail transportation can trace its history back to Europe in the 1500s, when miners used wheeled wagons on raised wooden tracks to bring coal or iron ore up from underground mines. They had found that horses could pull much larger loads on rails than they could on flat ground. This is because very little **friction** (resistance to motion) is produced when hard wheels roll on a hard surface.

Fig. 16-6. Our rail transportation system is a steel-wheel-on-steel-rail system, making it strong enough to transport large, heavy loads.

Regional and city rapid transit trains provide commuter service in some large cities. **Commuter service** is regular back-and-forth passenger service. For example, in New York City, many students travel to and from school on commuter trains. Fig. 16-7.

Freight Service

The railroad industry offers more freight (cargo) service than passenger service. Trains serve over 50,000 towns and cities in North America. There are two basic types of freight trains, unit trains and regular freight trains. Fig. 16-8.

A **unit train** provides efficient transportation. It carries the same type of freight in the same type of car to the same place time after time. Some unit trains are not owned by the railroad lines. The owners may pay the lines to pull their cars.

Regular freight trains offer a variety of short- and long-haul services. Cars are arranged on the train according to the type of cargo and final destination. The cars may be switched to several different trains before they are finally unloaded.

Fig. 16-7. Subway trains provide fast commuter service below the busy city streets.

Fig. 16-8. A unit train (A) is owned by a company and carries only that company's products. However, a regular freight train (B) may be made up of many different types of cars going to many different locations.

Rolling Stock—the Vehicles

Rail transportation vehicles are called **rolling stock**. There are three different groups of rolling stock:

- Engines
- Railroad cars
- Maintenance vehicles

Engines today are much safer and more comfortable than engines of the past. However, these modern locomotives still serve the same purpose—they pull the train. Most locomotives use diesel engines to turn electrical generators. The electricity produced then powers traction motors that turn the wheels. These combinations of engines and motors are called **diesel-electric locomotives**.

Railroad cars can transport many different forms of cargo. The most common car is the boxcar, a fully-enclosed car that can carry about anything that can be packed into it. Other cars are more specialized. For example, liquefied gas cars see only limited use.

Maintenance vehicles are used to keep rails clean and safe. Track inspection cars, brush-cutting machines, and hoist cars are examples of maintenance vehicles.

In the United States, all tracks are built to the same gage (distance between rails). This allows engines and cars from one line to interchange and roll on the tracks of any other line.

Facilities

Classification yards, shops, and terminals are facilities in rail transportation. They are connected by railroad tracks.

Classification yards are also called *switch yards*. This is where trains are "taken apart and put together." A train pulls into the yard and the cars are disconnected and sorted according to their destinations. The reclassified cars are rolled onto tracks with other cars going in the same direction. *Switch crews* and *dispatchers* are responsible for arranging the cars according to their final destinations. Fig. 16-9. Modern classification yards use computers to efficiently sort and classify cars. When the trains are made up, *road crews* couple (connect) multiple engines (more than one) to the train. Multiple engines are used to produce more power to pull large/long trains.

Each railroad line operates many repair and maintenance shops. Some are portable. These can be moved to where they are needed. Others are permanent. They are located at a particular place and cannot be moved.

In repair facilities, worn wheels are ground as round as new. Broken parts are replaced or rebuilt. Tons of grease and oil are applied to moving parts.

Terminals serve as meeting places for train crews. Business offices and communication facilities are housed in terminals. Two types of terminals are used, freight terminals and passenger terminals. *Freight terminals* are loading docks and storage places for cargo. At *passenger terminals*, people buy tickets, check their baggage, and wait for trains.

Apply Your Thinking Skills

1. Modern locomotives are powered by diesel-electric engines. What was used to power locomotives in the past?
2. Can you think of any benefits of having more passenger trains in the United States?

Fig. 16-9. A dispatcher keeps each train on the right track and makes sure the movement of freight is organized and continuous.

Water Transportation

Water transportation was the first major mode of moving people and cargo. Using the natural lakes and rivers, explorers were able to travel into areas where pack mules could not go.

Water transportation costs less than most other forms of transport. Cost is figured by the ton mile. A ton mile means moving one ton a distance of one mile. To take advantage of this low-cost transportation, a shipper must expect slower travel. Also, water transportation is limited to areas where there are navigable waterways. A waterway that is **navigable** is wide enough and deep enough to allow ships or boats to pass through.

Types of Water Transportation

There are two main types of water transportation companies, cargo lines and passenger lines.

Like rail transportation, most water transportation business involves moving cargo. Cargo movement that takes place on water within the United States is called *domestic inland shipping*. Foreign shipping is either *importing* (bringing in) goods from another country or *exporting* (sending out) goods to another country.

Passenger lines include luxury or cruise ship lines, paddlewheel riverboats that offer short recreational trips, and public transportation. Lake and river ferries are good examples of public water transportation.

Vehicles

Water vehicles are also called *vessels*. There are several different types of water transportation vehicles. The "workhorse" of domestic water transport does not even have an engine. It is a large floating box called a **barge** or lighter. (A barge is *lighter* than a ship.) Barges are tied securely together to form a *tow*, which towboats and tugboats push and pull through many of the major rivers, canals, and lakes in North America. Fig. 16-10. Barges can even be loaded onto ships and transported across oceans. Barges are used to ship bulk cargo such as petroleum products, coal, and grain.

Towboats and tugboats are relatively small vessels. They operate like railroad locomotives. These boats move barge tows and larger ships. A towboat has a flat *bow* (rhymes with "now"), or front end. It *pushes* barges. A tugboat has a pointed bow. It *pulls* barges. Fig. 16-11.

Fascinating Facts

The sail—a means of harnessing the natural power of the wind to efficiently power water vehicles—was used at least 5,000 years ago by Egyptians traveling on the Nile River. According to one theory, a sailor first thought of using the wind to power his boat when he noticed that he didn't have to row whenever a breeze caught his cloak!

Fig. 16-10. This cargo-filled barge is being moved as a tow.

Tugboat

Towboat

Fig. 16-11. Notice the difference in the bows of these "locomotives of the water."

Other types of ships include containerships (more about these in Chapter 17), tankers and supertankers, general cargo ships, and bulk carriers.

Cruise ships are like floating hotels. These passenger ships contain restaurants, recreation facilities, staterooms (bedrooms), water treatment plants, storage warehouses, and communication stations.

Some communities located on rivers or large lakes have large paddlewheel boats. Passengers are taken on short trips, usually lasting only a few hours. The trips may offer scenic tours, dinner, dancing, and other recreational activities. Ferries are used to transport people, and sometimes their cars, across rivers, lakes, and bays.

Waterways and Sea-lanes

Water transportation follows either inland waterways or sea-lanes. *Inland waterways* are navigable bodies of water such as rivers and lakes. *Channels* (deep paths) must be kept deep enough for large vessels, and these channels must be marked with buoys, lighthouses, and other markers. In the United States, the U.S. Army Corps of Engineers is responsible for channel maintenance. Locks and dams, needed to keep rivers navigable, are also built and maintained by the Corps of Engineers.

Sea-lanes are shipping routes across oceans. These do not require maintenance. The countries of the world agree in general on certain boundaries and locations of sea-lane routes.

Facilities

Major water transportation cargo facilities are ports, docks, and terminals. These places provide areas for loading and unloading equipment, needed services and supplies, cargo warehouses, and ship company offices. Passenger ships require the same facilities. Terminals built for passengers are more comfortable than those built for cargo.

Apply Your Thinking Skills

1. What makes water transportation cost less than other forms of transportation?
2. Why do tugboats and towboats have different shaped bows?

Highway Transportation

The most common mode of transportation is highway transportation. You are probably more familiar with this mode than you are with any other. Cargo and people are transported over millions of miles of highways, streets, and roads. The major advantage of highway transportation is the independence given the operator. As long as there is a road, a person can travel almost anywhere, anytime day or night. A major disadvantage is that highways all across the United States are becoming overcrowded, leading to traffic accidents, traffic jams, and air pollution.

Personal Transportation

People commonly use highway transportation for their own personal benefit. This is especially true in the United States. More people here use the highway for personal transportation than in any other country in the world.

Commercial Transportation

Commercial use of highways is very big business. This use is divided into two classes: "inter" and "intra." *Inter* means between. *Intra* means within. Commercial users of highways may travel *intercity* or *interstate*. This means they travel between cities or between states. They may also operate *intracity* or *intrastate*. That is within a city or within a state. If you were making a trip from Virginia to Arizona, how would your trip be classified? . . . That's right. Your trip would be classified as interstate.

Bus and taxi companies offer commercial passenger service. However, the most common commercial use of the highways is for freight service.

Types of Freight Operations

There are two types of freight operations:

- Motor freight carriers
- Owner-operators

Many companies own vehicles and hire people as drivers. As you travel the highway, you see many trucks with the same company name. These companies are *motor freight carriers*. Some carriers offer regularly scheduled pick-up and delivery.

In recent years, a new type of commercial trucking has become popular. The owner of a truck also operates it. He or she is an *owner-operator*. An owner-operator usually owns only one truck. This vehicle may be both office and home for the owner. Owner-operators hire out their vehicles and their driving services. They usually travel wherever the load needs to go.

Freight-Hauling Vehicles

A variety of vehicles are used in the trucking business. The cargo and the length of the trip determines the type of vehicle used. Three common types are:

- Single unit
- Tractor-semi-trailer rigs
- Trailer units

Single unit trucks are one-piece vehicles. They are also called *straight* trucks. The engine is in front, and the drive wheels are in the rear. The bed is permanently mounted over the drive wheels. Single units are used for carrying one type of cargo and for local hauling.

A **tractor-semi-trailer rig** is a combination of a tractor and a semi-trailer. Fig. 16-12. The *tractor* is the base unit that pulls the *semi-trailer*, which contains the cargo. They are called *semi* because the trailers have no front wheels. The rear wheels of the tractor support the front end of the trailer through a **"fifth wheel"** hook-up arrangement. Any

Fig. 16-12. This tractor-semi-trailer rig is a motor freight carrier.

semi-trailer can be hooked up to any tractor. Tractor-semi-trailer rigs are sometimes called "18-wheelers." (The tractor has a total of 10 wheels, and the trailer has two axles with 4 wheels on each axle.) These trucks can haul many different loads, depending upon the type of semi-trailer.

Sometimes *trailer units* are connected to the back of a semi-trailer to allow a single tractor to pull more freight. These units have both front and back wheels. Some states allow as many as two trailer units behind a tractor. Imagine the size of the engine that the tractor would need to efficiently pull that much! The total length could be over 140 feet long.

There are many different types of trailers and semi-trailers. A box shape called a *van body* is used most often.

Facilities

Commercial use of highway transportation continues to increase. The need for physical facilities is also increasing. Services must be provided before, during, and after trips. Cargo must be loaded, unloaded, and stored. Terminals and truck stops offer these services for both companies and independent operators.

Good road systems are necessary for a highway system to be effective. Roads originally were just paths that people and later animal-drawn vehicles traveled along. These roads were usually paved only in towns and cities. Paving is the covering of a road with a hard surface. Without paving, wheeled vehicles sank into the mud whenever it rained.

Stones, wood planks, bricks, and small crushed rocks were most often used for paving. When automobiles replaced slow-moving horse-drawn vehicles, much better roads were needed. Roads had to become wider to allow rapid travel in both directions. Better forms of paving became necessary to allow automobiles and trucks to travel safely at high speeds. Today, asphalt (a mixture of sand, crushed rock, and a tar-like substance) and concrete are most often used to pave our highways.

In the past, little planning was given to road construction. Roads were built to connect cities and towns as they were needed. As cars and trucks have become able to travel faster and faster, better and better roads have become necessary. The highway system of the United States now includes an interstate superhighway system that was designed for safe travel up to 70 miles per hour. Fig. 16-13.

There are millions of automobiles, motorcycles, and trucks using our highway systems every day. Highway designers and engineers must carefully plan how to try to keep all those vehicles moving smoothly and safely. As more and more people own automobiles and crowd onto the roads, designing these systems is becoming very difficult.

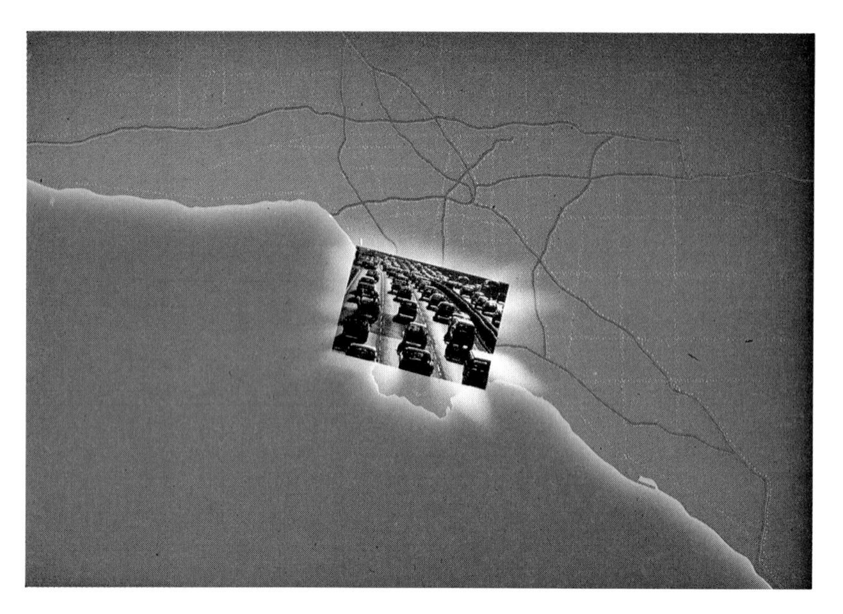

Fig. 16-13. The interstate highway system allows people and cargo to travel rapidly between all areas of the United States.

Highways are built and maintained by local and state governments using tax money. Taxes are collected on gasoline and diesel fuel by both the federal government and state governments. Some of the federal taxes are returned to states to help build and maintain road systems. Some states help pay for their highway systems with toll roads on which drivers must pay a certain fee, or toll, for each mile that they drive. The heavier the vehicle, the more the fee per mile. The fee is collected at toll booths, which may be automated or manned by a person collecting tolls.

Apply Your Thinking Skills

1. What forms of transportation do you think that people in other countries use most? Explain your answer.
2. Why are all weather roads necessary to our transportation system?

Pipeline Transportation

Pipeline transportation offers four special characteristics:

- It is the only mode in which the cargo moves while the vehicle stands still. Fig. 16-14.
- Most pipelines are buried in the ground, so they are unseen and quiet. Also, they cause no traffic congestion or accidents.
- Pipelines are laid out in straight lines across country. This decreases the transportation time.
- Theft from pipelines is difficult. Also, damage or contamination of cargo is rare.

Using Pipeline Transportation

Certain types of cargo, such as oil products, natural gas, coal, wood chips, grain, and gravel, can be transported very economically through pipelines. However, pipeline transportation is not as flexible as other modes. Service depends upon how close a customer is to a line.

Fig. 16-14. In pipeline transportation, the cargo moves and the vehicle stands still.

Cargo is shipped in batches. Solid materials are moved in a slurry. A **slurry** is a rough solution made by mixing liquid with solids that have been ground into small particles. The batch is put into the system. As it goes in, the amount is measured and recorded. The material is then pumped or pushed by compression (squeezing) through the lines.

Pipeline transportation is one-way. If products also need to flow in the opposite direction, two lines are installed.

Pipelines

There are over one million miles of pipelines crossing the United States and Canada. These lines range in size from two inches in diameter to fifteen feet in diameter. Most of these lines carry natural gas or petroleum products. Fig. 16-15.

Pipelines are made of either steel or plastic. There are three types of lines, gathering, transmission, and distribution.

Gathering pipelines are used to collect the cargo from the suppliers. These lines meet at central holding tanks and pumping stations.

Transmission pipelines are the main long-distance lines that transport the cargo. The batches end up at terminals and tank farms. Fig. 16-16.

Distribution pipelines deliver the cargo from the terminals to the customers. Small distribution lines probably bring water and natural gas right into the building where you live.

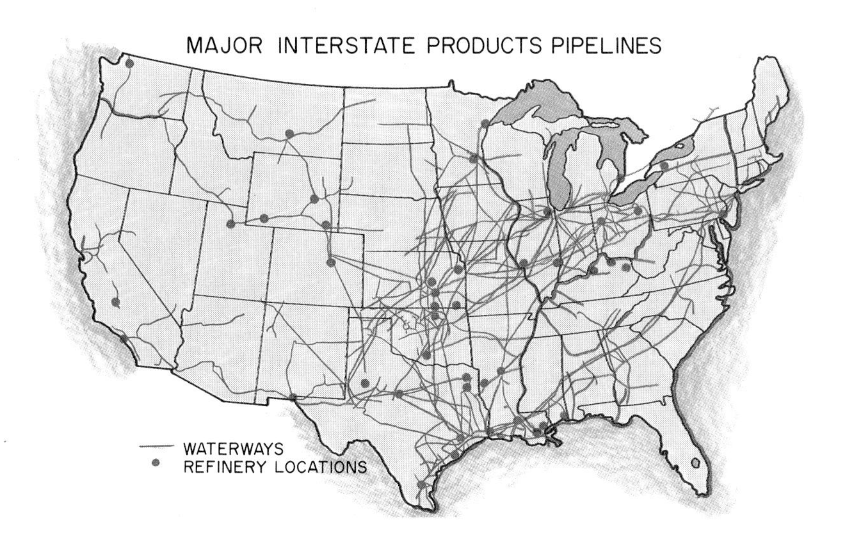

Fig. 16-15. Because you can't see them, it's difficult to imagine how many miles of pipelines are under the country.

Facilities

The facilities that service the pipelines are above ground. These include:

- Pumping stations
- Control stations
- Measuring stations
- Exchange stations

Pumping stations use pumps to move the cargo down a pipeline. Natural gas may be compressed up to 2000 pounds per square inch. It does not travel more than 15 miles per hour. Liquids such as crude oil travel 2 to 5 miles per hour. To keep cargo moving, pumping stations pump the batches every 30 to 150 miles along the pipeline. Fig. 16-17.

Control, measuring, and *exchange stations* are also located along pipelines. These facilities make sure the customers safely receive the correct size batches of their cargo.

Pipelines can become clogged. To prevent this, a scraper tool called a "pig" is regularly pushed through the pipeline. Pigs may be round like a ball or tubular shaped. Fig. 16-18. Some pigs have brushes on the outside. Others carry instruments that take readings on conditions inside the pipeline.

Fig. 16-16. Oil is stored in tank farms until it is needed.

Apply Your Thinking Skills

1. Why does pipeline transmission go in only one direction?
2. What might be some difficulties in transporting grain in a pipeline?

Fig. 16-17. Operators check the pumping stations, but most operations are remotely controlled by computers many miles away.

Fig. 16-18. Pipeline pigs are actually scrapers that are moved through the pipelines to clean them. Some pigs carry instruments that check the internal condition of the pipeline.

Investigating Your Environment
Environmental Invaders

Water vessels have been used for hundreds of years to transport people and cargo. Unfortunately, ships may also be responsible for transporting zebra mussels in their "ballast" water. Ballast is a heavy material placed in the bottom of a ship to make it stable. Often, sea water is pumped in for this purpose. Fish, mollusks, crabs, lamprey eels, and worms have all been transported from one part of the world to another in this way.

Zebra mussels traveled from eastern Europe to North American waters around 1985. Since then they have created disaster! They get into bodies of water and eat all the phytoplankton (tiny water plants). This starves out the other creatures who feed on these organisms.

Zebra mussels are able to attach themselves to any hard surface. They survive and reproduce well in intake pipes. Intake pipes take in water at industrial and water treatment plants. The constant flow of water in these pipes keeps food coming to the mussels all the time. They multiply so fast that they clog up the pipes. It has been estimated that a mother mussel will produce as many as 50,000 eggs a year.

Trying to get rid of the zebra mussels is definitely no easy task. Their larvae (mussels at an early, immature stage, before the changes and growth that come with adulthood) are so small that they slip right through filters on the intake pipes. Scraping or scrubbing them off works, but it is very expensive and they come right back. They have no known natural predators, although scientists are trying to find some microorganism that will prove toxic (poisonous) to them. Chlorine will kill them, but it cannot be widely used because it is harmful to the environment.

One scientist found that the mussels died when she put them in distilled water that was made to match the "recipe" of the EPA standards for "pure" water. The scientist finally determined that it was the potassium compounds in this water that killed the mussels. Unfortunately, the scientist found that it took fairly high concentrations of potassium to kill large numbers of zebra mussels. Such large doses also killed other mussels. Further tests are being done with potassium at lower concentrations because it was also found that low levels of potassium seem to prevent zebra mussels from attaching themselves to hard surfaces. This could lead to a method of preventing them from attaching to and clogging pipes.

Fig. IYE-16. Zebra mussels can attach themselves to any hard surface.

There is one positive note to this problem. Zebra mussels eat algae, simple green plants that float in and on water. Small amounts of algae can provide food for water life. However, large amounts of algae can use up much of the oxygen in the water, causing death for many of the creatures living there. The zebra mussel's appetite can help keep algae levels where they belong!

Take Action!

1. Hold a class debate on the following issue: "Resolved, that no chemicals shall be used to control pests."

Chapter 16 Review

Looking Back

Transportation is a large system made up of five smaller systems (subsystems): air, rail, water, highway, and pipeline systems.

Air transportation is fast, but relatively expensive. The three basic types of air transportation systems are: commercial, general, and military aviation.

Rail transportation is used to efficiently transport large, heavy loads. Most long distance rail transportation in the United States is freight rather than passenger service.

Water transportation is usually slow, but it is very inexpensive per ton mile. The two main types of water transportation companies are cargo lines and passenger lines.

In the U.S., highway transportation is the most common mode. It includes personal as well as commercial forms of transport.

Pipeline transportation is primarily used to transport liquids and gases. The three types of lines are: gathering, transmission, and distribution.

Review Questions

1. What is the main advantage of air transportation? Why is this so important?
2. What does V/STOL stand for?
3. What is the main difference between a dirigible and a blimp?
4. What is the FAA and what is its purpose?
5. What do the letters AMTRAK stand for? Why is this system important? Who owns it?
6. What is commuter service? Give two examples of different types of commuter service.
7. What is gage and why is it important for all railroad tracks in America to be the same gage?
8. What does ton mile mean? Does the ton mile cost more for water vessels or tractor-semi-trailer rigs?
9. Discuss the main advantage and disadvantage of highway transportation.
10. How is it possible to transport a solid in a pipeline? What is this called?

Discussion Questions

1. Can you explain in your own words how an airplane wing allows something heavier than air to fly?
2. Why do you think other countries use passenger railroad transportation more than the U.S. does? Do you think automobiles or trains are more efficient ways of transporting people?
3. What type of transportation system would you use to ship coal from Minneapolis, Minnesota, to Memphis, Tennessee? What type would you use to ship oil? Fresh fruit? An important document that must be signed? For each example, explain why you chose the transportation system you did.
4. Pipelines cannot cause traffic jams or accidents. Are there any accidents that can happen with pipelines?
5. List the five transportation systems. Next to each system list what you think are at least one advantage and one disadvantage of each system. (Consider important factors like speed, cost, safety, load size, and types of cargo.)

Chapter 16 Review

Cross-Curricular Activities

Language Arts 

1. Select a major city in a neighboring state. What would be the most economical means of transportation from your home to that city for two adults? Describe all means of transportation and detail the cost.
2. Outline the five modes of transportation described in this chapter. Use the complete sentence form for your outline.

Social Studies 

1. Choose any war (or major military conflict) between the Civil War and Operation Desert Storm. Do research, then briefly explain all the methods of transportation used by the military in that war. Is there one mode of transportation that seems to be most effective in helping a nation win a war?

Science 

1. The reason aircraft can fly is stated in Bernoulli's principle: Moving air creates low pressure, which, in turn, creates lift. You can demonstrate this principle with a thin, flexible strip of paper. Hold the paper horizontally by one end. (The other end will sag.) Blow an inch or two across and above the top of the unheld end of the strip. What happens? This occurs because the moving stream of air created when you blow forms a region of low pressure above the strip.
2. Archimedes' principle explains buoyancy—the basis of nearly all water transportation. Buoyancy is the result of the force that water or any other liquid exerts on an object. This force (an object's buoyancy) is equal to the weight of the water displaced by the object. If the force is greater than the weight of the object, the object floats. If it is less, the object sinks. Try immersing a variety of objects in water, from a paperclip to a piece of aluminum foil. In each case, tell which is greater: the weight of the object or the upward force of the water.

Math

1. Americans spend one billion hours a year stuck in traffic, which wastes two billion gallons of gasoline. Using current gasoline prices, figure out approximately how much money this wastes in one year.

Chapter 16 Technology Activity

Aerodynamics of Gliders and Airplanes

Overview

Many inventors in the late 1800s and early 1900s studied the theory (unproved principle) of flight. Some, like Otto Lilienthal of Germany, Clement Ader of France, Sir Hiram Maxim of England, and Octave Chanute and S. P. Langley in the United States, actually built and flew non-powered gliders based on their observations and discoveries. Probably the most famous aircraft inventors are Orville and Wilbur Wright of the United States.

All of these inventors believed that air would support the large wings of their gliders and hold the gliders aloft. After many experiments, they discovered that gliders and powered aircraft can remain aloft because of a set of scientific principles called aerodynamics. *Aerodynamics* is the science of air in motion interacting with a moving object. This interaction produces or results in a force. All aircraft in flight are constantly being pulled toward earth by gravity. Other aerodynamic forces acting on an aircraft are lift, thrust, and drag. Fig. A16-1. Gliders and powered aircraft are designed to use these forces to control flight. These forces are considerations in aerodynamic design. Fig. A16-2.

Goal

This activity is designed to help you examine the theory of flight. You will build two different types of simple, model gliders.

Equipment and Materials

8 1/2" x 11" typing paper or construction paper
paper clips
balsa wood (C grade) 3/16" x 3" x 14", 3/16" x 1/2" x 14 1/2", 1/16" x 3" x 8"
wood glue
modeler's glue or cement
scales (12")
triangles
modeling knives
pencils
sandpaper
1" x 1" x 4" blocks of wood
straight pins
tape measures
stopwatch

Procedure

Paper Glider

1. *Safety Note:* Before doing this activity, make sure you understand how to use the tools and materials safely. Have your teacher demonstrate their proper use. Follow all safety rules.
2. The first gliders you will make will be of a simple design made by folding an 8 1/2" x 11" piece of paper.
3. Plans are provided in Figs. A16-3 and A16-4. Make *two* gliders. Carefully fold

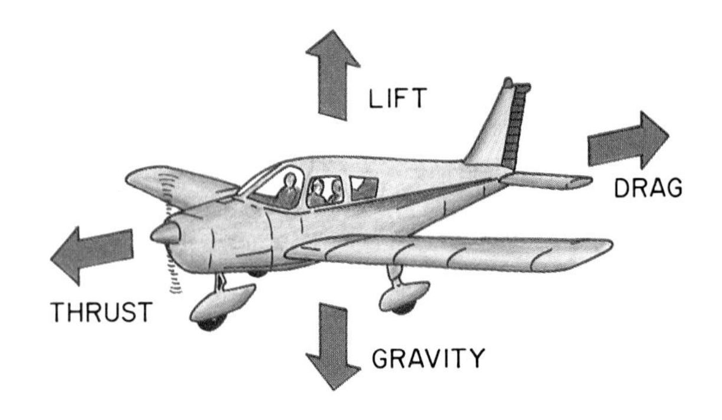

Fig. A16-1. The four aerodynamic forces that act upon an airplane in flight.

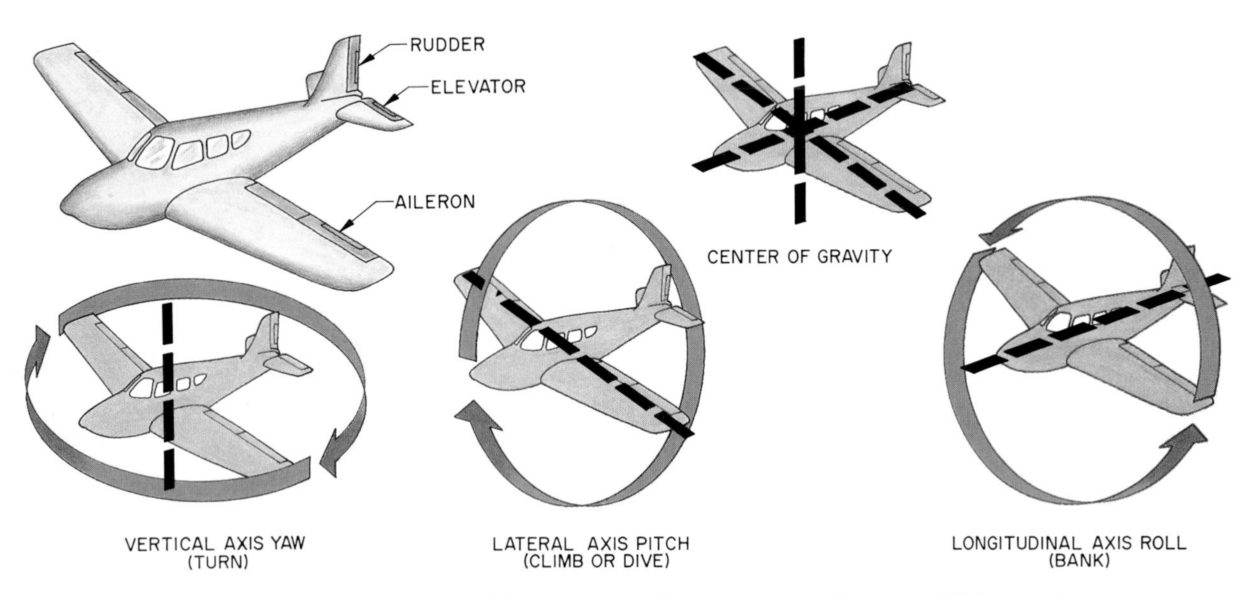

Fig. A16-2. Movable parts of an airplane—rudder, elevators, ailerons—are used to control flight. An airplane can turn in three different ways. It has three axes of rotation as shown. Each axis passes through the plane's center of gravity (center of total weight or balance point).

8 1/2" x 11" paper into the shape of the glider (Fig. A16-3).

4. One of the gliders is to be made for speed. The purpose of this glider is to show how sleek, narrow shapes "flow" through the air with little drag. The wings are shaped so there is enough lift to push on the wings, but not enough drag to slow the glider during its flight.
5. The second paper glider is to be made for maximum time aloft. To achieve more time aloft, the shape of the wings must be altered to catch more lift and use the drag over the wings more efficiently. Modify the wings of the second glider by folding them into a dihedral, as shown in Fig. A16-4. Place a paper clip on the bottom center of the glider as shown.
6. Test-fly your gliders in a large area such as outside or in a gymnasium. Time all flights with a stopwatch. Record the time aloft of each glider. Measure the distance each glider traveled in the air.

Wood Glider

7. The next model glider you will construct is made of balsa wood. Balsa wood is stronger than paper and weighs more. However, it is still light enough for use in the design of scale model gliders.
8. Begin construction by referring to the working drawings shown in Fig. A16-5 (p. 359). Begin with the wings. These are made from one piece of balsa wood 3/16" x 3" x 14". Cut the balsa wood to the shape and size shown.
9. Next, use sandpaper to shape the leading edge of the wings into an airfoil. The sectional view shows the airfoil shape. An airfoil enables the wings to develop a lower air pressure against the top surface of the wing, providing lift needed to keep the glider aloft.

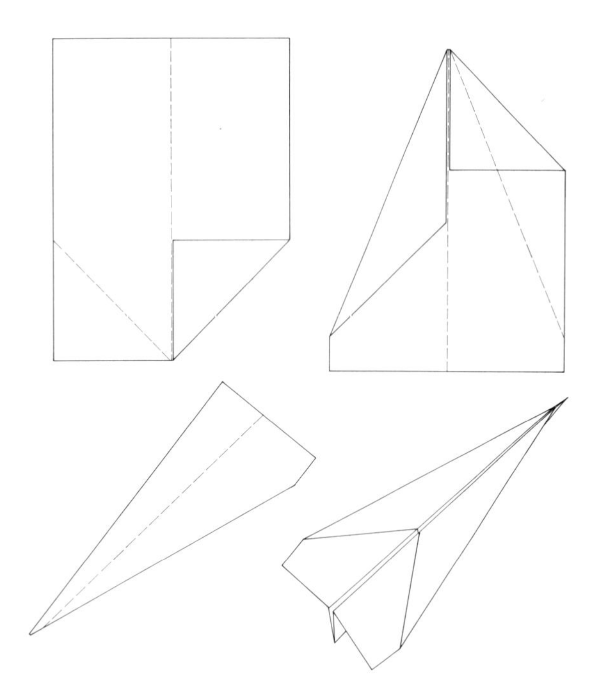
Fig. A16-3. Plans for making a paper glider.

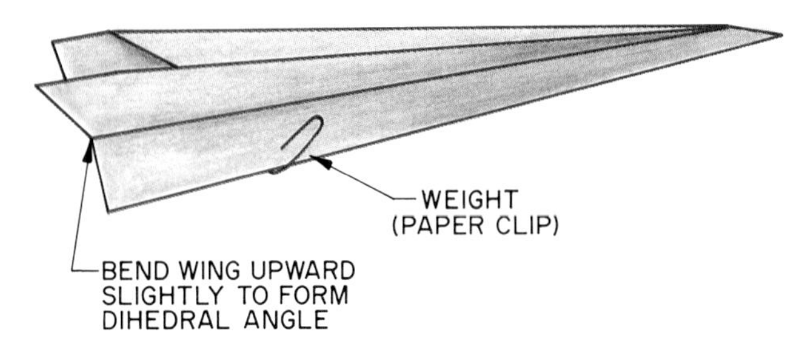

Fig. A16-4. Modified glider.

10. To increase stability, the wings must be formed into a dihedral angle. This is done by carefully cutting the 14" long wing in half, sanding a 7 1/2° angle on both sides of the wing along the cut line and gluing the wing back together. Sand from the bottom edge of the wing to the top edge. Assemble the dihedral airfoil wing according to the assembly detail drawing.
11. Use the 3/16" x 1/2" x 14 1/2" piece of balsa wood to construct the body or fuselage of your glider. Refer to the drawing of the fuselage in Fig. A16-5 for dimensions. (*Note:* Only the side view is shown.) Lightly sand the edges of the fuselage along its length. Remove the square edges only!
12. Make the tail section (horizontal stabilizer) from a 1/16" x 3 1/2" x 5 1/2" piece of balsa wood. Cut it to the dimensions shown in the drawing. Lightly sand the edges of the tail section.
13. Make a rudder as shown. The grain of the wood must run vertically through the rudder. Lightly sand the leading and trailing edges of the rudder. Do not sand the bottom. Leave it square for proper assembly.
14. Carefully assemble the glider as shown in Fig. A16-6. Make sure the wings and horizontal stabilizer are attached to the fuselage at a 90° angle. The leading edge of the wings should be 2 3/4" from the nose of the body. The rear stabilizer and rudder should be attached flush (even) with the end of the fuselage.
15. Test-fly your glider and record the results (time aloft and distance flown).

Evaluation

1. How much faster did the first paper glider go than the second paper glider? Did altering the shape of the wings change the aerodynamics of your second paper glider? Did the air currents or wind have any effect on the

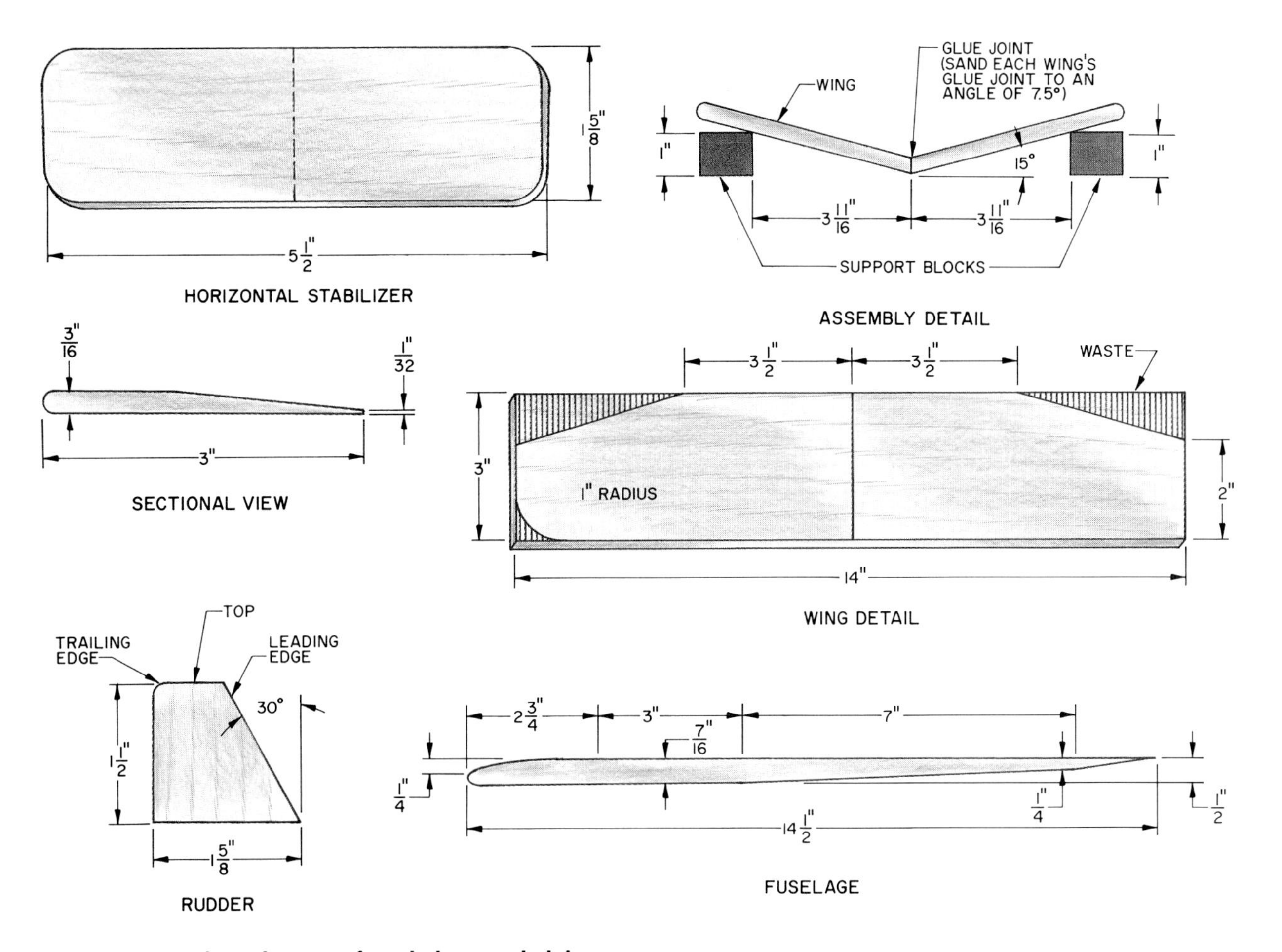

Fig. A16-5. Working drawings for a balsa wood glider.

performance of this glider? Try moving the paper clip on your second paper glider. How does this affect its flight?

2. How well did your wood glider fly? Did it appear to be balanced? What changes would make it fly better?

Credit:

Developed by Richard Seymour for the Center for Implementing Technology Education
Ball State University
Muncie, Indiana 47306

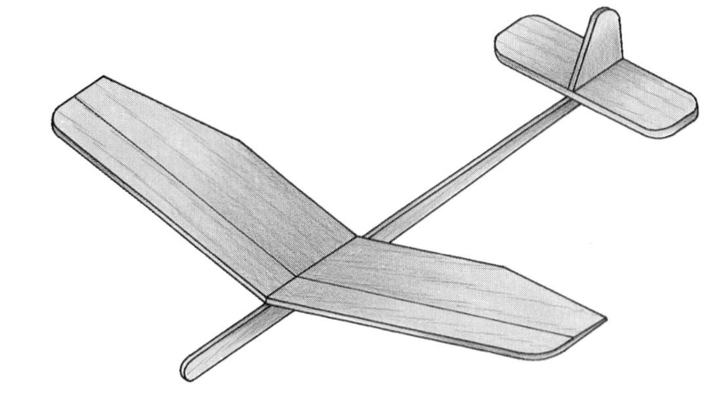

Fig. A16-6. Assembled glider.

Chapter 17

Intermodal Transportation

Looking Ahead

In this chapter, you will discover:

- what intermodal transportation is.
- how intermodal semi-trailer systems combine rail and highway transportation modes.
- what containerization is and what its advantages are.
- how liquid and solid bulk cargo are transported intermodally.
- how people can be transported on intermodal systems.
- why intermodal systems are efficient.

New Terms

containerization
container on flatcar (COFC)
containerships
escalator
intermodal transportation
piggyback
trailer on flatcar (TOFC)

Technology Focus

RR Stands for Roadrailer

If you could change the design of your vehicle to increase its miles per gallon (mpg) of fuel, would you? General Motors Corporation has. This has been done not only with its cars and trucks, but also with its fleet of railroad cars. In this case, the railroad cars are really railroad trailers. They're called *Roadrailers.*

A Roadrailer is an intermodal vehicle that stores and transports freight. A basic Roadrailer is a combination metal railcar and highway trailer. It is not quite as strong and heavy as a railroad boxcar. However, it is stronger and heavier than a highway semi-trailer. This container has both steel wheels to ride the rails and rubber tires to roll on the highways.

Changing from one mode of transportation to another takes only about four minutes. To make the change, a semi-tractor positions the trailer parallel over the rails. Then, either the truck driver or the railroad worker lowers the steel wheels. This raises the entire unit. The rubber wheels do not touch the ground or the rails. When the Roadrailer train reaches the rail terminal, the wheel change operation is reversed. Then a semi-tractor pulls the trailer onto the street or highway.

Using Roadrailer increases a train's fuel mileage. A trainload of trailers gets about 5 mpg per railcar. A train of Roadrailers gets about 20 mpg per Roadrailer. Why? There are two main reasons.

First, the Roadrailer is lighter than a railcar. Less energy is required to pull it. Second, Roadrailers are coupled together more closely than regular railcars. This makes the train more streamlined. It has less wind resistance.

Roadrailers reduce handling costs. A much smaller crew can move a trainload of Roadrailers than it takes to move, load, and unload the same number of highway trailers moved by "piggyback." In addition, no special equipment is needed.

There are some disadvantages to Roadrailers. For now, their use is limited to boxcar-type cargo. In addition, their lighter construction is not expected to last as long as heavy steel railcars. However, a Roadrailer costs so much less to move that it will still pay for itself. Roadrailers are expected to save General Motors millions of dollars every year.

Fig. TF-17. The "Roadrailer" is an intermodal vehicle that is a combination metal railroad car and highway trailer.

Advantages of Intermodal Transportation

In the past, transportation companies were very protective of their businesses. Every transportation company competed with every other transportation company. A truck company not only competed with other truck companies, but also with railroads, ship companies, and airlines as well. Today there is still competition, but things are quite different.

The service offered by each of the different modes is becoming more specialized. Companies cooperate. Often, railroads also own and operate trucks, ships, and barges. Airlines own trucklines and automobile rental agencies. Ship companies own railroads and trucking companies, and pipeline companies own fleets of highway tank trucks.

The process of combining transportation modes is called **intermodal transportation.** Fig. 17-1. A combination of two or more of the five modes of transportation (air, rail, water, highway, and sometimes pipeline) are used together to efficiently transport passengers and cargo. There is increased efficiency because much less time and labor are spent in loading and unloading cargo. This results in reduced shipping time and costs.

Apply Your Thinking Skills

1. Give an example of when it would be *necessary* to use intermodal transportation to transport cargo or passengers.

Fig. 17-1. During shipping, cargo is often transferred from one mode of transportation to another.

Intermodal Cargo Transportation

Let's take a look at intermodal cargo transportation and see how it works.

Trailer on Flatcar

Trailer on flatcar (TOFC) is an intermodal method of transporting cargo involving highway and rail modes. Trailers full of cargo are carried on flatcars.

TOFC is interesting to see. A trucking company uses tractors to move semi-trailers to the loading dock of the company wanting to ship the cargo. At the loading dock, the semi-trailers are carefully loaded and sealed by the shipper.

The trucking company moves the semi-trailers to a railroad yard. Here, they are loaded onto specially built flatcars. Fig. 17-2. The flatcars are connected to trains and are moved by the railroad to another yard close to the final destination. There, the semi-trailers are unloaded and connected to tractors. These units travel on the highways to the customers' loading docks. The semi-trailers are unsealed and off-loaded. (*Off-loaded* is another term for unloaded.)

Perhaps you know the common name for this popular intermodal method of transporting cargo. TOFC is often referred to as **piggyback**. Fig. 17-3. Use of piggyback or TOFC service has increased greatly since it was first introduced in 1950. There are even TOFC unit trains in operation. Have you ever seen a TOFC unit train? All the cars carry semi-trailers.

Fig. 17-2. A giant TOFC loader is used to load semi-trailers onto the flatcars.

Containerized Shipping

Containerization (or containerized shipping) is an efficient method of handling cargo. Cargo is loaded into large containers before it is transported. Containerization is used by all modes but pipeline.

Fig. 17-3. Piggyback service is an efficient and economical intermodal operation.

The most popular container used by the transportation industry is basically a large metal box. The *standard* (most typical) size box is eight feet high, eight feet wide, and forty feet long. The frame of the box is well-supported. The corners are strongly reinforced. There are holes in the corners to provide a means of grasping the box during loading, securing, and unloading operations.

Containerization has created new forms of intermodal transportation. The same standard size container is used in several different ways by the highway, rail, and water transportation modes. Fig. 17-4. A smaller container is used in air transportation.

Highway Mode

Containers can be made into semi-trailers. This is done by placing them on a frame that has eight tire wheels and a fifth wheel. Each "semi-trailer" can then be driven to its destination or lifted onto railroad flatcars and moved piggyback style.

Rail Mode

Containers can also be loaded directly upon railroad flatcars. When they are fastened directly to the flatcars, the method of transportation is called **COFC** for **container on flatcar**.

Fig. 17-4. A standard container can be carried on a highway truck frame (A), a containership (B), or a special rail flatcar (C).

Water Mode

These same standard containers are also loaded onto barges and ships. Many ocean-going ships are specially designed and built to carry containers. These are called **containerships**. As you can see, containers really do get around.

Air Mode

The airlines use containers that are shaped to fit into different models of aircraft. These are smaller and lighter than the standard containers. They also fit into standard highway mode vehicles. Some of these air-mode containers have round ends and some have square ends.

Like the larger containers, air-mode containers can be loaded and sealed by the shipper. They are transported by truck to and from airports. Loading containers into aircraft and unloading them requires special equipment. The containers are moved into the plane through the cargo hatch. Fig. 17-5.

Fig. 17-5. Containers are loaded into the hatch of a plane.

Advantages of Containerization

There are many advantages to containerized shipping. The most important advantage is less handling of cargo. Fig. 17-6. For example, let's compare the typical transport of a product, first without a container and then in a container.

Suppose you own a factory in Peoria, Illinois. Your company produces special automobile transmissions (speed-changing gears). You have just received a large order for your product from a company in Germany. How will you ship it?

Way 1. The shipping department decides not to use containers. Products will travel as break bulk.

Fig. 17-6. Containerized shipping permits less piece-by-piece handling of cargo.

Workers crate each transmission separately. The crates are loaded, one at a time, into a semi-trailer or trailer unit. A trucking company moves the trailers to a railyard. TOFC service is not available. The transmissions have to be unloaded from the truck and reloaded into a boxcar. The boxcar is placed in a train and moved to the seaport. The transmissions are unloaded from the boxcar and again placed into a truck.

The truck delivers the transmissions to the overseas or export dock. There, the transmissions are unloaded from the truck and placed in a warehouse. When enough for a full shipload is gathered in the warehouse, the ship comes for the cargo. Your transmissions are loaded into the holds (storage areas) of the ship.

When the ship arrives at a port in Europe, the transmissions are again unloaded and packed, one at a time, into a railroad car. When they arrive in Germany, they are loaded into a truck and delivered to the automobile company. The automobile company unloads the truck.

Using the break-bulk method of cargo transportation to send the transmissions is not efficient. Too much loading and unloading are required.

Way 2. The shipping department decides to ship the transmissions in a standard container.

An empty container fastened to a highway wheel frame is delivered to your factory. The transmissions are placed in plastic bags and then packed into the container. Shipping crates are not needed. Special racks within the container hold the transmissions in place. After the container is completely loaded, the workers lock and seal the entire container.

A trucking company picks up the container and moves it to a railroad yard. There, the entire container is placed on a flatcar and moved to the export dock. The container is stacked on the dock with other standard containers. When the ship arrives, the container is hoisted onto the ship and transported across the ocean.

In Europe, the locked and sealed container is unloaded. It is placed on a railroad car or on a wheel frame to be hauled by a tractor-semi-trailer rig. When the container arrives at the German automobile factory, the customer breaks your seal and unloads the transmissions.

Do you see the reduced amount of handling? Using a container saves a great deal of labor and money. Also, there is less chance for theft. The chances for damage to cargo from dropping, bumping, or exposure to bad weather are also decreased.

Containerization is a very efficient form of transportation. It has become a favorite intermodal method of shipping cargo.

Moving Liquids

It is not practical to build pipelines to every city and town. By using rail or highway modes, pipeline companies are able to expand their range of service to almost anywhere. Some tank truck cargo companies actually call themselves *"rolling* (or highway) *pipelines."*

Oil is often shipped by rolling pipeline. First the crude or refined oil products are shipped by underground pipelines. When the product reaches a terminal, it may be temporarily stored in large tanks or pumped directly into railroad tank cars or tank semi-trailers. Fig. 17-7. When the cargo is delivered to the final destination, pumps are again used to unload the oil.

Oil is not the only product that is moved by this type of intermodal transportation. Big newspapers and printing companies buy ink in great quantities. Their most efficient delivery system uses pumps and tanks to transfer the ink from one mode to another.

Fig. 17-7. Refined oil products are being loaded into highway tank trucks for transport to customers.

Have you seen milk tank trucks? These transport milk from the farm to the dairy. The milk is never exposed to air. Milking machines collect milk from the cows. The milk travels through pipes to holding tanks. From there, it is piped into tank trucks. At the dairy, it is pumped through hoses and pipelines. That's efficient cow-to-dairy transportation!

Moving Coal and Gravel

Another form of intermodal transportation combines on-site conveyors, trailers, and railroad cars. In the northwestern United States, coal and gravel are mined for customers in other parts of the United States. As these materials are gathered, they are loaded onto very long conveyors. A *conveyor* is a device that moves materials over a fixed path. In this case, the material is moved across rugged land. Some of these conveyors are over a mile in length. Finally, the conveyor dumps its load into trucks or railroad cars. Fig. 17-8.

In rail transportation, hopper cars are used in unit trains. Sometimes the trains don't even completely stop moving while they are being loaded. The engine slowly moves the cars of the train under the conveyor. The conveyor loads the cars so fast that the train can keep moving.

Fascinating Facts

The first U.S. petroleum pipeline, built in 1865 near Titusville, Pennsylvania, was five miles long. That pipeline could move 800 barrels of oil a day from an oilfield to a railroad.

Fig. 17-8. This conveyor is carrying coal from the mine to the train, where it quickly dumps its load into the awaiting hopper cars.

The hopper-car unit trains travel to the customer's location. There they are unloaded. There are two main methods of unloading cars. In one method, the cars are moved over elevated sections of track that have open spaces between the rails and the ties. Hopper doors on the bottom of each car are opened. The load is dropped onto a conveyor. The conveyor carries the material to a storage area and dumps it.

Another method of unloading the railroad cars involves rolling each car individually onto a section of mechanical track. Special clamps lock the car onto the mechanical track. The track and car rotate together until the car is upside down. Fig. 17-9. The load dumps out all at once. The mechanical track section continues the circular motion until the car is upright. Then the empty car is moved ahead and replaced by the next full car. The procedure is repeated until all cars are unloaded. The train must stop and start as each car is uncoupled, unloaded, and moved on. However, this is still a very fast method of unloading cargo.

Other Intermodal Cargo Transportation

Other forms of intermodal cargo transportation include the vacuum-operated transfer of flour or grain between ships, railcars, and trucks. This system is much like cleaning with a vacuum cleaner.

Military equipment comes in many odd shapes and sizes. The United States military must have an even broader system of intermodal transportation. Fig. 17-10. A container of supplies may be shipped by highway, rail, water, and air and then finally parachuted into an inaccessible (hard to reach) camp from a large airplane.

Fig. 17-9. An unloader dumps the entire railcar load at one time.

Fig. 17-10. Military vehicles must be able to accommodate cargo of all sizes and shapes.

Engineers and designers continue to think of new methods of handling materials. Use of intermodal transportation will increase and become even more efficient.

Apply Your Thinking Skills

1. Would it be a problem if containers were not "standard size"? Why?
2. Why is it not practical to build pipelines to all towns and cities?

Intermodal Passenger Transportation

People also use intermodal transportation when they travel. Let's see how a person named Bill traveled from his home in New York City to the Superdome in New Orleans, Louisiana.

First, Bill walked to a bus stop. From there, the bus (highway mode) carried him to a city transit train station. Bill boarded the train (rail mode) and was transported across town to a station near the airport. He took a taxi (highway mode) to the airport. Inside the airport, a moving walkway (on-site) transported him to his departure gate. Fig. 17-11.

When it was time to board the plane, Bill walked down a covered telescoping ramp called a *jet way*, right into the plane (air mode). Fig. 17-12. The first stop was at the Dallas-Fort Worth airport in Texas. Bill was scheduled to change planes there. He traveled from one gate to another by a computer-controlled passenger vehicle that travels on special tracks (rail mode). Fig. 17-13. At the next gate, he boarded another plane for the final part of his trip to New Orleans.

When he arrived in New Orleans, Bill walked out of the plane through another jet way. He rode another on-site method of transportation, a moving stairway called an **escalator**, to the next level. From the escalator, he walked to a bus stop. The bus delivered him to the entrance ramp of the Superdome.

Fig. 17-11. A moving walkway is a conveyor device used for transporting people. It is also called a moving sidewalk.

Fig. 17-12. Passengers walk through a covered, telescoping ramp called a jet way when boarding or leaving an airplane.

Fascinating Facts

The Otis Elevator Company installed the first escalator back in 1900 at the Paris Exposition. In 1901, this escalator was moved and reinstalled in a Gimbel Brothers' department store in Philadelphia, Pennsylvania, where it was used until 1939.

Bill's trip used many different modes of transportation. Just think, he will rely on all of them again when he returns home.

Comparing Types of Intermodal Services

Passenger and cargo intermodal transportation are much alike. The major difference is that passengers are usually responsible for making the connecting links between modes. The transfer of cargo between modes is the responsibility of the transportation company.

Fig. 17-13. A computer-controlled automatic passenger vehicle that travels on special tracks.

Apply Your Thinking Skills

1. List all the vehicles that Bill traveled in or on during his journey from New York City to the Superdome. Discuss the convenience each provided.
2. Could we use pipelines to transport people?

Global Perspective
Progress on Two Wheels

Do you dream of someday owning a car? To many people around the world, the automobile has become a symbol of progress. However, according to research done for Worldwatch Institute, many reasons exist for taking a new look at the bicycle, not just as a means of exercise or recreation, but as a means of transportation.

The bicycle holds great promise for people in Third World countries. Third World countries are those whose technology is relatively underdeveloped. Most of these countries are poor. Even if a person there could afford to buy a car, fuel is scarce and few roads exist. Today, in many places, people must walk wherever they need to go, carrying heavy loads on their backs, shoulders, and even their heads. Bikes can ease the load and even enable individuals to transport larger loads by attaching a trailer to the bike.

Governments can do a great deal to encourage bicycle riding. They can have roads suitable for biking constructed. These roads are easier and cheaper to build than roads designed for motorized vehicles because they can be narrower and do not require paving.

Government encouragement of bicycle use can also lead to the development of industries to produce bicycles. For example, about 50 years ago, India began a bicycle industry by first importing parts, assembling them, and then selling the completed bicycles. Then they began producing frames in existing workplaces. They gradually expanded and built factories to manufacture the parts themselves. Today, India is one of the world's major producers of bicycles.

Using bicycles can also have a positive effect on industrialized countries. Major problems faced by these societies include air pollution, traffic jams, and dependence on oil as fuel. Motorized vehicles are major contributors to these problems. Adopting alternative means of transportation could help remedy these problems. Bicycle riding could be encouraged for short trips. Perhaps intermodal transportation, combining bike riding with other modes of transportation, could be used. For example, bicycle roads could be built and safe parking areas provided for commuters near mass transit stations.

Bicycles can give millions of people in underdeveloped areas new mobility. They may also provide industrialized nations with at least a partial solution to existing problems.

Extend Your Knowledge

1. Another vehicle that is operated by human power is the ricksha (rickshaw). Write a description of a ricksha. Is there more than one kind? In what countries have they been used?

Fig. GP-17. In some countries, bicycles are a popular choice of transportation because they are more affordable than automobiles.

Chapter 17 Review

Looking Back

Intermodal transportation is the process of using more than one transportation mode to move people or cargo. The increased efficiency of intermodal systems contributes to lower transportation costs and time.

Trailer on flatcar companies combine highway and rail modes to ship cargo. Semitrailers are loaded onto rail flatcars for part of their journey.

Containerization is the use of special containers for shipping cargo. The containers are filled with cargo at the start of a journey and emptied only when they reach the final destination. Containerization greatly reduces the amount of handling needed, saving much labor and money. It also reduces losses due to damage or theft.

Liquids are frequently shipped by intermodal systems. A combination of pipelines, tank trucks, and railroad tank cars may be used to ship large quantities of oil, milk, and ink.

Solid bulk cargos, such as coal and gravel, are often moved by conveyors, trailers, and railroad cars.

People commonly use intermodal transportation when they need to get from place to place.

Review Questions

1. What are the two main advantages of intermodal transportation?
2. Define and briefly describe TOFC. What is the common name for this method of transportation?
3. Describe containerization in the highway mode of transportation.
4. Define and briefly describe COFC.
5. In what ways are containers used in the air mode of transportation different from those used in the highway, rail, and water modes?
6. List four advantages of containerization.
7. What is the main advantage for pipeline companies in using intermodal transportation?
8. What is a "rolling pipeline"?
9. Briefly describe the intermodal transportation used to move coal from the mine to the customer.
10. What is the major difference between intermodal passenger and intermodal cargo transportation?

Discussion Questions

1. Why would industrial companies be interested in buying different types of transportation systems?
2. If companies using various transportation modes did not cooperate, would intermodal transportation be practical?
3. Do you think it would be practical to develop airplanes large enough to carry standard size containers? What about lighter-than-air aircraft?
4. Early American settlers traveling across the United States often used intermodal transportation systems. What were those systems?
5. Can you think of any disadvantages to intermodal transportation systems?

Chapter 17 Review

Cross-Curricular Activities

Language Arts 

1. Interview a grandparent, parent, and peer about their preference in transportation from home to a nearby large city. Compare their answers. Which mode would you choose? Develop your findings into a brief essay.
2. Interview delivery persons at your local grocery store. Where do they pick up their cargo? How did it arrive at that point? How many deliveries do they make in one day? Present your findings to the class in an informative speech.

Social Studies 

1. Review the concept of containerization. Find out how spent fuel from America's nuclear power plants is moved away from the plant sites. What type of safeguards do they provide for the public's safety? Try to find out which states, if any, move their fuel out of state. What types of precautions do they take to protect the public?
2. Illustrate intermodal transportation by finding products that are shipped via several different forms of transportation from one state to another. After you select a product and all the ways it is transported, draw or trace a map of the United States and illustrate that product being moved from one state to another. Show the route the product takes, and diagram the mode of transportation in the states in which the product travels.

Science 

1. One problem in transferring loads made up of particles, such as sand or gravel, is that, at first, particles slide through an opening. After a time, however, no more particles fall because the material has developed a stable slope and no more particles will slide down. The angle at which this happens is known as the *angle of repose*. Using small samples of sand, sugar, flour, and other particulate materials, find the angle of repose for each by pouring the material until you form a pile with the steepest possible sides. Stick an index card through the pile, and trace its slope on the card. The angle formed between the sloping line and the bottom of the card is the angle of repose.

Math

Horsepower is a unit that is used to express the power (rate of doing work) of an engine. One horsepower is defined as 550 foot-pounds of work per second. One foot-pound is the work needed to lift 1 pound a distance of 1 foot. If an engine lifts a 550-pound object a distance of 4 feet in 1 second, it is working at a rate of 2,200 foot-pounds per second (550 x 4 ÷ 1). This engine is delivering 4 horsepower (2,200 ÷ 550).

1. An engine can lift 1,650 pounds a distance of 2 feet in 1 second. How many horsepower is it delivering?
2. How many horsepower would the engine be delivering if it took 6 seconds to lift the same weight 1 foot?

Chapter 17 Technology Activity

Passage to the Southern Land

Overview

Suppose you have a rich uncle who has invited you to visit him. He will pay for your trip, but you have to make the arrangements. It's going to be a long journey. Your uncle lives in Sydney, Australia. You'll need to use several modes of transportation to get there.

Goal

The goal of this activity is to plan a trip from your home to Sydney, Australia. Fig. A17-1. You will need to determine what modes of transportation to use, the time required for each part of the journey, and the cost.

Equipment and Materials

pencil
paper
a world map or atlas

Procedure

1. On the world map or atlas, locate your community. Then find Sydney. What is the approximate distance between the two?
2. The longest part of your journey will involve crossing the Pacific Ocean. Decide whether you will travel by airplane or ship. Look up articles about Australia in travel magazines or contact a travel agency to learn what the common routes are, how long the journey takes, and what it costs.

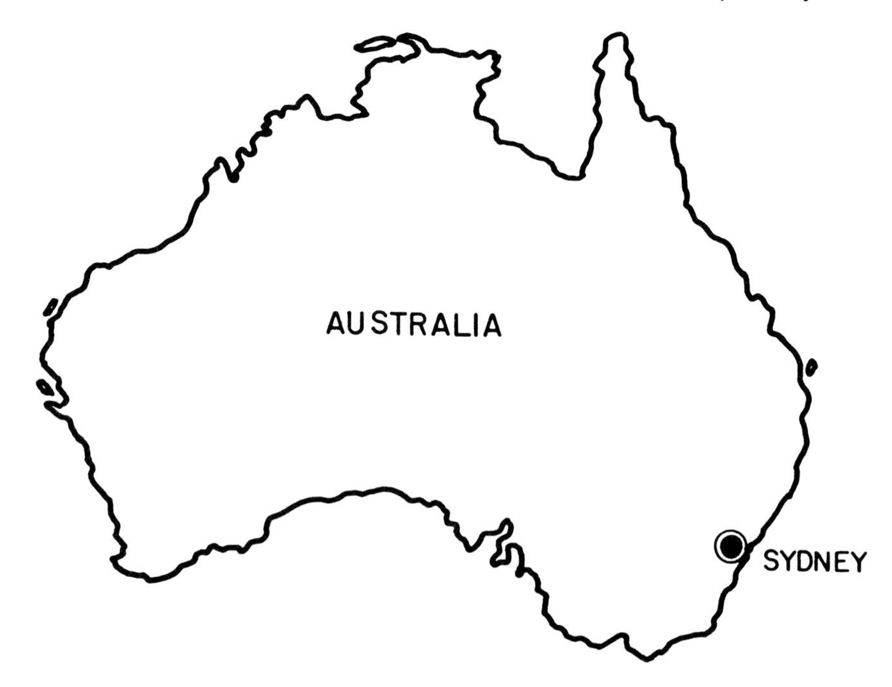

Fig. A17-1. Here is where Sydney is located in Australia.

Chapter 17 Technology Activity

3. Once you have learned where you can catch a ship or plane to Sydney, you'll need to figure out how to get to that departure point. Is it close enough for someone to drive you there? Will you need to fly to the departure point? Could you perhaps take a train? Gather information about time and costs.
4. Make a table similar to the example shown in Fig. A17-2. Write down each part of the journey, from leaving your house to the arrival in Sydney. For each part, list the mode of transportation, the time it will take, and the cost. (Your uncle is generous, but he does like to know where his money is going.)

Evaluation

1. How many modes of transportation will you use to reach Sydney?
2. How long will your journey take?
3. How much will it cost your uncle?
4. What were your reasons for choosing the modes of transportation that you did? Now that you see the entire time and cost for the trip, do you want to change to any different modes? Why or why not?

FROM	TO	MODE OF TRANSPORTATION	TIME	COST

Fig. A17-2. This is how your table should be set up on your notebook paper.

Chapter 18

Power in Transportation

Looking Ahead

In this chapter, you will discover:

- what a prime mover is.
- the two main types of engines and how they operate.
- the different kinds of motions engines produce.
- the advantages and disadvantages of the different engine types and designs.
- the systems common to most engines and how they work.
- how two-stroke cycle and four-stroke cycle engines operate.
- the two types of motors, how they operate, and their advantages and disadvantages.
- what types of power are used for different types of transportation vehicles.

New Terms

combustion
engine
external-combustion engine
four-stroke cycle engine
horsepower
hydraulics
internal-combustion engine
linear motion
motor
power
prime mover
reciprocating motion
rotary motion
RPM
two-stroke cycle engine

Technology Focus

Automated Guided Vehicles (AGVs)

One of the newest types of on-site transportation is the automated guided vehicle system. In this type of system, computer-controlled vehicles are used to carry parts and products around manufacturing plants, warehouses, and large mail-order stores.

Automated guided vehicles (AGVs) are usually four to eight feet long. They may be equipped in a number of different ways to carry different loads. A robot-equipped AGV can even load and unload itself.

An AGV can go practically anyplace it is programmed to go. There are four basic ways to keep an AGV on course. The most common, but least flexible, method is wire guidance. An antenna on the AGV "homes in" on the magnetic field surrounding a wire that is placed in the floor.

Many AGVs use optical guidance systems. These "read" painted lines. Different routes are painted different colors. The vehicles are programmed to follow certain colors.

Infrared signals can be used to lead AGVs about the workplace. This type of signal is much like the waves in your microwave oven.

Another method of guiding AGVs involves the use of laser ranging and direction finding. This type of AGV has an on-board computer that uses the information obtained by the lasers to direct the vehicle's movement.

Most on-site AGVs are operated by electric motors powered by batteries. They move around a plant quietly, creating few environmental problems. Some units are programmed to locate battery charge connections to "feed" themselves when they are not working.

AGVs are the "feet and legs" of industrial robots. At the beginning of 1987, there were over 15,000 AGVs in use throughout the world. More and more systems are being installed every day. General Motors alone is currently installing systems that will use about 1,500 AGVs. AGV systems are becoming increasingly popular because of their flexibility and efficiency.

Fig. TF-18. Automated guided vehicles (AGVs) such as the one shown here are being used more and more in industry these days to carry around different parts and products.

Prime Movers

Prime movers supply the **power** (the use of energy to create movement) in transportation. A **prime mover** is a basic engine or motor.

The terms *engine* and *motor* are often used to mean the same thing. This is not wrong, but it is possible to be more exact. In this book, we will use the following definitions:

- An **engine** is a machine that produces its own energy from fuel.
- A **motor** is a machine that uses energy supplied by another source.

Fig. 18-1.

An engine creates its own energy from fuel (usually gasoline).

An electric motor gets its energy from an outside source (such as electrified cables or batteries).

Fig. 18-1. Examples of an engine and a motor.

Many different types of engines and motors are used in transportation. In this chapter, we will examine the most common ones. This will help you develop a basic understanding of how energy becomes the power that keeps the transportation system working.

Apply Your Thinking Skills

1. What things do you use in your everyday life that are powered by engines? What are some things you use every day that are powered by motors?

Internal-Combustion Engines

Most engines burn fuel to create heat. The heat creates pressure. The engine uses the pressure to create mechanical force or motion. There are two basic types of engines: internal-combustion engines and external-combustion engines. (**Combustion** means burning.)

Internal-combustion engines are the most common engines used in transportation. In an **internal-combustion engine**, fuel is burned within the engine. *Within* is a key word. Internal means within or inside.

Types of Motion

An internal-combustion engine can produce three different kinds of motion. These are:

- Reciprocating motion
- Rotary motion
- Linear motion

Fig. 18-2.

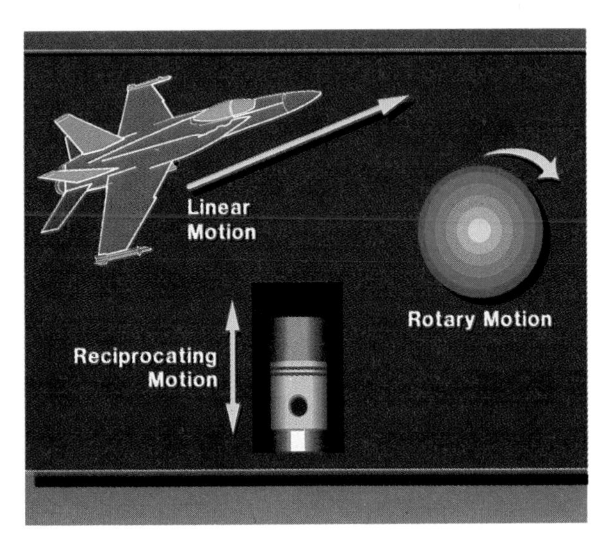

Fig. 18-2. Three types of motion can be produced by internal-combustion engines. Reciprocating motion is a back-and-forth or up-and-down motion. Rotary motion is a circular motion. Linear motion is motion in a straight line.

Reciprocating Motion

Most transportation vehicles are powered by reciprocating internal-combustion engines. These include most highway and rail vehicles and smaller air and water vehicles.

A reciprocating engine uses pistons, connecting rods, and a crankshaft. During operation, the **reciprocating motion** (up-and-down or back-and-forth motion) of a piston is changed into the circular or rotating motion of a crankshaft. Fig. 18-3.

Not all of the energy used by the piston becomes crankshaft power. Some power is lost while the engine is operating. This is due to the *friction* of the mechanical connections. The parts rub together and create resistance to motion between the parts.

The speed of the engine is determined by the size of the parts. The larger the piston (weight and diameter), connecting rods, and crankshaft, the slower the speed of movement. Very large reciprocating internal-combustion engines used in train locomotives or on ships may turn at only 100 to 400 **RPM** (**R**evolutions of the crankshaft **P**er **M**inute). Very small reciprocating internal-combustion engines, like those used in model airplanes, may turn at well over 20,000 RPM.

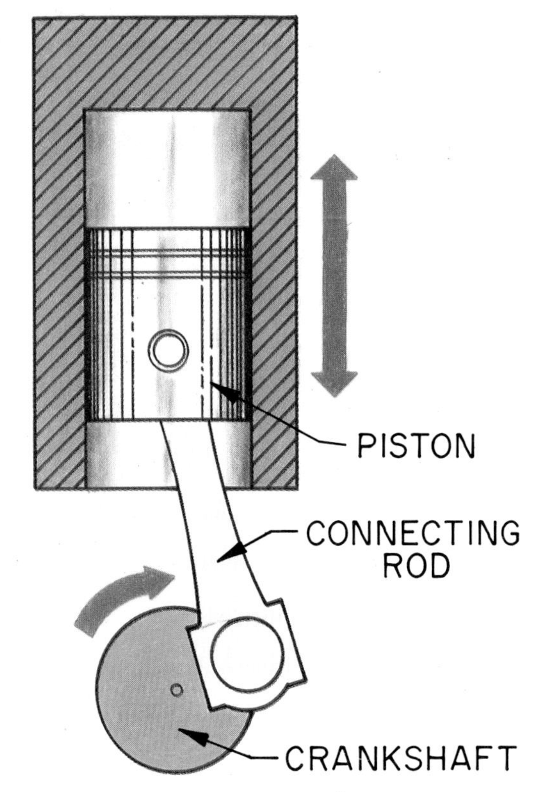

Fig. 18-3. In a reciprocating engine, the piston moves up and down, turning the crankshaft in a circular motion.

The major advantages of reciprocating internal-combustion engines include the following:

- The RPM of the engine can be quickly and easily controlled. This makes them well-suited for highway stop-and-go vehicle movement.

- The cost of manufacturing a reliable engine is low. Also, the cost of rebuilding or repairing is low compared to the cost of repair for other types of internal-combustion engines.

There are also disadvantages of this type of engine:

- Only a limited amount of energy can be economically produced by a reciprocating internal-combustion engine.
- The amount of horsepower that can be provided by each piston is limited. **Horsepower** is a unit of measurement of power. One horsepower equals the force needed to raise 550 pounds at the rate of one foot per second. High-horsepower engines usually have many pistons. Fig. 18-4. They have more moving parts than smaller engines and they weigh more. These factors make large engines more expensive to produce.
- The faster the RPM of the engine, the shorter its usable life span.
- Reciprocating-motion internal-combustion engines use quick-burning fuels. These fuels are more expensive than the slow-burning fuels used in jet engines.

Fig. 18-4. A modern automotive piston engine is quite powerful. Power is a measurement of how quickly work can be done. Power is measured in horsepower.

Rotary Motion

Rotary-motion engines produce more horsepower than reciprocating engines. This is partly because there is less mass. *Mass* is the size and weight. There is also less friction.

Rotary motion is a circular motion. Like all internal-combustion engines, rotary-motion prime movers turn because exploding fuels form expanding gases that push against the blades of a turbine or rotor. A *turbine* is a wheel with evenly spaced blades or fins attached to it. A *rotor* is the triangle-shaped part of a rotary engine that revolves in a specially shaped combustion chamber. Fig. 18-5. The turbine or rotor is usually mounted on the same straight shaft as the device that is being powered.

Rotary-motion engines work best where the same RPM is held for extended periods of time. However, some automobile manufacturers are using small internal-combustion rotary-motion engines for stop-and-go road use.

The following are some advantages of rotary-motion internal-combustion engines:

- There are fewer moving parts to maintain or cause friction than in a reciprocating engine.
- Horsepower can be increased by increasing the size of the rotor or turbine. No additional parts are required.
- A wide variety of fuels can be burned to produce the expanding gases.
- A constant RPM can be maintained for very long periods of time without damage to the engine parts.

Some disadvantages of rotary-motion engines are the following:

- Quick changes in RPM cannot be made efficiently.
- The cost of manufacturing and rebuilding the engines is greater than it is for reciprocating engines.
- Because of the high RPM attained, the rotor or turbine must be kept well-lubricated and in perfect balance.

Linear Motion

Linear-motion engines are generally referred to as *jet* or *rocket engines*. **Linear motion** is motion in a straight line. The power from a linear-motion engine must be pointing in the exact opposite direction as the transportation vehicle will be going. Fig. 18-6.

In a jet engine, power is created by the internal combustion of the fuel. Fig. 18-7. The more gas that is expanded, the faster the vehicle will go. The expanding gas is aimed behind the vehicle. The force of the gases escaping rearward makes the vehicle move in the opposite direction—forward. This resulting forward force is called *thrust*.

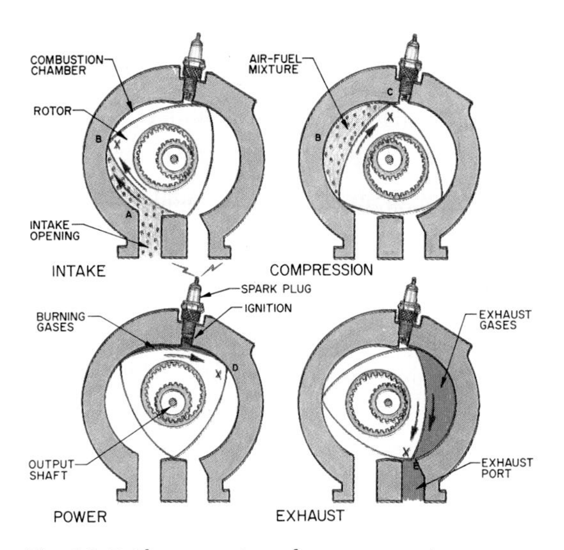

Fig. 18-5. The operation of a rotary engine.

Fig. 18-6. A jet engine produces linear motion. Powerful jet engines can propel aircraft at very high speeds.

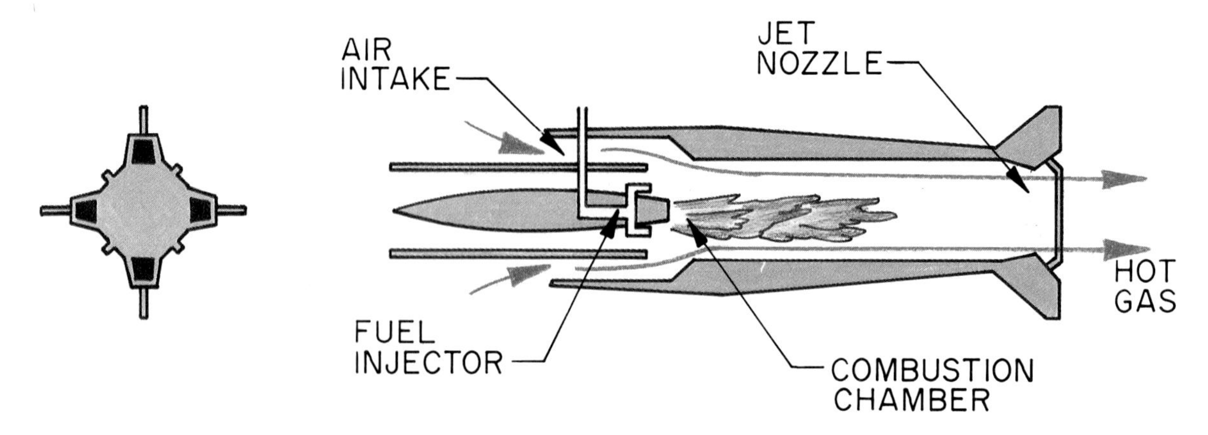

Fig. 18-7. Operation of a simple jet engine.

Linear-motion internal-combustion engines are used on large aircraft, rockets, and spacecraft.

Rocket engines work much like jet engines. All engines need to mix oxygen with fuel for combustion since nothing can burn without oxygen. The main difference between jet and rocket engines is that jet engines take in air (which is part oxygen) from the atmosphere. Since rocket engines are usually designed for use in spacecraft outside earth's atmosphere, a rocket engine must carry all of its own oxygen and fuel supply. This supply may be either solid or liquids that are carried as two separate chemicals. When the chemicals are combined, combustion takes place and hot gases expand through a nozzle to produce thrust like a jet. A liquid fuel rocket engine may be turned on and off. Once a solid fuel rocket is started, combustion will continue until all the chemicals are used.

There are some advantages of the linear-motion internal-combustion engine:

- Very high horsepower or thrust can be developed.
- Several different types of solid or liquid fuels can be burned to make the expanding gases.
- There are no moving parts needed to transfer the power to the vehicle.

Some disadvantages of the linear-motion internal-combustion engine include the following:

- Very high manufacturing and rebuilding cost. Some rocket engines are built to be used only once.
- No directional change of thrust or motion is possible. The force is generated in one direction only.

Systems in an Engine

To understand how engines work, let's examine one example of an internal-combustion engine—a single-cylinder small engine. Other types of internal-combustion engines are more complex but operate in basically the same way.

Engines have six systems:

- Mechanical system
- Lubrication system
- Fuel system
- Ignition system
- Starting system
- Cooling system

The *mechanical system* contains the piston, connecting rod, and crankshaft. Fig. 18-8.

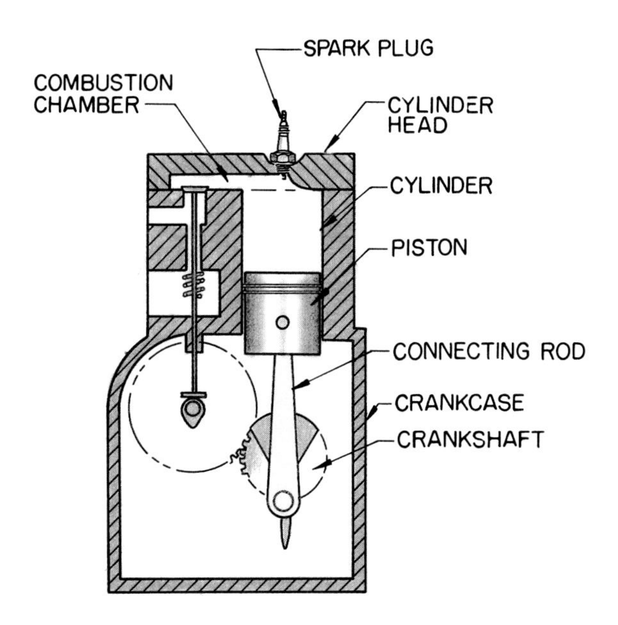

Fig. 18-8. Basic framework of an engine.

The reciprocating motion of the piston is changed into the rotary motion of the crankshaft.

The function of the *lubrication system* is to decrease friction. Lubricating (oiling) parts makes them slippery. Lubricating is done in basically three ways. On some engines, an oil pump pushes oil through channels to the moving parts. Other engines rely on simple splash-and-dipper devices to oil the parts. Certain engines are lubricated by oil that is mixed into the fuel and supplied by the fuel system.

The *fuel system* supplies fuel to the cylinder. It also mixes the fuel with air before it enters the cylinder for combustion, and it carries away exhaust. Parts of a fuel system include:

- an air cleaner to prevent dirt from entering the engine.
- a fuel tank.
- a *carburetor* or fuel mixer to mix the air and fuel accurately to ensure efficient engine operation.
- *intake valves* or ports—mechanical devices that allow the air-fuel mixture to flow into the cylinder.
- *exhaust valves* or ports and pipes to allow spent gases to escape from the engine.
- a muffler to quiet the engine noise.

Fig. 18-9.

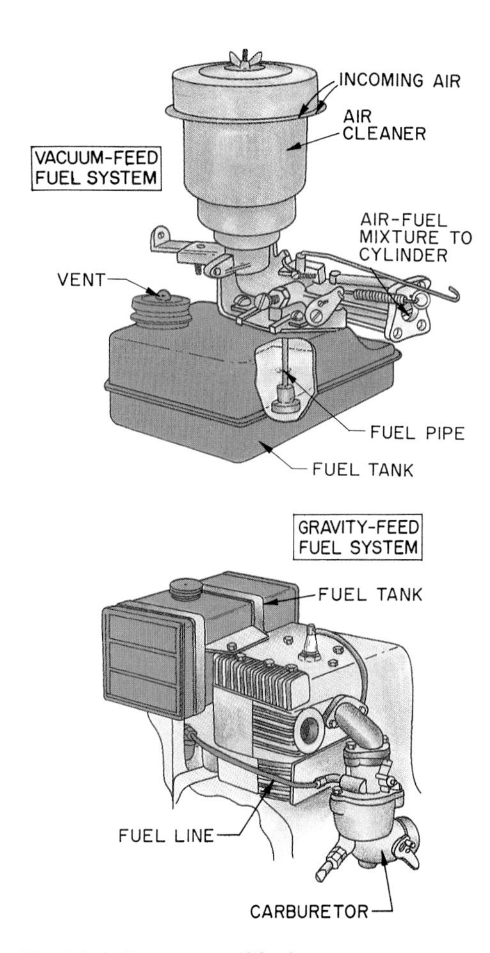

Fig. 18-9. Two types of fuel systems.

The *ignition system's* job is to ignite the air-fuel mixture at the correct moment. The ignition system includes:

- *spark plugs* to supply a spark to begin combustion.
- a *coil* to increase the spark plug's spark.
- *breaker points* and *condenser* or other electronic devices that switch the electricity flow.
- wires for the electricity to flow along.
- a *generator* or *magneto*—devices that produce electricity from motion of the engine.

If the mechanical, lubrication, fuel, and ignition systems are all working correctly, the engine can be started.

The *starting system* may be as simple as a rope on a pulley connected to the *flywheel* (the part in a small engine that is set into motion by the action of pulling a starting rope). It may also be as complex as a starting motor that works off a storage battery. Most small engines have recoil (rewind) rope starters. Fig. 18-10. The starting system must stop operating once the engine begins running.

While the engine is running, heat from combustion and friction must be kept under control. This is the job of the *cooling system*. Small engines are usually air-cooled. An air-cooling system includes a fan (usually part of the flywheel) and metal shrouds (covers). The shrouds contain and direct the cooling air movement. Fig. 18-11.

Larger engines are usually water-cooled. This type of system is more complex. It contains a water pump, radiator, hoses, and a water jacket. The water jacket is usually built into the crankcase and cylinder head.

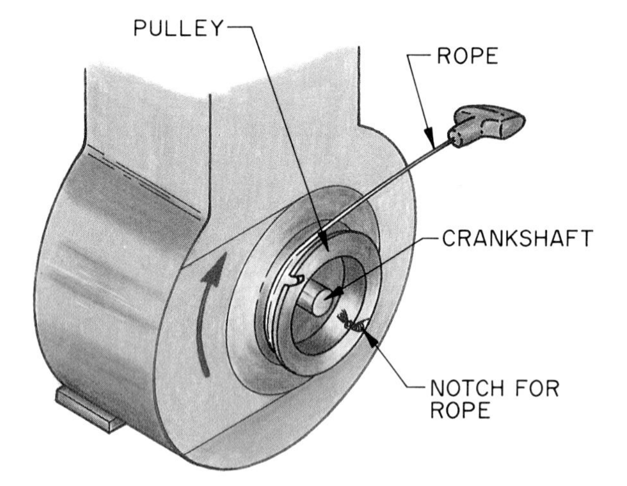

Fig. 18-10. With a rope starter, the operator turns the crankshaft by pulling the rope. This starts the engine, which then keeps the crankshaft turning.

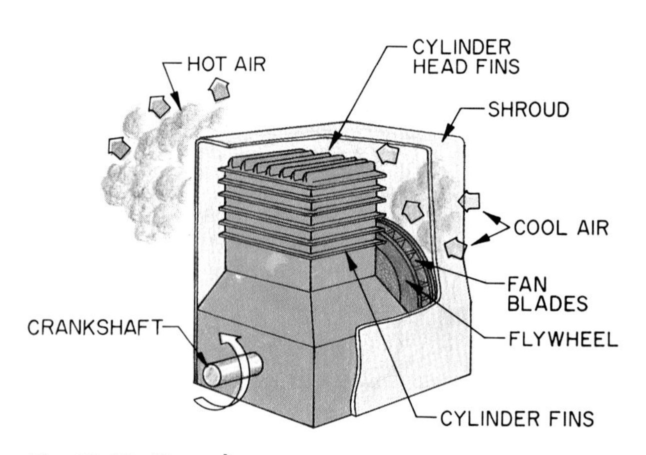

Fig. 18-11. Air-cooling system.

Engine Operation

Internal-combustion engines can be either two-stroke cycle or four-stroke cycle designs. Most engines use a four-stroke cycle. A **four-stroke cycle engine** has four *strokes* of the piston for one combustion *cycle*. Fig. 18-12.

Stroke 1: The air-fuel mixture is drawn into the cylinder (intake stroke).

Stroke 2: The mixture is compressed (squeezed into a smaller space) (compression stroke).

Stroke 3: The *spark plug* fires, and the burned gases expand and force the piston down (power stroke).

Stroke 4: the piston comes back up, pushing the spent gases out of the cylinder (exhaust stroke).

These four strokes together make one *cycle*. While the engine is operating, the cycles repeat over and over again. The name of this engine type is usually shortened to "four cycle" or "four stroke."

Fascinating Facts

Back in 1860, the French inventor Jean Étienne Lenoir developed a one-cylinder internal-combustion engine that used street-lighting gas as fuel. A few years later, he put the engine in a vehicle and was able to travel six miles in about two hours.

A traveling salesman, Nikolaus August Otto, read about Lenoir's invention in a newspaper. He built a successful four-cylinder gas engine in 1876. This invention helped pave the road to the development of the automobile and the airplane.

The type of engine that became popular in automobiles was built in Germany in 1885. Working independently, Gottlieb Daimler and Karl Benz both developed successful high-speed four-cycle gasoline engines. Daimler first attached his to a boat, then to a bicycle, and then, in 1887, to a four-wheeled vehicle. Benz installed his engine in a three-wheeled carriage. The Daimler and Benz companies later merged and, in 1926, made their first Mercedes-Benz automobile.

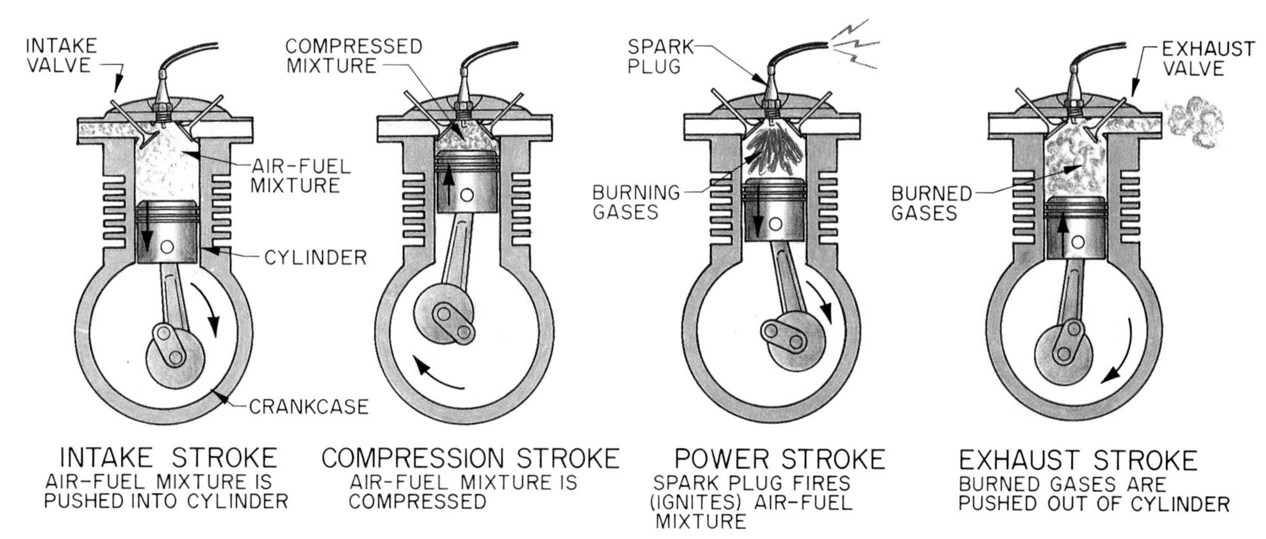

Fig. 18-12. Operation of a four-stroke cycle engine.

Two-stroke cycle engines (usually called "two cycle" or "two stroke") complete the four operations in only two piston strokes. Fig. 18-13. During the first stroke, the piston goes up. It allows the air-fuel mixture to enter the cylinder and compresses it at the same time. The spark plug fires when the piston is at the top. This pushes the piston down for the power stroke. When the piston is at the bottom of the stroke, the exhaust is let out of the cylinder.

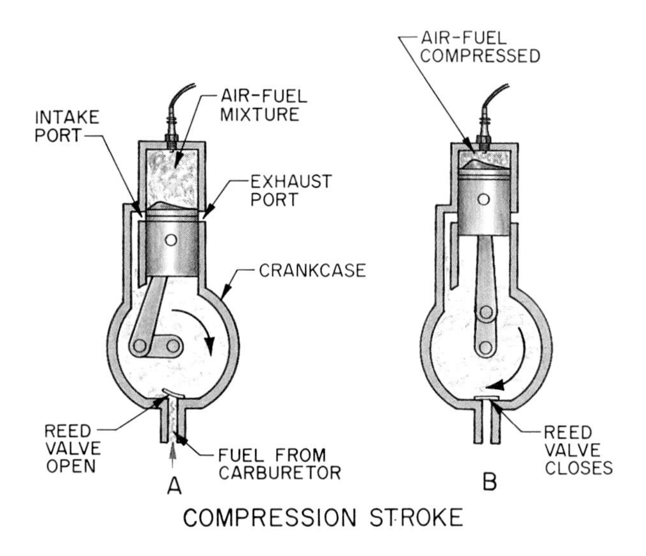

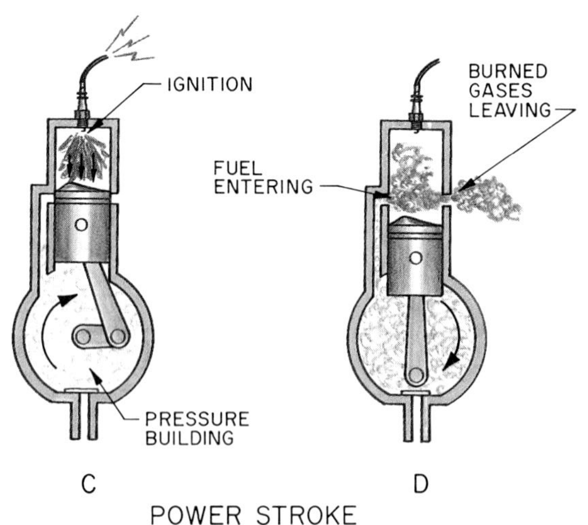

Fig. 18-13. Operation of a two-stroke cycle engine.

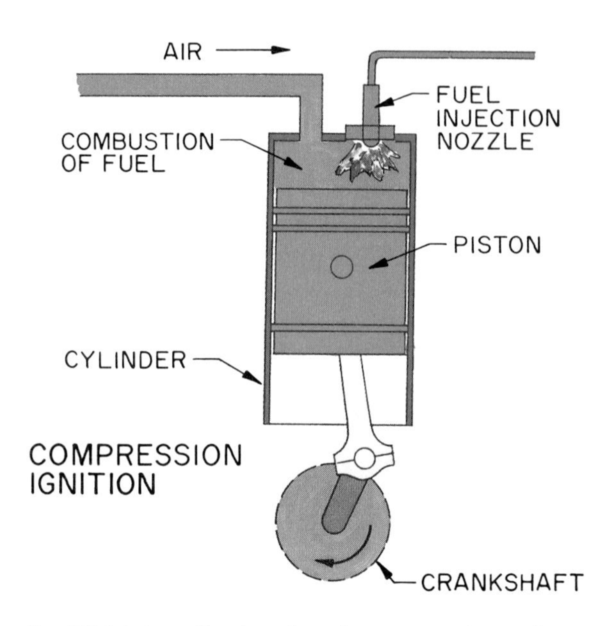

Fig. 18-14. In a diesel engine, the compression cycle squeezes the air-fuel mixture together very strongly. The added compression creates heat that causes the air-fuel mixture to burn.

This design allows the two-stroke cycle engines to have fewer moving parts. The shorter cycle and fewer parts allow the two-stroke engine to run at higher RPMs than four-stroke engines. A disadvantage of the two-stroke engine is that it usually produces more pollution than a four-stroke engine.

A special type of two- or four-stroke cycle engine, called a diesel, does not need a spark plug. Fig. 18-14. Diesel engines last a long time and are very good at pulling heavy loads. They are used in trucks, cars, locomotives, and many other types of vehicles.

Apply Your Thinking Skills

1. What would happen to an engine if you didn't keep it properly lubricated?

External-Combustion Engines

An external-combustion engine was one of the original manufactured devices that provided power for transportation vehicles. Today, the principle of external combustion is still used to a great extent in prime movers for large ships and railroad locomotives.

In **external-combustion engines**, the fuel is burned *outside* the engine. Like internal-combustion engines, external-combustion engines can provide reciprocating and rotary motion. However, the advantages and disadvantages of these engines are different when the energy source is external. External-combustion engines are not used to produce linear motion.

Almost all of the external-combustion engines currently used in transportation are steam-driven. The basic procedure for providing steam is simple. Water is heated inside a closed container called a boiler. The water becomes high-pressure steam. This steam is used to drive pistons or turn turbine blades to produce mechanical energy. Fig. 18-15.

The steam engines used in transportation vehicles operate as a closed system. After steam has been used, it is allowed to cool. The steam condenses into water. The water is reheated and again becomes steam. This process keeps repeating. In a closed system, there is little noise or water contamination.

In transportation vehicles, the size of the steam boiler must be appropriate for the size of the vehicle. Using a heavy boiler in a compact automobile would not be efficient.

The heat needed to boil the water can come from many sources. Early transportation vehicles burned coal and wood. Some still do, but most modern vehicles burn petroleum products. Some modern military vessels use nuclear fuel to boil the water. Because very little nuclear fuel is needed, these vessels do not need to refuel often.

Fig. 18-15. Operation of a steam engine.

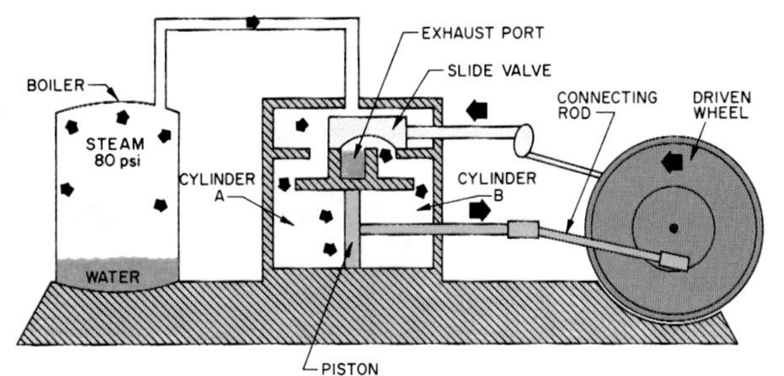

Steam moves from the boiler into cylinder A. It pushes against the piston causing the piston to move to the right.

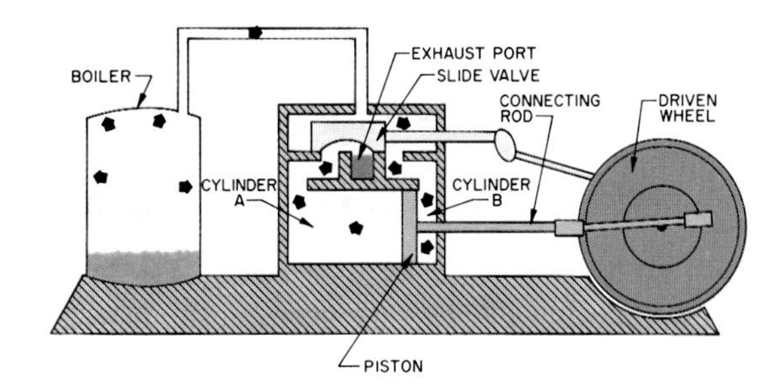

The slide valve cuts off the steam supply to cylinder A. The steam enters cylinder B and pushes against the other side of the piston. The piston moves to the left, the slide valve opens, and the cycle is repeated.

Advantages of an external-combustion steam-operated engine include the following:

- Almost any type of fuel can be used to produce the steam.
- The boiler and the engine can be placed in different physical arrangements.
- Many different designs of steam-driven engines are available to serve specific needs.

There are several disadvantages to external-combustion steam-operated engines:

- Boilers require a great deal of space. Their use is limited to large vehicles.
- Even when handled carefully, steam is dangerous.
- The steam and water cause metal to deteriorate (break down). Continuous care and preventive maintenance are necessary.

Apply Your Thinking Skills

1. What are some of the safety hazards of using steam to power engines?

Motors

Electric- and fluid-powered motors are used to move transportation vehicles. Most on-site transportation devices are powered by motors.

Electric Motors

For many years, electric motors have powered urban mass transportation commuter vehicles. The oldest operating electric street car line runs a daily schedule on tracks in New Orleans. Electric power is supplied to each car's motors by an overhead electrified cable. Other lines may use a "hot" third rail on the ground. Fig. 18-16.

Fig. 18-16. Many trains use electricity as their power source.

Rubber-wheeled city buses once were powered by overhead wires. Most of these have been replaced with diesel-engine buses.

Electrically powered passenger and freight trains are relatively common in the northeastern section of the United States and all across Europe. In addition, several large cities have electrically powered mass transportation systems. Washington, D.C., and Atlanta, Georgia, are two examples. Can you name others?

As batteries are improved, more and more electrically powered vehicles will be operated on streets and highways. Small cars that operate efficiently on batteries for short distances are available now. Perhaps in the near future, we will be able to buy solar-powered electric vehicles.

There are many advantages to electrically powered transportation vehicles. The most important ones are:

- Except for pollution caused by electrical generating plants, they do not pollute the environment.
- Operating costs are comparatively low.
- They require less maintenance and repair work than other power systems.

- Little or no noise is produced by electrically powered equipment.
- Electric motors will generally run in either direction. No transmissions (gears) are needed.

Vehicles that operate on electricity have one big disadvantage. Supplying electricity to each vehicle is often difficult and inconvenient. This problem is especially serious for vehicles that are operated independently, such as cars and trucks. However, a train on a track does not change its route. Electricity can be supplied more easily.

Fluid-Powered Motors

Fluid-powered motors can develop considerable horsepower, but not much speed. They are used in heavy equipment and other slow-moving vehicles that require a lot of power.

Fluid-powered motors are controlled by increasing or decreasing the pressure of the fluid. Using the reactions of fluid under pressure to develop motion is called **hydraulics**. A fluid pump produces the pressurized fluid. It can be powered by almost any rotating power source. Most fluid-powered vehicles use internal-combustion engines to power the fluid pumps.

Like electric motors, most fluid-powered motors will also reverse. They run efficiently in either direction.

Apply Your Thinking Skills

1. What would be some problems for solar-powered electric cars?
2. Do you think electric cars would be cheaper or more expensive to build than gasoline cars? Why?

Power Used for Transportation Vehicles

Different types of power are used for different modes of transportation. Let's look at the kinds of power that are used for highway transportation, air transportation, rail transportation, water transportation, and pipeline transportation.

Highway Transportation

Highway transportation generally uses reciprocating internal-combustion engines. The two main types of engines used to power different modes of highway transportation are gasoline engines and diesel engines.

Gasoline Engines

As you learned earlier in this chapter, internal-combustion engines can be either two-stroke cycle or four-stroke cycle designs. Four-stroke cycle gasoline engines are used to power nearly all automobiles and large motorcycles that you see on the highway today. Gasoline engines are also used to power many of the trucks and buses traveling down the highway.

Diesel Engines

Since diesel engines are good at pulling heavy loads, they are used in many huge freight trucks and large buses. Some cars are powered by diesel engines because this type of engine is more fuel-efficient than a gasoline engine. However, because of various problems with the small diesel engines used in cars, they have not become as popular in the United States as manufacturers once thought they would be.

Air Transportation

Propeller-driven airplanes use reciprocating internal-combustion engines. Jet planes have linear-motion internal-combustion engines that can develop huge amounts of power for their size and weight. Spacecraft and guided missiles are powered by internal-combustion rocket engines. Rocket engines carry their own oxygen and fuel supply. They are the most powerful internal-combustion engines. Fig. 18-17.

Rail Transportation

Locomotives can be classified according to how they are powered. These classifications include steam locomotives, diesel-electric locomotives, electric locomotives, gas-turbine electric locomotives, and diesel-hydraulic locomotives.

Fig. 18-17. Rocket engines are the most powerful internal-combustion engines.

Fascinating Facts

An English mining engineer, Richard Trevithick, invented a steam-powered carriage in 1801. When his first model broke down and then went up in flames, he developed an improved version. He drove friends around London, racing along at eight miles an hour. He had hoped that people would like his carriage and that they'd consider it a convenient way to travel. However, there was little interest.

Trevithick decided to put his carriage on rails, and he built the first steam locomotive in 1804. That locomotive could pull five wagons with 70 passengers and 10 tons of iron at five miles per hour.

Steam Locomotives

Steam-powered railroad locomotives were once very important to rail transportation. They are still used in many parts of the world. However, few remain in use in the United States.

Steam engines used in railroad transportation combine both reciprocating- and rotary-motion principles. Have you seen movies that showed the old railroad steam engines? The "chug-chug" sound was made when steam was released from the *reciprocal* (piston-driven) *engines.*

Some modern railroad locomotives use steam in a *rotary-engine* arrangement. Steam is produced and piped into a chamber. It is directed at the fins of a turbine rotor. The pressure of the steam turns the rotor in a circular direction. The more steam, the faster the rotor turns. Other designs use gas turbines to turn the rotors.

The rotor is connected to an electrical generator. The rotating generator produces the electricity needed to power electric motors. These electric motors turn the axles and wheels of the locomotive.

Diesel-Electric Locomotives

Most modern locomotives use diesel-electric power. These locomotives work just like the rotary-engine-powered models except that a diesel engine is used to produce the electricity. The diesel engine turns an electric generator. The generator produces the electricity needed to power the electric motors. Diesel-electric engines have several advantages over steam engines. Diesel-electric engines are very efficient and long-lasting. They need much less space than steam engines, so the locomotives can be smaller. Diesel-electric engines also accelerate faster than steam engines. Fig. 18-18.

Electric Locomotives

The operation of electric locomotives is very similar to diesel-electric locomotives. The one big difference is that an electric locomotive gets its power from wires suspended above the track or from an electrified third rail instead of producing its own electric power. A subway train is an example of an electric locomotive.

Gas-Turbine Electric Locomotives

A few trains use gas turbines to generate power. In gas-turbine electric locomotives, the force of hot gases runs turbines. These turbines operate electric generators. The power produced by these generators runs the trains.

Diesel-Hydraulic Locomotives

Diesel-hydraulic locomotives use diesel engines to drive a torque converter instead of a generator. (Torque is any force that produces or tends to produce rotation or torsion.) The torque converter, which includes a pump and a turbine, uses fluids under hydraulic pressure to transmit and regulate power received from the diesel engine. The pump forces oil against the

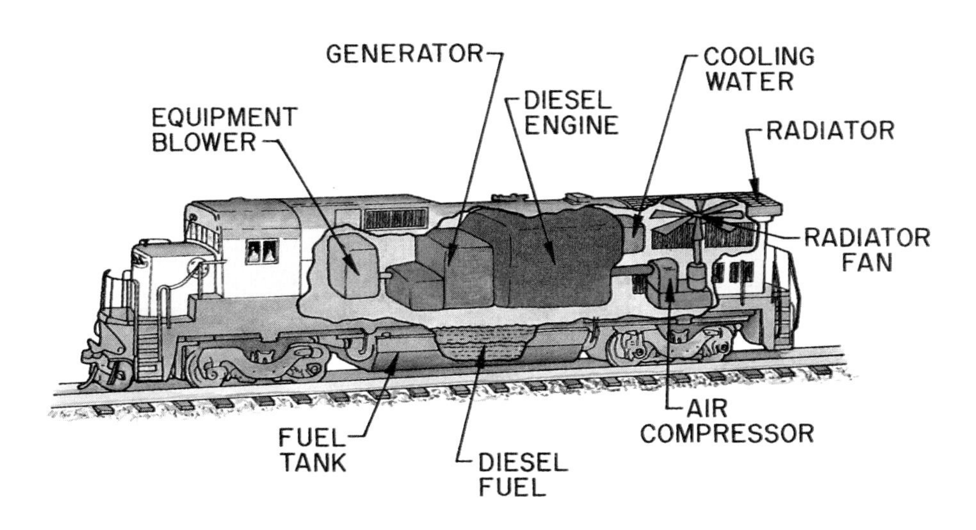

Fig. 18-18. Even though alternate forms of power are being found for locomotives, diesel-electric engines still power most modern locomotives that are on the tracks today.

blades of the turbine, and this action causes the turbine to rotate and to drive a system of gears and shafts that moves the wheels of the locomotive. Diesel-hydraulic locomotives are not used in the United States, but they are widely used in other countries, such as Germany.

Water Transportation

Even though water transportation is the slowest form of transportation, it is also inexpensive, safe, and fairly flexible. As you learned earlier, water transportation is used to transport both passengers and cargo. As international trade continues to grow, the use of water transportation to move cargo will also increase. Fig. 18-19. Let's look at the two major types of power that are used for water transportation.

Steam Engines

Reciprocal and rotary engines are both used to power water transportation vehicles.

Fig. 18-19. Today, water transportation is used more for cargo transport than it is for passengers.

Ships have room for large boilers and various methods of creating heat. Many ships use nuclear reactors to boil water and produce steam.

Gas-Turbine Engines

Gas-turbine engines are used to power many of the different types of ships in the water today. As you have already learned, hot gases turn the turbines in this type of engine. The revolving turbines turn the propeller shaft that powers the ship.

Pipeline Transportation

You learned in Chapter 16 that pipelines move material in the form of gases, liquids, or liquids and solids mixed into a slurry. The power for pipelines is produced by pumps that are turned by electric motors. Pumps may be of several different types, but they all perform the same job—"pushing" the pipeline contents along the pipeline. The pumps add pressure to the pipeline in almost the same way that a bicycle pump adds pressure to a tire. The pressure "pushes" the material along the pipeline just like a piece of paper shoots out of a straw when you blow into the straw.

If there were no friction in a pipeline, only one pump at the beginning of the line would be necessary. However, since there is friction between the contents and the pipeline walls, pressure must be added to the pipeline regularly. Pumping stations along the pipeline continuously add pressure to replace what is lost to friction.

Apply Your Thinking Skills

1. Why do you think there are so few steam-powered locomotives in use in the United States today?

Investigating Your Environment
Are Cars Driving Us to Disaster?

Our world has come to depend on cars, but gasoline-engine automobiles cause many problems. Every year more than 200,000 people die in accidents and millions more are seriously injured. Car exhaust causes a lot of pollution. It contains carbon monoxide, carbon dioxide, nitrous oxides, lead, and hydrocarbons. These poisons contaminate our soil, water, and air. They contribute to acid rain and the depletion of the ozone. A car's air conditioning system gives off chlorofluorocarbons (CFCs). CFCs break down the ozone layer. Even after a car is discarded, CFCs are released when the car is crushed. Discarded batteries, tires, and motor oil pile up in our landfills, causing further water and soil poisoning. Our demand for petroleum products to fuel our automobiles has led to problems. Oil spills have polluted miles of shorelines and killed thousands of fish, birds, otters, and other types of wildlife.

With so many cars, traffic jams are often a problem. Billions of gallons of fuel are wasted by cars stuck in traffic jams. Building more roads does not help. More roads draw more cars, which in turn creates the demand for more roads.

Since 1970, cars have been equipped with devices that remove some pollutants from exhaust, and the Environmental Protection Agency (EPA) has ordered that the amounts of lead in gasoline be reduced. Newer cars are lighter and more aerodynamic, so they use less fuel. Computer-controlled engines reduce emissions and waste less energy.

Engines using other types of fuels, such as methanol or compressed natural gas, are imperfect solutions. Engines that are powered with methanol emit carcinogenic (cancer-causing) formaldehyde. Compressed natural gas engines are bulky and subject to leaks that contribute to global warming. Battery-operated electric cars have been designed, but they have to be recharged frequently, and some type of fossil fuel must be burned to make the electricity. Solar-powered cars have also been designed, but they are not practical at this time.

Fig. IYE-18. Automobiles produce many undesirable outputs when they are operated. Air and noise pollution are two undesirable outputs from automobile usage.

Though automobile technology continues to improve, mass transit may be the best way to reduce problems caused by cars. Streetcars, subways, railroads, rail systems, and even car pools use less energy per person and can greatly reduce pollution and congestion.

Take Action!

1. Prepare an advertising campaign (slogans, radio and TV ads, posters, etc.) to encourage people to use mass transit and/or bicycles rather than automobiles.

Chapter 18 Review

Looking Back

Power is using energy to create movement. Engines and motors are prime movers. They supply power in transportation.

There are two basic types of engines: internal- and external-combustion engines.

The three types of motion produced by internal-combustion engines are reciprocating, rotary, and linear.

Engines have the following six systems: mechanical, lubrication, fuel, ignition, starting, and cooling systems.

Internal-combustion engines operate in either two- or four-stroke cycles. These engines must both complete four operations: intake, compression, power, and exhaust.

Steam engines are the most common external-combustion engines. They are used in large ships and locomotives.

Motors used in transportation systems are either electric- or fluid-powered. Electric motors are frequently used to power mass transportation vehicles. Fluid-powered motors apply the principles of hydraulics to operate heavy equipment.

There are different types of power used for different types of transportation.

Review Questions

1. Explain the difference between an engine and a motor.
2. What is the major difference between internal-combustion and external-combustion engines?
3. Describe a reciprocating-motion engine. Give two advantages and two disadvantages of this engine. What vehicles are powered by this type of engine?
4. What is rotary motion? Tell how rotary engines work. Give two advantages and two disadvantages of this engine.
5. What is linear motion? Describe a linear-motion engine. What vehicles use this type of engine? Give two advantages and two disadvantages of this engine.
6. List the six systems usually found in a small gasoline engine. Briefly explain the job of each system.
7. Name and describe the four strokes that make up one full cycle of a four-stroke engine.
8. Briefly describe how the power is provided for the most common type of external-combustion engine.
9. Give four advantages of electrically powered transportation vehicles. What is the disadvantage of these vehicles?
10. Describe hydraulic power. Give one advantage and one disadvantage of motors using this type power.

Discussion Questions

1. Do you think we will see more rotary engines in the future? Why or why not?
2. Why are most automobiles powered by four-stroke cycle rather than two-stroke cycle engines?
3. Why would it be difficult and impractical to develop large electric-powered trucks?
4. Discuss the effects of friction on engines and on pipeline transportation. Are there any times when friction is necessary to vehicles? If so, when?
5. If you were going to design a "perfect" transportation vehicle, what type of engine would you put in it? Why?

Chapter 18 Review

Cross-Curricular Activities

Language Arts

1. A cleaner environment is a universal concern. In a persuasive essay, present the advantages of owning and operating an electric car.
2. Make an outline to explain the six basic systems in an engine.

Social Studies

1. Find out when the internal-combustion engine was developed. Tell how this invention affected transportation as we know it and what effect it had on the oil industry.
2. Who was responsible for the first steam engine? Find out on what other invention he based his invention. Then list some developments that evolved from his invention.

Science

1. Friction can be troublesome when energy is lost because parts of an engine rub together. Rub the flat surfaces of two jar lids together briskly, and note how much heat (friction) is produced. Now oil them, and rub the two lids together again. Is the friction increased or decreased? Why is the lubrication system such an important part of an engine?
2. External- and internal-combustion engines work because expanding gases exert force. You can demonstrate this simply by raising a book off the table with a balloon. Place the book on the balloon and blow up the balloon. What stroke of a four-stroke cycle engine is represented here? What part of the engine is represented by the book? What event is represented by your blowing air into the balloon?

Math

Current flow in an electrical circuit is calculated by using Ohm's law, which states the following:

$$I = \frac{E}{R}$$

The **I** is the amperage, or strength of the current (measured in amps); the **E** is the voltage (measured in volts); and the **R** is the resistance (measured in ohms).

1. Using a meter, you have determined that a device on an electrical circuit is drawing 2 amps when 120 volts is applied. What is the device's resistance?
2. If you apply 24 volts to a circuit that contains 24,000 ohms of resistance, what will the amperage of the current be?

Chapter 18 Technology Activity

Building a Cardboard Hovercraft

Overview

Hovercraft, which are also known as ground-effects machines and air-cushion vehicles (ACVs), are vehicles that support themselves a small distance above the ground or water surface on a cushion of air forced downward by vertical fans. This invisible cushion of air eliminates almost all friction between the hovercraft and the ground or water. Gas-turbine engines provide the power for the vertical fans. Propellers drive a hovercraft forward, and most of these vehicles have rudders for steering.

Hovercraft can transport both passengers and freight. They are used for surface transportation tasks that cannot be accomplished efficiently by automobiles, trucks, buses, trains, or boats. Hovercraft are also ideal for transportation in arctic waters, where winter ice often makes waterways impassable for conventional ships and where the ice is not solid or smooth enough for the passage of trucks or other vehicles.

In this activity, you will design and build a hovercraft using cardboard and a balloon. After your vehicle is built, you will test it to see if it will lift off the surface.

Goal

To design, build, and operate a hovercraft using cardboard and a balloon.

Equipment and Materials

10" or 12" balloon
8" x 8" piece of cardboard (corrugated)
8 1/2" x 11" piece of paper (heavy)
plastic clothes bag (clear) or Glad wrap
plastic lid from coffee can
1/8" x 3/8" pop rivet
washer for pop rivet
compass
protractor
masking tape
glue

Procedure

1. *Safety Note:* Before doing this activity, make sure you understand how to use the tools and materials safely. Have your teacher demonstrate their proper use. Follow all safety rules.
2. Find the center of the 8" x 8" piece of cardboard. (Use the diagonal method of drawing a line from corner to corner.)
3. Using the compass, draw a circle around the center of the cardboard. (The larger the circle, the better your hovercraft will work.)
4. Using a protractor, divide the circle into 4, 5, 6, 7, or 8 equal parts. Then draw lines from the center to these points.
5. Cut a 2" x 4 3/4" piece of heavy paper. Fold and glue a 1/4" hem on one of the long (4 3/4") sides. Fold and glue the hem again. Then on the other edge, make 3/8" cuts down the side. Now, make a cylinder with the hemmed edge to the outside. Glue and tape this cylinder together. Fig. A18-1.

Chapter 18 Technology Activity

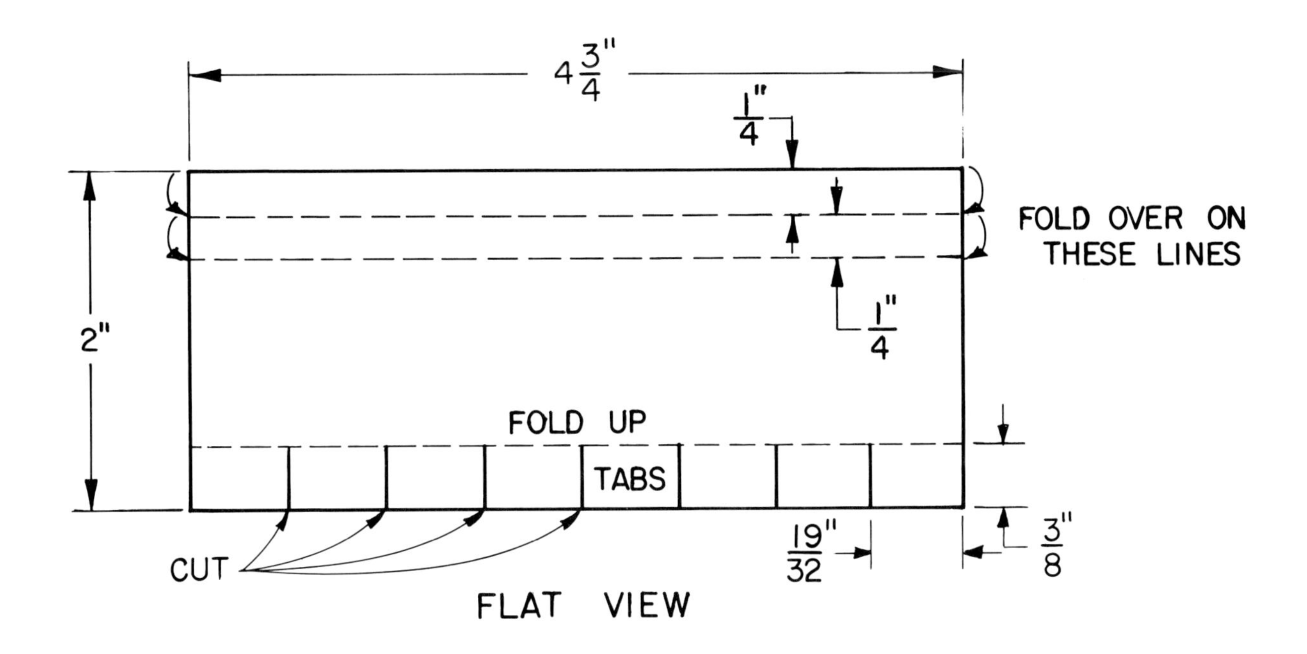

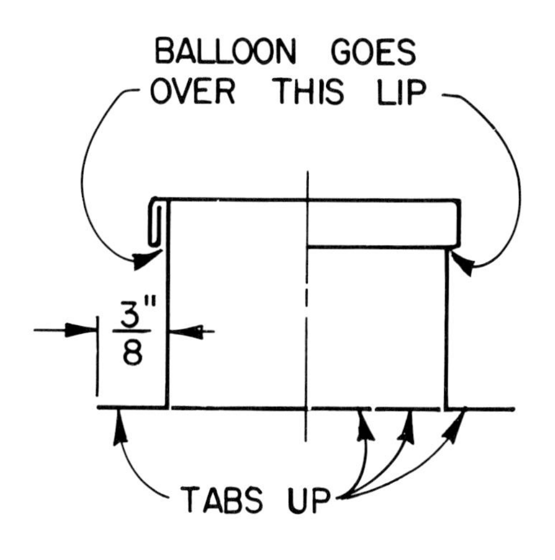

Fig. A18-1. Plans for making the cylinder.

Chapter 18 Technology Activity

6. Cut two 1" x 1 1/2" pieces of corrugated cardboard and make them fit together inside the cylinder. Fig. A18-2.
7. Cut a hole 2" from the center of the 8" cardboard disk so that the cylinder made in Step 5 will fit into it. Fig. A18-3.
8. Glue and tape the edge of the cylinder with the 3/8" tabs on it into the hole. Fold one cut under the disk and another on top.
9. Use the plastic clothes bag or plastic wrap to make a circle 1" larger than the cardboard circle made in Step 3.
10. Stretch the plastic bag over the cardboard's side with the lines drawn on it and tape securely all the way around the edge.
11. On the bottom side of the plastic, cut 1/2" holes about 2" from the center of the cardboard on the lines you drew in Step 4.

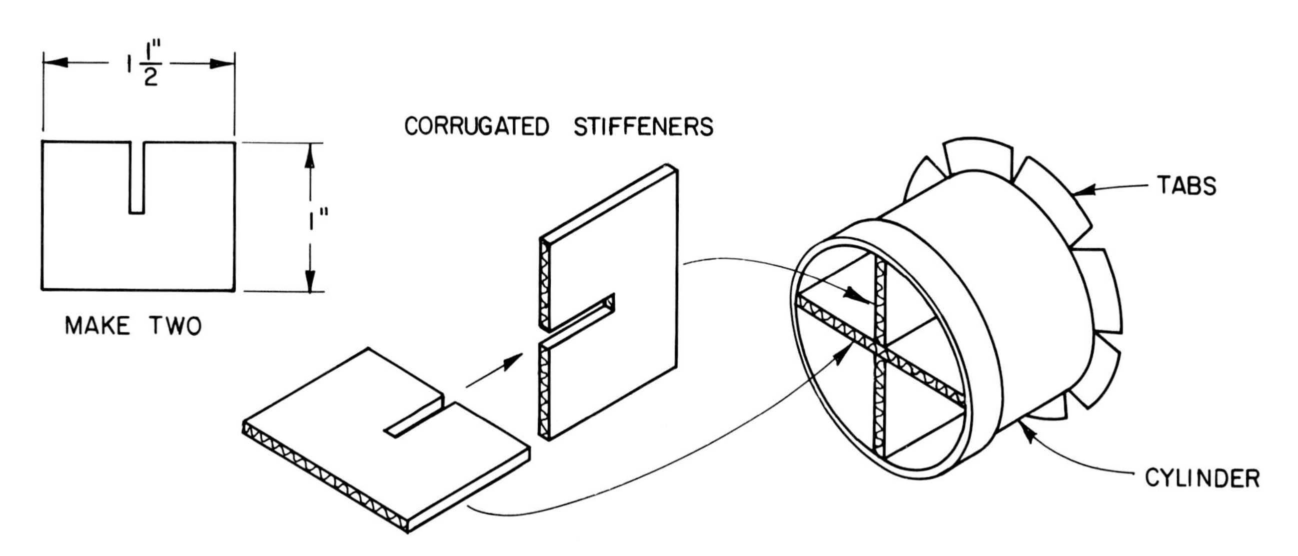

Fig. A18-2. Cutting and fitting two pieces of corrugated cardboard inside the cylinder.

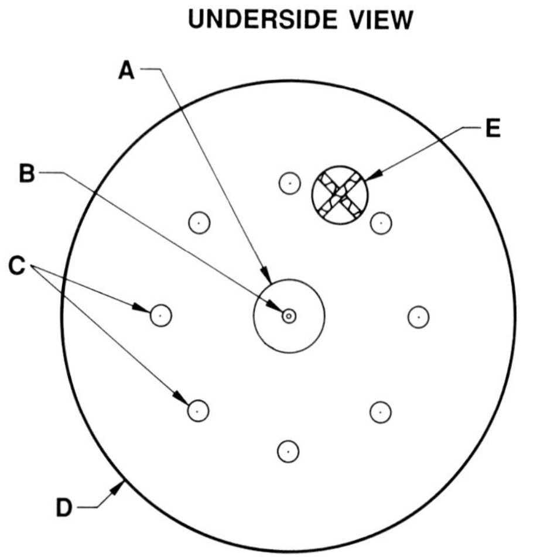

Fig. A18-3. The underside view of the cardboard hovercraft.

A PLASTIC CIRCLE FROM COFFEE CAN LID
B POP RIVET
C AIR HOLES CUT IN PLASTIC BAG
D CORRUGATED CARDBOARD CIRCLE
E HOLE FOR CARDBOARD DISK

Chapter 18 Technology Activity

12. Cut a 1" circle from the plastic coffee can lid. Then, using the pop rivet and washer, fasten the 1" plastic circle to the bottom of the plastic-covered cardboard disk with the pop rivet inserted from the bottom and the washer on top of the plastic-covered cardboard. Fig. A18-3.
13. Cut a 4" x 11" piece of heavy paper and tape it on top of the disk 1" from its edge. This is to support the balloon.
14. Let the glue dry for 12 hours.
15. Blow the balloon up, twist its bottom, and slip it over the cylinder. Then, untwist the balloon's bottom and watch the disk float. Figure A18-4 shows what your finished hovercraft should look like.

Evaluation

1. How well did your cardboard hovercraft float?
2. How well did your classmates' hovercrafts float? Take a look at the ones that worked the best (hopefully, yours is one of them). Why did they float so well?
3. If you could design another cardboard hovercraft, what would you do differently to make it work better?

Credit: Paul Simmons and Robert Campbell

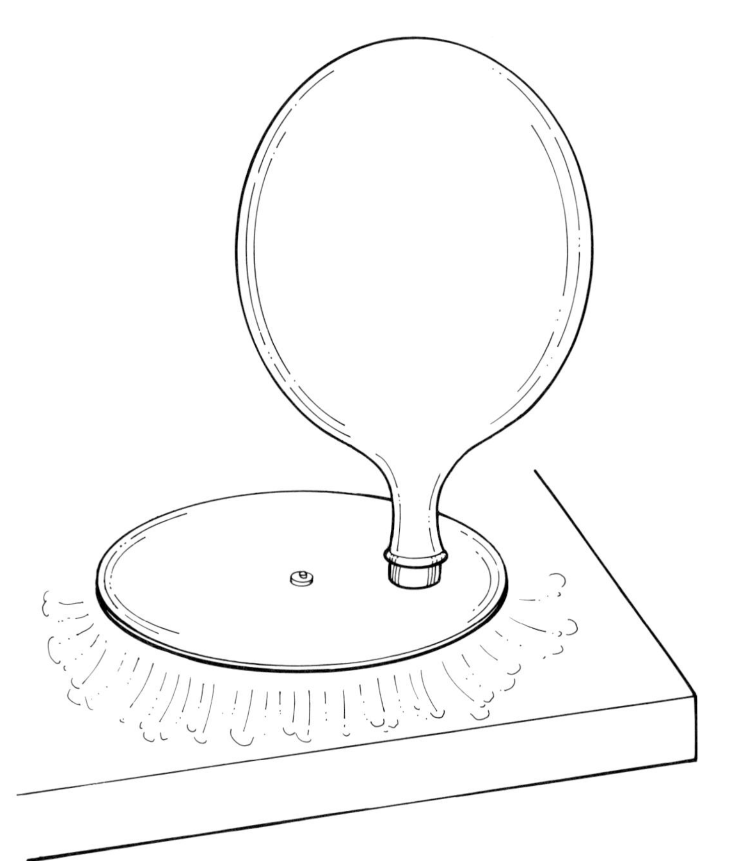

Fig. A18-4. The finished cardboard hovercraft.

Chapter 19

Trends in Transportation Technology

Looking Ahead

In this chapter, you will discover:

- what aerodynamics is and how it influences vehicle design.
- some new materials that are being used in vehicles.
- how electronics is revolutionizing vehicle operations.
- some alternative energy sources that affect vehicle designs and operations.
- the types of space transportation systems and how they work.
- that satellites are launched by space transportation systems and can have impacts on our transportation systems on earth.
- some current and possible future uses of biotechnology in transportation systems.

New Terms

aerodynamic drag
aerodynamics
booster rockets
ceramics
composite
downsizing
gasohol
maglev systems
methane gas
methanol
National Aeronautics and Space Administration (NASA)
payload
photovoltaic cells

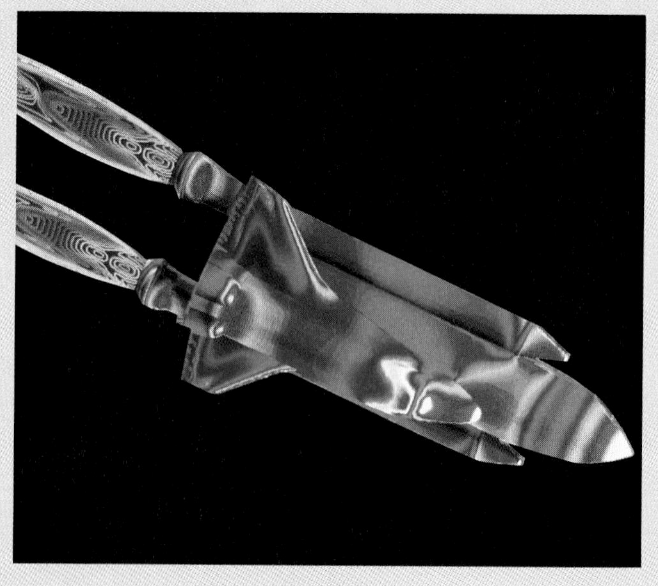

Technology Focus

The Flying Car

It's not too early to order your twenty-first century flying automobile. A $5,000 deposit will reserve a Moller 400, which could be cruising the skies at 355 mph by the year 2000. The Moller 400 is a vertical takeoff and landing vehicle (VTOL). A prototype of this volantor (a term meaning something that hops nimbly and quickly about) has already been flown and has demonstrated its practicality. It can take off and land safely in about the same amount of space as a helicopter. One of its eight engines could fail and the vehicle could still land or take off. All but two could fail in flight, and it could still glide until it could land safely. Each engine has only three moving parts, so chances of engine failure are small.

Paul Moller's first hurdle in developing a working volantor was to develop a quiet, low-weight, high-powered engine. He bought technology information from a manufacturer of snowmobile engines. He made more than 70 changes in the design before he developed an engine with the power-to-weight ratio needed for vertical takeoff.

The Moller 400 engines are 150-horsepower Wankel engines with two spinning rotors that act as pistons. Two engines are mounted inside each of four nacelles (streamlined enclosures). Fans on each engine propel the engine's thrust against louvered vanes. The vanes direct the air downward for takeoff. In flight, the vanes alter position to change the direction of thrust for forward motion.

Three computers do most of the work of piloting. The pilot has a throttle to select altitude and rate of climb and a joystick to guide the vehicle in the desired direction. The computers do all the rest.

Paul Moller sees his Moller 400 as the first step in achieving his dream for the future—an entire personal transportation system in the skies.

Fig. TF-19. Shown here is a model of the Moller 400 that Paul Moller has been working on for over twenty-five years.

Vehicle Designs and Operations

New developments in transportation technology are making today's vehicles more efficient, more economical, and safer than vehicles in the past. Further developments hold great promise for the future. Most of the changes being made involve:

- Aerodynamics
- Size and materials
- Electronics

Aerodynamics

Aerodynamics—say the word slowly to yourself—AIR-oh-die-NAM-iks. Do you hear the word "air"? "Dynamics" refers to forces and how they affect motion. **Aerodynamics** is a science that deals with the interaction of air and moving objects. That's what vehicles are—moving objects.

Air tends to resist or slow movement. Vehicles meet this resistance as they are moved through air. The faster they travel, the greater the resistance. A strong, power-robbing force is created. It is called **aerodynamic drag**. Design engineers look for ways to reduce this drag force. Doing so eases the work load of the engine and thus improves fuel efficiency.

Highway Vehicles

Much progress has been made in reducing the aerodynamic drag on highway vehicles. The results are impressive. Not only are the newer cars and trucks more fuel-efficient, they are also safer and easier to handle. This is because the aerodynamic design also makes vehicles more stable when hit by crosswinds. An additional bonus is wind noise is also reduced.

Shape is the critical factor. Engineers describe an aerodynamically efficient shape as "slippery." They try to design vehicles that will "slip" smoothly through the air. Fig. 19-1. Air should glide smoothly over, under, and around a vehicle. Even a mirror or door handle that sticks out can disturb the air flow and increase drag.

Aircraft

Engineers are currently developing new aircraft designs. The basic wing (cross-section) *shape* is not likely to be changed. However, the length, positioning, angle, and other factors about wings may be changed in relation to the total design. A major goal is to improve lift and further reduce drag. Fig. 19-2.

Size and Materials

Lightweight vehicles are generally more fuel-efficient than heavy vehicles. One way to achieve weight reduction is to design and produce smaller vehicles. In industry, this is referred to as **downsizing**. Today, most personal transportation vehicles are smaller

Fig. 19-1. This car has an aerodynamically efficient shape. It was designed to "slip" through the wind with as little aerodynamic drag as possible.

Fig. 19-2. This is a possible design for a hypersonic aircraft that would be able to take off from a runway and fly into orbit.

and slower than vehicles built in the recent past. If consumer demand does not change, this trend is likely to continue.

Today, trucking companies often buy smaller commercial vehicles. These cost less to buy than large, heavy vehicles. They are also cheaper to operate. In doing this, companies are exchanging speed and larger loads for smaller money investments.

Design engineers are looking for ways to reduce weight without reducing size. Often, in a new model, lightweight materials such as aluminum or plastics are used instead of heavier materials such as steel.

Plastics

Many new, smaller aircraft are made primarily of special fiber-reinforced plastics. They are light, but strong. Parts of larger planes are also made of these materials. Engineers continue to look for other ways in which they may be used.

Plastics have long been used in car and truck interiors. This trend continues. Now, superior new kinds of plastics are replacing those used previously. It is expected that more and more plastics will be used to form *exterior* parts. A study done by the University of Michigan indicates that up to 70% of new cars will have plastic body panels by the year 2000.

Ceramics

Ceramics are materials, such as earthenware and porcelain, made from nonmetallic minerals that have been fired at high temperatures. These materials are used in spacecraft because they can withstand extreme temperatures and because they insulate well. These same qualities will probably prove useful in other vehicles. Some new experimental engines include ceramic parts. Fig. 19-3.

Fig. 19-3. This experimental engine contains mostly ceramic parts.

One vehicle engineers hope to be able to use ceramics in is trucks. Diesel engines used in trucks produce great amounts of heat. Air passes through the truck's radiator. This air helps cool the water in the radiator. Large amounts of air are needed. The front of a truck is built large and flat like a "wall" to accommodate the airflow. This shape increases aerodynamic drag and reduces fuel efficiency.

In order to give trucks a more aero-efficient shape, the engine must be redesigned. Some engine makers are looking for ways to use ceramics to replace metal parts or to protect them from the heat. If they are successful, air-cooling will not be necessary. Profiles of trucks can be changed. Fuel efficiency should increase for three reasons:

- Aerodynamic shape
- Lighter weight (redesign or elimination of radiator)
- Less energy lost in the form of heat

Before ceramics can be widely used, one major problem must be overcome. New production techniques are needed. Ceramic materials are so hard that special diamond-coated tools must be used to cut them. Another method is to develop special shaping molds.

Another drawback of ceramics is that they are brittle. A part with even a small blemish may fracture (break) when put under stress.

Composites

A **composite** is a new material made by combining two or more materials. Each component of a composite retains its own properties, but the resulting material has more desirable qualities. Composites are being used more and more today to make parts for different types of transportation. For example, fiberglass reinforced with a thermosetting resin—material that has "set" after being heated and cannot be easily reshaped—is now being used for boat hulls and automobile bodies. Kevlar™, which is a composite that is very difficult to cut, is being used to make reinforcing belts for tires. Carbon/graphite composites are now being used to make lightweight bicycle frames and to produce several new types of aircraft. The use of composites in transportation systems will increase as different materials are combined and found to be lighter, safer, and longer-lasting.

Electronics

Computers, microprocessors, and other electronic devices are revolutionizing vehicle operations. Probably the first use of electronics in vehicles was stereo radios and tape decks. However, engineers soon found that electronic controls could be used to reduce the amount of air pollutants in vehicle exhaust. The controls precisely regulated the air-fuel mixtures in the engines. Not only did this reduce waste, it also improved engine performance.

Engineers had learned that electronics could be more efficient than mechanical means. They began to look for other ways to use electronics. Fig. 19-4.

Today, electronics is used in many vehicle subsystems, such as brakes and power steering. The role of electronics is expected to expand rapidly in the next few years. The goal is described as "an engine with a brain." A computer will control more and more systems operations.

Fiber Optics

The use of fiber optics is expected to increase. The optical fibers used in vehicles are usually thin strands of plastic. They carry signals in the form of light. These signals may be used to control such items as power windows and door locks.

Fig. 19-4. This automobile dashboard uses liquid crystal display (LCD) technology. Tiny liquid crystals that act as light-controlling cells are put into flexible film panels. A wide variety of graphic displays are possible. Gages and switches can be located on the same panel.

Signals from several controls can be multiplexed (merged or combined) on fibers. A pair of optical fibers can replace many wires. This saves space and weight. It also greatly simplifies the wiring system. Fig. 19-5.

Eventually, fiber optics may be used to link electronic operations in engines. However, the fibers currently available cannot take the heat produced by today's engines. Remember, however, that engines may be redesigned, and new materials may be developed. Advances are made every day in technology.

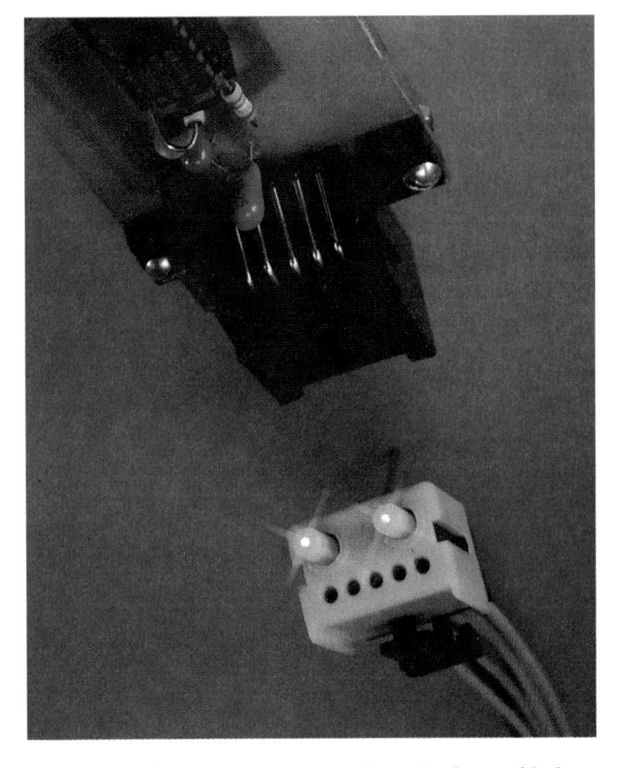

Fig. 19-5. Fiber optics carry signals in the form of light. Using fiber optics can greatly simplify wiring systems.

Apply Your Thinking Skills

1. What are some features on new cars that you have seen that make them more aero-efficient?
2. Can you think of any advantages, other than weight reduction, of using plastic body panels on cars? Any disadvantages?

Alternative Power Sources

One of the biggest problems facing transportation is finding power sources that are in plentiful supply and do not harm the environment.

Natural Power Sources

The oldest methods of transportation relied upon natural power sources. Wind-powered sailing ships traveled the seas long before engines were developed. Today, modern technology is finding new ways to use wind to power ocean-going ships. Fig. 19-6.

Wind energy is also being turned into electrical energy. Some small sailing yachts have windmills and electrical generators installed on their masts. Even when winds are light, the electricity produced by these units is enough to charge the batteries of the boat. The same designs, on a larger scale, are used to provide supplemental (additional) electricity on larger ships.

The sun is a potential source of natural energy for transportation use. Ways to use its power are being developed now. **Photovoltaic cells** are devices that can change the energy from the sun's light (solar energy) into electricity. These have already been used to power experimental vehicles. Some vehicles use the power from the cells to directly power an engine. More often, the electricity from the photovoltaic cells does not go directly to a motor. Instead, the solar-generated electricity is used to charge high-capacity batteries. Fig. 19-7. The photovoltaic cells can charge the battery whenever the sun shines on them. This allows the vehicle to be used even when the sun is not shining.

Fig. 19-6. A "Daughter of the Wind." This ship, called the *Alcyone*, uses computer-controlled turbosails to capture the power of the wind. Turbosails are economical to operate; they save energy; and they produce almost four times as much thrust as the best sails available today.

Fig. 19-7. Electrically powered cars can benefit from an array of on-board solar panels. The *Destiny 2000* has a solar array embedded in the hood and trunk which contributes 250 additional watts to its batteries when exposed to sunlight.

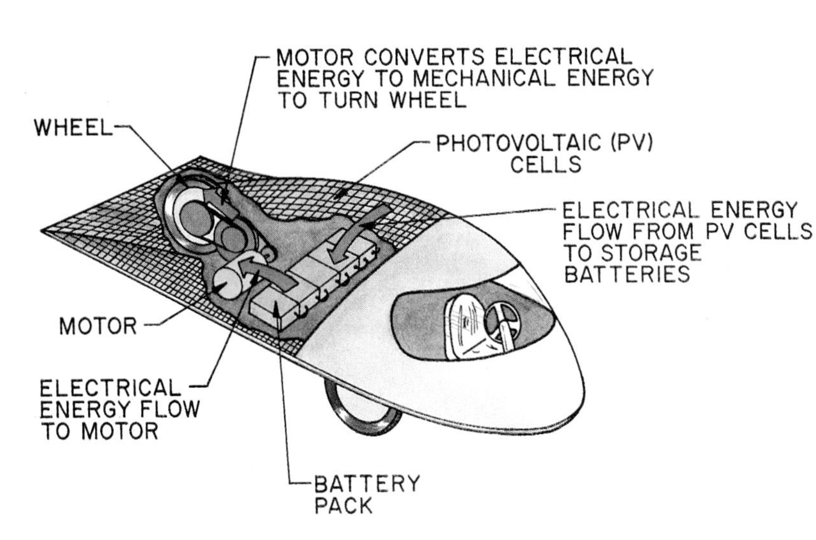

Fig. 19-8. The GM *Sunraycer* uses its photovoltaic cells to charge special batteries. The batteries then provide electrical power to the motor and wheels.

The charged batteries also produce a more constant flow of electricity to the motor. This makes it much easier to maintain constant speeds and gives the vehicle longer range. Fig. 19-8.

Currently, the use of wind and solar-produced electricity is limited by battery capacities. Most batteries that are powerful enough to store enough electricity to power vehicles are either very heavy or very expensive. Several types of new battery designs that may soon solve this problem are now under development. However, the future for use of energy from natural sources in transportation holds great promise.

Fascinating Facts

Computer programs are available today for some servicing operations. The knowledge of many automotive service specialists is worked into a computer analysis program. With such a program, a computer can efficiently diagnose a vehicle's problems. It examines various factors about the vehicle's operating systems. Then it "makes decisions" based on the input of the specialists. Such technology is called a *smart system* (also known as an expert system). In the near future, similar systems may actually be built into vehicles.

Alternative Fuels

The use of alternative fuels in internal combustion engines is very possible in the near future. Alternative fuels would be fuels other than the fossil fuels (gasoline, diesel fuel, oil, natural gas, propane, and coal) that currently power most transportation vehicles.

One of the most promising of these fuels is hydrogen gas. Hydrogen can be burned in internal combustion engines much like current fuels but with several advantages. The waste product of burning (combusting) hydrogen is water vapor. Because of this, little or no pollution would be produced from hydrogen-powered vehicles. Hydrogen is one of the most plentiful elements on earth, so there would be little danger of running out of the supply. The main disadvantage is that hydrogen gas is very explosive. This makes it difficult and expensive to build safe hydrogen fuel tanks for vehicles. Many scientists and engineers feel that these two problems will be solved in the near future.

Other alternative fuels include those produced with biotechnology. You'll learn about these later in this chapter.

Maglev Systems

Maglev systems are rail systems that operate on the scientific principle that like poles of a magnet repel each other. The word maglev is short for *mag*netically *lev*itated. (Levitate means to rise or float in the air.) Magnets in the maglev guideway (rails) repel magnets of like polarity on the bottom of the maglev vehicle. Fig. 19-9. This action causes the train to levitate above the guideway, creating a nearly frictionless riding surface. Magnets are also involved in the vehicle's propulsion. Changing the polarity of the magnets on the train and the guideway at the proper moments speeds up or slows down the train.

Because there is so little friction and the vehicles are aerodynamically designed, the trains can easily be accelerated to speeds over 250 mph. They glide quietly, smoothly, and swiftly along the guideway using relatively little energy.

Fig. 19-9. A maglev train can reach speeds higher than 250 mph.

A maglev system is very different from our present "steel-wheel-on-steel-rail" system. In the United States, studies are being done to see whether such a system would be practical and economical.

Stirling Cycle Engines

The *Stirling cycle engine* is an experimental external-combustion engine that makes use of the scientific principle that gases expand when heated and contract when cooled. In internal-combustion engines, combustion of the fuel/air mixture causes the mixture to expand in a cylinder. That expansion pushes upon a piston and does work. The Stirling cycle engine has a closed cylinder with a gas (usually helium or hydrogen) sealed into it. The gas is not burned. It constantly flows back and forth in the cylinder between a hot space (heated by an external combustion source) and a cold space. Thus, it is heated and cooled in a repeating cycle. (A device called a regenerator helps heat the gas at the beginning of the cycle and cools it by absorbing the heat at the end of the cycle.) This alternate heating and cooling process causes the gas first to expand and then to contract. This causes a repeating cycle of pressure changes that cause a power piston to move back and forth inside the cylinder. A rod connects this power piston to a crankshaft. The crankshaft converts the power piston's *reciprocating* (back-and-forth) motion to the rotary (rotating on an axis, like a wheel) motion of the drive shaft. This action, in turn, causes a displacer piston to move back and forth and force the gas through the regenerator.

One major advantage of the Stirling cycle engine is the fact that it would pollute much less than current internal-combustion engines. Since the Stirling cycle engine does not involve burning a fuel-air mixture, no burned waste products would be exhausted into the air. Another advantage is that the engine is very efficient, so it would need very little energy to operate. The main disadvantage is that it is a very complex, expensive engine to build. So far, this disadvantage has kept the Stirling engine from coming into general use. Experimental Stirling engines and designs are being tested, and we may soon see many types of vehicles running on these promising engines.

Apply Your Thinking Skills

1. Which of the alternative power sources do you think is the most practical for use on a widespread basis? Explain your answer.
2. Why would maglev trains be able to travel at much higher speeds than trains in common use today?

Transportation in Space

The future of space transportation has been discussed ever since 1865. That was the year that the French author Jules Verne wrote *From the Earth to the Moon*. It was the first science fiction novel written about space travel. Most people considered Verne's idea of a person traveling through space to the moon to be ridiculous. They were not proved wrong for almost 100 years. In July 1958, the **National Aeronautics and Space Administration (NASA)** was founded to plan and operate the U.S. space program. In 1969, Neil Armstrong and Edwin Aldrin landed on the moon in an Apollo spacecraft.

Space transportation is still very new. The future of human space travel is still almost as hard to predict now as in Jules Verne's time. Space is a very hostile and dangerous place for humans. An unprotected human would die almost instantly in space. There are still seldom more than a handful of people in space at any one time. However, that may change soon. The first true spaceships, space shuttles, make frequent trips into space and back. America is planning to build a permanent space station within the next decade and also to explore the surface of Mars.

There are two main types of space transportation systems—manned and unmanned. Manned systems carry humans into space. Unmanned systems are all systems without human crews or passengers. Manned space flight is still much less common than unmanned flights. Manned spacecraft use two basic designs of spacecraft: nose-cone-mounted spacecraft and space shuttles. These types of spacecraft must provide everything necessary to keep their passengers alive in the vacuum of space.

Nose-cone-mounted spacecraft were the first type used to transport humans into space. These spacecraft are sealed containers put onto the nose of a booster rocket and launched into space. **Booster rockets** are rockets used to push a payload into space. A **payload** is anything transported into space. When the rocket reaches space, the spacecraft separates from the booster rocket. Fig. 19-10. These nose-cone-type spacecraft have no wings and have no ability to fly through the atmosphere on their own.

Space shuttles can be called the first true spaceships because they take off, maneuver in space, and fly back through the atmosphere for landing. Shuttles have their own powerful engines, but they still need the help of a booster rocket to get into space. Fig. 19-11. Each shuttle can be used over and over again. They are versatile and powerful

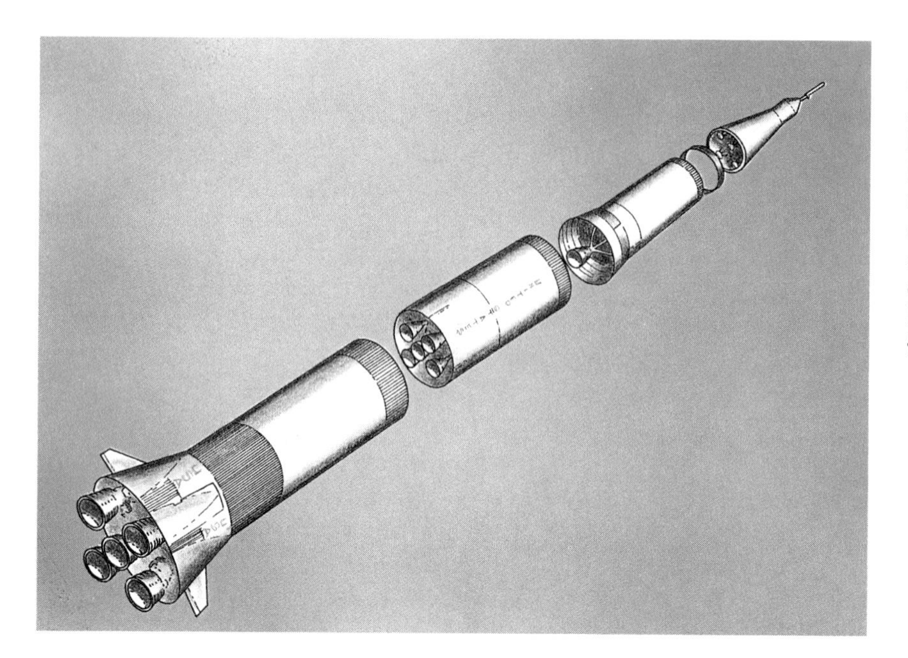

Fig. 19-10. Nose-cone-mounted spacecraft (like this Apollo vehicle) sit on top of a booster rocket. Once the booster rocket has pushed the nose-cone-mounted spacecraft high enough, they separate. The spacecraft continues into orbit and the booster rocket burns up while falling back to earth.

enough to be called "space trucks." Fig. 19-12. Most shuttle flight missions include launching satellites. Fig. 19-13.

Unmanned systems use nose-cone-type vehicles. These nose cones are usually just hollow containers. Once in orbit, these containers open and the payload is maneuvered into position with tiny rocket engines. These vehicles and their booster rockets can be used only once.

The most common cargo on unmanned space transportation vehicles is satellites. There are now hundreds of satellites in orbit around the earth. Most of them are communication satellites. Fig. 19-14. Some are programmed to send information to receivers on earth. Others receive microwave signals from ground stations or vehicles. Then they transmit these signals to other stations or vehicles at other locations. Some are for special purposes like photographing weather systems. Others are military "spy" satellites.

Fig. 19-11. The shuttle *Atlantis* on its launch pad. The large red middle tank contains fuel to power the shuttle's main engines during launch. Reusable booster rockets are mounted on both sides of the fuel tank.

Fig. 19-12. The large cargo bay and robot arm allow the space shuttle to deliver large payloads into orbit.

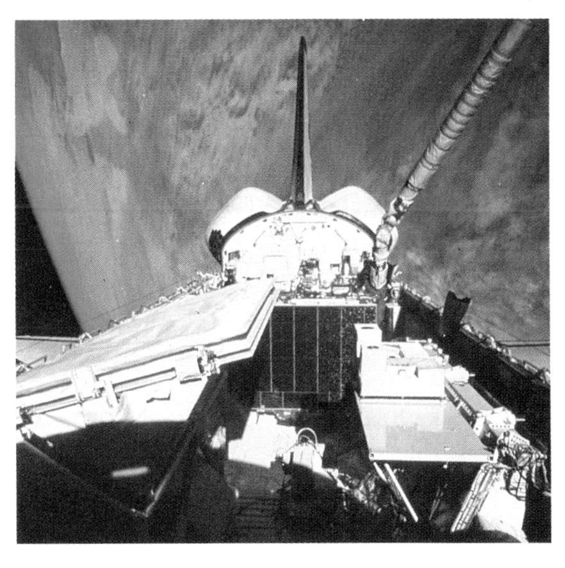

Fig. 19-13. The cargo bay of the shuttle opens to allow satellites to be lifted out.

Some of the communication satellites will have an impact on transportation systems here on earth and within the earth's atmosphere. Using communication satellites, trucking companies will be able to keep a continuous *log* (record) of each truck's location. It would be possible to monitor truck functions as well. For example, a terminal manager would be able to check the refrigeration unit on a truck that is carrying meat products while the truck is on the highway. Truck drivers will also be able to talk to their companies or to each other from any part of the country. Air traffic controllers will be able to accurately monitor aircraft on long-distance flights. This will improve the efficiency and safety of air travel. Routes can be more precise. There will be less danger of encountering (meeting) other aircraft.

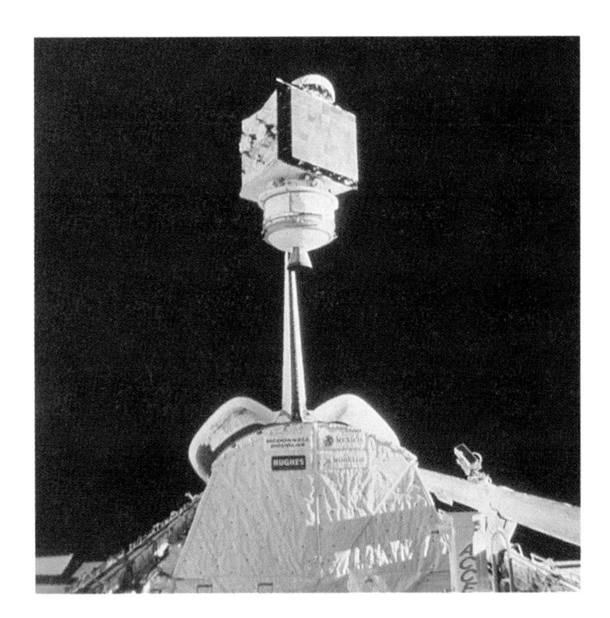

Fig. 19-14. A modern communication satellite. The large photovoltaic cells provide all the electrical power used by the satellite while in orbit.

Unmanned space transportation systems also include robotic exploration vehicles like the *Voyager*. Fig. 19-15.

Manned and unmanned space transportation is by far the most expensive transportation system. It costs thousands and sometimes millions of dollars to put just one pound of payload into space. The future of space transportation will really be determined by how much money governments and corporations are willing to spend to explore space. Many people feel that things like mining and manufacturing in space will make space transportation worth the cost. Others feel that space transportation will always remain too expensive.

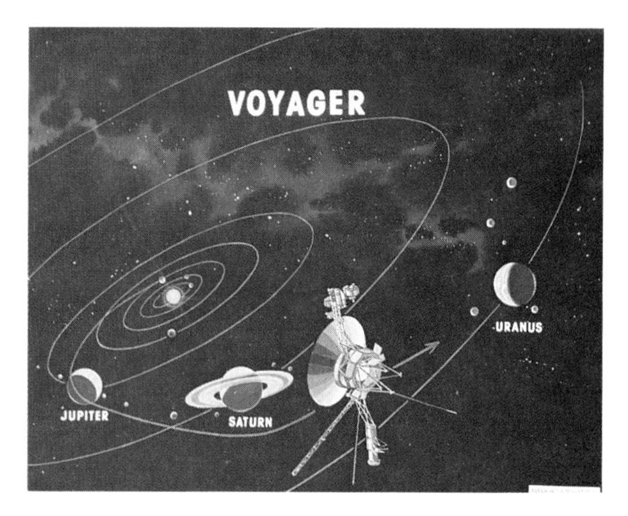

Fig. 19-15. The robot spacecraft *Voyager* visited several planets in our solar system. It sent back photographs that are helping scientists understand more about our solar system.

Apply Your Thinking Skills

1. How do space transportation systems affect our ability to use space?

Transportation and Biotechnology

Currently, the most frequent use of biotechnology in transportation systems is in the production of fuels. In the United States, corn is converted into methanol. **Methanol** is a chemical that can be used as fuel in internal combustion engines. Usually it is mixed with gasoline to make a mixture called **gasohol**. Gasohol can be used in automobiles just like straight gasoline, and it produces less pollution than straight gasoline. Because the methanol in gasohol is derived from corn, gasohol production also provides a new market that helps keep farms profitable. In some countries, other crops are used to produce fuels. The government of Brazil is attempting to produce enough alcohol from sugar cane to fuel the country's automobiles.

Methane gas is a waste product given off when plant or animal waste decays. It is a gas very similar to natural gas and propane. Propane is already used to power many vehicles. Some large garbage dumps already collect this methane. Experiments have shown that methane may work well in powering vehicles.

Some scientists think that the future will see many uses of biotechnology in transportation systems. Biotechnology may help us be able to "grow" new superstrong lightweight materials for vehicles. "Oil eating" microorganisms may help clean up oil spills.

Apply Your Thinking Skills

1. Do you think food shortages may someday occur if a crop that was originally grown only as a source of food, such as corn, is also commonly used as a source of fuel?
2. Can you think of any possible dangers in using methane gas to power vehicles?

Global Perspective
Flying Low

Fast-moving land vehicles are sometimes colorfully described as "flying low." New train systems—some in operation now and others being developed—involve trains that move at high speeds. Most of today's high-speed-rail (HSR) systems are wheel-on-rail systems. However, some of the systems in use or being tested are truly not touching a surface.

A very advanced HSR system has been developed in France. It is called *train à grand vitesse* (TGV), which means "train of great speed." The train, called *Atlantique*, runs between the cities of Paris and Le Mans. It can reach speeds up to nearly 300 mph, but generally cruises around 185 mph. It is aerodynamically designed and utilizes the latest in electronic equipment. An important feature of the train is that it runs on tracks that are standard European gage (distance between rails). Thus, it can be operated in any country in Europe.

Another HSR train that runs on standard tracks has been developed in Sweden. Called the *X000*, it runs between the cities of Gothenburg and Stockholm. Its top speed is about 125 mph, but cruising speed is less than 100 mph. The train is particularly suited to running on winding mountain tracks. It has rubber cushioning under the axles and a special system that allows each set of wheels to turn with the curve of the track.

In a maglev system, there is no physical contact between the train and the guideway. Maglev trains are actually "flying low."

The maglev system being developed in Germany is called the *Transrapid*. On the test rack near the city of Lathan, the train has reached speeds up to 230 mph. When construction of a line between Essen and Bonn is completed in the late 1990s, speeds up to 250 mph are expected.

Another maglev system expected to be completed in the late 1990s is in Japan. The Japanese National Railways are currently operating trial runs on a test track near the city of Miyazaki. Speeds over 300 mph have been reached, but operating speeds are expected to be between 150 and 200 mph.

Other countries around the world are recognizing the potential of high-speed-train travel and are developing their own system. Watch for more of the trains that "fly low."

Extend Your Knowledge

1. Since the early 1980s, the NASA program has been working to design and build an aircraft that will be able to take off from a runway and fly directly into space. Find information about this or other research projects presently under way to develop new types of vehicles. Share your findings.

Fig. GP-19. The *Atlantique* is a high-speed-rail (HRS) system that runs in France and can reach speeds of nearly 300 mph.

Chapter 19 Review

Looking Back

Transportation is continually changing. Aerodynamic design is making our vehicles sleeker and more fuel efficient. New materials, light yet strong, are being developed. Advances in electronic technology are revolutionizing vehicle systems operations. The use of alternative power sources is being explored. Vehicles powered by solar energy or hydrogen fuel may soon be practical. High-speed maglev rail systems are being developed that use the power of magnetism to propel them. Tests and experiments are being done with Stirling cycle engines to develop practical engines that are efficient and nonpolluting.

The two types of space transportation are manned and unmanned. The two basic designs of spacecraft are nose-cone-mounted and shuttles.

Satellites are put into orbit by both manned and unmanned spacecraft. These satellites are a very important part of our communication system. They also have impacts on transportation systems.

Fuels produced by biotechnology are already being used in vehicles. We may see other uses of biotechnology in transportation in the future.

Review Questions

1. Why is it important to reduce aerodynamic drag on vehicles?
2. Why are more and more lightweight materials being used in vehicles?
3. What two characteristics of ceramics make them desirable for use in engines? What are two drawbacks of using ceramic materials for engines?
4. List two advantages and one disadvantage of using fiber optics in vehicles.
5. What are photovoltaic cells?
6. What are the advantages of hydrogen as a fuel? The disadvantages?
7. Describe a maglev system. What factors enable it to be a high-speed transportation system?
8. What is a payload? What is the most common payload launched into space?
9. What are the main advantages of a space shuttle over a nose-cone-mounted spacecraft?
10. Discuss three impacts that communication satellites will have on transportation systems.

Discussion Questions

1. Since lighter vehicles get better gas mileage, why aren't all cars made small and light?
2. How could improved battery technology be very important to transportation systems?
3. Do you think it is more important to develop vehicles and engines that are more fuel efficient or to develop alternative power sources?
4. Should our government spend tax money to study space transportation systems?
5. Describe what you think it would be like to live in a permanent space station. Include any advantages and disadvantages of such a life.

Chapter 19 Review

Cross-Curricular Activities

Language Arts 

1. In a creative essay, describe the ideal transportation of the future.
2. For the next three days, keep a list of various types of transportation that are affecting your life. Then write a brief essay describing how transportation influences your daily life.

Social Studies 

1. Why do car manufacturers have to be concerned with aerodynamic designs and fuel efficient autos? What are some specific ways to make cars more fuel efficient other than the design changes discussed in this chapter. Research these questions on your own and then share the information with the rest of the class.
2. Find out how deregulation benefitted the car and truck industry. Give some specific examples of laws that regulated the car and truck industry that have been lifted. Why were these laws enacted in the first place, and what caused legislators to finally lift them?

Science 

1. Aerodynamic design makes air glide smoothly over, around, and under a vehicle. You can demonstrate this effect with a lighted candle, a 2-liter soda bottle, and a rectangle of cardboard with dimensions similar to those of the bottle (approximately 4 inches by 9 inches). Placing the cardboard on end with its full surface facing you, hold it between your lips and the candle. Try to blow the candle out. What happens? Now try blowing the candle out with the bottle instead of the cardboard in front of it. What happens? How does the bottle's shape make this possible?

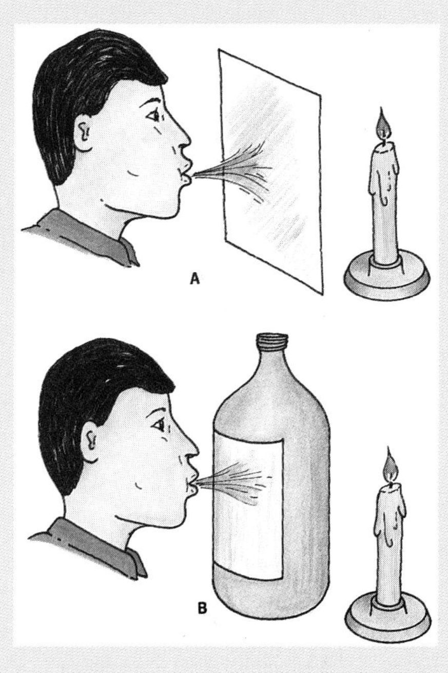

Math

1. Megan is working in a factory's shipping department. She has to pack 24 boxes, each measuring 3" x 4" x 12", into a crate. The crate has inside measurements of 18" x 8" x 24". Can Megan fit all 24 boxes into the crate?
2. Doug works at the same factory. He has to pack 144 boxes that measure 6" x 6" x 2". What size shipping crate will Doug need? (There are several correct answers. Choose the ones that are most practical.)

Chapter 19 Technology Activity

Solar-Powered Vehicle

Overview

Most of today's vehicles use petroleum products for fuel. As you know, experiments are being done to find alternative power sources that are in plentiful supply and do not harm the environment. In this activity, you will build a vehicle that uses solar energy as its power source.

This energy must be converted to electricity and transmitted to an electric motor, which generates mechanical energy. The mechanical energy causes the wheels to rotate and move the vehicle. You'll also need indicators, which provide feedback on fuel levels, speed, etc.

Goal

To design, build, and operate a solar-powered vehicle.

Equipment and Materials

two-inch thick Styrofoam™ polystyrene
razor knife (pointed craft-type)
coping saw
fine sandpaper
solar cells (2 to 10)
1/4" nut driver
22-gage solid wire (blue and red)
wire strippers (set to twenty-two gage)
voltmeter
toggle switch
rotary selector switch
one-inch diameter dowel rod
Stanley Center Square
axle, front (1/8" diameter bare welding rod)
plastic straws (for axle housing)
wheels (1/4" slices of 1"-diameter dowel rod with a 1/8" hole in the center
drill press with 1/8" bit (to drill wheels)
gear motor (1)
Liquid Nails® construction adhesive
1/4" paneling scraps (for mounting motor and front axle assembly)
#4 x 1/2" combination-head sheet metal screws (for attaching things to paneling)
#1 Phillips screwdriver
latex or acrylic paint
one-inch foam brush

Procedure

1. *Safety Note:* Use all tools and equipment carefully and safely. Also, make sure that you know how to wire properly. Ask your teacher for help if you are in doubt about any procedure.
2. Your teacher will divide your class into groups and tell you where the tools and supplies are located.
3. Please read this entire activity before performing any of it.

Developing the Propulsion System

4. Carefully read the assembly instruction card that comes with the gear motor. Assemble your motor so that it produces the greatest mechanical advantage.
5. Carefully study Fig. A19-1 to make sure that you know how to connect the solar cells in series, parallel, and series-parallel formations.
6. Using the wire stripper, cut an eight-inch red and an eight-inch blue piece of 22-gage wire for each of your solar cells. Then strip 1/2" of the insulation from

Chapter 19 Technology Activity

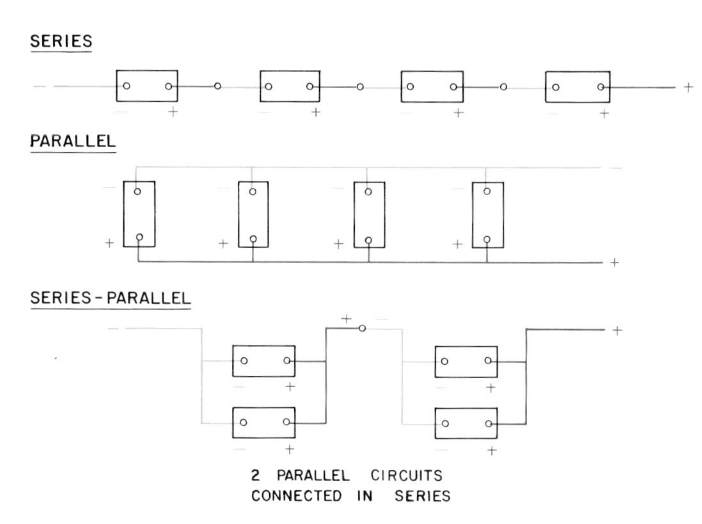

Fig. A19-1. The formation of series, parallel, and series-parallel wiring.

each end of each wire. Bend one end of each wire into a loop.

7. Using the 1/4" nut driver, gently (using only your index finger and thumb on the handle to tighten the nuts) attach the loop of one red wire to each plus (positive) terminal of each solar cell. Similarly, attach the loop of one blue wire to each minus (negative) terminal of each solar cell.
8. Wire your solar cells in series, parallel, and series-parallel to determine which configuration of cells provides the best voltage for your motor. Just twist the wire ends together to make the connections. Use a voltmeter to make your measurements.
9. Using this information, choose an electrical configuration for your solar cells and gear motor. Make a note of your decision for later.

Developing the Structure

10. The solar cells need to be on the top, front, back, or sides of the vehicle so that *all* solar cells in any one series will have maximum illumination. A shaded or blocked solar cell will act like a 5,000 ohm resistor, effectively switching the power off.
11. Sketch what you would like your vehicle to look like. Keep in mind the angle of the sun at the time of day you intend to run your vehicle. A solar cell works best when the sun's rays are perpendicular (at right angles) to the face of the cell. Now, draw the placement of the solar cells on your sketch and dimension your drawing. Lightweight, stable, aerodynamic vehicles have been found to work best.
12. Have your teacher approve your drawing and dimensions.
13. Obtain the necessary tools and supplies.
14. Build your structure (vehicle body) from two-inch thick Styrofoam. Use Liquid Nails® construction adhesive to attach anything to the structure. See examples in Fig. A19-2.

Developing Control and Suspension

15. Suggested methods for controlling your vehicle's on/off state, speed, and direction are shown in Fig. A19-3.
16. Suggested methods for making and mounting your vehicle's front wheels are shown in Fig. A19-4. Methods for mounting the wheels on the gear motor and then mounting that assembly on the structure are shown in Fig. A19-5.

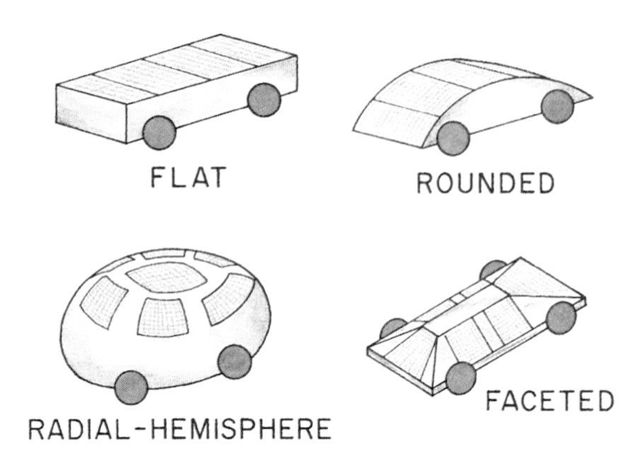

Fig. A19-2. Here are possible shapes for your solar-powered vehicle.

Assembling Your Vehicle

17. Use the Phillips screwdriver and screws to mount the gear motor to a small piece of paneling.
18. Use the Liquid Nails® adhesive to mount the motor assembly to the structure. For a more streamlined design that will have less wind resistance, you may want to recess your motor assembly.
19. Use the Liquid Nails® adhesive to mount the steering system to the structure. To aid in streamlining, you may want to recess your steering assembly.
20. Assemble the solar cells in the configuration you chose earlier (Step 9).
21. Mount the solar cells on the structure using staples made from scraps of twenty-two gage electrical wire.
22. Wire and mount the toggle switch and rotary switch if you have decided to use them.

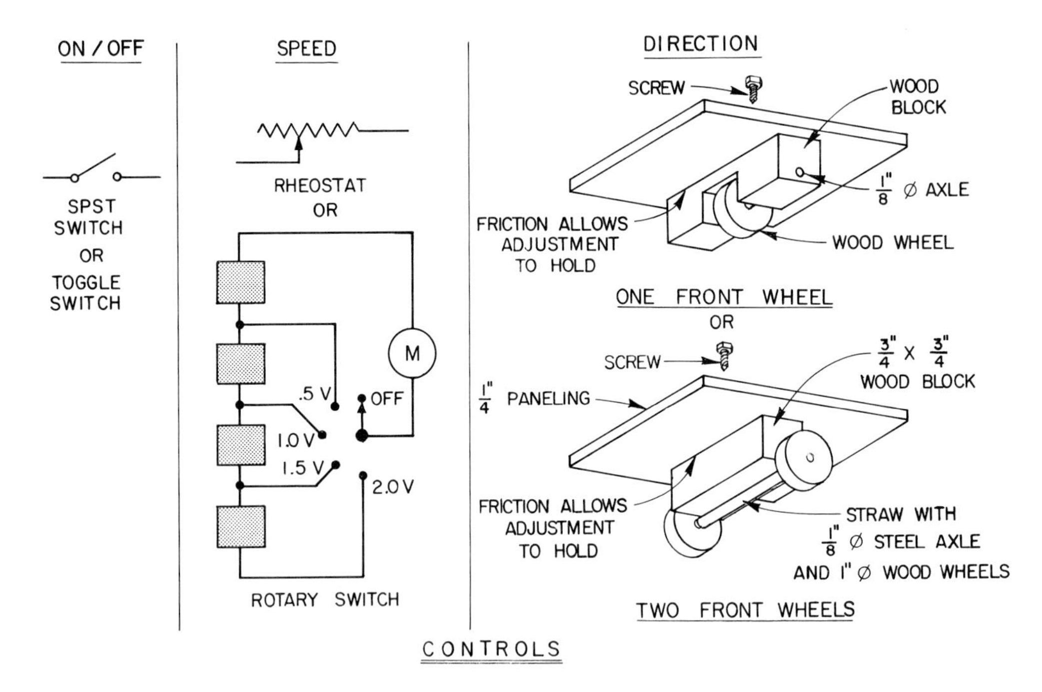

Fig. A19-3. Suggested methods for controlling your vehicle's on/off state, speed, and direction.

Chapter 19 Technology Activity

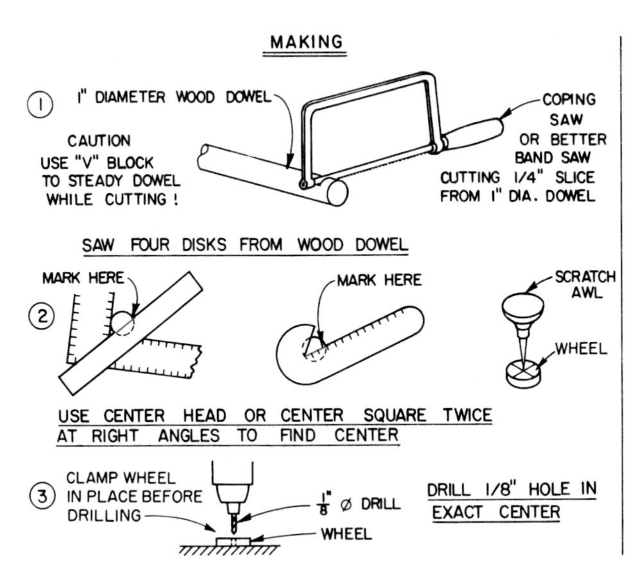

Fig. A19-4. Making and mounting your vehicle's front wheels.

The Support System

23. You are the support system for your vehicle. You must troubleshoot problems, obtain any extra parts that are needed, and repair your vehicle when it breaks down.

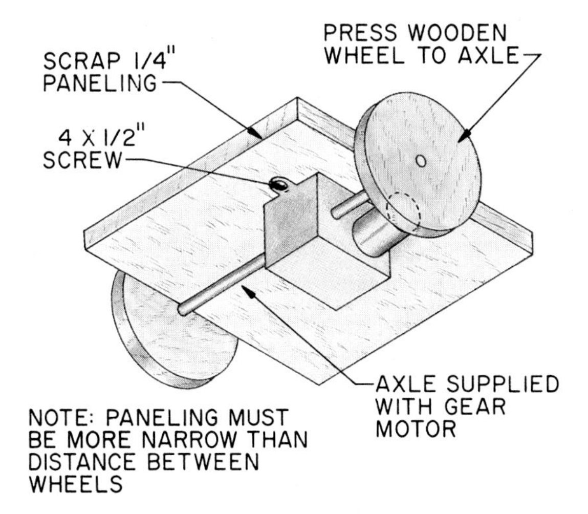

Fig. A19-5. Mounting the wheels on the gear motor and then mounting that assembly on the structure.

24. Use the foam brush and paint to decorate your project. Make sure you don't get paint in any of the working parts.

25. Test your vehicle. If necessary, make appropriate changes in wiring, gearing, switching, solar arrangement, etc.

Evaluation

1. How does the angle of the sun's rays, with respect to the surface of the solar cells, affect energy output?
2. Do you think that an aerodynamic shape is an important consideration when designing a vehicle? Why?
3. Using the system model (see Chapter 2), analyze your vehicle as a whole and its power system. If you could design another solar-powered vehicle, what would you do differently?

Credit: Daniel B. Stout

Section IV Activity

Designing a Transportation System

Overview

A local supermarket has decided to do away with its shopping carts. Instead, it is seeking a more efficient, less cumbersome means of carrying a shopper's purchases from aisle to aisle. You have been asked to design this method of conveyance.

STEP 1: State the Problem (Design Brief)

If you have done any grocery shopping, you know the problems of using shopping carts. They are noisy; they create dangerous and unattractive clutter in parking lots; they can be difficult to maneuver, especially for older customers. See Figs. SCIV-1 and SCIV-2. However, you also know you need some system of transporting items from the shelves to the check-out lines. Your goal is to devise a plan that will fulfill this basic need while eliminating the drawbacks of the system now used.

On a clean sheet of paper, type or write neatly a design brief. This is a short statement of your problem or task, including some of the major points you must consider in finding a solution. For example, your task here is to design a better way of carrying groceries. Your considerations might include the system's cost, and how readily shoppers will accept it.

Fig. SC IV-1. Cluttered parking lots are one reason why supermarkets could use a different method of carrying groceries.

STEP 2: Collect Information

These are some ways in which you might gather the information needed to design your system:

- Interview shoppers. Ask them what qualities they would like to see in a new means of grocery transportation.
- Visit places where vehicle conveyance, such as conveyor belts or monorails, are an important part of business.
- Inspect several supermarkets to determine the width of the aisles, the height of the shelves, and the types of items sold in each section.

STEP 3: Develop Alternative Solutions

On the same sheet of paper as your design brief, list several possible solutions. These might include:

- A wire-basket cart mounted on a monorail track running up and down each aisle.
- A corps of "grocery attendants" who deliver items from their respective aisles to the check-out lines.
- An above- or below-ground conveyor belt that moves groceries from a warehouse area to the check-out counters.

STEP 4: Select the Best Solution

From your list of ideas, choose the best one. These are some of the points to consider as you make your decision:

- How much will the system cost, not only to install, but to operate?
- Will shoppers be able to understand and use it easily?
- Will it provide access to items in all parts of the store?
- Will it transport items safely, without damaging them?
- Will it eliminate the old problems without creating new, equally undesirable ones?

STEP 5: Implement the Solution

Make a drawing or write an explanation of your proposal, including as many details and specifics as possible. Make a model or give a demonstration of your plan.

STEP 6: Evaluate the Solution

Present your solution to your teacher and classmates. Discuss its strengths and weaknesses. Revise your plan, taking these suggestions into account.

Fig. SC IV-2. Shopping carts also cause problems inside supermarkets.

Section V

Construction Technology

The building process begins by considering purposes, possibilities, and resources. This approach to needs and impacts is appropriate for projects as diverse as homes, office towers, and public works.

Technology Time Line

3000 B.C.	2000 B.C.	1000 B.C.	1	1000	1100	1200

2750 B.C. Construction of Stonehenge begins in England.

2700 B.C. Eygptians begin building pyramids.

A.D. 50 Glass is first used for windows.

70-82 The Colosseum is built in Rome.

214 The Great Wall of China is begun.

326 The first St. Peter's Basilica in Rome is begun.

650 The first windmills are built in Persia to crush grain.

1100 Windmills are improved and used in Holland to pump water from the lowlands.

An actual construction project starts as a series of questions, discussions, and plans. The builder's drawings and specifications are a record of those plans. After being analyzed to generate materials lists, the drawings help manage the labor that forms those materials into a finished structure.

1632-53 Taj Mahal is built as a tomb for Indian ruler Shah Jahan's deceased wife.

1784 Benjamin Franklin invents bifocals.

1825 The Erie Canal is opened, linking Lake Erie and the Hudson River.

1839 Charles Goodyear makes rubber stronger through a process called "vulcanization."

1859 In London, the bell known as "Big Ben" tolls for the first time.

1300 | 1400 | 1500 | 1600 | 1700 | 1800 | 1900

1350 The Leaning Tower of Pisa is completed.

1883 The Brooklyn Bridge is completed. Its span remained the longest in the world until 1903.
1884 The first skyscraper, the Home Insurance Building, 10 stories tall, is opened in Chicago.
1885 The Washington Monument is dedicated.
1886 The Statue of Liberty is dedicated on Liberty Island in New York Harbor.

1889 The Eiffel Tower is opened for the World's Fair.

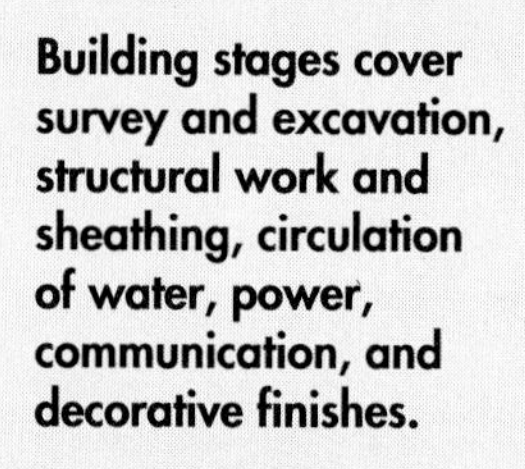

Building stages cover survey and excavation, structural work and sheathing, circulation of water, power, communication, and decorative finishes.

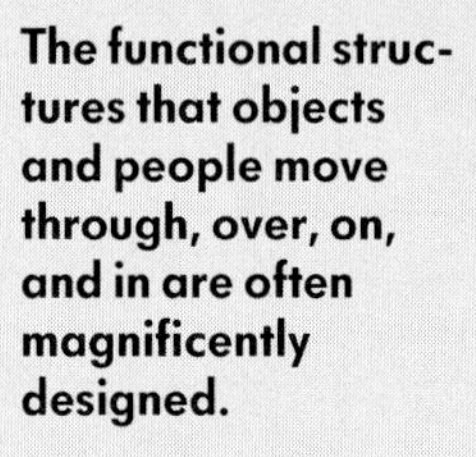

The functional structures that objects and people move through, over, on, and in are often magnificently designed.

Technology Time Line

1900 | 1910 | 1920 | 1930 | 1940

1914 The Panama Canal is completed.

1931 The 102-story Empire State Building is completed and is the world's tallest building at the time.

1933 Alcatraz becomes a federal prison and is believed to be "escape-proof."

1936 Hoover Dam is completed.

1937 The Golden Gate Bridge is opened for the first time.

1938 Owens-Corning begins mass-producing fiberglas

Construction methods for factories and homes have been changed for building high in the air, under water, and below ground. Now they must change as well to build a space station in micro-gravity conditions.

New trends in technology: bridges with prefabricated supports have shorter construction periods, and sports arenas with suspended-cable roofs allow unobstructed viewing.

1963-65 Vehicle Assembly Building is constructed in Cape Canaveral, FL. Built for constructing spacecraft, it is the world's most spacious building.

1965 The Gateway Arch in St. Louis is completed and is the largest monument in the world at the time, standing 630 ft. high.

1974 The Sears Tower in Chicago is completed and is the tallest building in the world.

1976 The Canadian National Tower in Toronto, Canada, is the tallest freestanding structure in the world at 1,815 feet.

1990 131 feet below the English channel, workers from France and England shake hands after completing one phase of the "Chunnel," a tunnel connecting France and Britain.

1950 | 1960 | 1970 | 1980 | 1990

1959 Frank Lloyd Wright, one of America's most influential architects because of his unique style, dies at the age of 92.

1968 The fourth Madison Square Garden is opened in New York.

1969 John Hancock Center opens. It is the second largest building in the world at the time (1,350 ft. high).

1982 The Vietnam War Memorial is opened in Washington, D.C.

Chapter 20

Construction Systems

Looking Ahead

In this chapter, you will discover:

- the importance of construction.
- the four basic types of construction.
- how the seven resources are used as inputs in construction systems.
- what kinds of structural materials are used in construction systems.
- how processes, outputs, and feedback work in a construction system.

New Terms

architects
building codes
commercial construction
construction industry
foundation
industrial construction
public works construction
residential construction
site
structural materials
superstructure

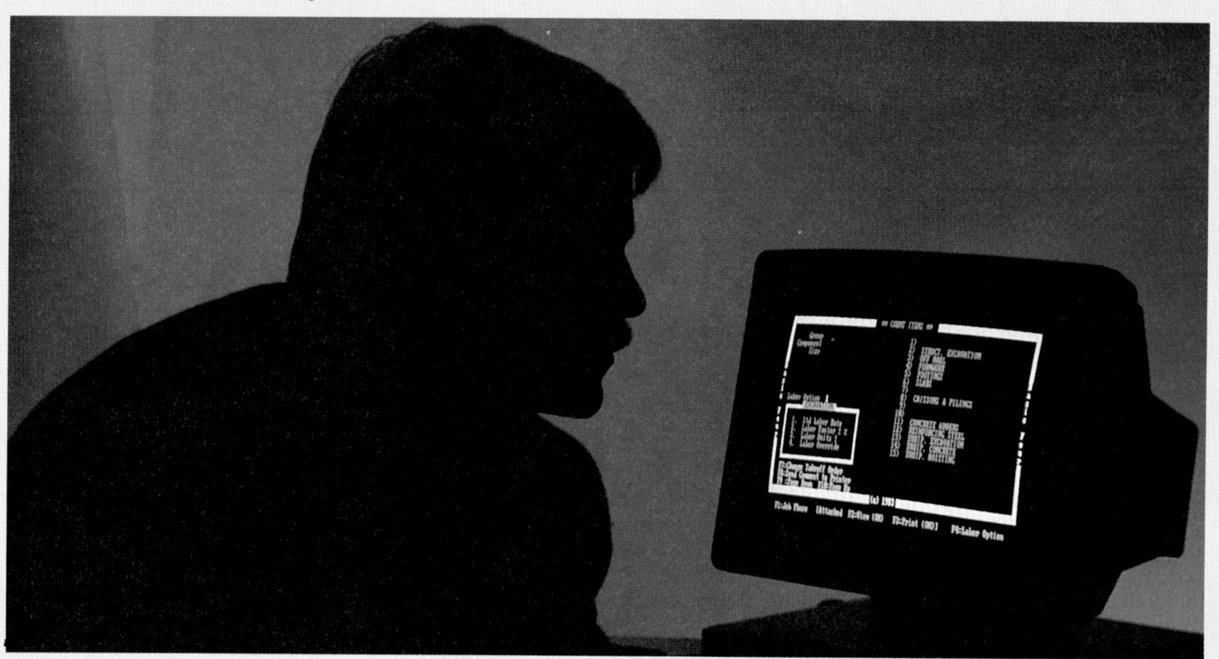

Technology Focus

The Remarkable Geodesic Dome

R. Buckminster (Bucky) Fuller (1895-1983) was an inventive genius. He had no architect's license. He had no college degree. However, he profoundly influenced modern architecture. The invention for which he is most famous is the *geodesic dome*. This design has been described as "the most significant structural innovation of the twentieth century."

A geodesic dome is a rounded or curved structure. It consists of a framework of short, lightweight tubes or struts. These are usually made of aluminum or steel. They are ingeniously arranged to form a repeating and connected series of geometric shapes called tetrahedrons. The framework is usually covered with heavy plastic sheeting, although sometimes a more rigid material is used. Rigid plastic or glass panels or aluminum sheeting may be chosen, depending upon the building's intended use or purpose.

The domes are relatively easy and inexpensive to construct. They are also very strong and practical. Many of Fuller's domes are already being used as homes, factories, and offices. There is even a geodesic dome convention center and botanical garden.

A geodesic dome can enclose an enormous amount of space without internal vertical supports such as columns or posts. For example, the Union Dome in Baton Rouge, Louisiana, is ten stories high. It has an unobstructed floor area that is 384 feet in diameter.

The most famous geodesic dome is really a sphere (ball shape). This is the impressive EPCOT center, near Disney World in Florida. It is 180 feet in diameter and 18 stories high.

Fuller's domes are indeed remarkable. Someday, a whole city may be covered by one dome. The environment would be completely controlled and spring-like. It would be free of snow, wind, rain, and pollution. The influence of Fuller's design may extend well into the future. The engineering and architectural principles used in geodesic dome design may someday be applied to the construction of space colonies.

Fig. TF-20. Geodesic domes may be the most significant structural innovation of the 20th century.

What Is Construction?

Construction is the process of producing buildings or other structures. Even though construction projects can be as varied as bridges, houses, skyscrapers, or geodesic domes, most structures have two major parts: the foundation and the superstructure. The **foundation** is the part of the structure that rests upon the earth and supports the superstructure. Foundations are usually made from concrete. The **superstructure** is the part of the building or other structure that rests on the foundation. In buildings, the superstructure consists of everything from the first floor up.

The production system that produces constructed products is the **construction industry**. It offers many opportunities for different types of work. The skills required to work in construction are varied. They all have something to do with building. That, in a word, is exactly what construction is—building.

Our Constructed World

Generally, when people think about construction, they think about the building of houses. Construction has helped us meet our need for shelter. It has also helped us meet many other human needs as well.

For example, constructed products include buildings and other structures where we can:

- Work
- Play
- Study and learn
- Shop
- Receive health care
- Worship

Because of construction, we have a network of highways, bridges, airports, and tunnels. These enable us to travel freely about the country.

Certain buildings have been so beautifully designed and constructed that they are considered great works of art. Other structures have become important historical or cultural symbols. Fig. 20-1.

Fig. 20-1. Structures do more than meet our basic needs. For example, the Lincoln Memorial is a symbol of our great respect for our sixteenth President. Thousands upon thousands of people visit this structure every year.

Fascinating Facts

More than 4,500 years ago, Egyptians built a spectacular 480 feet high monument—The Great Pyramid. Designed as a tomb for King Khufu, the pyramid contains 2,300,000 blocks of limestone, weighing 5,000 pounds each. The base alone covers an area that's larger than ten football fields!

Types of Construction

There are four major types of construction:

- Residential
- Industrial
- Public works
- Commercial

Residential construction refers to building structures in which people live. Most residential structures are single-family or private homes. Fig. 20-2. However, residential construction also includes the building of small multifamily units. *Multi* means more than one. Multifamily units have two or more apartments or dwelling areas.

Duplexes and town houses are multifamily units. A *duplex* consists of two apartments. These are usually side by side under a single roof. A *town house* is a single two-story unit, but several of these are built side by side to form a single long building. Town houses are sometimes called *row* houses.

Most residential construction is done by fairly small construction companies. Usually, common materials (such as wood and brick) and basic building techniques are used.

Industrial construction includes the building and remodeling of factories and other industrial structures. This type of construction is usually planned by specialized engineering firms. Industrial structures are usually built by large construction firms that have many employees.

Public works construction is building structures intended for public use or benefit. This type of construction includes large projects such as dams, highways, bridges, tunnels, sewer systems, airports, hospitals,

Fig. 20-2. Single-family homes such as this are one type of residential construction.

Fascinating Facts

The Great Wall of China, the largest public works construction project ever undertaken, extends more than 1,500 miles and took hundreds of years to complete. Ch'in Emperor Shih Huang Ti started the project back in 214 B.C. by building walls to join then current fortifications and watchtowers in order to protect China's northern border.

About 25 feet high, the wall is 26 feet wide at the base and 16 feet wide at the top. It was built with almost 400 million cubic yards of material—enough to build 120 pyramids or a wall 6 feet high around the world at the equator.

schools, and parks. Fig. 20-3. The projects are usually funded by federal, state, or local taxes.

Public works construction projects are built by large construction firms. Most of these have hundreds of employees and much heavy equipment.

Commercial construction consists of building structures used for business. Supermarkets, shopping malls, restaurants, and office buildings are examples of commercial construction. Fig. 20-4. Commercial projects are usually large-scale construction projects that involve millions of dollars and many workers.

The building materials and techniques used in commercial and industrial construction are somewhat different from those used in residential construction. These structures often have steel frames and many concrete parts.

Apply Your Thinking Skills

1. Your community includes many different kinds of construction. For each type of construction (residential, industrial, public works, commercial), make a list of five actual structures in your community that fit into that category.

Fig. 20-3. Schools, like North Texas State University, are a type of public works construction.

Fig. 20-4. This commercial structure provides space for many business offices.

Parts of the Construction System

Like other technological systems, a construction system has inputs, processes, outputs, and feedback. These all work together to achieve the goal of constructing a finished structure. Fig. 20-5.

Inputs

The inputs include the seven technological resources that will be used in building the structure:

- People
- Information
- Materials
- Tools and machines
- Energy
- Capital
- Time

People

The building of structures requires the skills, hard work, and cooperation of many people. **Architects** design structures and develop the plans for building them. Workers directly involved in the actual construction process operate equipment such as bulldozers; install wiring and plumbing; lay bricks and blocks; and measure, saw, and fasten boards. Others provide needed materials and equipment. There must also be people to oversee the workers, monitor progress, order materials, schedule work and deliveries, and inspect the structure. Fig. 20-6 through 20-8.

Information

Many kinds of information are needed for construction systems. For example, when architects decide what kind of structure to design, they collect information about the needs of the people who will be using that structure. If it is a school, for example, how many students will attend it, what are their ages, and do they have any special needs? Will physical education facilities be needed, and if so, what kind? Is an auditorium

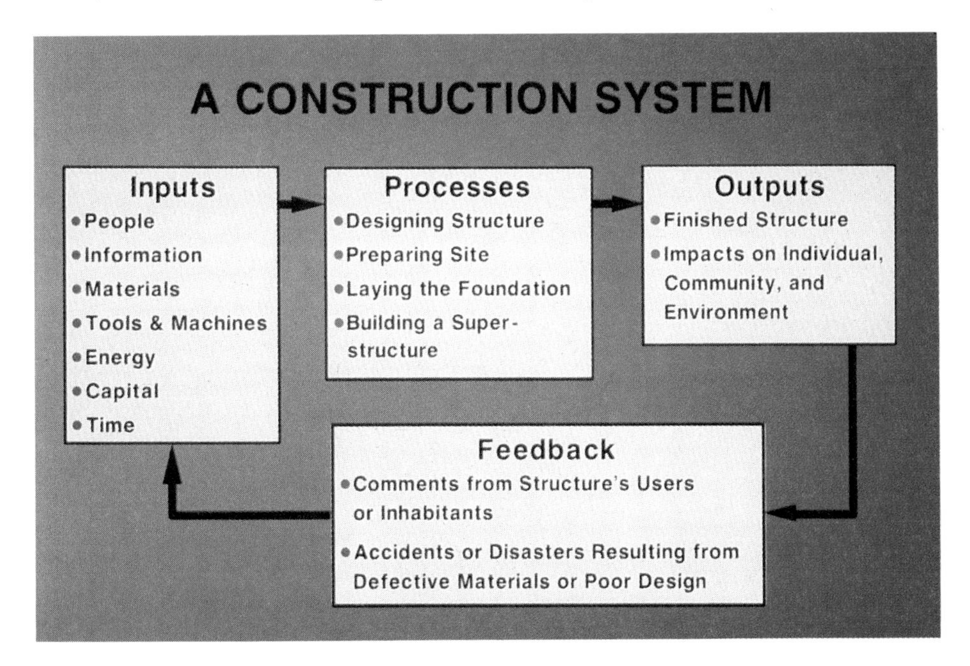

Fig. 20-5. The parts of a construction system are input, processes, output, and feedback.

Fig. 20-6. Architects design buildings.

Fig. 20-7. This construction worker, who's laying concrete blocks for a new home's basement, is directly involved in the construction process.

Fig. 20-8. Plumbing as well as electrical systems must be checked by a building inspector.

needed? A cafeteria? A lab or shop? All of this information is necessary for the architect to design a good school for these particular students.

Information must also be gathered about the construction site itself. The **site** is the land on which a project will be constructed. Information must be gathered about whether the land is flat or hilly, whether the soil provides good drainage, and whether the soil can support the structure. For buildings, information must be gathered about lot size and the exact boundaries of the lot.

Information about any laws or codes regulating construction must also be gathered so the structure can meet these requirements. For example, **building codes** specify the methods and materials that can or must be used for each aspect of construction.

Materials

Many different kinds of building materials are available. The size and nature of the project under construction determine what is needed.

Structural materials are those used to support heavy loads or to hold the structure rigid. They are chosen for strength and stiffness. Wood, steel, aluminum, concrete, and masonry are often used as structural materials.

Wood has many practical advantages as a structural material. It is relatively strong and stands up fairly well to bad weather. Wood is readily available in many sizes and shapes and in a variety of structural panels, such as plywood and particleboard. It can be cut and shaped. It can be fastened together with nails, screws, bolts, or adhesives such as glue.

Wood has some disadvantages, though. For example, wood may burn, warp, split, or rot. This depends upon the conditions to which it is exposed. Untreated, it can be attacked by termites and other insects.

Steel is made by combining iron with small amounts of carbon. It is an outstanding structural material. Its major advantages are that it's very strong and rigid (stiff). Fig. 20-9. Steel can be formed into various structural shapes, such as I-beams or girders, plates, sheets, and rods. Steel can also be formed into wire, which can be woven into extremely strong rope or cable. Steel units can be welded together or fastened with bolts or rivets.

Steel also has disadvantages as a structural material. For example, ordinary steels may rust or corrode. These must be protected by either being painted or embedded (enclosed) in concrete. Also, when steel is exposed to temperatures above 700°F (371°C), it will rapidly lose its strength. It should be covered with a fireproof material such as concrete, especially in high-rise buildings.

Fig. 20-9. A steel frame is used when constructing most large buildings.

Aluminum is a metal used when light weight and resistance to corrosion are important. Like steel, aluminum can be formed into a variety of shapes. For example, it can be formed into I-beams or girders, rods, and wire. It can also be rolled to form plates and foil. Aluminum parts can be welded or fastened together with rivets or bolts.

Concrete is a mixture of sand, gravel, water, and cement. When properly cured, it is a very strong material. *Curing* is not simply "drying." It is a chemical reaction that makes the concrete hard and strong. It can be used to build such structures as bridges, tunnels, highways, and large buildings.

Concrete has many advantages as a structural material. Besides being strong, it is moldable. It can be poured into molds to form almost any kind of shape. It may be made into blocks. Huge panels or entire walls can be made of concrete. Separate concrete units can be joined to form structures, such as high-rise buildings.

Concrete also has some disadvantages as a structural material. It is heavy. In addition, it must be prepared properly, poured carefully, and finished and cured correctly. If even one of these processes is not done correctly, the concrete may crack or crumble. Fig. 20-10.

Fig. 20-10. Placing concrete. All processes must be done carefully in order for the concrete to become hard, strong, and durable.

Masonry is a broad term that includes both natural materials (such as stone) and manufactured products (such as bricks and concrete blocks). The material selected depends on the type of project. Most masonry structures are held together by some kind of cement mortar or other adhesive.

Masonry has the advantage of being fireproof and is also a sturdy structural material. It is usually more costly than wood, however, and the bricks or stones must be laid well. This means a lot of skilled labor is involved. In addition, the mortar that holds the brick or stone together tends to crumble over time and must be replaced periodically. Maintaining the mortar is called *repointing*.

Other materials used in construction include roofing materials, vinyl siding, insulation, wallboard, plaster, electrical wiring and lighting, plumbing supplies and fixtures, and heating and cooling systems.

Tools and Machines

There are four basic types of tools and machines used in construction: hand tools, portable tools, light equipment, and heavy equipment.

Hand tools are simple tools powered by humans. Hand tools commonly used in construction include hammers, saws, screwdrivers, shovels, and wrenches.

Portable power tools are tools that are powered by electricity or air and are small enough to carry. Examples include electrically powered saws and drills and air-powered nailers and staplers. Fig. 20-11.

Light equipment is equipment that can be moved about fairly easily. Ladders, sawhorses, and wheelbarrows are commonly used. Some light equipment, such as surveying equipment, is fairly complex. Fig. 20-12.

Large and powerful equipment is referred to as *heavy equipment*. Cranes, bulldozers, and trucks are examples of heavy equipment. Fig. 20-13.

Energy

Energy, of course, will be used to build a structure. Petroleum and electrical power run the many machines and tools that construction teams use, such as bulldozers, power drills, and cranes. Human energy is used every time a worker drives a nail or an electrician installs wiring.

For most types of structures, once the structure is complete, energy will still be needed to run the internal systems. Can you imagine your house without a furnace or electrical power? It would be difficult to drive over the Brooklyn Bridge or Golden Gate Bridge at night without all the lights marking the outline of the bridge. Energy is an input to construction systems both when structures are being built and when they are being used.

Capital

The land on which a structure is built is a very important part of capital in construction. Land is very expensive to buy and a valuable asset to own. In addition, many hundreds of millions of dollars are spent on equipment, tools, materials, and other supplies. Large sums are also spent on workers' wages.

Fig. 20-11. Some portable power tools such as this pneumatic nailer are powered by air.

Fig. 20-12. Some light equipment can be complex. This surveyor is using a transit to accurately measure distances and angles so he can determine the exact property lines.

Fig. 20-13. Heavy equipment, such as this crane, can accomplish a great deal in a short period of time.

Time

Time is a critical resource in construction technology. Scheduling is of the utmost importance. Materials must be delivered so that they are available as they are needed. The jobs that need to be done are dependent on one another, so if one job is not done on time, it will affect the schedule of many other jobs. For example, the ground must be cleared and prepared before construction can begin. Bad weather can cause serious delays in this process, as well as other outdoor construction processes. If the ground is not prepared, workers cannot build the foundation. The frame cannot be constructed if the foundation has not been completed. Electricians cannot install wiring and roofers cannot install roofs until the framing is complete. Construction companies do not want to pay employees who are waiting for their phase of work to begin. In addition, construction contracts often include financial penalties if construction is not complete by a certain date or bonuses if it is completed before a certain date.

Processes

The processes of construction include everything from choosing and preparing a site to turning over a finished building to its new owner. Fig. 20-14. Perhaps you have watched the gradual progress of a construction project, such as a new home or a shopping mall, in your own community. If so, you have seen a number of construction processes going on. The next chapters in this section will examine some of these construction processes in more detail.

Output

The main output of a construction system is, of course, the finished structure. Depending on the purpose of the particular system, this can range anywhere from a backyard treehouse, to a new dam or bridge, to a whole new development of residential or commercial buildings. Output also includes any impacts on society and the environment.

Suppose a city builds a new skating arena. Perhaps people will begin to use the arena rather than nearby ski areas for recreation. The arena may encourage skating professionals to move to the community and set up a skating school. These are some of the possible social impacts.

Construction systems also have impacts on the environment. This is one of the reasons they need to be monitored so closely from the very beginnings of the projects. Sometimes a proposed project may present

Fig. 20-14. Planning a project, putting up a frame, and finishing a structure are all processes of a construction system.

possible threats to natural features and wildlife. The benefits of the project may not be worth the potential harm. Even if the project doesn't damage the environment, it may change it in undesirable ways. For example, people might not want to build an industrial factory in the middle of the town's business section.

Controversies often arise over proposed construction projects and the impacts they might have on the surrounding environment and community. As in any other technological system, the advantages and disadvantages of a proposed project need to be carefully weighed.

Feedback

Feedback on construction systems is important. Even though people try to plan the best possible structures, they sometimes overlook important factors. People using a building, for example, might realize there is not enough ventilation, and the ventilation system will need to be improved.

A dramatic kind of feedback was provided during the Tacoma Narrows Bridge accident. The Tacoma Narrows Bridge, built in Washington state, was a suspension bridge that opened on July 1, 1940. On November 7, 1940, the bridge began to sway and twist violently. At 11:00 a.m., a 600-foot length of the main span broke off and collapsed. (Fortunately, the bridge had been closed to traffic early in the day.)

One of the main reasons for the disaster turned out to be the proportion of the vertical depth of the span to the length of the span. Fig. 20-15. The thin span was too flexible to withstand the motions set in action by the wind.

The collapse of the bridge was feedback that resulted in major structural changes. When the bridge was rebuilt, the vertical depth of the span was increased from 8 to 33 feet.

Apply Your Thinking Skills

1. Describe the resources that were probably used to build the building in which you live.

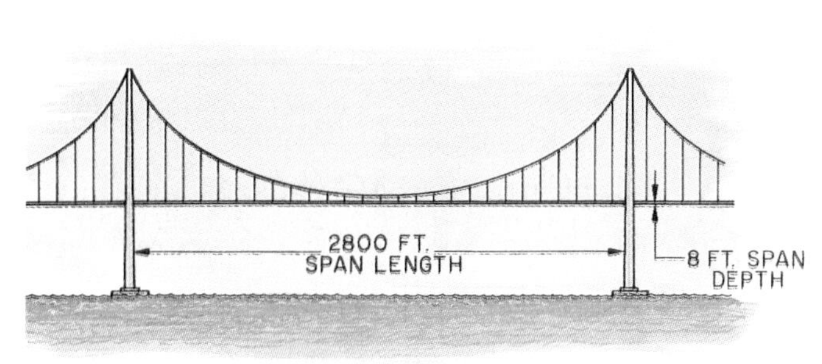

Fig. 20-15. The small vertical depth of the bridge span is one of the factors that led to the Tacoma Narrows Bridge catastrophe.

Investigating Your Environment

Can We Have Construction without Destruction?

Materials used in residential construction include wood, concrete, brick, asphalt, and various metals. These materials come from the land. Wood is harvested from trees. Concrete, brick, asphalt, and metals are made from materials that are quarried or mined.

Harvesting wood can be harmful to the environment. A tree's roots act as natural anchors, holding soil in place. Without trees, precious soils are blown or washed away, making the land unsuitable for growing other types of vegetation. Trees supply oxygen and they provide shelter and food for animals. Though the harvested trees are often replaced with new trees or other vegetation, these are often not native to that area. In addition, it takes many years for trees to grow. All of this upsets the ecosystem of an area. *Ecosystems* are the unique situations of interdependent plant and animal communities found in nature. Deserts and jungles are examples of ecosystems.

Fig. IYE-20. Recycled wood can now be used to build homes. This practice helps decrease the number of trees that are cut down for residential construction.

Quarrying and mining can also be harmful to the environment. Quarrying is the process of taking stone blocks, sand, and gravel from the land. Mining is the process of taking metals from the land. There are many ways to carry out these processes, but essentially it involves some type of digging. Mining and quarrying not only ruin the natural beauty of an area, but they disrupt the ecosystem. As is the case with logging, mining and quarrying often make the land unfit for anything to grow.

With new technology, houses can now be built out of recycled materials. A company in British Columbia has glued, pressed, and then microwaved strips of scrap wood to produce a wood similar to new wood. New types of cement made from scrap stone, wood, and plastic are available. A company in Illinois makes nails by melting down scrap metal.

Another way of recycling is re-using. Millions of board feet of lumber from houses that have been torn down get tossed into landfills. Much of this wood is still good, but building codes require that lumber have a grade stamp. Codes need to be changed so old wood may be evaluated and given a stamp of approval, too.

Take Action!

1. Are there any companies in your community that salvage old building materials? Find out and report to the class.

Chapter 20 Review

Looking Back

Construction is the process of producing structures. The four main types of construction include residential, industrial, public works, and commercial construction.

Construction systems include inputs, processes, outputs, and feedback. The inputs include the people, information, materials, tools and machines, energy, capital, and time needed to design, plan, and complete the structure. Processes include activities such as choosing and preparing a site, laying the foundation, and building a superstructure. The output includes the finished structure plus any impacts that structure has on society and the environment. Feedback provides the construction system with information about any problems with the structure itself, such as inferior quality materials or poor design features. Feedback also provides the system with information about any negative impacts on individuals, the community, or the environment.

Review Questions

1. Name and describe the two major parts of a structure.
2. Define the four types of construction and give examples of each.
3. How are the building materials used in commercial and industrial construction different from those used in residential construction?
4. What are architects?
5. Define site. What information about the site must be gathered?
6. What are building codes?
7. What are structural materials and what qualities do they need to have?
8. Name the five structural materials commonly used in construction and list any advantages or disadvantages they have.
9. Name the four basic types of tools and machines used in construction, and give at least three examples of each.
10. Briefly discuss how important scheduling is to construction systems.

Discussion Questions

1. Why might some people object to certain construction projects?
2. Choose a building or other structure in your community or a nearby community that interests you. What features or qualities of this structure interest you and why? What impacts has the structure had on the community?
3. Describe an ecosystem in the area of the country where you live. Has construction or the harvesting of construction materials had any effects on this ecosystem? If so, what is being done or can be done to reverse these effects?
4. How do building codes benefit consumers?
5. If you could choose a new construction project for your community, what would it be? Explain why you would choose that particular project.

Chapter 20 Review

Cross-Curricular Activities

Language Arts 

1. Select a famous landmark and research its construction. Encyclopedias may help provide material.
2. In a three-minute speech to classmates, describe a famous landmark without identifying it by name. Then have your classmates guess the landmark you just described.

Social Studies 

1. Look up Louis Sullivan. What did he develop? How did Sullivan's development have an impact on the construction business? Find out how and why Sullivan came up with this development and how it benefits cities.

Science 

1. Many hand tools are based on one or more of the six simple machines. These simple machines make it possible to apply mechanical energy at one point and deliver it to another point. Demonstrate this principle by using a prying tool to pry open a container top such as a paint-can lid. Where do you apply the force? How does the simple machine (in this case, a lever) deliver the force? In what way is the force changed?
2. You can observe the changes that take place in concrete construction through observing the hardening of plaster of paris, or molding plaster. Using a plastic container, mix molding plaster with water according to label directions. You will see that it first forms a *colloidal gel*—a liquid mixture in which the particles are mixed evenly throughout and do not settle to the bottom. As time passes, heat is given off by the reaction between gypsum (the major ingredient of the plaster) and water. Next, crystals grow and interlock, forming a hard and solid mass, just as they do in concrete.

Math

When you borrow money for some purpose, the money that you borrow is called the *principal*. You are charged a fee for using someone else's money. This fee is referred to as *interest*. The interest figured only on the original principal is referred to as *simple interest*. The *annual interest rate* is the percentage of the principal that must be used to calculate the interest based on one year.

The formula for calculating simple interest is:

INTEREST = PRINCIPAL x RATE x TIME

1. Ian borrows $1,200 to replace a garage roof. If Ian borrows the money for one year at a rate of 10.5%, how much money will he pay in interest?
2. Maria borrowed $860 for one year to install floor covering in two rooms. If she pays $77.40 in interest, what was the interest rate?

Chapter 20 Technology Activity

Identifying Types of Construction in Your Community

Overview

After completing this activity, you will know the major types of construction underway or recently completed in your community.

Goal

To identify the different types of construction that are underway or have been recently completed in your community.

Equipment and Materials

one 35mm camera to take color slides
sufficient film to take one or more slides of each building being constructed in your community
a map of your community

Procedure

1. If you live in a small community where little or no construction is underway or has been recently completed, you may need to adopt a community for this activity. If you live in a large community with a great deal of construction underway, it may be necessary to divide the class into teams.

Fig. A20-1. This office building is an example of a commercial construction project.

Chapter 20 Technology Activity

Each team would be responsible for a different section of the community. At least four categories of construction should be identified in the study: residential, industrial, public works, and commercial. The construction may be new, or it may be a renovation, or an addition. See Fig. A20-1 for one example of a type of construction project.

2. Locate the construction sites throughout your community. Mark these sites on your map. Code each of the construction sites using the categories you have identified in Step 1.
3. Photographically record each site in at least one stage of construction. In some situations, it is best to have the permission of the owner before taking the photographs.
4. If possible, prepare a series of photographs of one or more construction sites showing various stages of construction. These photographs could record the construction of an entire building, the construction of an addition such as a house deck, or the renovation of a bathroom or kitchen.
5. If possible, talk to the general contractor on at least one project. Find out what kind of services had to be employed to help complete the project. It is not important to have the names of the organizations providing this help unless the general contractor volunteers the information.
6. Organize the information you have obtained during this study. Make a brief presentation to your class, summarizing your findings.

Evaluation

1. Were you able to find each of the four categories of construction in your community?
2. Which of the four categories was the easiest to find? Which was the hardest?

Chapter 21

Planning Construction

Looking Ahead

In this chapter, you will discover:

- how construction projects are initiated (started).
- that construction projects are subject to regulations.
- factors to consider when selecting a site.
- the steps in the design process.
- the types of working drawings and what they show.
- what specifications are and why they are important.

New Terms

city planners
elevations
engineers
floor plan
power of eminent domain
private sector
public sector
scale drawings
site plan
specifications
topography
utilities
zoning laws

Technology Focus

Maria Dolores Marquez

Maria Dolores Marquez is a landscape architect in San Diego, California, where she is an associate in the landscape architecture and planning firm of Burton Associates. She specializes in the design, project management, and construction administration of residential, office, and recreational projects. Dolores has worked in the landscape architecture and planning industry for twelve years, but she didn't know she wanted to be a landscape architect until junior high school.

In school, Dolores wasn't very interested in any of her classes. When she took a class in arts and crafts, however, it changed her life. "In that class," Dolores fondly remembers, "we did everything." She especially enjoyed designing a house, drawing plans, and building a model of it. "That was the one I really got into," says Dolores. "From that point on, I wanted to be an architect."

In high school, she took a drafting class from an architect. Dolores tried to learn as much as she could about an architect's career from him. He advised her to try landscape architecture because it was new and because there would be a great demand for landscape architects.

Dolores followed her teacher's advice and attended California Polytechnic at San Luis Obispo, California, where she studied design, biology, and plant science. She was most interested in the design courses. Dolores says she "really enjoys being creative."

Dolores served as the project manager for the Escondido, California, Civic Center project and the Mira Mesa, California, Community Center. These projects have won various awards.

Dolores' favorite projects are designing and directing smaller projects, where she can fully exercise her creative talents. Most developers are looking for creative designers. She accomplishes this by designing walkways, entryways, patios, pools, and fountains that harmonize with the landscape and with the house. "My favorite part of any design is selecting the plant material," says Dolores.

Dolores is fulfilling her life's ambition in her work. Her dream began with that one important arts and crafts class in junior high school.

Fig. TF-21. Maria Dolores Marquez has fulfilled her life's ambition by becoming a landscape architect.

Initiating Construction

Most buildings and other structures are built for ordinary people. These people make up the **private sector** (part) of our economy. As you learned in Chapter 20, there are four types of construction—residential, commercial, industrial, and public works. The private sector is directly responsible for three of those types. For example, a family may need a home (residential). A person in business may need a store or a warehouse (commercial). A company may need a factory or other facility (industrial). Private funds are used to pay for the design and construction of these projects.

The public sector of our economy is responsible for public works construction. This **public sector** includes municipal (city), county, state, and federal governments. People are appointed or hired by the government to serve on boards or in agencies, bureaus, departments, or commissions. These people are responsible for initiating (beginning) construction projects such as highways, post offices, and fire stations. Tax money is used to pay design and construction costs. Fig. 21-1.

Community Planning

Like the building of a house or other structure, the "building" of a community must be carefully planned so that it can be designed to best meet the community's needs. This planning is usually done by **city planners**. Fig. 21-2. These people have studied all aspects of community development. Many larger communities have planners on their permanent staffs. Smaller communities usually hire planners on a temporary basis. These consultants work closely with city, county, and state officials and various governmental agencies. In the course of their work, they study:

- the size and character of the population
- the economy of the community
- the nature and quantity of natural resources such as oil, gas, water, timber, and farmland
- transportation facilities
- educational facilities
- history and culture of the area

After learning all they can about a community, planners identify areas of potential growth. They also identify potential problems that might limit future

Fig. 21-1. Making decisions about public works construction is a big responsibility. These people must account to the taxpayers for spending tax money wisely.

Fig. 21-2. A city planner studies a community and analyzes needs.

Fascinating Facts

Back in the 1880s, inventor and businessman George M. Pullman built America's first planned "industrial village." Pullman, famous for his railway sleeping car, built his model "suburban factory town" on a 4,000-acre site 13 miles south of Chicago.

Pullman, Illinois, not only had a factory to build railroad cars but also had housing for workers and their families. The town, which became part of Chicago in 1889, also had a hotel, a theater, a library, and a farmers' market.

growth and development. They work to find solutions to these. Finally planners make recommendations for future community development. Fig. 21-3.

What will be best for the community? Citizen representatives and elected officials sit on planning commissions and boards. These people study all recommendations or

Fig. 21-3. A planner presents development ideas to the planning board. Interested citizens attend these sessions and make their wishes and opinions known. All of us must work together to build good communities.

plans carefully. Before deciding whether to initiate any construction, they consider the potential impacts on:

- the lives and property of the people living in the area
- the local economy
- the level of employment
- property values
- taxes
- the health, safety, and general welfare of the people in the area

Apply Your Thinking Skills

1. What type of new construction might benefit your community? Describe the positive impacts it would have on the community. Would it have any negative impacts?

Controlling Construction

No matter what type of construction is desired, or who desires it, it must meet the requirements and standards set up by the community in which it is to be built.

Communities are divided into residential, commercial, and industrial zones. Communities have boards of appointed officials that set up special **zoning laws** that tell what kinds of structures can be built in specific parts of the community. These laws may also specify such things as:

- maximum property size
- maximum height of a building
- the number of families that can occupy a house
- the number of parking spaces a commercial building must provide

These laws are designed to protect homeowners from traffic, noise, and other environmental problems.

In addition to zoning requirements, all structures must meet certain building codes. As you learned in Chapter 20, *building codes* are local and state laws that specify the methods and materials that can or must be used for every aspect of construction. To make sure each structure is constructed according to these building codes, the structure must be inspected throughout the construction process and when construction is completed. You will learn more about building inspection in Chapter 22.

Apply Your Thinking Skills

1. Why are zoning laws important to community development?

Site Selection and Acquisition

Two basic decisions must be made before construction can begin. One is choosing the best site. The other is choosing the best design. These two decisions are not usually made independently. The design may be influenced by the nature of the site. The site choice must be suitable for the design. Once the best site has been chosen, it must be acquired.

Selecting a Site

In addition to being suitable for the basic design of the structure, the site must also meet a number of other important criteria. Although specific criteria will vary from structure to structure, the same basic factors must be considered:

- Location. Is the site in the city or the country? Is it near roads or highways? For stores and restaurants, is it an area where there will be a lot of potential customers?
- Size. Is the site large enough? If not, can an adjoining site be acquired?

- Shape. Is the site long and narrow, short and wide, pie-shaped, or L-shaped? Will the planned structure fit well on the site?
- Topography. **Topography** refers to the site's surface features. Is there a lake or stream? Are there hills, gullies, large rocks, or trees? What is the nature of the soil? Is it dry sand or wet clay? Fig. 21-4.
- Utilities. **Utilities** are services such as electricity, natural gas, and telephone. Are these available at the site? How much will it cost to have them installed? Also, if the property is outside the city limits, what kind of water and waste disposal systems—if any—are available?
- Zoning. Will zoning laws permit the type of planned structure to be built there?
- Cost. Is the price of the site reasonable and affordable?

Fig. 21-4. Falling Water, a home designed by the famous architect Frank Lloyd Wright, makes use of the natural topography of its site.

Fascinating Facts

The choice of a site for a home can become a way to express a personal vision, an ideal, or a belief.

In 1767, when Thomas Jefferson was 24, he designed a house that would show his vision for the new nation. He named the completed house Monticello. The 35-room red brick mansion was built on a hilltop near Charlottesville, Virginia—facing westward. The site reflected Jefferson's belief in the country's unlimited possibilities.

Acquiring the Site

Once the site has been selected, the land or property must be acquired.

The simplest way to acquire land is to buy it from the owner. That is *direct purchase*. Doing this may not be as easy as it seems. For example, the land may be prime (the best) farmland. Owners may consider it too fertile (rich) to be used for buildings or other structures. The land may be located in or near a major city. Land there is scarce, and is usually very expensive. Quite often, the owner of a piece of land may simply not want to sell. He or she may ask for more money than the buyer is able or willing to pay.

The right to own property is precious. However, sometimes land may be needed for public purposes. For example, federal, state, or local government officials may decide to make certain public improvements. Perhaps a new road, school, dam, or public housing is needed. A landowner may refuse to sell. If there are no acceptable alternatives, the government can take legal steps to force the owner to sell by exercising its power of eminent domain. The **power of eminent domain** is a law that states the government has the right to take private property for public use. (The rights or needs of all come before the rights or needs of one.)

The legal process for taking over land that an owner has refused to sell is called *condemnation*. The government must prove that the property is needed. If it does, then the owner has to sell. However, he or she must be paid a fair price for the land. Individual rights must be protected.

Apply Your Thinking Skills

1. If you were building your own dream home, what kind of site would you choose? Why?

The Design Process

In this section, we will look briefly at the design process used for residential and commercial structures. Similar steps are followed for all types of construction.

Identifying Specific Needs

Suppose a family needs a new home. They may decide to have one *custom-built*. This means the house will be planned and built especially for them. To begin the construction process, they hire an architect. An *architect* is a person who designs structures and develops the plans for building them. The family chooses someone who is experienced in designing custom-built houses.

A meeting is held between the architect and the family. Fig. 21-5. At the meeting, the architect asks questions to identify specific needs or problems:

- How large is the family? What is the age and sex of each member? This may determine how many bedrooms and bathrooms the house should contain.
- Does either spouse work at home? If so, what kind of work does he or she do? This may determine whether there is a need for an office or den. Perhaps there should be a separate entrance, soundproofing, or special wiring for telephones, computers, or other equipment.
- What are the family's special interests or hobbies? For example, if a family member enjoys woodworking, there may be a need for a workshop.
- Does the family entertain a lot? There may be a need for a large family room, a formal dining room, or a guest room.

Fig. 21-5. Identifying the needs of the client is the beginning of the design process.

- How much money can the family afford to spend? Money is the single most important factor. The amount available influences nearly all other decisions.
- Where is the family going to live? Suppose they plan to live in town. Land there may be scarce and lot sizes rather small. The architect may recommend a two- or three-story house. If, on the other hand, the family plans to live in the country, a sprawling, one-story house may be preferred.

A similar procedure is followed in other types of construction. For example, suppose a company needed a new office building. Then company representatives would meet with the architect. Questions would be asked to identify company needs. For example:

- How many people will be working in the new building?
- What types of work will be done?
- Is the company planning to expand in the near future?

Fig. 21-6. After determining client needs, the architect begins making preliminary plans.

Developing Preliminary Ideas

After the architect has learned about the family or company, he or she begins jotting down preliminary (beginning) ideas and plans. Fig. 21-6. First, rough sketches of several possible designs are made. Some may be thrown away almost immediately. Others may be saved and reviewed later. As more facts are gathered, certain of these ideas may begin to show greater promise. The best ones are considered carefully. Soon a more nearly finished (though still preliminary) design is developed.

Refining Ideas and Analyzing the Plan

Architects review and revise preliminary design ideas. They consider:

- The special needs and desires of the client.
- The money available.
- The site on which the structure will be built.

As they refine their ideas, they may prepare renderings or models. This depends on the type of project and the people making the decisions.

In commercial construction, one or more engineers may be asked to analyze the preliminary plans. **Engineers** figure out how a structure will be built and what structural materials will be used. For example, they determine whether a certain material is strong enough for its planned use. Basically, engineers deal with matters related to the function and strength of structures. Questions they need to consider might include:

- What is the soil type; how much weight can it carry? Will special supports be needed?
- What are climate conditions? Are high winds or heavy snows common? Are temperatures extremely hot, or do they vary greatly?
- Will the structure serve its planned purpose as it is currently designed? For example, will a sports stadium efficiently accommodate large crowds?
- Could different materials or construction techniques be used to reduce costs and yet maintain quality?

Answers to these and other engineering questions help the architect make *renderings* of the refined ideas. These help clients visualize what the structure will look like. Fig. 21-7. Sometimes models are made. However, this is usually done only when a presentation must be made before a group. Fig. 21-8.

Fig. 21-7. Renderings are refined drawings. Colorful renderings help clients visualize the appearance of planned structures.

Preparing Final Plans and Specifications

The architect's refined preliminary plans are reviewed by the client and the engineers. When plans are approved, final drawings and specifications are prepared.

Working Drawings

The *working drawings* contain the information needed to construct a project. They are drawn to scale. In **scale drawings**, a small measurement is used to represent a large measurement. For example, one-fourth inch (1/4") may represent one foot (1'). Measurements given on the drawings are for the actual size, however.

There are four types of architectural working drawings:

- Site plan. The **site plan** shows where the structure will be located on the lot. Boundaries, roads, and utilities are included. Fig. 21-9.
- Floor plan. The **floor plan** shows the locations of rooms, walls, windows, doors, stairs, and other features. Fig. 21-10.
- Elevations. **Elevations** are drawings that show the finished appearance of the outside of the structure, as viewed from the ground level. A separate elevation is made for each side of the structure. Fig. 21-11.

Fig. 21-8. Models are useful in helping people understand complicated or unusual construction projects.

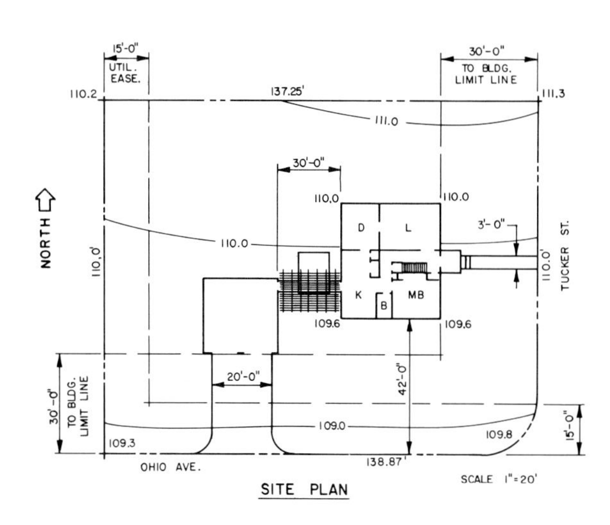

Fig. 21-9. The site plan. Notice that the scale used for the drawing is shown at the bottom.

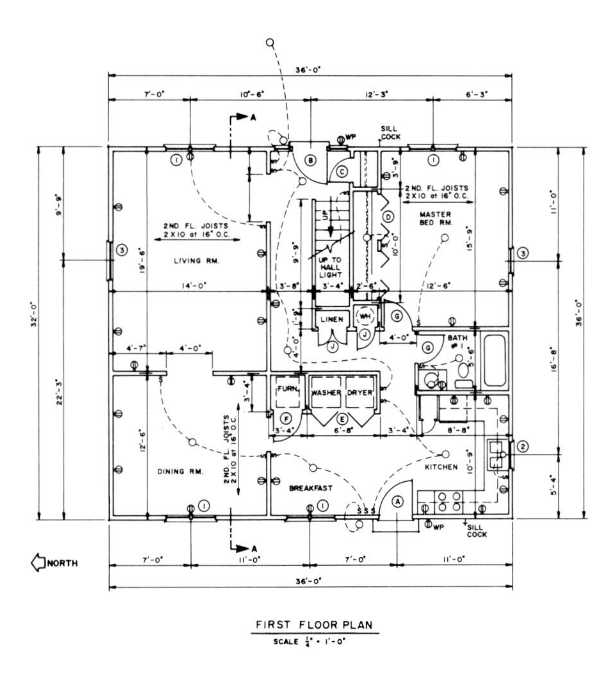

Fig. 21-10. Notice this floor plan is labeled "First Floor." A separate floor plan is made for each story of a building.

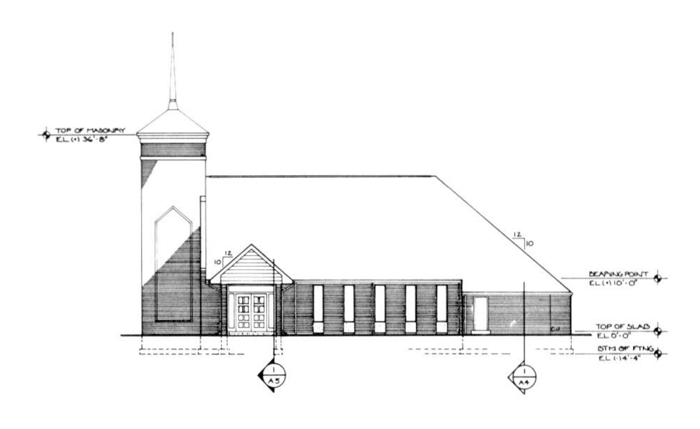

Fig. 21-11. This elevation shows the front view of a church. Other elevations must be drawn to show the sides and the rear of the church.

- Detail drawings. These are special drawings of any features that cannot be shown clearly on floor plans or elevations, or that require more information to be constructed. Fig. 21-12.

Specifications

In addition to all the working drawings, a set of specifications must be prepared. In architecture, **specifications** are written details about what materials are to be used for a project, as well as the standards and government regulations that must be followed. They describe or list the size, number, type, and (if appropriate) model number and color of every item to be included in the finished building. Construction details and materials that could not be shown on the drawings are given in the specifications. Fig. 21-13.

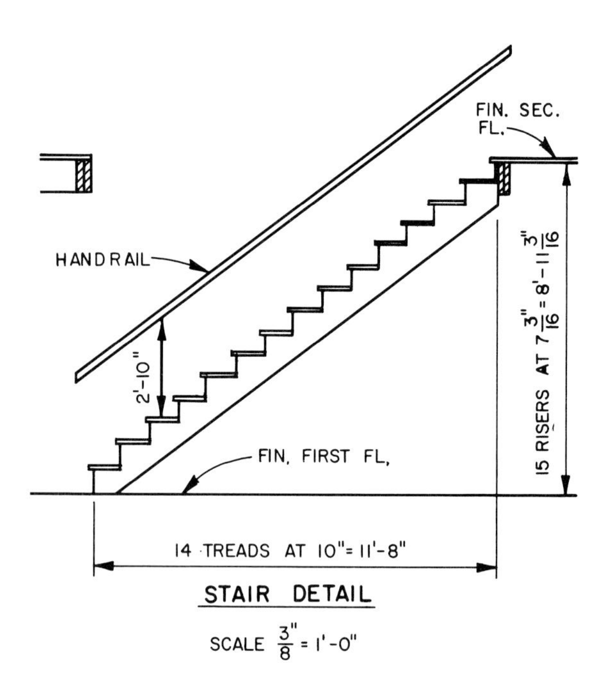

Fig. 21-12. This is a detail drawing of a staircase that could not be shown in enough detail on the floor plan. Notice all the additional information given on this drawing.

Division 4: Carpentry and Millwork

Sec. 1. Scope. This division includes the furnishing and installation of all carpentry, millwork, and related items required to complete this work as indicated on the drawings and/or hereinafter specified.
Sec. 2. Materials.

a. Rough lumber shall be Framing Lumber — No. 3 Grade Southern Pine.
b. Exterior millwork. Horizontal siding shall be 6" bevel redwood.
c. Sheathing shall be 5/8" aluminum-foil-backed foam as made by Dow Chemical.
d. Attic insulation shall be 12" fiberglass batts installed in accordance with manufacturer's instructions.
e. Interior trim shall be of pine. Pattern selected by owner.
f. Interior doors shall be as indicated on door schedule.
g. Exterior door frames shall be 1 3/4" thick, rabbeted with 1 1/8" outside casings of white pine.
h. Interior door frames shall be 7/8" thick pine.
i. Double-hung sash shall be Andersen or equal.
j. Kitchen cabinets shall be birch. Countertops Formica or equal.
k. Closets and wardrobes shall have 7/8" by 12" shelving and one clothes pole running length of space.
l. Living room, hall, bedroom floors to be straight-line oak; select grade, finished natural; kitchen vinyl composition tile over plywood underlayment; bathroom, ceramic tile set in rubber into squares, cemented to plywood sub-floor; entrance hall imitation slate.

Fig. 21-13. Sample specifications for work to be done on the project. Each part of the project is specified in detail.

Specifications are extremely important. They provide owners with an accurate description of the materials and services they are buying. Contracting firms use specifications when calculating costs and when building the structure. The specifications serve as a guide or set of instructions.

Apply Your Thinking Skills

1. Interview two different members of your class. After identifying the needs of their particular families, describe how the houses you might design for each of their families would be different.
2. Suppose you were going to design a child's treehouse. What questions would you ask in order to identify specific needs?

Global Perspective
Born from a Mountaintop

There is emerging in Osaka Bay, 400 kilometers (about 250 miles) from Tokyo, an island where none existed before. When construction of this artificial island is complete, it will hold an airport capable of handling over 150,000 takeoffs and landings and accommodating 30 million travelers each year.

The site chosen was 5 kilometers (3 miles) offshore, where the water was about 18 meters (59 feet) deep. To keep fill materials (rocks and soil) from being washed away, a sea wall was constructed to surround the 4 kilometer (2 1/2 mile) long and 1.2 kilometer (3/4 mile) wide island. Then workers began placing fill within the enclosure. Ultimately, over 168 million cubic meters (220 million cubic yards) of fill will be used. To obtain this fill, contractors are cutting off the tops of three mountains. The rock from these mountaintops is crushed into gravel, hauled to the site, and dumped.

The settlement of landfill used to form the island creates problems for the construction of the terminal building. To cope with it, jacks will be installed in the foundation. A monitoring system will keep track of the level of the floors and adjustments will be made with the jacks as needed to keep the building level and secure.

Although the design is basically simple, the airport should be impressive and easy to use. One side of the terminal building will face the single runway and the planes. This side will be made of stainless steel and glass and will be aerodynamically shaped, in much the same way as a boomerang is shaped. The other side will face the railway and access roads from the mainland. Trees and other greenery will be planted on that side, extending into a greenhouse that will run the entire length of the building.

Fig. GP-21. Kansai International Airport is being built on an artificial island three miles from the mainland.

The cost of the project is expected to be 1.4 trillion yen (10 billion dollars) or more. Construction is expected to be completed in 1994. Then this island made from mountaintops will be a functioning international airport—Kansai International Airport.

Extend Your Knowledge

1. Select a structure in your area that you find interesting and write a brief history of its construction. What type of structure is it? When and by whom was it designed and built? Describe any unique design or structural features. What unusual construction problems, if any, were faced by the builders, and how were these problems resolved? Include other facts that you find interesting and informative.

Chapter 21 Review

Looking Back

Construction may be initiated by the private sector or the public sector. A city planner may study a community and analyze its needs, then make recommendations to planning commissions about areas of potential growth and development. These recommendations are studied carefully before any decisions are made to initiate construction.

Zoning laws and building codes are drawn up to make sure that any structures built will be in the best interest of the community.

There are many factors to be considered when selecting a site. The site is usually acquired through direct purchase. In some cases, the government may have to exercise its power of eminent domain to purchase property needed for public works.

Architects and engineers design and plan structures. The design process involves a series of steps. The first is identifying specific needs. Then preliminary ideas are jotted down and the best ones are analyzed and refined. Finally, working drawings and specifications are prepared.

Review Questions

1. Briefly identify the private and public sectors of our economy. Which one would be responsible for having a new school built? Where does the public sector get the money to pay for design and construction costs?
2. Define zoning laws and give three examples of things they might specify.
3. Name at least five factors that should be considered when selecting a home's site.
4. Describe the power of eminent domain. When is it necessary to exercise this power?
5. Give five examples of questions an architect might ask to identify the needs of the family for whom he or she is designing a home.
6. Describe the role of the engineer in the design process.
7. What is shown on a site plan?
8. What is shown on a floor plan?
9. Define elevations.
10. What are specifications?

Discussion Questions

1. Suppose you are on the planning commission for your community. What would you recommend about a proposed shopping mall to be built near your school? Explain your reasoning.
2. Again supposing you are on the planning commission for your community, what would you recommend for a clothing manufacturing industry that would employ 5,000 people? Again, explain your reasoning.
3. Explain why it is important for an architect to identify the needs of a client before designing a structure.
4. Why are specifications such an important part of the plan for a construction project?
5. In what kinds of situations do you feel it is justifiable to exercise the power of eminent domain? How would you feel if it were your house that was going to be taken away by this means?

Chapter 21 Review

Cross-Curricular Activities

Language Arts 

1. Invite an architect, contractor, plumber, electrician, or anyone in the construction field to class for an informal discussion about the construction business. Be sure to follow up with thank-you notes after the visit.
2. Ask a parent which room in the house he or she would like to have remodeled. Ask for specific details. Make a sketch of the room, and write a brief description of the major changes that would be made.

Social Studies 

1. Make a two-column chart. On the top of one column place the words "Private Sector." Label the other column "Public Sector." Then make a trip to your county seat's commercial district. Make a list of all the names of the buildings in that area of town. Label each as either a private or public sector building, and list each in the appropriate column.

Science 

1. Some structures are designed to lean. That is, they are not meant to be at 90° to the ground. Such structures must be carefully designed. The location of the center of gravity—where the structure's weight can be considered to be concentrated for the purposes of determining its stability—is vital to the design. If the center of gravity is too far from the place where the structure is supported (its base), the structure will become unstable and fall. You can demonstrate this effect with an empty quart-size milk or juice carton. See how far you can tilt it before it falls over. Now open the top, and drop in a large lump of clay. Now how far can you lean it before it falls? You have lowered the center of gravity by adding the weight of the clay, making the carton stable at a greater angle.

Math 

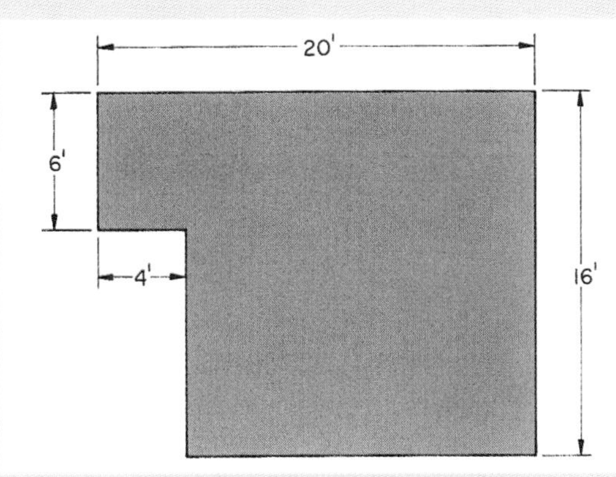

1. Amy measured a room in her basement and determined the sizes as shown above. If all dimensions are in feet, what is the total area of the room's floor?
2. If Amy chooses a floor covering that costs $1.62 per square foot, what will be her total cost for the floor covering?

Chapter 21 Technology Activity

Preparing an Architectural Rendering

Overview

After completing this activity, you will know some of the ways in which you can make sketches of buildings look more realistic.

Goal

To learn about architectural renderings and to develop a notebook of your own renderings.

Equipment and Materials

magazines showing sketches and floor plans of houses and other buildings
white drawing paper with a slight texture
No. 2 and No. 4 pencils
kneaded eraser

Procedure

1. Develop a collection of illustrations showing the many ways in which drawings of houses and other buildings can be drawn to show them in a more realistic manner. Group these illustrations according to the manner in

Fig. A21-1. This is a hand-drawn architectural rendering.

which they were drawn. Adding detail to a simple line drawing, using shading and color, is termed *rendering* the illustration. Your illustration categories will include those completed entirely by pencil. Others may be completed entirely with ink. Still others will include a combination of pencil, ink, watercolor, markers, and special dyes. See Figs. A21-1 and A21-2 for examples of renderings.

2. Study your collection of illustrations. Then prepare a simple pencil sketch to show a grouping of shrubs and trees. Do not add many details to the leaves and branches. The purpose of the illustration is to suggest shrubs and trees and not detract from the details of the building that you really want to emphasize. Also study the illustrations to note how depth can be added by making some of the plants darker than others.
3. To improve your drawing skills and test your ability to observe, you can copy illustrations from your collection. Always work on a clean, smooth surface. When you begin shading large areas, any debris under the paper will appear as a dark spot on your drawing.

Fig. A21-2. This is a computer-generated architectural rendering.

Chapter 21 Technology Activity

Knowing this may help you select surfaces that will actually give you the texture you desire on your drawing. It is also important not to slide your hand across the pencil drawing. This will cause smudged areas. When drawing in areas where there are already pencil lines on the paper, most illustrators rest their hand on a clean piece of paper.

4. You may want to extend your sketching by going outside to draw a small section of a building. It is not necessary to draw the entire building. The skills you develop will help you develop future house designs and plan the best landscaping layout.
5. In Fig. A21-3, Figures A and B have been provided to assist you in further developing your skills. Many of the details of the buildings have been eliminated. Enlarge the size of these sketches to fill a sheet of 8 1/2" x 11" paper.
6. Use a No. 2 and No. 4 pencil to add detail to the building and landscaping around the building. It may be interesting to see how the members of the class start with the same general sketch and finish with a number of variations at the end of the class.
7. Develop a notebook of examples of your own work that shows ways to illustrate bricks, blocks, siding, shingles, trees, shrubs, people, and many other objects revealed in an architectural rendering.
8. Note that computers are currently being used to complete architectural renderings. This is time consuming, even for a skilled computer operator. Individual hand sketches are still a quick way to show a client what the project will look like before construction begins. Such sketches help you and the client make last-minute changes that will save time and money during the construction process.

Evaluation

1. How well did your different renderings turn out? Did your classmates' renderings turn out a lot different from yours?
2. Did you enjoy preparing these renderings? Would you be interested in this type of work for a career?

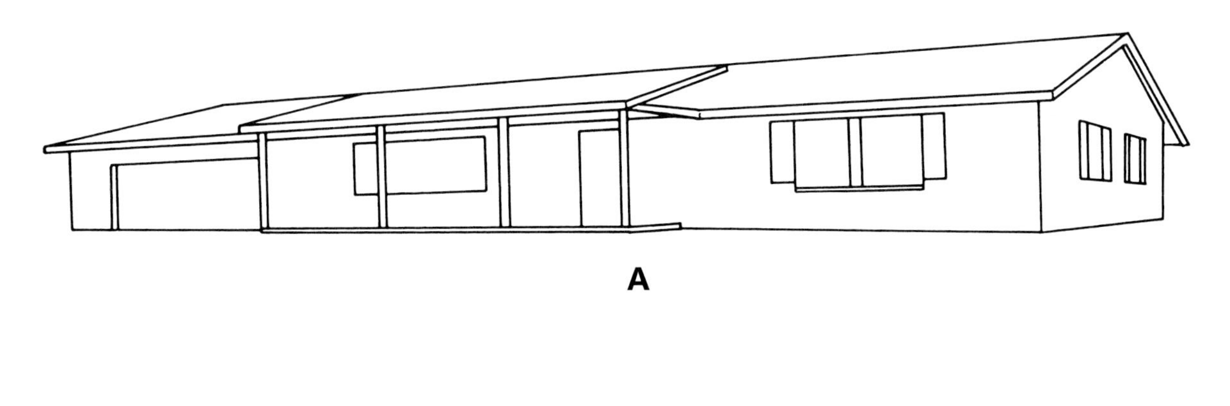

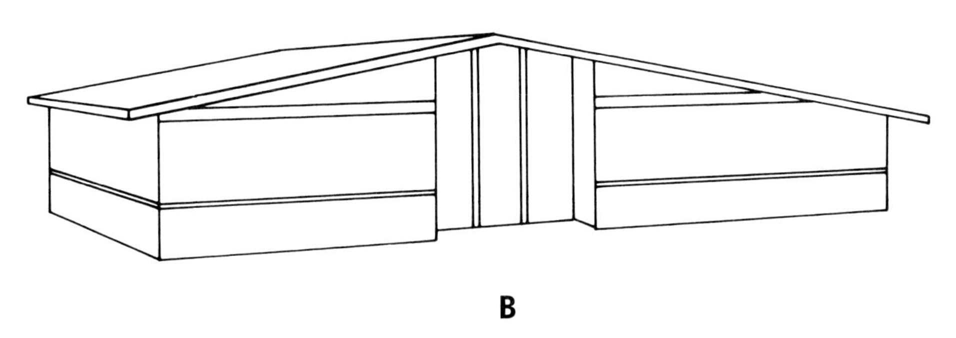

Fig. A21-3. Here are two examples of buildings that you can enlarge and use to make renderings.

Chapter 22

Managing Construction

Looking Ahead

In this chapter, you will discover:

- who manages construction projects.
- how a contractor is selected.
- what schedules are and why they are important in construction.
- how construction is monitored.
- the importance of quality and quality control in construction.
- when and how inspections are made and why they are important.

New Terms

bid
contractor
monitor
Occupational Safety and Health Administration (OSHA)
schedule
subcontractors

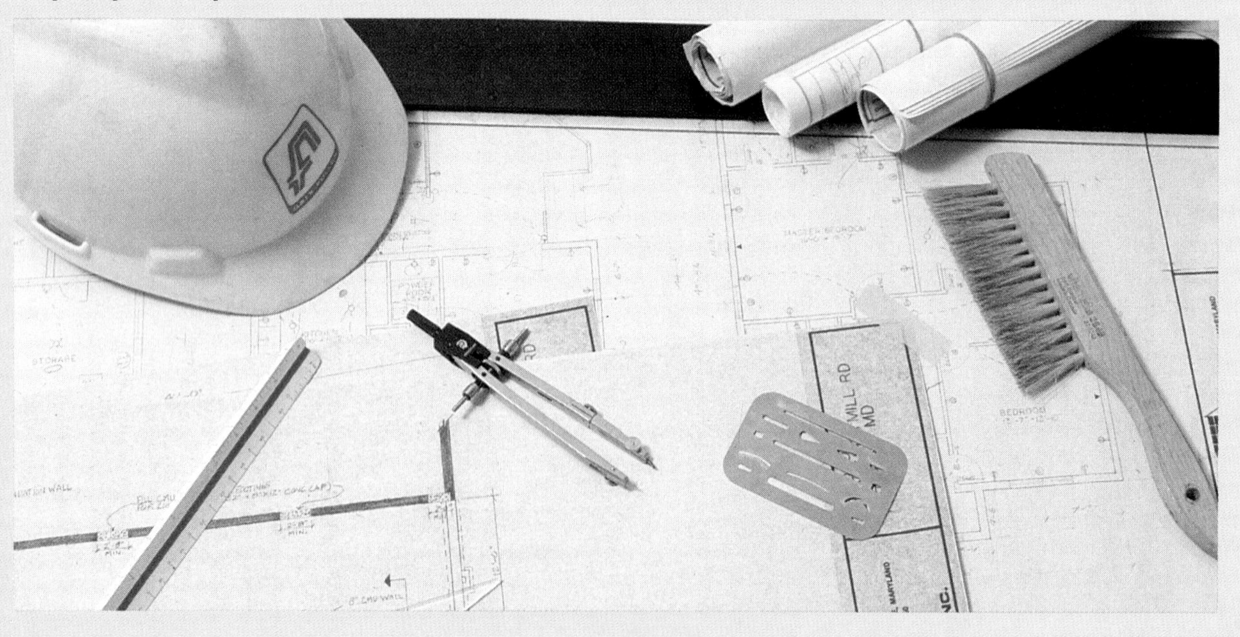

Technology Focus

Managing from Afar: Building the Brooklyn Bridge

John Augustus Roebling designed the Brooklyn Bridge, which is a suspension bridge that spans the East River and connects Brooklyn and Manhattan in New York City. He was a successful pioneer in the design and construction of suspension bridges. He began working on the project in 1867. For more than two years, he worked tirelessly, designing the bridge and surveying the site. Just before beginning construction, he was seriously injured. He developed tetanus. He died on July 22, 1869. His son, Washington Augustus Roebling, then became chief engineer of the project.

The Brooklyn Bridge was the first bridge to be built on a foundation of pneumatic caissons (pronounced: new-MAT-ik KAY-sahns). These are very tall cylinder-like devices. They are open at the bottom and partially filled with compressed air.

To build the bridge's foundation, caissons were placed on the bottom of the river. Workers (or "sandhogs") entered a caisson at the top. They climbed down a long, narrow ladder to the open bottom or work chamber, which was filled with compressed air. Inside the caisson, they were able to excavate deep into the floor of the river. As the workers dug deeper and deeper, the caisson was lowered, until it came to rest on solid rock. Then, the caisson was firmly and permanently anchored to the bottom of the river and completely filled with concrete.

During the early work on the bridge, Washington Roebling spent a lot of time in this compressed-air atmosphere at the bottom of the caissons. In 1872, after more than 12 hours in the work chamber, he became ill. He had developed "caisson's disease," or the "bends." It permanently damaged his health. He was never again able to visit the construction site. Instead, he was confined to his home in Brooklyn. However, he didn't quit.

Fig. TF-22. During construction of the Brooklyn Bridge, Washington Roebling developed the bends and had to manage the project from his home in Brooklyn until the bridge was done eleven years later.

For the next eleven years, Roebling supervised the building of the bridge by watching the work through a telescope. His wife carried his instructions to the bridge site each day as the work progressed. Finally, fourteen years after it was begun, the Brooklyn Bridge was completed. It has now been in continuous use for more than 100 years.

Who Manages Construction Projects?

Management procedures may vary according to the size and type of construction project. Generally, an architect or engineer oversees the project. Fig. 22-1. He or she is responsible for seeing that the owner's wishes are carried out. This person supervises the work done by the contractor. A **contractor** is a person who owns and operates a construction company. Contractors are responsible for the actual building of projects. They must follow the plans and specifications developed by the architects or engineers. Fig. 22-2.

The contractor is chosen after planning is complete. The person or group planning the project announces or advertises the plans.

Fig. 22-2. Contractors must be sure that the plans are carried out down to the finest details.

Fig. 22-1. An architect or an engineer oversees the construction project.

Qualified contractors submit bids for the contract or the right to build the project. A **bid** is a price quote for how much a contractor will charge for a project. The amount equals the total of the contractor's best estimate of what it will cost to build the project according to the owner's plans and specifications, plus the amount of profit the contractor hopes to make. Generally, the contractor who submits the lowest bid is awarded the contract. Fig. 22-3. Usually, the project must be completed by a certain date. The contractor must control costs and yet make sure that high-quality work is done.

Contractors may also hire subcontractors. **Subcontractors** specialize in certain types of construction work. Fig. 22-4. For example, the electrical system in a building is usually installed by a subcontractor.

Fig. 22-3. Some contractors hire an estimator. This person estimates the cost of materials, labor, equipment, and other items for a project. The contractor bases the bid on this estimate.

Apply Your Thinking Skills

1. Usually, the project is awarded to the contractor who submits the lowest bid, but this is not always the case. Why might a client choose a contractor who gave a slightly higher bid?

Scheduling and Monitoring Construction

All the workers, materials, and equipment needed to complete a construction project must come together at the right times. To accomplish this, the contractor must prepare schedules. A **schedule** is a plan of action that lists what must be done, in what order it must be done, when it must be done, and, often, who must do it. Fig. 22-5. Of course, schedules must have some degree of flexibility and allow time for unexpected events like bad weather or design changes.

Fig. 22-4. Plumbing is usually installed by a subcontractor.

Scheduling

The three main schedules that govern (control) a construction project are:

- Work schedule
- Materials delivery schedule
- Financial schedule

Work Schedule

Before a useful work schedule can be prepared, the contractor must first analyze the job. That is, he or she must know and understand every phase of the total project. A contractor must know *what* must be done and *where, when,* and *how* it should be done. Also, the contractor must decide *who* should do each job. By studying both the drawings and the specifications, a contractor can determine:

- The number of workers needed
- The skills or crafts needed
- The time required for each process
- The order or sequence of jobs

Using this information, the contractor can then prepare a written work schedule.

Fig. 22-5. This construction manager uses a bar chart to help schedule construction tasks in the right order.

Materials Delivery Schedule

Suppose a contractor was hired to build a brick house. Then, on the day the bricks were to be laid, only the bricklayers show up—no bricks.

To avoid such problems, the contractor prepares a materials delivery schedule. This is a list of every kind of material needed to complete the building. It includes the quantity, style, color, and price of each material. It also shows where and when the materials must be delivered.

Financial Schedule

A contractor pays wages to workers on a regular basis. Materials and supplies must also be purchased. This means a regular and dependable source of money is needed.

How and when a contractor will be paid is worked out in advance with the owner (person or persons who initiated the project).

Fascinating Facts

Construction projects may take days, years, or even decades! The construction of India's Taj Mahal, one of the world's most beautiful tombs, took from 1632 until 1653—and required 20,000 workers! A modern skyscraper can be completed within a year or two. A prefabricated building, which is a building that is made of components made in a factory and shipped to the building site, can be set up within just a few days.

Generally, contractors are paid certain amounts as the work progresses. The amount is usually a percentage of the total price of the job. However, what if money that was expected does not arrive in time? Suppose no arrangements were made? Then the contractor could be in trouble.

To avoid these problems, the architect or engineer prepares a financial schedule. It lists amounts and dates for payment. The financial schedule is reviewed and approved by the owner and, possibly, the bank.

Monitoring

Careful scheduling alone is not enough to assure a successful operation. In order to make sure that all the terms of the contract are properly met, the project must be carefully monitored by the contractor or those especially hired to do the monitoring. To **monitor** a construction project means to watch over and check it.

Monitoring Materials

A purchasing agent (buyer) is responsible for obtaining the right materials at the right price. The materials must be the proper kind and the correct quantity. They must be reasonably priced. The purchasing agent works closely with the various materials suppliers. The purchasing agent also prepares and monitors the materials delivery schedule.

The purchasing agent checks all materials as they are received. Fig. 22-6. Materials must be:

- Those that were ordered
- Delivered to the correct site
- Delivered on time
- In good repair
- Of acceptable quality
- The quantity ordered

All of these requirements must be met each time materials are purchased, delivered, and paid for.

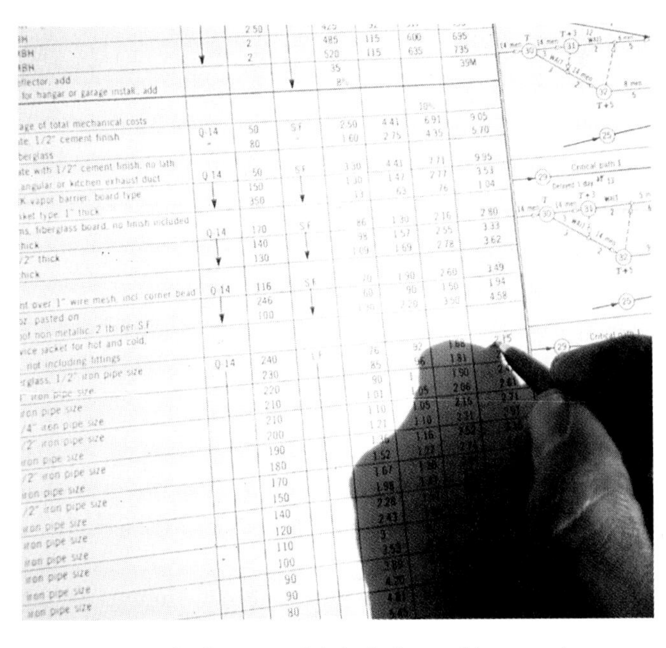

Fig. 22-6. Monitoring materials includes making certain the right quantity was delivered to the site.

Monitoring Job Progress

Keeping a construction project on schedule is important. This requires careful supervision and close monitoring of the work schedule. Suppose bad weather interrupts work. The contractor must find ways to make up for time lost. If needed materials are not on hand, the purchasing agent must notify the supplier at once. Careful monitoring of job progress is vital to the success of the project. Fig. 22-7.

Fig. 22-7. The contractor checks to make certain the work is getting done on schedule.

Monitoring Quality

Quality is extremely important. Most contractors take pride in work well done. They also want the structure to be safe. There are other things to consider as well. For example, suppose the quality of the work does not meet the standards described in the original contract. Then the owner may not be required to pay for the work that was done. The unacceptable work would have to be repaired or replaced. Making extensive repairs is very costly, and could put the entire project way behind schedule. The contractor may have to bear the cost of the repairs. In addition to all of this, contractors are often legally liable (responsible) for quality. For example, if a bridge should later collapse, the contractor may be sued.

Contractors monitor the quality of both the materials and the work. Quality is vital.

Apply Your Thinking Skills

1. The purchasing agent must make sure that materials are of acceptable quality. Discuss some of the possible advantages and disadvantages of using materials of higher quality than required by the building codes.

Inspecting Construction

All work done must meet the requirements of the contract and the standards set by building codes. To ensure this, the structure is *inspected* carefully. This means it is examined carefully by someone who knows what the correct results should be and what conditions should be met.

Inspections are usually done on a regular basis. This way problems can be spotted early. Corrections can be made without losing much time.

What Is Inspected?

It is usually not necessary to inspect all parts of a project. Trying to do so would be difficult, expensive, and time-consuming. Inspections are usually limited to the more critical parts of the project. For example, inspectors routinely check structural, electrical, and mechanical elements.

They check the quality and appropriateness of the materials being used. They make sure work is being done properly. In short, inspectors are concerned with the three principal (main) elements of any construction project:

- Materials
- Methods
- Quality

Quality cannot be added or attached to a structure. It must be part of it from the beginning. *Everyone* involved in the project must be committed to quality. It must be present in the design of the structure, in the engineering, in the materials used, and in the way the work was done. That's what inspectors look for. They make certain that each detail of the construction process has met or exceeded (gone beyond) the standards required for quality construction.

Who Inspects?

The types of inspectors depend upon the nature of the project. Inspections may be conducted by local building inspectors or quality-control specialists. Sometimes insurance agents or bank representatives inspect projects. All projects are inspected by representatives of the government. Fig. 22-8.

All levels of government—city, state, and federal—have set safety regulations. The **Occupational Safety and Health Administration (OSHA)** has been established by the federal government. It is part of the U.S. Department of Labor. OSHA sets safety standards. Its representatives visit sites to see that those standards are met.

Fig. 22-8. Inspectors make sure that safety standards are met.

Apply Your Thinking Skills

1. Inspections are done periodically as a project progresses. What problems might occur if all of it is put off until the end of a project?

Investigating Your Environment

Sick Buildings

The construction industry involves the talents of many people to ensure high quality. Nevertheless, it has been found that many buildings and homes can be hazardous to your health. Some have been built with hazardous materials, others are harmful because of their design.

Lead was once widely used in construction for plumbing systems and roofs. It was also used in most paints. However, it was discovered that lead can be absorbed by the body and accumulate in one's system, causing a gradual poisoning. Lead poisoning can cause serious illness and brain damage. Materials containing lead are still found in many older buildings.

Asbestos is an inorganic, fibrous material that was once used to make roofing, insulation, cement, and floor materials. When asbestos particles are inhaled, they are likely to cause lung cancer or other lung diseases. The Environmental Protection Agency (EPA) has proposed a total ban on the use of asbestos by the year 1996. Federal laws have been passed that make it mandatory that all asbestos be removed from or sealed off in all government buildings, including schools.

Construction materials and new construction techniques have contributed to a problem called "sick building syndrome." Materials such as paneling, carpeting, plywood, and paint release fumes or fibers into the air. Cleaning agents, pesticides, and glue also emit fumes into the air. Since today's buildings are tightly sealed to reduce heating and cooling costs, these fumes and fibers are trapped inside, causing indoor air pollution. This pollution can cause people to become ill.

The most dangerous source of indoor pollution is probably radon gas. Breathing this colorless, odorless radioactive gas can increase one's risks of lung cancer. Radon gas is naturally present in the soil. It is more concentrated in some areas of the country than others. Radon gas can pass through tiny holes or cracks in a structure's base. Radon levels are usually higher in basements. Newer buildings, with their airtight structure, are especially vulnerable to radon problems.

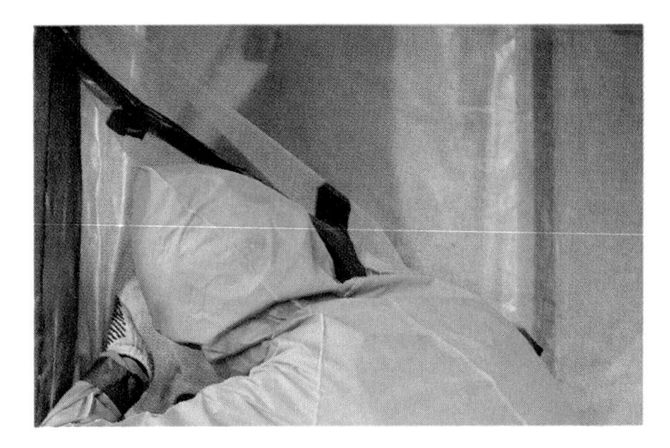

Fig. IYE-22. The removal of asbestos in buildings can be dangerous. Workers must take special safety precautions to avoid inhaling this material into their lungs.

Fortunately, most indoor pollution problems are easily detected and easily treated. The fumes need to be released or diluted with fresh air. Keeping air ducts clean greatly improves ventilation. Ventilation systems can be modified to be air exchangers. Architects and engineers should design buildings with air exchange in mind. Simply opening a window and turning on a fan can greatly reduce the amount of toxins (poisons) in homes and other buildings.

Take Action!

1. Research how asbestos removal is done in buildings and how much this process costs. Also, find out what happens to the asbestos after it is removed from a building. Write a short report on the information you obtain.

Chapter 22 Review

Looking Back

Construction projects must be well managed. Architects oversee most building construction sites. Engineers work with the architect. Contractors are responsible for the actual construction of the project. They may hire subcontractors to do parts of it.

Contractors study plans and make bids on projects.

To make sure all the terms of the contract are met—that the project is built as designed, according to specifications, and completed on time—careful scheduling and monitoring are necessary.

Various people, including government representatives, periodically inspect construction to be sure that everything is done properly. Quality is a vital element of construction.

Review Questions

1. Briefly describe the responsibilities of a contractor.
2. What is a bid? How is the figure for a bid arrived at?
3. Define and give two examples of subcontractors.
4. What information does a contractor need to consider when making a work schedule?
5. What is the purpose of a materials delivery schedule?
6. Why is a financial schedule necessary?
7. What are the six things a buyer checks for when materials are received?
8. What does the term "inspection" mean? Why are structures inspected?
9. Briefly discuss the three general types of things that inspectors check on a construction project.
10. What does the acronym OSHA stand for? What does OSHA do?

Discussion Questions

1. What traits would a contractor need to do a good job?
2. What kinds of individuals might be responsible for inspecting construction projects?
3. A contractor must be able to make accurate estimates of job expenses. What are the risks if the estimates are way off?
4. Discuss the need for allowing time for unexpected events in construction schedules.
5. After reading the "Investigating Your Environment" article about sick buildings, what advice would you give to someone thinking about buying an older home? What advice would you give to someone planning to build a new home?

Chapter 22 Review

Cross-Curricular Activities

Language Arts 

1. As a class, write a business letter to your local building commission or zoning body and request a list of building codes for your neighborhood. Are there any unusual codes for your neighborhood? Have a class discussion on the information you receive.
2. Discuss with your classmates the questions that you might consider asking when hiring a contractor. What qualifications would you expect this person to have? What resources would be available to you for finding out this type of information?

Social Studies 

1. Reread the description at the beginning of the chapter of how the Brooklyn Bridge was built. Research to find a new bridge that was built somewhere in the United States in the last decade. Identify the bridge and compare and contrast the building of that bridge with the construction of the Brooklyn Bridge.

Science 

1. One consideration in ensuring quality of construction is making sure surfaces that should be horizontal are perfectly horizontal. For this purpose, construction workers use a tool called a level. It has a fluid-filled tube with an air bubble trapped inside. Because air is less dense than the fluid, it floats to the top of the middle of the tube. You can make a simple level. Obtain a cylindrical glass container, such as an empty olive jar, with a tightly-fitting lid. Fill it with water almost to the rim, and fasten on the lid. Place it on its side on a flat surface. Where is the air bubble? Now place it on a sloping surface. What happens to the air bubble? How could the device be used to determine whether a vertical surface is really vertical?
2. Until recently, caissons were often used in construction jobs under water. You can demonstrate how a caisson works with a drinking glass. Hold the glass upside down. (Hold it straight; do not tilt it.) Then plunge the glass directly downward into a deep container of water, such as an aquarium or laundry tub. The air inside will prevent water from entering, even though the bottom is open.

Math

Matt has his own home-repair service. When he is working on a job, he charges $12.50 per hour for labor.

1. Matt is currently bidding on a job where he estimates that the materials and other expenses will amount to $475. He also has figured that it will take him 6 hours to complete the job. What bid will Matt submit for this job?
2. On another job, Matt can undertake several tasks. He figures that it will cost him $6 in materials and 2 hours labor to fix a sink. Matt also figures that he can repair some broken furniture in 4 hours with no appreciable materials cost. How much should Matt charge for this job?

Chapter 22 Technology Activity

Bidding and Estimating a Construction Project

Overview

After completing this activity, you will know how to bid and estimate a small construction project.

Goal

To bid and estimate the cost of materials for constructing a storage barn.

Equipment and Materials

pencil
notebook paper

Procedure

1. For this activity, your job is to carefully and accurately estimate the cost of the materials needed to construct the gambrel roof storage barn shown in Figs. A22-1, A22-2, and A22-3.
2. Carefully study the drawings and specifications provided for this structure. Refer to the construction specifications listed in the next column.
3. Complete a bid estimate form similar to that shown in Fig. A22-4 by filling in the appropriate spaces. The description column requires that each part of the structure be identified by name (*e.g.*, stud, joist, roof, sheathing, etc.).
4. In the size column, identify the size of one part.
5. In the quantity column, identify how many of each part will be needed for the structure.

10' x 12' Storage Barn
Construction Specifications

Floor joists: 2" x 6" x 10' treated lumber—
16" on center (O.C.)
Subflooring: 3/4" x 4' x 8' plywood or
oriented-strand board (OSB)
Studs: 2" x 4"—16" O.C.
Rafters: 2" x 4"—24" O.C.
Wall sheathing: 5/8" APA texture 1-11
(special rated siding 303)
with shiplapped edges and
4" or 8" O.C. grooves
Roof sheathing: 3/8" or 1/2" 4' x 8'plywood
or waferboard
Roofing: A.) 15# building felt
B.) 3-tab fiberglass shingles
Ridge board: 2" x 6" x 12' pine
Eaves: 2 required—2" x 6" x 12' pine
Nails: 16d box 3/4" galvanized roofing nails
8d box 1 1/2" ring shank floor nails
8d galvanized (for wall sheathing)
Drip edge required on all roof edges
Gussets for rafters: 3/8" plywood or 3" x 6"
truss fasteners
Door hinges: 6" T-hinge
Corner trim: 1" x 6" cedar
Paint: as needed

6. List the cost of one part in the unit cost column.
7. Determine the extension cost by multiplying the unit cost times the quantity required.
8. After all materials have been identified and their costs determined, add the prices in the extension column to determine the total material cost of the project.

Chapter 22 Technology Activity

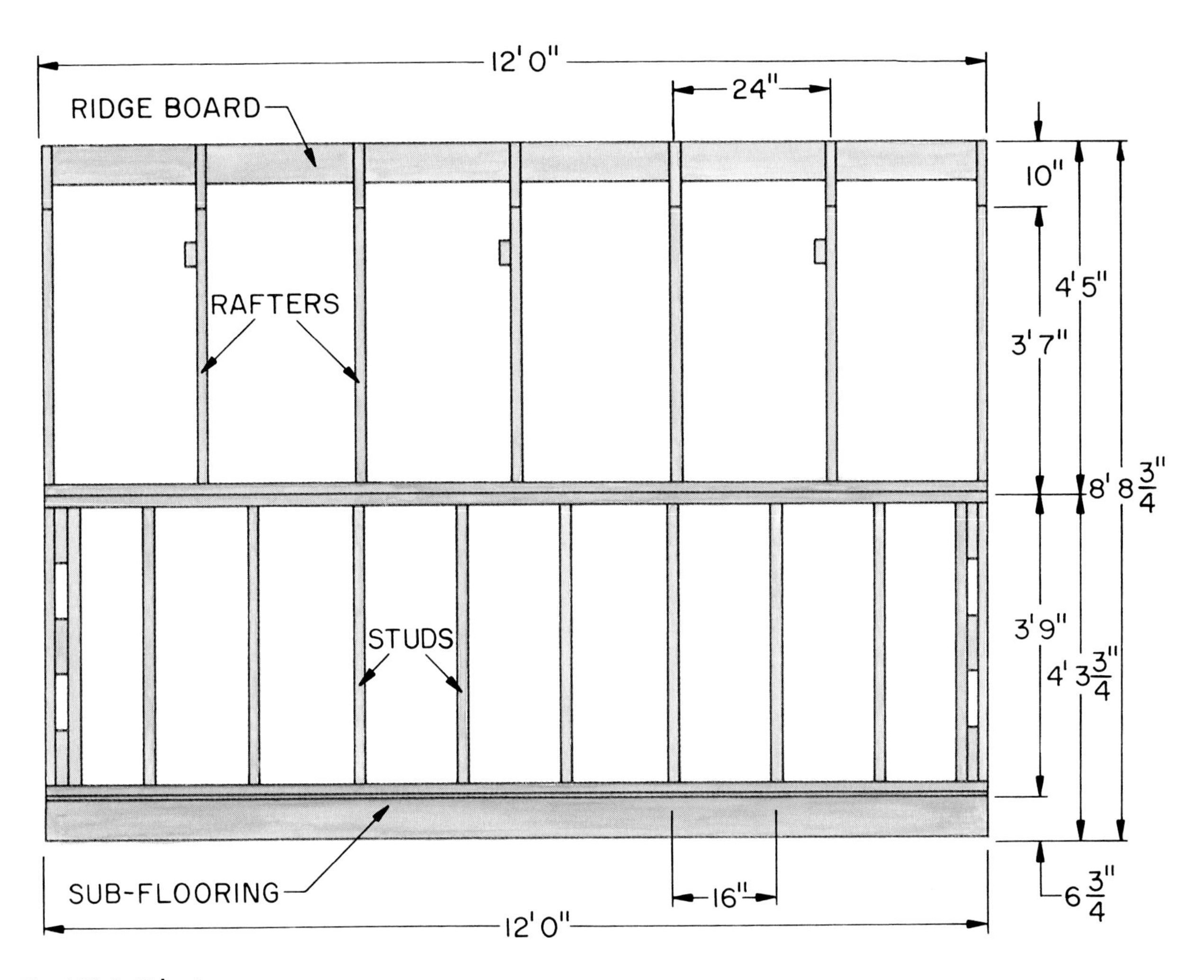

Fig. A22-1. Side view.

Chapter 22 Technology Activity

Fig. A22-2. Back view.

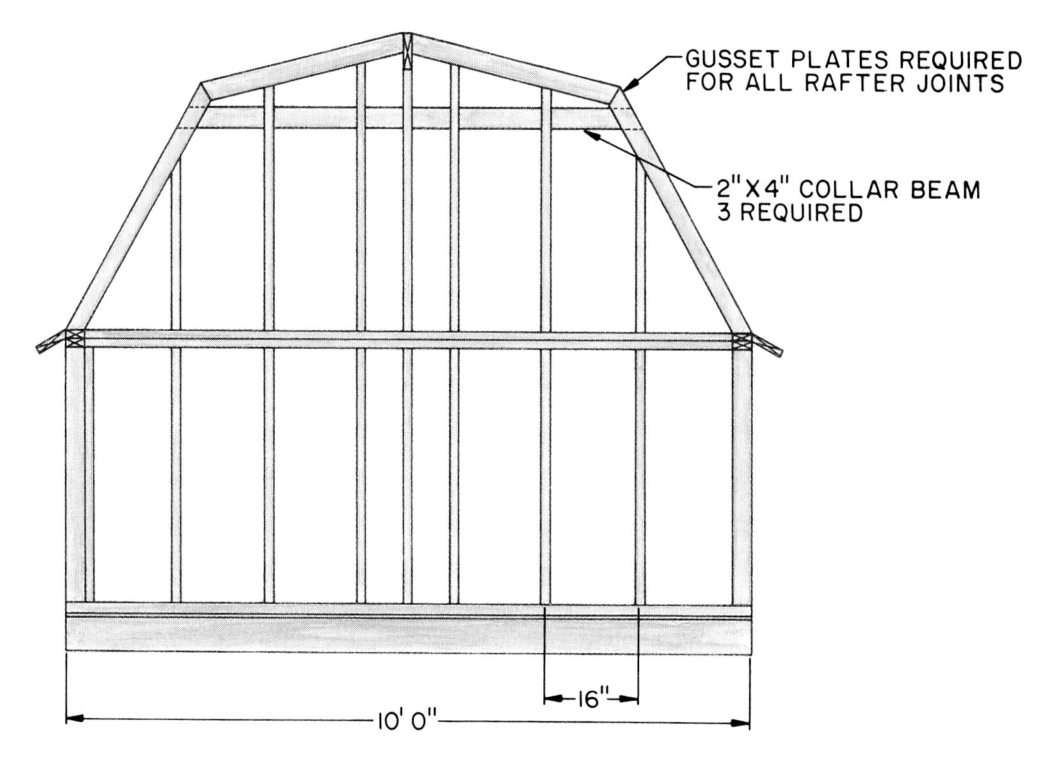

Fig. A22-3. Front view.

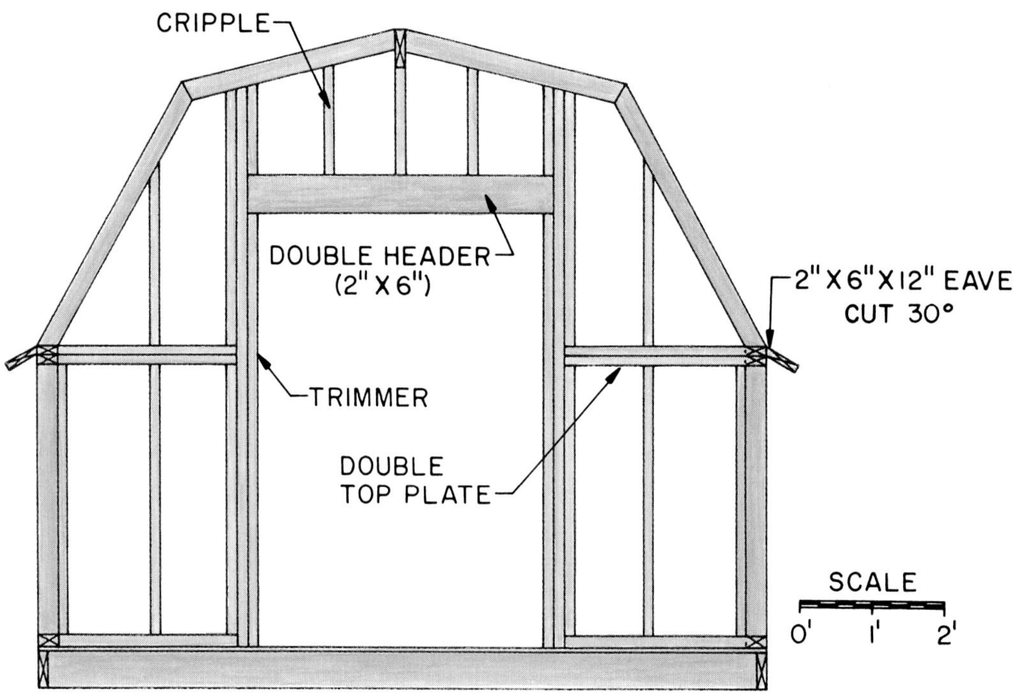

Chapter 22 Technology Activity

	Construction Bid Estimate Form				
PROJECT DESCRIPTION ______			PREPARED BY: ______		
No.	DESCRIPTION	SIZE	QUANTITY	UNIT COST	EXTENSION COST
1					
2					
3					
4					
5					
Total					

Fig. A22-4. Example of a construction bid estimate form.

Evaluation

1. How long did it take you to complete this activity? Did you come up with the same materials costs that your classmates did?
2. Did you enjoy this type of work? Would you be interested in doing this type of work for a career?

Chapter 23

Constructing Homes and Other Buildings

Looking Ahead

In this chapter, you will discover:

- what is involved in preparing a construction site for a project.
- how foundations or substructures are constructed.
- how superstructures are built.
- how interiors are finished.
- the nature and importance of finish work and landscaping.
- how the project is transferred to the owner.

New Terms

batter boards
demolition
earthmoving
excavating
footing
insulation
joists
landscaping
load-bearing wall structures
sheathing
studs
subfloor
surveying
transit
trusses

Technology Focus

NAWIC

Women are breaking down the barriers to employment in the construction industry. Traditionally, construction has been a man's world. Only 10 percent of the construction work force are women. However, their numbers are increasing. The U.S. Census Bureau predicts that 42 percent of all new entrants in the construction work force during the 1990s will be women. The construction industry will need 200,000 new workers each year. The National Association of Women in Construction (NAWIC) is preparing to assist women gain employment in this male-dominated field.

NAWIC is sponsoring training in pre-apprenticeship programs. Students get instruction in the skills necessary for working in the construction industry. They study blueprint reading, math skills, and safety. They also receive hands-on training in four building trades. The NAWIC Education Foundation also offers other courses, including a home-study course in construction management. This course can lead to becoming a Certified Construction Associate.

One of the main reasons women don't choose to enter the construction industry is that they are not aware of its many career opportunities. NAWIC has begun working with elementary schools by sponsoring a national block-building competition. Winners receive prizes of U.S. Savings Bonds ranging from $100 to $2,500 to use as a down payment towards their education.

The organization is also sponsoring Career Days to acquaint junior and high school students with the career opportunities in construction. The 9,000 members of NAWIC believe that if they can reach one out of every ten graduating high school students, then they can secure the female workers they need to maintain a quality work force.

Women who are already in other areas of the work force are also being contacted by NAWIC. The construction industry has traditionally provided higher salaries than other industries. NAWIC believes that there are many clerical workers who would thrive on the challenges and rewards of the construction industry.

Fig. TF-23. These days, more and more construction workers are women.

Preparing the Construction Site

Construction of a project begins with preparing the site. First, the site is surveyed to establish the property lines. Next, the site is cleared of anything that might interfere with construction. Then the building's position on the site is laid out.

Surveying the Site

The first step in preparing the site is to measure and mark the site in order to identify the boundaries of the property. This is called **surveying**. Surveyors use a **transit** to measure horizontal and vertical angles. They use a chain to measure horizontal distances. Once all measurements have been made and recorded, a stake is driven into the ground at each corner of the property to mark the boundaries. Fig. 23-1.

Clearing the Site

Next, the site must be cleared of anything in the way of new construction. This might include trees, old structures, rocks, and/or excess dirt.

Trees the owner wants to keep are marked. Then bulldozers can be used to remove unwanted trees.

Demolition

Demolition processes are used to rid the site of old buildings. **Demolition** involves destroying a structure by tearing it down (wrecking) or blasting it down with explosives. A crane that swings a large, steel wrecking ball against the structure may be used to demolish large structures. A bulldozer may be used to demolish small structures. Tall buildings, smokestacks, and similar structures can be demolished quickly by blasting. Fig. 23-2.

Fig. 23-1. Surveying is a very precise land-measuring process.

Fig. 23-2. The demolition of a building can be done with a crane that has a wrecking ball attached to it.

The rubble (pieces of concrete, brick, and other building materials) that remains must then be cleared away. Fig. 23-3. Most of it is usually hauled away in dump trucks. Some of it may be bulldozed to low areas of the site and used as fill.

Not everything that is removed from a construction site is destroyed. For example, signs, trees, and even whole houses can sometimes be *salvaged* (saved) and relocated. Even when a building is demolished, there is often much that can be saved. Doors, windows, lighting fixtures, and cabinets are examples of items that are often salvaged.

Fig. 23-3. Once an old structure is demolished, the rubble must be removed.

Earthmoving

In **earthmoving**, excess earth and rock are cleared away and the remaining earth is leveled and smoothed. This is done with heavy and powerful equipment, such as bulldozers and graders.

Occasionally, swampy areas or ponds must be drained or pumped dry. Then the area is filled in or covered with a deep layer of earth or sand. Also, small creeks or streams may need to be blocked or redirected.

Laying Out the Site

Laying out the site is the process of identifying and marking the exact location of the structure on the property. The site plan shows how far the building will be from the edges of the property. Using the site plan as a guide, workers take measurements from the stakes surveyors placed at each corner of the property to the corners and edges of the proposed building. They mark these new locations with stakes.

The boundaries of the proposed building are usually marked using batter boards. **Batter boards** are boards held horizontally by stakes driven into the ground. String is used to connect batter boards at opposite sides of the building. The batter boards are placed four to five feet outside the building's boundaries. The attached strings cross directly over the boundary stakes at the corners of the building. Fig. 23-4. The corner stakes can then be removed, and the batter boards are used to guide excavation.

Apply Your Thinking Skills

1. Why must survey work be accurate?

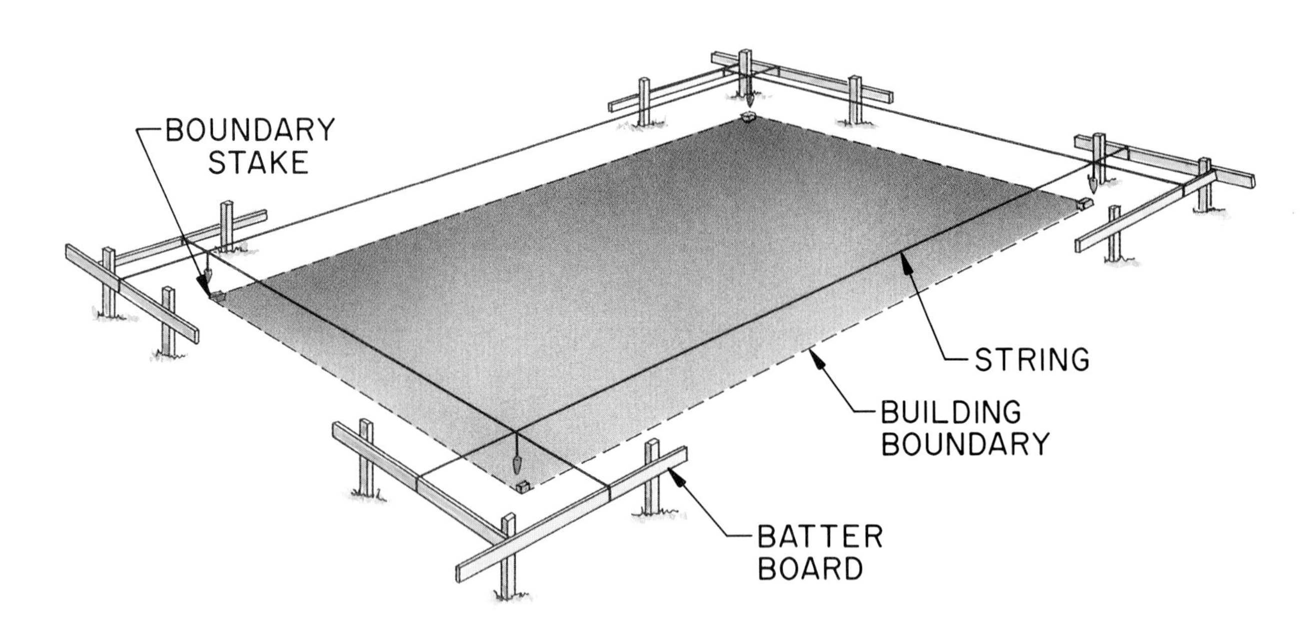

Fig. 23-4. Batter boards and string are used to lay out the building's location. They are also used to guide excavation.

The Foundation

As you learned in Chapter 20, the foundation is designed to support the superstructure.

Excavating for the Foundation

Excavating, or digging, for the foundation is done by heavy equipment such as backhoes, front-end loaders, and trenchers. Fig. 23-5. The size, shape, and depth of the excavation depends upon the design of the building. For example, a ranch-style house may be designed to rest on a simple concrete slab. It will require a wide but shallow area below the surface of the ground. On the other hand, a two-story house with a full basement will require a wide but deep opening. The excavation for a skyscraper may extend straight down, deep into the ground.

Fascinating Facts

Biltmore, the dream home of George Washington Vanderbilt II, was built during the years 1891-95 on a 125,000-acre estate in the Blue Ridge Mountains of North Carolina. Designed by architect Richard Morris Hunt, the 250-room mansion became the largest private house ever built in the United States.

Vanderbilt himself was involved in the design, helped lumberjacks clear the site, and labored along with the workers in the construction of his mansion.

The soil may be checked by an engineer to make sure it will be able to support the structure. If the soil is soft or very loose, it may have to be *compacted* (packed down to make it firm). Soil can also be made firm sometimes by adding certain chemicals to it.

Fig. 23-5. Heavy equipment like this backhoe is used to excavate for the foundation.

Parts of the Foundation

The two important parts of the foundation are the footings and the walls.

The **footing** is the part of the structure below the foundation wall that *distributes* the structure's weight. Fig. 23-6. The footing is usually twice as wide as the foundation wall so it can distribute the weight over a wider area. Footings are made of reinforced concrete.

The foundation walls *transmit* the weight of the superstructure to the footing. Foundation walls may be made from concrete blocks or from poured concrete that has been reinforced with steel. In buildings with basements, the foundation walls become the basement walls. Refer to Fig. 23-6.

Apply Your Thinking Skills

1. Why is concrete a good building material for foundations?

Building the Superstructure

Once the foundation is finished, work can begin on the superstructure.

Types of Superstructures

The superstructure of a building is usually one of two types: framed or load-bearing wall structures.

Framed Structures

A framed building has a main "skeleton" or framework that supports the weight of the building. The framework consists of various structural members and other supports fastened together. These give the building its particular shape. The members may be concrete, steel, or wood.

Most houses have a wood-frame structure. The frames of the walls, floors, and roofs are made from the following framing members:

- **Studs**. These are evenly spaced vertical boards that form the frame of exterior and interior walls. Fig. 23-7.

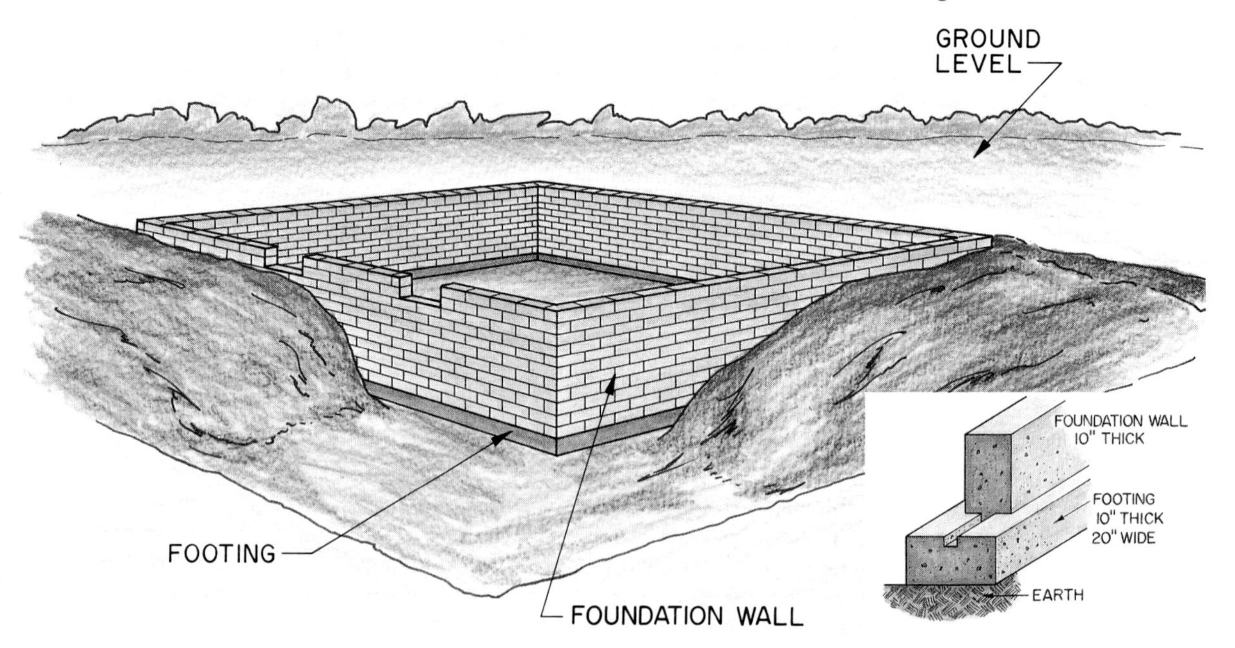

Fig. 23-6. The main parts of the foundation are the footings and the foundation walls.

Fig. 23-7. ***Studs*** **of 2" x 4" or 2" x 6" lumber are used to frame the walls.** ***Trusses*** **(preassembled triangular frames) are used to frame the roof.**

- **Joists**. These are evenly spaced horizontal boards that form the frames that support floors and ceilings. Fig. 23-8 (p. 484).
- **Trusses**. These are preassembled triangular frames that are used to frame the roof. See Fig. 23-7 again.

These members may be nailed, screwed, or bolted together.

A steel frame is used for large industrial or commercial structures, such as office buildings. Fig. 23-9. The steel for these frames is prepared in a fabricating shop. Then it is delivered to the site. Ironworkers assemble and erect the steel framing members on the site according to the building plan. The steel parts are bolted, riveted, or welded together to make a rigid frame.

Load-Bearing Wall Structures

In **load-bearing wall structures**, the heavy walls support the weight of the structure. There is no frame. Bearing walls are usually made of concrete blocks, poured concrete, or precast concrete panels. Fig. 23-10. Bearing-wall construction is best suited for low buildings of one or two stories.

Enclosing the Superstructure

The wall and roof frames of most frame buildings are first enclosed with layers of sheathing. **Sheathing** is a layer of material, such as plywood or insulating board, that is placed between the framing and the finished exterior. Doors and windows are installed. Decorative finish materials such as wood paneling, vinyl, aluminum or wood siding, stone, or brick are placed over the wall sheathing. Fig. 23-11. The sheathing over the roof frame is first covered with roofing felt. Then roofing materials such as asphalt or wood shingles or tiles are applied.

Installing Floors

A **subfloor**, consisting of sheets of plywood, is nailed or glued to the floor joists. See Fig. 23-8 again (p. 484). The

Fig. 23-8. Parallel horizontal framing members called *joists* are used to support the floor. This worker is laying a plywood subfloor over the joists.

Fig. 23-9. A steel frame is used for most large buildings.

Fig. 23-10. For poured-concrete walls, concrete is pumped through hoses and poured into forms. The forms are the molds that hold the concrete until it hardens.

Fig. 23-11. Finish siding is applied over sheathing.

subfloor serves as a base for the finish flooring. It also provides a surface for workers to walk on when completing other parts of the building. An additional layer of plywood, called *underlayment*, is nailed or glued to the subfloor before the finish floor is applied.

Installing Utilities

Utilities is a term used to refer to various service systems in a building. They include:

- Electrical systems
- Plumbing systems
- Heating, ventilating, and air conditioning (HVAC) systems

Utilities are installed in two stages. First they are *roughed in*. Parts such as wires, pipes, and ductwork are placed within the walls, floors, and ceilings before the interior surfaces are enclosed. Ductwork for HVAC is usually installed first, then piping, and finally, wiring. Fig. 23-12 (p. 486).

After the interior walls are enclosed, the utilities are *finished*. Such things as light switches, plumbing fixtures, and temperature controls are installed.

Installing Insulation

Insulation is material that is applied to walls and ceilings to help keep heat from penetrating the building in the summer and cold from penetrating in the winter. This helps make the house more energy-efficient. Insulation comes in various forms (Fig. 23-13):

- Batts or blankets. These are thick fiberglass sheets or rolls, with a paper or foil backing, designed to fit snugly between framing members.
- Rigid panels. These are large sheets of plastic foam or natural fibers.
- Loose fill. Loose fill is fibrous or granular material that is blown into place using a special hose.

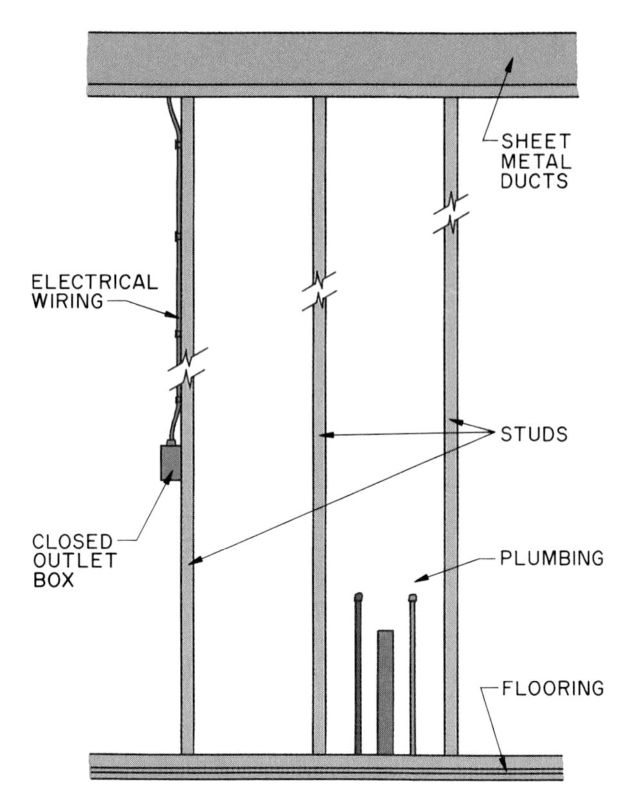

Fig. 23-12. Utilities must be roughed in before interior walls can be finished.

Finishing the Interior

After the utilities are roughed in and the insulation is installed, the interior is ready to be finished.

Ceilings and Walls

Most walls and ceilings are first enclosed with drywall. *Drywall* is a general term used for plasterboard or wallboard. It is a heavy, rigid sheet material. Drywall is nailed directly to the wall studs and ceiling joists. It may also be fastened with adhesives. Holes must be cut in the drywall for electrical outlets, light switches, and lighting fixtures.

A filler is placed into the dents or spaces or seams between the drywall panels. Fig. 23-14. These areas are then taped. Another layer of filler is applied. When the filler is dry, the surfaces are sanded smooth. Now the walls and ceilings can be finished with wallpaper or paint. Paneling and tile are also popular wall finishes.

Floor Coverings

Finish flooring is usually installed over the underlayment after the walls and ceilings have been finished. This is done to avoid

Fig. 23-13. Various types of insulation can be used to help keep a building warmer in winter and cooler in summer.

Fig. 23-14. This worker is filling in the spaces between the drywall sheets.

damage to the flooring. Wall-to-wall carpeting, sheet vinyl, and vinyl or asphalt tiles are commonly used for floor covering. Wood, ceramic tile, and flagstone are also popular. Materials are installed in different ways. Special fasteners or adhesives may be required.

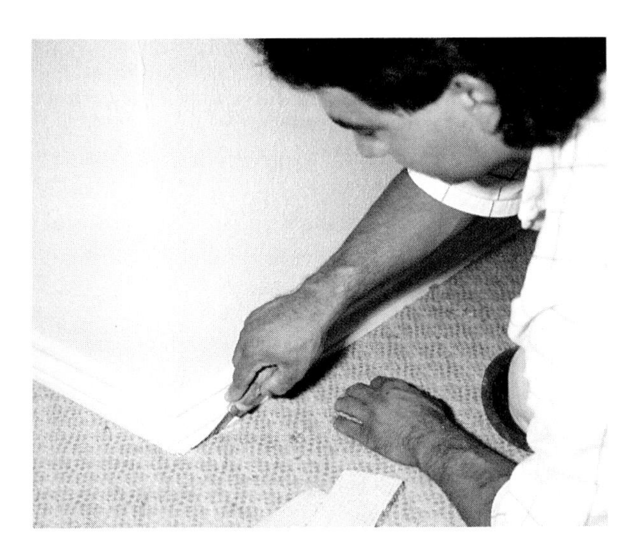

Fig. 23-15. Trim is used to cover joints where ceilings, walls, and floors meet. It gives the room a finished appearance.

Fig. 23-16. Installing accessories, such as this light fixture, is the last step in finishing the inside of a structure.

Trim and Other Finish Work

Trim is the woodwork, baseboards, and moldings used to cover edges and the joints where the ceilings, walls, and floors meet. Trim is also applied around doors and windows. Most trim used in homes is wood trim. Plastic and metal trims are also available. Fig. 23-15.

Room doors and closet doors are put in place next. Then hardware is installed. Hardware includes items such as doorknobs and towel bars.

Installation of *accessories* is the last indoor finishing task. Kitchen and bathroom cabinets are major accessories. Plumbing and electrical fixtures are also major accessories. Fig. 23-16. Shelving, countertops, and other built-ins are also installed as part of finish work.

Apply Your Thinking Skills

1. Explain why steel framing is usually used in the construction of tall buildings.
2. Explain what you would see if you took a cross-section of a floor.

Post-Construction

Once construction is complete, the site can be finished. After all this is done, ownership of the completed project can be transferred to the client.

Finishing the Site

The site as well as the structure must be finished. The two major types of outdoor finishing that need to be done are paving and landscaping.

Paving

Several areas around a home or other building need to be paved. Driveways, walkways, parking lots, and patios are examples. These are usually made of concrete. Their locations, sizes, and shapes are indicated on the working drawings. The builder marks or stakes out these areas on the ground. Forms that will hold the concrete while it cures may be set up. Then the earth is carefully leveled and compacted. Finally, the concrete is poured and finished.

Landscaping

Landscaping begins after all the debris from construction has been removed. **Landscaping** includes changing the natural features of a site to make it more attractive. It includes shaping and smoothing the earth with earthmoving equipment as well as planting trees, shrubs, grass, and other vegetation. Landscaping is done according to a landscaping plan. This plan shows how the finished site should look. Fig. 23-17.

Transferring the Project

The last step in the construction process is formally transferring the completed project to the owner. After construction is complete, the project is given a final inspection. This is done to make certain that the terms of the original contract and specifications have been fulfilled. The quality of the work must also be acceptable.

A number of important legal matters must be attended to. All outstanding bills for the construction of the project must be paid by the contractor. There must be no claims of money owed against the property.

After all requirements have been satisfied, the final payment is made to the contractor. Then the keys to the building are given to the new owner. The new owner now takes over full responsibility for the property.

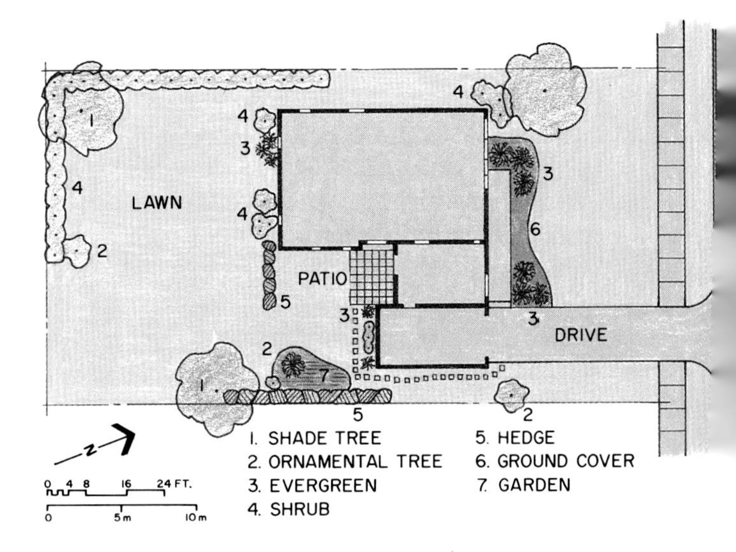

Fig. 23-17. Landscape plans are prepared by landscape architects or contractors.

Apply Your Thinking Skills

1. Think about the landscaping that surrounds your home. Then discuss how you might landscape the area if you were the landscape architect.

Global Perspective

Whole Lotta Shakin . . .

Earthquake—a fearful phenomenon, even terrifying! The earth, the one thing we consider most stable and dependable, shakes beneath our feet.

Earthquakes occur in countries around the world. In recent times, the United States, Armenia, Mexico, Iran, Chile, and the Philippines have all experienced disastrous earthquakes.

What causes earthquakes? The earth's crust is broken into pieces called plates. These plates "float"on the molten interior of the earth. When movement of a crustal plate occurs, there is an earthquake.

During an earthquake, constructed products receive the worst damage. It is the breaking and collapsing of structures that causes the greatest loss of life. Building design, construction techniques, and construction materials can be used in ways that can help make buildings resistant to earthquake damage.

To keep damage to a building at a minimum during an earthquake, the design should be simple, box-like, and symmetrical (balanced). Construction techniques should bind the members of the building together. Components not interwoven into the whole structure (such as stairs) could break away and fall during an earthquake. Columns and walls should be simply made and extend from the foundation to the roof. Buildings should be relatively flexible. Materials such as wood or steel should be used instead of the more brittle concrete. Brick buildings also tend to split apart.

Several earthquake protection systems are being developed and used in construction. One important type is called "base isolation." In this type of system, shock absorbers (such as rubber bearings or springs) are placed between the foundation and the building itself to absorb the movement of the ground.

Another type of system involves active control. Computer-controlled weights or cables are moved to adjust the position of the building to earth movements.

Earthquake-resistant measures do add to the cost of buildings. However, they may well be worth the price when the earth begins to shake.

Extend Your Knowledge

1. Designing structures for safety is important. Safety on construction sites is also a major concern. Think about the activities that go on during the construction of a building. Make a list of safety rules workers should follow to avoid accidents and injuries.

Fig. GP-23. Constructed products receive the worst damage during an earthquake.

Chapter 23 Review

Looking Back

The first step in constructing a building is preparing the site. This involves surveying the site, clearing away unwanted objects, leveling and smoothing the site, and laying out the site.

Next, the excavation is done for the foundation, which supports the superstructure. The footings and walls of the foundation are then constructed.

The superstructure of a building is the part above the ground. Superstructures may have wood or steel frames or may be load-bearing-wall structures. Once the structure is framed, it is enclosed, the subfloor is laid, the utilities are roughed in, and insulation is installed. After this is all completed, the interior can be finished.

The last step in completing the project is finishing the site. This involves paving and landscaping. When everything is done, the project is transferred to the owner.

Review Questions

1. What is surveying?
2. Describe two demolition techniques.
3. What does it mean to "lay out the site"?
4. Name and describe the two main parts of the foundation. Tell the function of each.
5. Name and describe the three types of frame members used in a wood-frame house.
6. Describe a load-bearing wall structure.
7. What is sheathing?
8. What are the three types of utilities discussed in the text? Briefly describe the two stages in which utilities are installed.
9. What is the purpose of insulation? Name and describe two types of insulation.
10. Briefly describe what is done during landscaping.

Discussion Questions

1. What might be some disadvantages of a wood frame for a house?
2. Discuss why load-bearing wall structures aren't used for high-rise buildings.
3. Sometimes people who are having a new home built want to do some of the work themselves. Doing this saves them money and gives them a sense of accomplishment and satisfaction. If you were having a new home built, what jobs would you do?
4. What are some of the important things a landscape architect needs to consider when developing a landscape plan?
5. Suppose you were going to redecorate your room. What would it look like? How would you select the materials needed to do the job?

Chapter 23 Review

Cross-Curricular Activities

Language Arts 

1. In an essay, explain what elements must be considered when preparing a site for construction.
2. Outline all the steps needed to construct a new home in your community. Phrases may be used in place of complete sentences.

Social Studies 

1. Do research on a recent earthquake. Find out about the amount of damage it did to structures, including highways and bridges. How many people were injured and how many were killed? Had any measures been taken to prevent earthquake damage?

Science 

1. The insulation of homes and other structures is necessary because heat moves through materials. Some materials, such as metals and glass, are good *conductors* (transmitters) of heat. (Heat passes through them easily.) Others, like wood and air, are poor conductors. You can see this for yourself. Put your hand on various surfaces at room temperature: metal, paper, plastic, glass. Some will feel cool because they conduct the heat of your hand quickly away.
2. The Leaning Tower of Pisa is in danger because its center of gravity is too far from its point of support. (The center of gravity is the point in any object where its weight is concentrated.) Stand with your back against a wall. Keeping your heels against the wall, try to touch the floor. You will be unable to do this because your center of gravity moves too far from your point of support. Your position becomes unstable. Like the Leaning Tower, you are in danger of falling over.

Math X÷

Slope is an angular measurement used to indicate the steepness of surfaces such as a roof, a hillside, or another type of incline. The formula that is used to calculate slope is RISE ÷ RUN. Rise is the height of the surface, and run is the horizontal length.

1. A roof has a rise of 4" and a run of 12". What is its slope?
2. If another roof has a run of 12" and a slope of 1/2 (6/12), what is its rise?

Chapter 23 Technology Activity

Load-Bearing Exterior Wall Construction

Overview

The exterior walls of the superstructure of a building are designed and constructed to protect the interior portions from the elements (sun, wind, rain, and snow). Often, they are also designed to support the weight of the building, including the floor, interior walls, and roof. Walls that support the weight of the superstructure are called load-bearing walls.

In this activity, you will build a model superstructure of a building. It will be similar to the defense towers built in southern Europe during the eleventh and twelfth centuries. These defense towers were constructed of natural materials that were readily available. For example, some were made of stone. Others were made of bricks made from mud. The tower walls had to be thick to support their own heavy weight.

Goal

To design and build a model tower in five days.

Equipment and Materials

graph paper
hammer
string
table knives and spoons (to be used as trowels)
8 oz. measuring cup
32 oz. (or larger) cans for mixing mortar
mortar boards—8" x 8" galvanized sheet metal
handsaw
stone or gravel, 3/4" to 1 1/2" dia.
masonry cement
masonry sand
wood strips for floor joists and roof planks, 1/8" x 1/2" x length required
1/4" wood dowels for roof rafters
one plywood (AC) board, 1/2" x 8" x 12" for each tower building site
5 common nails, 4d, for each tower
wire brads, 1 1/2" x 18 (or other suitable size)
wood glue
paint and other finish materials

Procedure

1. *Safety Note:* Before doing this activity, make sure you understand how to use the tools and materials safely. Have your teacher demonstrate their proper use. Follow all safety rules.
2. Ask your school's media resource person if information is available about the defense towers built in southern Europe during the eleventh and twelfth centuries. If so, make one or more sketches of such towers.
3. Create your own tower design. Sketch three or four tower designs on 1/4" graph paper. Do not make the base of your tower larger than 6" x 6". Include an interior space in your design.

Day One

4. Lay out the perimeter (outer limits) of your tower base on the 1/2" x 8" x 12" building site board. Place strings as shown in Fig. A23-1.
5. Lay out your tower base within the marked perimeter. Remove the strings.

Chapter 23 Technology Activity

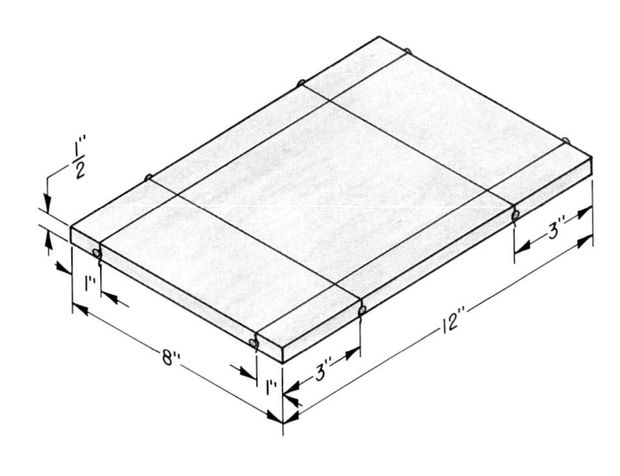

5 - 4d NAILS

TOP VIEW

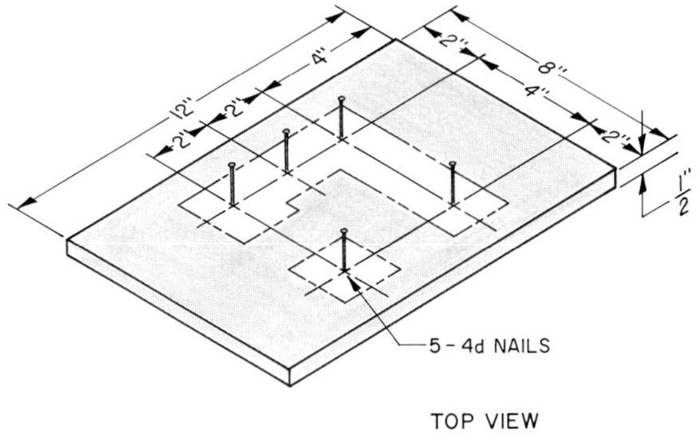

Fig. A23-2. Structure anchors.

Fig. A23-1. Locating the structure.

6. Drive the 4d nails into the building site board as shown in Fig. A23-2. (Your tower design might require additional 4d nails, but do not use too many.) Drive the nails in about 3/8".
7. Select stones and gravel to be used on your tower. Place them at your work station.
8. Mix the mortar in a can. Use the following ratio: 1 part masonry cement to 3 parts masonry sand. (A "part" can be any amount, but this amount must be the same for each ingredient. Your teacher will tell you how much to use.) Add enough water to make the mortar workable. Stir the mixture as you add the water. Do not add too much water!
9. Refer to Fig. A23-3 throughout the construction process. Place the mortar mixture to a height of 4" on your building site board. Do not allow the mortar to slump (slide) outside the base lines that you laid out earlier. Place a door on the back side of your tower. Place a window on one of the other sides. Add some stones or gravel to the outside of the walls.
10. Cut and evenly space some 1/8" x 1/2" floor joists in the wet mortar.
11. Clean the container in which you mixed the mortar. Clean your work area.

Day Two

12. Mix a new batch of mortar.
13. Gradually build the second floor walls up another 4". Reduce the thickness of the walls by 1/2" as you build them up. Put a window in one wall.
14. Cut and evenly space some 1/8" x 1/2" floor joists.
15. Clean the container in which you mixed the mortar. Clean your work area.

Chapter 23 Technology Activity

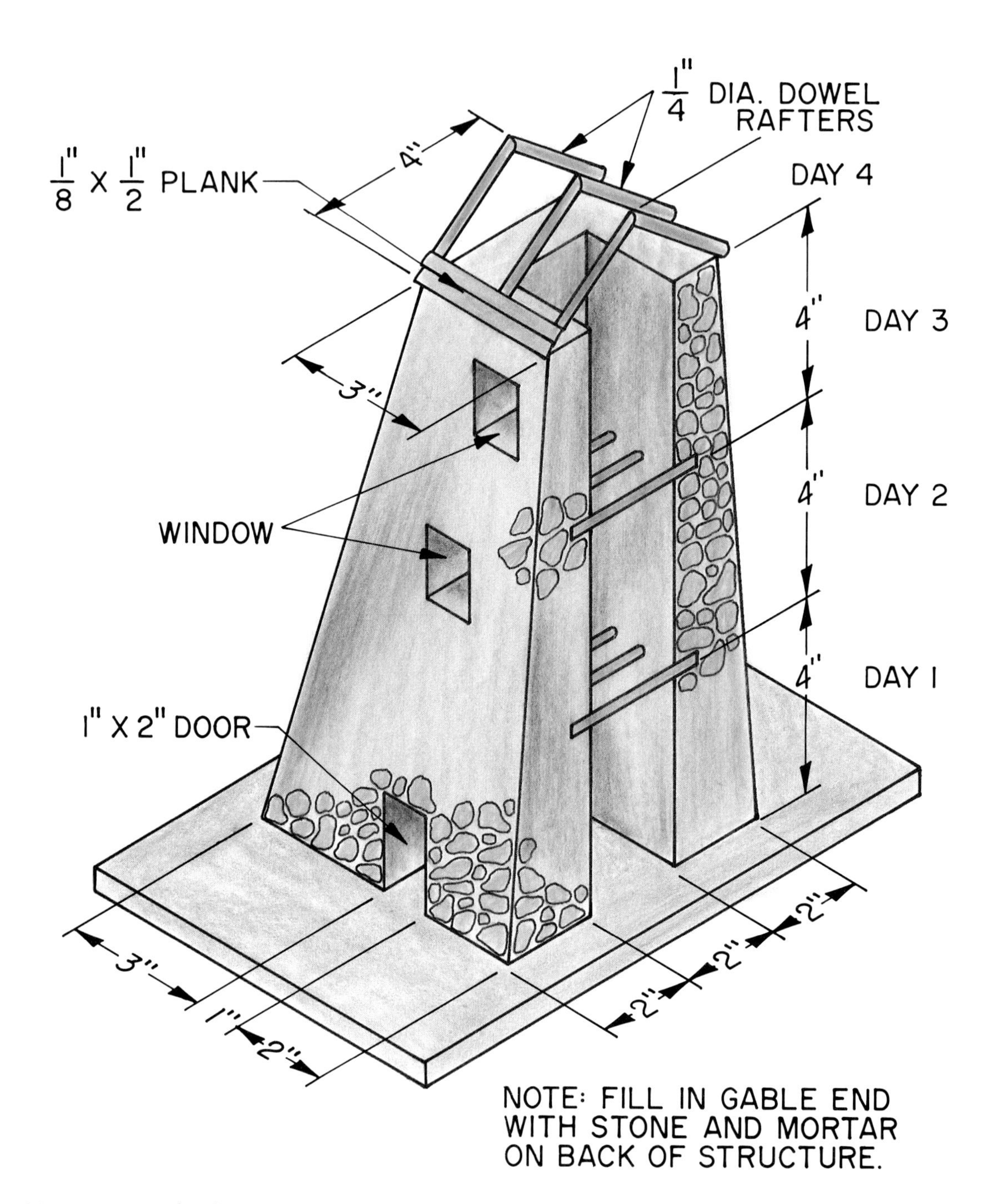

Fig. A23-3. Structure details.

Chapter 23 Technology Activity

Day Three

16. Mix a new batch of mortar.
17. Build the third floor the same way you built the second floor.
18. Be sure to reduce the wall thickness another 1/2" as you build it 4" higher. Add another window on one side of the tower.
19. Smooth the top of the tower with a trowel (knife).
20. Clean the container in which you mixed the mortar. Clean your work area.

Day Four

21. Mix a new batch of mortar.
22. Enclose the gable end of the tower with stone and mortar.
23. Cut and place the 1/4" dowel rafters. Set the ends of the rafters in mortar and glue them at the ridge.
24. Cut and glue the 1/8" x 1/2" roof planks on the rafters.
25. Clean the container in which you mixed the mortar. Clean your work area.

Day Five

26. Paint and finish the structure for display.

Evaluation

1. What problems, if any, did you encounter while building your tower?
2. How well did your tower turn out in comparison with your classmates' towers?
3. If you could design and build another tower, what would you do differently?

Credit:

Developed by Richard Henak for the Center for Implementing Technology Education
Ball State University
Muncie, Indiana 47306

Chapter 24

Other Construction Projects

Looking Ahead

In this chapter, you will discover:

- how highways are constructed.
- the types of bridges and how they are constructed.
- how dams are constructed.
- how canals are used and constructed.
- how tunnels are constructed.
- how pipelines are constructed.
- how air-supported structures are built.

New Terms

abutments
arch bridge
beam bridge
cable-stayed bridge
cantilever bridge
cofferdams
lock
movable bridge
reservoir
suspension bridge
truss bridge

Technology Focus

Hoover Dam—a Hydroelectric Giant

Dams are structures built across rivers to control or block the flow of water. Hoover Dam stands in the Black Canyon of the Colorado River, between the states of Arizona and Nevada. It was originally named Boulder Dam for nearby Boulder City, Nevada. However, it was renamed in honor of Herbert Hoover, who was President of the United States when the dam was under construction.

Hoover Dam is one of the tallest (or highest) dams in the world. It is also one of the largest concrete structures ever built. The dam is 726 feet tall, 1244 feet wide, 45 feet thick at the top, and 660 feet thick at the bottom. It contains more than four million (4,400,000) cubic yards of concrete. That's enough concrete to pave a standard highway (16 feet wide) from New York City to San Francisco, California—a distance of 2930 miles.

The dam was designed primarily to generate electricity and to provide a source of drinking and irrigation water. It has been an extraordinary success. The power plant that is located at the base of the dam has a capacity of over one million (1,249,800) kilowatts. (Remember, hydroelectric energy is produced by the force of water falling from a dam onto a turbine that drives an electric generator.) Much of the electricity is carried by power lines to Los Angeles, California, 250 miles away.

Lake Mead is the *reservoir* (lake) that was created behind the dam when the flow of the river was blocked. It is approximately 115 miles long and 589 feet deep. This huge lake contains more than ten trillion (10,000,000,000,000) gallons of water. It is one of the largest artificial lakes in the world.

Water from the reservoir is used to irrigate more than one million (1,000,000) acres of farmland in Arizona, Nevada, and California. Irrigation has more than doubled agricultural production. The value of the crops produced runs into hundreds of millions of dollars each year.

The reservoir also provides many millions of gallons of drinking water for a number of cities in southern California. The water travels through a 240-mile-long aqueduct, itself a large-scale construction project.

Hoover Dam is truly a most extraordinary engineering achievement. This hydroelectric giant greatly affects the area and the people around it.

Fig. TF-24. The Hoover Dam, which is one of the world's tallest dams, generates large amounts of electricity and provides a source of drinking and irrigation water.

Other Types of Construction Projects

Construction involves many projects besides buildings. You see at least some of these every day. In this chapter, we'll take a look at some major types of construction projects that are not buildings. These include:

- Highways
- Bridges
- Dams
- Canals
- Tunnels
- Pipelines
- Air-supported structures

Apply Your Thinking Skills

1. What are some structures in your community or a nearby community that fit into the above categories? Can you think of any other structures that do not fit in any of these categories?

Highway Construction

Highway construction includes the building of any type of highway, street, or road. Before actual construction begins, the land must be thoroughly surveyed to determine the best route for the road. After this is determined, a surveying team marks the edges and centerline of the planned highway with stakes.

Next, the soil must be prepared for construction. Trees, roots, and rocks must be removed. Fig. 24-1. The ground must be leveled and graded. Earth must be removed from areas that are too high, and low spots must be filled in. Then the soil must be compacted (firmly packed) to prepare it for paving.

Next the roadbed is prepared. The two major types of roadbed are flexible and rigid. A flexible roadbed has a thick gravel subbase. The subbase spreads the load on the highway into the soil below the highway. This is usually covered with a concrete base. The top layer is an asphalt mixture. Fig. 24-2.

Fig. 24-1. This bulldozer is preparing the ground so a new highway can be constructed. Building a road requires extensive earthmoving.

Fig. 24-2. A paver spreads the paving material and forms the pavement at the same time.

A rigid roadbed is made by placing steel bars on a sand base, then pouring concrete over the bars. The resulting steel-reinforced concrete slab can spread the weight of traffic over a large area.

To complete the road, lights, pavement markings, signs, and possibly traffic signals are added.

Apply Your Thinking Skills

1. What type of roadbed, rigid or flexible, lies beneath the road in front of your school? In front of your home?

Bridges

Bridges are structures built to allow people and vehicles to pass over obstacles. An obstacle is something that is in the way. When we think of bridges, we most often think of those built over water. However, bridges may also extend over valleys, highways, or railroad tracks.

A bridge *span* is the distance between supports. The word "span" is also used to mean "extend across," such as "the bridge spans the Spoon River."

Substructure of Bridges

The substructure of a bridge consists of:

- Abutments
- Piers
- Piles (sometimes)

See Fig. 24-3 (p. 500).

Abutments are the supports at the ends of a bridge. They not only support the bridge itself, but also the earth at the ends of the bridge. Abutments are usually made of reinforced concrete. They are sometimes referred to as end piers.

Piers are vertical structural supports. They are used when more than one span is needed. They are positioned to keep the bridge from sagging.

Supports for a bridge must rest on a solid surface. When the earth under the bridge is not solid, piles are used. *Piles* are wood,

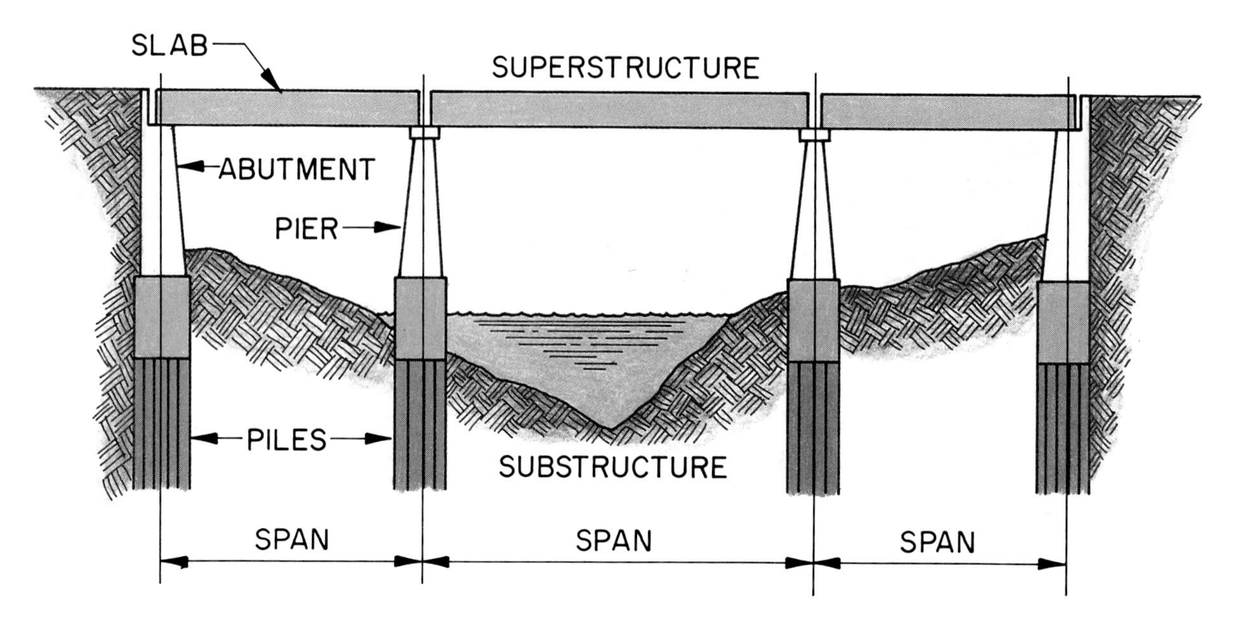

Fig. 24-3. The parts of a bridge structure.

metal, or concrete members. They are driven down into the earth to a solid base. Piers and abutments are placed on top of them.

Types of Bridge Superstructures

The type of bridge constructed depends on how long the bridge must be and on the weight it must support. There are seven common types of bridges (Fig. 24-4):

- **Beam bridge**. Piers (beams) support spans of concrete slabs reinforced with steel girders.
- **Arch bridge**. The load of the bridge is transferred to the arch, which is supported on each end by an abutment. Single or multiple arches may be used.
- **Truss bridge**. As you learned in Chapter 23, a truss is a triangular framework. Trusses may be used above or below the roadway to support the bridge. Trusses are also used in combination with other bridges, such as suspension bridges, to give them additional support.
- **Cantilever bridge**. Beams called cantilevers extend from the ends of the bridge. They are connected by a section called a suspended span. To remember this kind of bridge, think of two diving boards at opposite sides of a pool being connected by placing a board across the top of them.
- **Suspension bridge**. These have two tall towers that support main cables that run the entire length of the bridge. The cables are secured by heavy concrete anchorages at each end. Suspender cables dropped from the main cables are attached to the roadway. These bridges are used to span long distances. Probably the most famous suspension bridge is the Golden Gate Bridge in San Francisco.
- **Cable-stayed bridge**. These are similar to suspension bridges except the cables are connected directly to the roadway. To date, most cable-stayed bridges have been built outside the U.S. This is because some of our engineers are concerned about the strength and durability of these bridges.

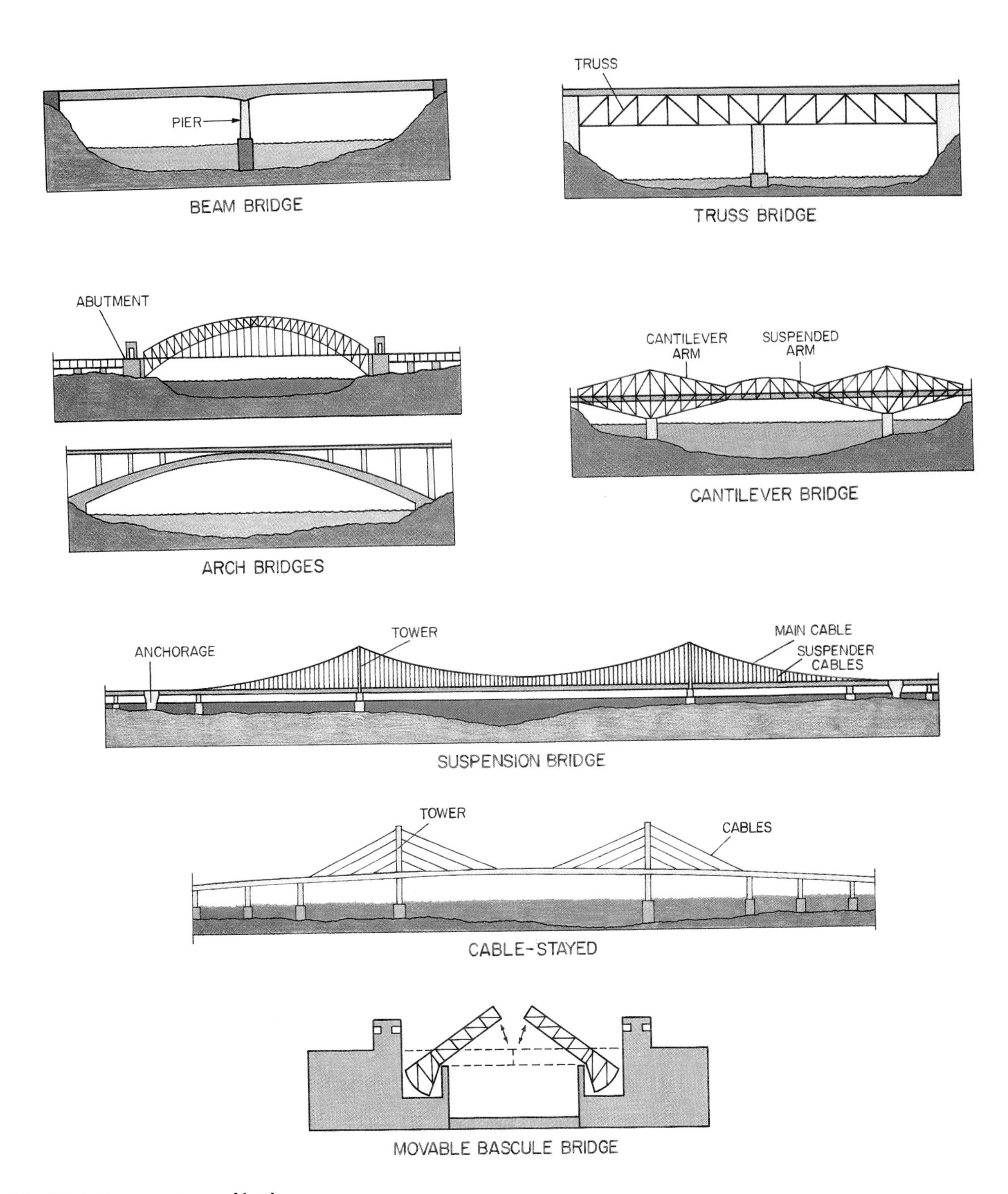

Fig. 24-4. Common types of bridges.

- **Movable bridges.** These bridges are designed so that a portion of the roadway can be moved to allow large water vessels to pass underneath. Bascule bridges open by tilting upward. Lift bridges have a section of roadway that moves up between towers. Swing bridges have a section that moves sideways.

Apply Your Thinking Skills

1. Think of at least three bridges in your community or a nearby community. What kind of superstructure does each of these bridges have?

Dams

A dam is a structure that is placed across a river to control or block the flow of water. The water that collects behind the dam creates a reservoir. A **reservoir** is a lake in which water is stored for use. That is one of the main reasons for building a dam—to provide a dependable water supply for nearby communities. As a bonus, the lake can also be used for recreation. The other main reason for constructing a dam is to collect water to power the water turbines in a hydroelectric power station at the base of the dam. Fig. 24-5.

Dams may be made of earth, concrete, steel, masonry, or wood. Usually a combination of materials is used.

The three main parts of a dam are its:

- Embankment
- Outlet works
- Spillway

The *embankment* blocks the flow of water. *Outlet works* are used to control the flow of water through or around the dam. When water is needed downstream, gates of the outlet works are opened to allow water to flow through. A power plant may be part of the outlet works. The *spillway* acts as a safety valve that allows excess water to bypass the dam when the reservoir becomes too full due to flooding. If water could not bypass the dam, the dam would break.

Constructing a large dam is a complicated process. Temporary watertight walls, called **cofferdams**, must be built to keep the construction site dry. As work progresses, another cofferdam is built farther out in the river, and the first cofferdam is removed. The dam is built in carefully planned stages. Fig. 24-6.

Fascinating Facts

Louisiana's Lake Pontchartrain Causeway became the world's longest bridge when it opened in 1956. Extending about 29 miles—nearly 24 of them over water—it connects New Orleans with Mandeville to the north. A twin span, which opened in 1969, is 228 feet longer than the original one.

Fig. 24-5. A dam's reservoir provides a dependable water supply for communities. It can also be used for recreation and to supply power to water turbines in a hydroelectric power station.

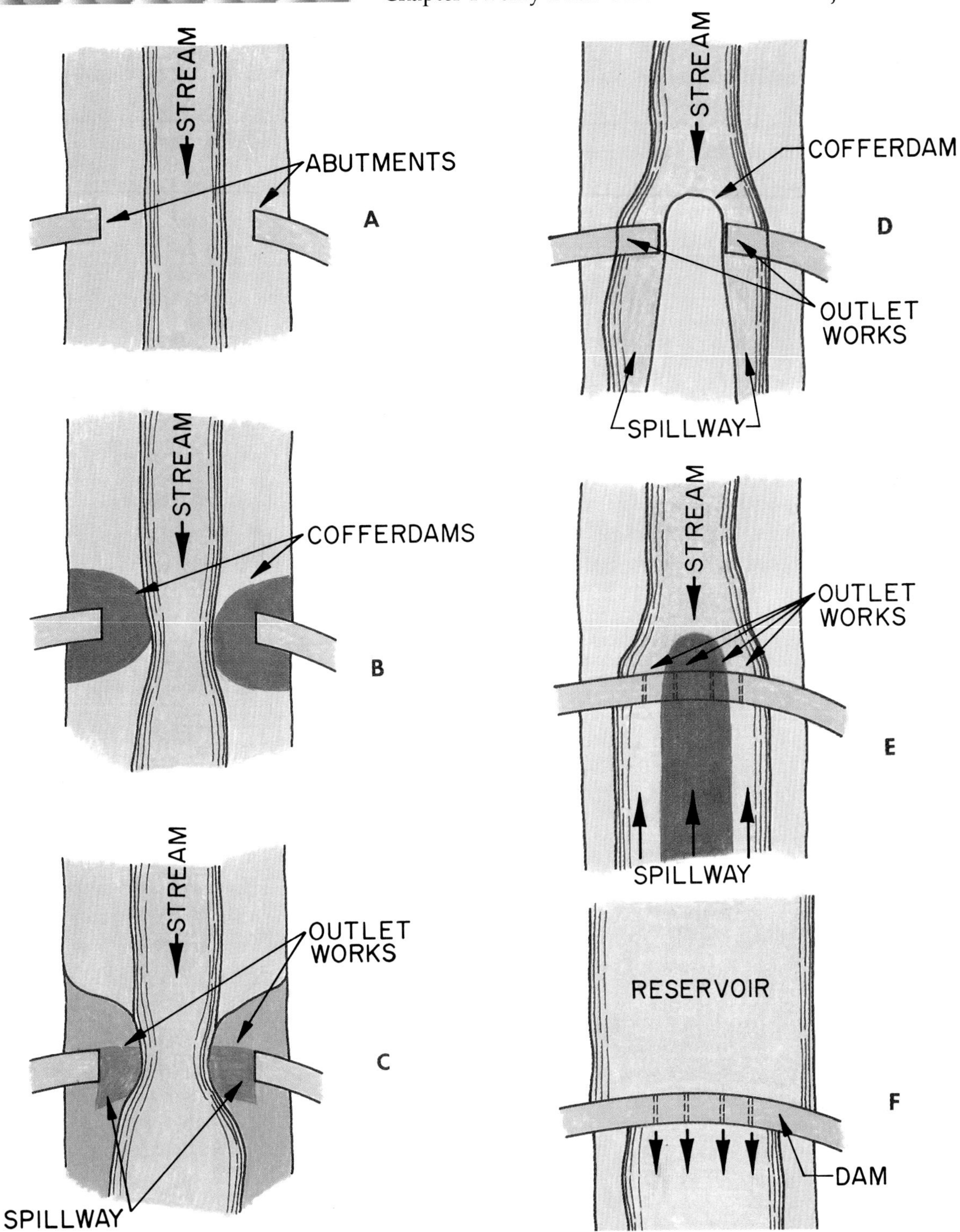

Fig. 24-6. Stages in building a dam.
A. Building abutments.
B. Building the first cofferdams.
C. Building the first part of the outlet works and spillway.
D. Building another cofferdam and channeling water through the outlet works.
E. Completing the outlet works and spillway.
F. Completed dam.

Apply Your Thinking Skills

1. The wide reservoirs that are created when dams are built may flood the property of homeowners or businesses that had been near the waterway being dammed. How do you think these people should be compensated?

Fascinating Facts

In August 1914, a passenger-cargo ship traveled from the Atlantic to the Pacific Ocean—two oceans previously divided by land. This voyage marked the official completion of the world's most famous engineering marvel—the Panama Canal. The 51-mile-long canal—built by thousands of men who had to cut and clear the way through jungle, swamps, and hills—fulfilled a dream first proposed by explorers four centuries earlier.

Ships previously had to travel around Cape Horn on the southern tip of South America to get from New York to San Francisco. This was a trip of 13,000 miles. The opening of the canal cut the length of the trip to 5,200 miles!

Canals

Canals are artificial waterways that are built for irrigation or navigation. *Irrigation* canals carry water from a place where water is plentiful to another place where water is needed. These are constructed to supply water to land that otherwise could not be used to grow crops. Fig. 24-7. *Navigation* canals connect two bodies of water. Navigation canals may also be constructed when a river has a portion that either bends too much or is too shallow to navigate. The canals allow ships to bypass those parts of the river.

Fig. 24-7. These irrigation canals help make more land available for farming.

Construction of a canal requires a great deal of earthmoving. Once the excavation is complete, clay, concrete, or asphalt is usually used to line the canal. This prevents leaks and erosion (washing away of the soil). Canals may also require the construction of locks if the two waterways being connected are at different elevations. A **lock** is an enclosed part of the canal that is equipped with a gate. The level of water within the lock can be changed in order to raise or lower ships from one water level to another. For example, the Welland Canal, which connects Lake Erie and Lake Ontario, contains eight locks along its twenty-eight-mile route to help ships change levels safely.

Apply Your Thinking Skills

1. Constructing a canal often involves clearing away a great deal of brush and trees, much earthmoving, and rerouting of water. What effects do you think all of this has on wildlife in the area?

Tunnels

A tunnel is an underground passageway. Tunnels are built to allow people, vehicles, or materials to pass through or under an obstruction. For example, tunnels may be built under busy city streets for subways. They may be built under rivers or through mountains for railroads or highways. A tunnel may also be built to carry water around a dam. There are three common types of tunnels: earth, immersed, and rock.

Earth tunnels are constructed in soil or sand. Because sand and soil can be unstable, these tunnels are hazardous to build. As earth tunnels are dug, concrete sections can be installed to prevent collapse.

For immersed tunnels, pre-manufactured sections are floated to the tunnel site. Here they are sunk into trenches that have been scooped out at the bottom of the waterway. The sections are then connected to form the tunnel.

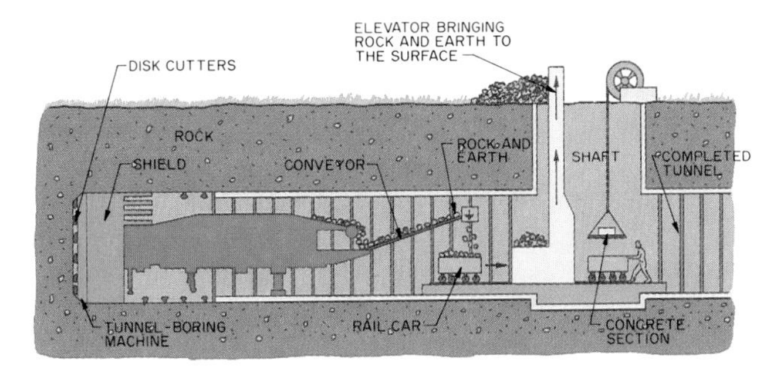

Fig. 24-8. Giant boring machines use disk cutters to cut tunnels through soft rock. A conveyor system and elevator transport the rock and earth that is cut away up to the surface.

In rock tunnels, material is removed by blasting or by using giant boring machines. Fig. 24-8. The three parallel tunnels of the Chunnel, being constructed under the English Channel to connect England and France, are being cut by giant boring machines. Two of the tunnels will be used for trains that will carry passengers and vehicles. A smaller central tunnel will be used for maintenance. Fig. 24-9.

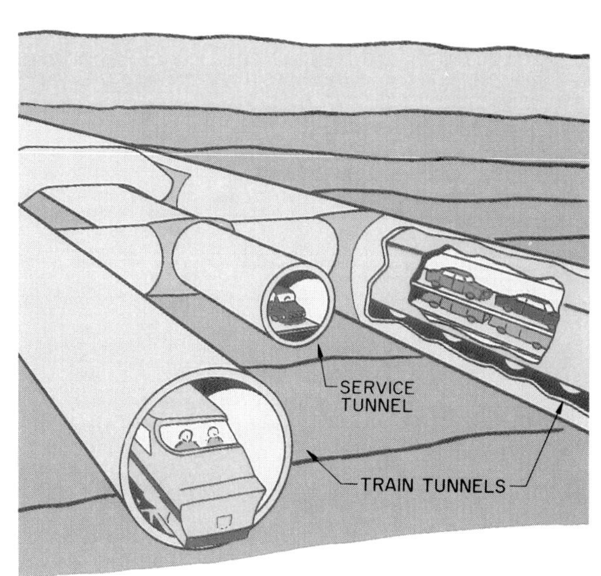

Fig. 24-9. The 31-mile-long Chunnel (the English Channel tunnel) will allow trains carrying passengers and vehicles to travel quickly and comfortably between England and France.

Apply Your Thinking Skills

1. What impacts do you think completion of the Chunnel might have on businesses and industries in France and England?
2. What impacts do you think the Chunnel will have on the lives of the people who live in these countries?

Pipelines

As you learned in Chapter 16, pipelines are an efficient way to transport products such as crude oil, refined petroleum, and natural gas. Most pipelines are buried underground. There are above-ground pumping stations along the pipelines that are used to maintain the pressure needed to keep the product moving.

Once the route for a pipeline has been surveyed and marked, backhoes and trenchers dig the trenches that will hold the pipe. Sections of pipe one to four feet in diameter are manufactured in factories. These are transported to the site and laid next to the trench. Then the sections of pipe are welded together. Fig. 24-10. Cranes then lift the pipe and place it in the trench. Fig. 24-11. Before being covered with soil, the joined sections must be inspected for leaks. This is usually done with X-ray or ultrasound equipment. After any necessary repairs are made, the trenches holding the pipeline are covered. The earth must be

Fig. 24-10. Pipefitters weld sections of pipe together.

Fig. 24-11. Cranes with special pipe-holding devices lift the pipe into place.

packed down all around the pipe. Otherwise, the ground may later "settle," causing the position of the pipes to shift. This could cause damage to the welded joints, resulting in leaks. A tamper may be used to pack the earth as the trench is gradually filled. A roller or the wheels of a heavy tractor may be used to finish tamping and leveling the surface.

Apply Your Thinking Skills

1. The 800-mile-long Trans-Alaska Pipeline, which carries crude oil from north Alaska to Alaska's southern coast, carries 25 percent of the petroleum used in the United States. What effects do you think the construction and continued operation and maintenance of this pipeline have on the people and the economy of this state?

Air-Supported Structures

Air-supported structures are a specialized type of structure whose main advantage is that they can be set up and taken down easily. Their portability is one of their greatest assets. For this reason, they are often used for exhibition buildings. These exhibition buildings can be easily transported, set up, and taken down to transport again when an organization tours from one location to another. Fig. 24-12 (p. 508).

An air-supported structure is made up of four elements:

- a structural membrane
- a means of supporting the membrane
- a means of anchoring the membrane to the ground
- a way in and out of the building

The structural membrane is usually made of a strong, impermeable, synthetic (human-made) material. The membrane, which is

Fig. 24-12. Air-supported structures can be used to provide shelter for sporting events and exhibitions.

usually translucent, is supported by introducing an air supply that will create an uplift of the membrane, distributing tension evenly through the structure. An electrically driven fan usually blows air into the structure for this purpose. All parts of the membrane must be kept under tension to withstand external air pressure. To accomplish this, the fan keeps the air pressure within the structure at a level high enough to counteract the air pressure outside the structure.

Because the air within the structure creates an uplift, the structure must be suitably anchored to keep it from pulling away from the ground. The membrane is usually attached to a rigid concrete base (buried underground) to keep it firmly anchored at ground level.

Getting in and out of an air-supported structure is different from getting in and out of conventional buildings. If a traditional door were used in an air-supported structure, air would leak out of the building every time someone opened it. Also, if the door opened inward, people would have to push hard to overcome the internal pressure. If it opened outward, the internal pressure would make the door pop open with extra force. One method of solving this problem is to use an air lock. To create an air lock, two sets of doors are used. People enter through an outer set, which is then closed before opening the inner set. The air lock this creates provides an opportunity for pressure to be equalized when people are going in and out.

Apply Your Thinking Skills

1. What part of a traditional building corresponds to the air supply in an air-supported structure?

Investigating Your Environment

The Environmental Impact of Construction

Pollution isn't the only way technology can harm the environment. In 1969, Congress passed the National Environmental Policy Act (NEPA). NEPA requires government agencies to do an environmental impact study before building large-scale projects. An *environmental impact study* determines the possible effects that the project will have on the environment. The report of the results is called an environmental impact statement.

The Glen Canyon Dam on the Colorado River was built before environmental studies were required. It provides "peak power" to utilities in the Southwest. This means it is used to supply additional power at times when demand is high. This causes the river to rise and fall as much as 13 feet a day. These wild swings in the water level have increased fishing accidents, destroyed prehistoric ruins, stranded trout, and wiped out four species of fish. A simple solution is to operate the dam evenly and not just at peak hours.

Eight federal dams have been built on the lower Columbia and Snake Rivers. These rivers were once passageways for 16 million young salmon making their way out to sea. Now they die in dam turbines and in slack water because there's no current to carry them where they need to go.

Dams aren't the only human-made structures that have imposed a threat to nature and the environment. Sometimes, even construction projects built with the best of intentions can cause harm. In Mississippi, a new highway was built to replace a highway that ran through the marsh habitats of the sandhill crane. The new highway was built away from these habitats. Unfortunately, the new highway, a wide band of dry land, made the situation worse. It became the perfect passageway for the cranes' predators. Raccoons, crows, and coyotes can now move easily and quickly along this dry route, preying on the cranes and their eggs.

Fig. IYE-24. This "salmon ladder" allows salmon to bypass this dam, which is blocking their regular migrating route.

Take Action!

1. Land development is a controversial issue in many communities. Hold a class debate on the issue: "Resolved, that there shall be no new construction on land that is not already zoned for commercial or residential use."
2. Conduct your own environmental impact study. Select an area of new construction (in your community, if possible) and research its effect on the ecosystem.

Chapter 24 Review

Looking Back

There are many types of construction projects besides traditional buildings. Highways, bridges, dams, canals, tunnels, pipelines, and air-supported structures are all constructed projects.

Surveying, earthmoving, and paving are the major construction processes used to build roads, streets, and highways.

Bridges are built over obstacles such as water or railroad tracks. Types of bridges include beam, arch, truss, cantilever, suspension, cable-stayed, and movable.

Dams help control the flow of water. They provide a dependable water supply, help control flooding, and provide the falling water needed to produce hydroelectricity.

Canals are artificial waterways built for navigation or irrigation. Canals may also require the construction of locks.

Tunnels provide ways to move through or under obstacles. The three main types of tunnels are earth, immersed, and rock.

Most pipelines are laid in trenches. Usually, only the pumping stations are above ground. The sections of pipe are joined, tested for leaks, placed in trenches, then covered with dirt.

Air-supported structures are a unique kind of structure in which a structural membrane is supported by air pressure. Their main advantage is portability.

Review Questions

1. Describe the two types of roadbed used in highway construction.
2. Name and briefly describe the three parts of a bridge's substructure.
3. Describe a truss bridge.
4. Name and describe one type of movable bridge.
5. What is a reservoir? Name two uses for a reservoir.
6. What is a cofferdam?
7. Describe the function of an irrigation canal. Describe two uses of navigation canals.
8. Describe a lock and tell how it is used.
9. What is the Chunnel?
10. Why won't a conventional door work in an air-supported structure? Describe the type of entrance/exit system needed in these structures.

Discussion Questions

1. What impacts might occur if a major highway that goes directly through the center of a town were to be replaced with a new highway that would bypass the town?
2. What factors do you think need to be considered when deciding on the type of bridge to build across a waterway?
3. What factors would you consider if you had to decide whether or not to build a dam in a specific location?
4. Discuss some negative and positive impacts of various types of construction (highways, dams, etc.) on farming.
5. The 800-mile-long Trans-Alaska Pipeline carries crude oil from north Alaska to the state's southern coast. Think about Alaska's climate and topography (surface features, such as forests, lakes, etc.). Then discuss some problems that might have occurred when constructing this pipeline.

Chapter 24 Review

Cross-Curricular Activities

Language Arts 

1. Research and write a report on the construction of the Chunnel.
2. Write a report on the history of tunneling, with emphasis on how tunneling was done before specialized machines were developed.

Social Studies 

1. Find out the following information on the Trans-Alaska Pipeline: when and where was it built; how much did it cost; how was it built; and what problems were encountered during its construction? Then discuss how this pipeline has benefitted the U.S. Do you think the pipeline could have been built in a better location? Explain your answer.
2. The Johnstown, Pennsylvania, flood of 1889 was the result of a dam giving way. Research and write a report about the flood, as well as its aftermaths. Was a new structure built to replace this dam? If so, what design changes were made to prevent a recurrence of this tragedy?

Science 

1. A cantilever bridge has one or more spans that extend beyond the supports. It works because the center of gravity of the suspended span always remains close enough to the support to provide stability. You can build a model cantilever bridge with a dozen books of the same size. Make two piles of six books each, placing them side by side. The top book of each pile represents the span. Pull lower books out in a stepped fashion so that the books on the bottom are as far as they can be from each other without the structure toppling (see below).

Math X÷

1. Walt wants to install a patio that will measure 13 1/2' x 24'. If he intends to pour the concrete 4" thick, how many cubic yards of concrete will he need? (Hint: there are 27 cubic feet in a cubic yard.)
2. An adjusting screw will move 5/64" for each full turn. How far will it move in 6 turns?

Chapter 24 Technology Activity

Designing, Building, and Testing a Model Bridge

Overview

Before a superstructure can be constructed on a site, it must be carefully designed and analyzed. The people responsible for these jobs are members of a technological team. A technological team is a group of highly qualified and educated people. Each has specialized training in certain areas of technology. The team consists of engineers, scientists, technicians, and skilled craftspersons.

In this activity, you will follow the problem-solving process to construct a model bridge and test its strength. You will do some of the jobs that would be done by members of a technological team. For example, as an engineer, you will try to solve a construction design problem. As a craftsperson, you will draw and build a scale model of your chosen design. As a technician, you will be responsible for determining if your design is sound and workable.

Goal

To design, build, and test a model bridge.

Equipment and Materials

1/4" graph paper
notebook paper
6 balsa or basswood strips (3/32" x 1/4" x length)
white polyvinyl resin glue
cardboard
straight pins
drafting instruments
modeling knives
utility knives
blueprinter or other duplicating machine
size C transparent drafting paper
test apparatus (See Fig. A24-4)
scales to weigh models in ounces
bathroom scales to weigh loads in pounds
lead shot in two 64 fluid ounce cans

Procedure

1. *Safety Note:* Before doing this activity, make sure you understand how to use the tools and materials safely. Have your teacher demonstrate their proper use. Follow all safety rules.
2. Your teacher will assign this activity to be done in groups of two or individually.
3. You are to design, draw, construct, and test a scale model truss bridge with beam supports. (*Beams* are horizontal structural members.) This bridge will be designed for vehicular traffic only (highway or rail vehicles). Any truss design that meets the following conditions may be used:

- The bridge must be designed within the size limits given by your teacher. The span of the bridge will be the same for all students.
- The bridge will not have piers or other intermediate vertical supports under the roadbed.

Chapter 24 Technology Activity

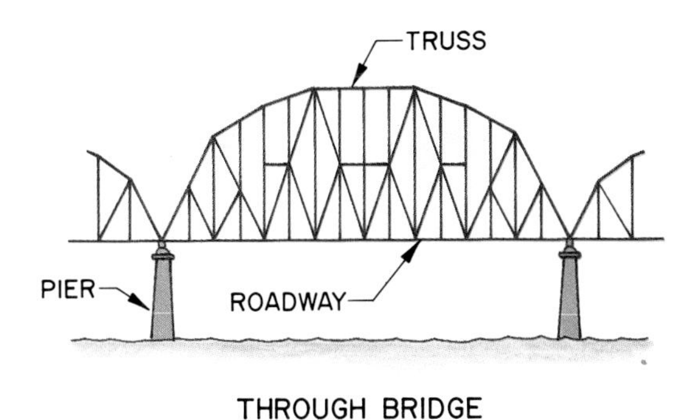

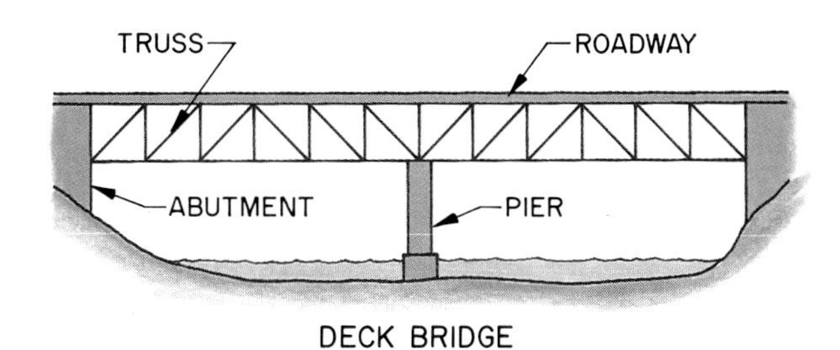

Fig. A24-1. A truss bridge may be a through bridge or a deck bridge.

- The bridge design must be a truss design on beams. Figs. A24-1 and A24-2.
- The bridge must be constructed for the greatest amount of strength using the least amount of materials.
- All scale-model bridges will be constructed of either balsa wood or basswood strips as assigned by your teacher.
- The scale-model bridge must be assembled using only white polyvinyl resin glue at the joints.
- No more than six pieces of wood (3/32″x 1/4″ x width of bridge) are allowed on the roadbed portion of the bridge.

4. Using this text and information in your school's media center, collect information about truss types, bridge design, and strength of materials.

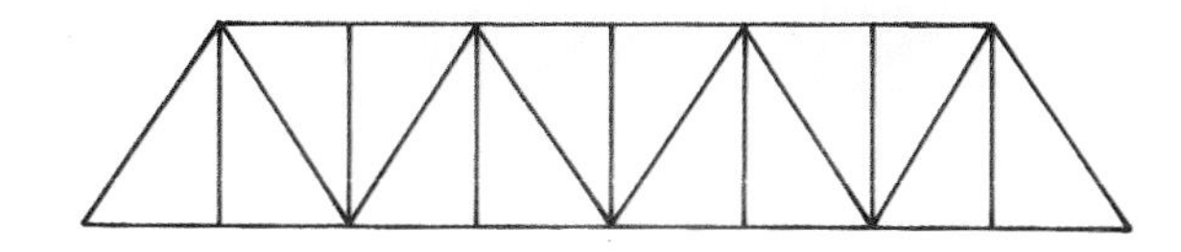

A. WARREN TRUSS

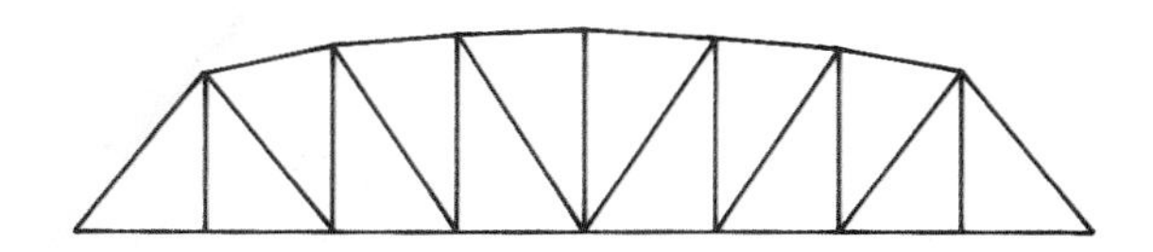

B. PRATT TRUSS

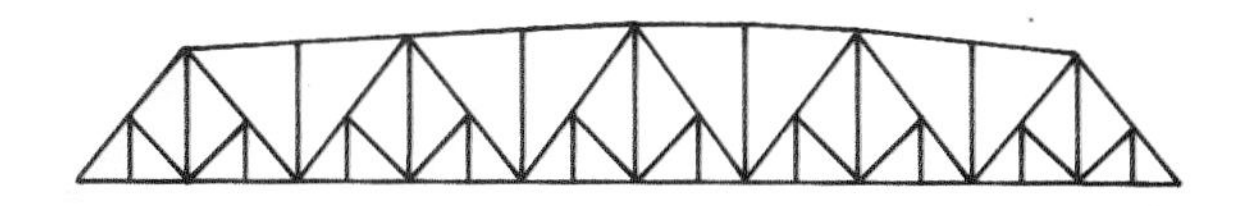

C. SUBDIVIDED WARREN TRUSS

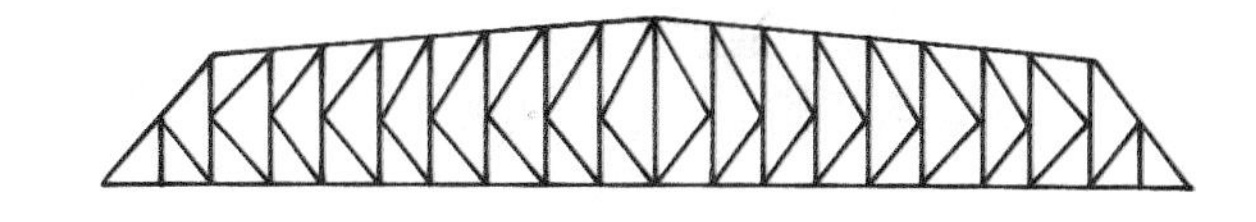

D. K-TRUSS

Fig. A24-2. Types of trusses.

Chapter 24 Technology Activity

5. After gathering information about truss bridge designs, brainstorm ideas (with your partner, if assigned). Sketch your design ideas on 1/4" graph paper.
6. Select your best design.
7. The soundness of your design can be determined mathematically. Engineers use special terms to describe the results. Your bridge design may be unstable, statically determinate, or statically indeterminate. An *unstable* structure is unacceptable. It has too few members and must be redesigned. A *statically indeterminate* design has more members or supports than necessary. It should be redesigned. A *statically determinate* design is in equilibrium (balance). It has the minimum number of structural members to support its own weight and the load it must carry. A design that is statically determinate is a good design and is acceptable.

A. You can figure statical determinacy. Use the formula $k = 2j - r$.

- k = minimum number of necessary members in the truss
- j = number of joints in the truss
- r = number of reaction components at the exterior supports necessary for stability of the truss ($r = 3$ for all of your designs)
- m = number of actual members in the truss

(1) To figure your bridge's statical determinacy, count the number of joints (j) on one side of your bridge. Write this on a sheet of paper.

(2) Next, count the members (m) of your bridge. A *member* is each part between the joints. Write the number of members on your paper.

(3) Using the formula stated above, find the minimum number of necessary members (k).

B. Judge your results as follows:

- $m < k$—bridge design is unstable (< means "less than")
- $m > k$—bridge design is statically indeterminate (> means "greater than")
- $m = k$—bridge design is statically determinate

8. Redesign your bridge, if necessary. (Repeat steps 5, 6, and 7.)
9. Make a scale working drawing of your design on size C drafting paper using the scale 1/4" = 1'0".

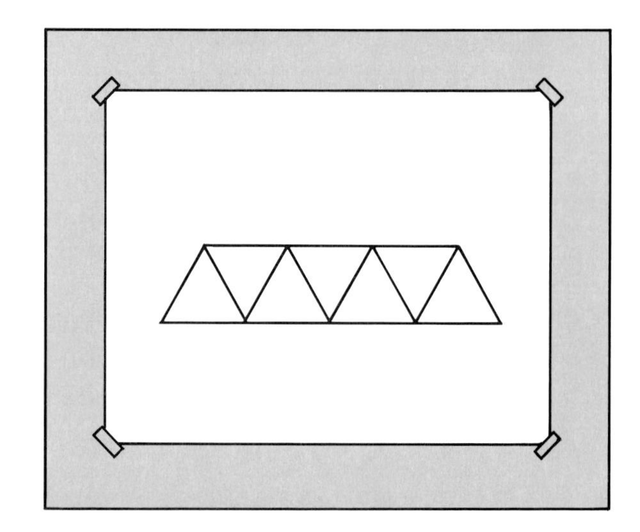

Fig. A24-3. Tape your drawing onto a piece of cardboard.

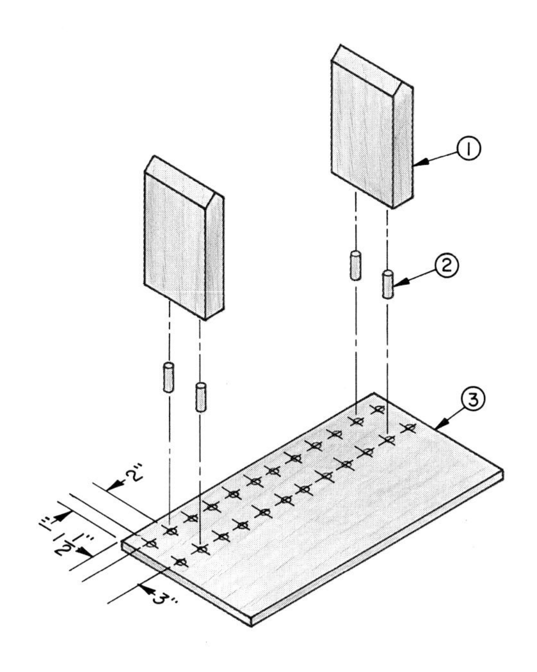

BILL OF MATERIALS

Part*	Quant.	Material	Size
1	2	Maple	1 3/4" x 6" x 10"
2	4	1/2" Dowels	2" long
3	1	Maple	3/4" x 12 1/2" x 24"
4	1	3/8" Dowel	6" long
5	2	Screw Eyes	#6
6	2	Postage String	10"
7	4	S-Hooks	(Size as needed)
8	1	Box	4" x 8" x 3"

*Numbers in this column correspond to circled numbers on the drawing.

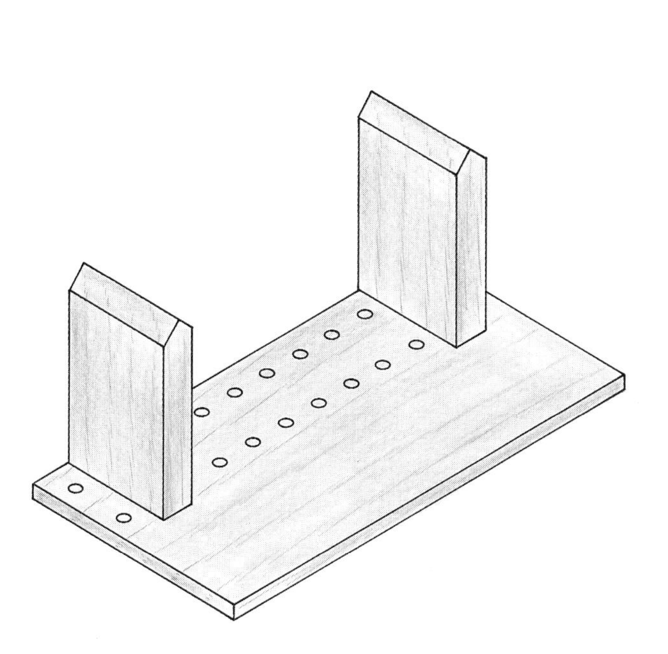

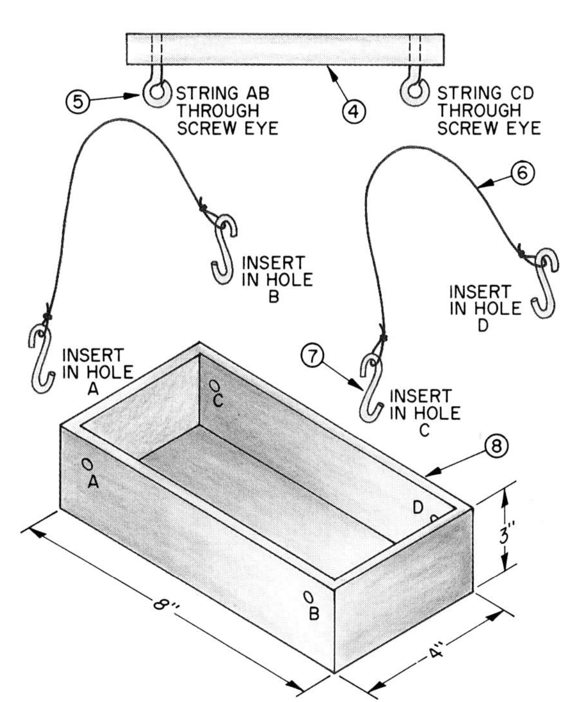

Fig. A24-4. Plans for making a test apparatus.

10. Using a duplicating machine, make a copy of your drawing. Turn in the original to your teacher.
11. Prepare a bill of materials for your bridge.
12. Tape the copy of your drawing on a piece of cardboard that is slightly larger than your drawing. Fig. A24-3.
13. Construct the sides of your scale-model bridge by cutting the truss members to size according to your bill of materials. Place the cut members over their respective locations on the drawing. Use the white glue to assemble the bridge. Use straight pins to hold the truss onto the drawing.
14. After assembling the two sides, glue the roadbed, top cross-members, and sides together to complete the bridge.
15. Weigh the model bridge and record its weight on a separate sheet of paper.

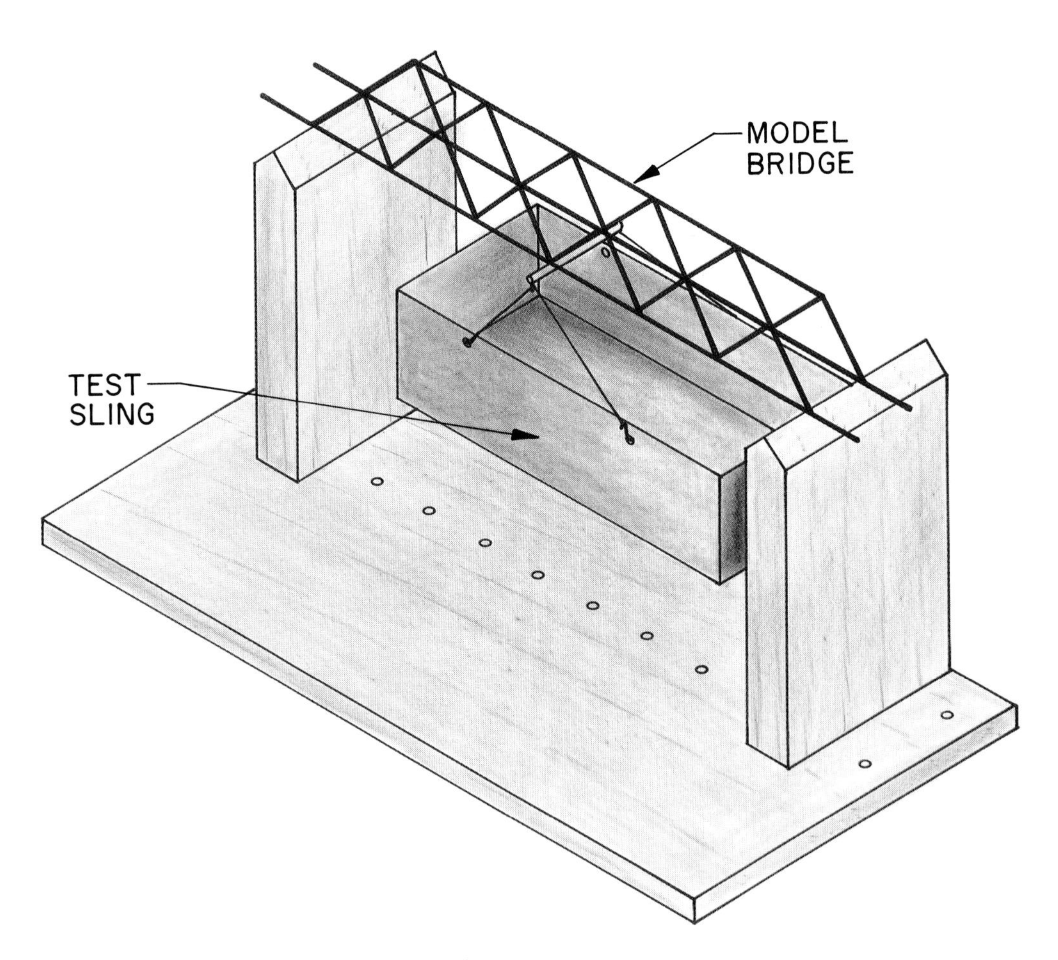

Fig. A24-5. Setup for testing the strength of your model bridge.

16. Test the strength of the bridge by load-testing it on the test apparatus. (Figure A24-4 provides information for constructing the apparatus.) This test is destructive. It will destroy your model.

To load-test your bridge, place it on the test apparatus as shown in Fig. A24-5. Place the test sling under your bridge and fasten it as shown. Next, slowly pour lead shot into the test sling. Be careful not to spill any of the shot. Continue adding to the load until the bridge fails. Weigh the shot you poured into the test sling. Record the failure weight on the same paper on which you recorded the bridge weight.

17. Determine the efficiency rating of your bridge design. Using the same sheet of paper, complete these two formulas:

(1) *Failure weight/length of material ratio*
A = Failure weight, in ounces
B = Total length of material used, in inches
Efficiency rating = (A ÷ B) x 100

(2) *Structure efficiency*
A = Failure weight, in ounces
B = Bridge weight, in ounces
Efficiency rating = (A ÷ B) x 100

18. Evaluate your results. The greater your efficiency ratings, the more efficient your structure.

Evaluation

1. At what weight did your model bridge collapse? Did most of your classmates' model bridges hold more or less weight than your bridge? Why?
2. If you could design and build another model bridge, what would you do differently?

Credit:
Developed by David Watson and Patti Farrar for the Center for Implementing Technology Education
Ball State University
Muncie, Indiana 47306

Chapter 25

Trends in Construction Technology

Looking Ahead

In this chapter, you will discover:

- some of the innovative tools and equipment used in construction.
- new materials that have been developed and how they are used in construction.
- advances made in design and construction techniques.
- how biotechnology is related to construction.
- some interesting facts about construction in space.

New Terms

adhesives
ergonomics
geotextiles
modular construction
panelized construction
prefabricated
smart building

Technology Focus

Living in the Future

Julie ducked her head under the pillow when she heard her mother patiently repeating over and over, "Julie, it's time to get up for school. Today is Monday, September 21, 2020, and it is seven a.m."

Julie was still dreaming about taking the flying car she hoped to get for graduation to a rock concert. She slept deeply until the shutters on her windows, controlled by light sensors, opened to let in the morning light. Suddenly awake, Julie shut off the alarm that simulated her mother's voice. She couldn't argue with a talking clock. It would persist until its sensors indicated she was on her feet.

"Straighten your covers and convert to a day lounge sofa," Julie commanded her voice-activated bed. Then, she thought about what she needed to do that day. She checked her portable computer for her assignments. "Darn!" she exclaimed. "I have to finish my report on Dad's new space station before school." Julie's father was part of the construction crew for the orbiting space relay station. He had sent Julie electronic photographs over the weekend while they talked on the vision telephone.

Julie decided to have breakfast before doing her schoolwork. The lights came on as she entered the kitchen. She selected a quick breakfast from the touch panel on the family electronic menu display. Her meal was produced within a few seconds. Julie finished eating and threw the scraps into the recycler.

Julie settled comfortably into her ergonomically designed study unit and logged into the school computer with the modem on her portable computer. She quickly dictated a few paragraphs about the problems of unfolding the third story of the space station. Her spoken words were entered into the school's mainframe computer. Julie remembered the electronic photographs her dad had sent. She resized them and inserted them into her report. She clicked "print." As she left the house, the home security alarm was activated and the lights went out.

When Julie reached her microbiology classroom, she found all the walls were gone. Now, Julie's lab was situated at the rear of the school grounds, next to the pond they were analyzing. "I sure hope the state college will have modern smart card readers for my daily schedule," she thought as she picked up her laser-printed research paper at the copy center and headed off to microbiology.

Fig. TF-25. In the future, computers will control all major systems in "smart" homes. The home's occupants may communicate with the computer by touch screen or voice input.

Developments in Construction

Many of the basic skills and techniques used in construction have been in use for hundreds of years. However, there have been, and continue to be, many new and dramatic developments in construction. New tools, equipment, and materials have been developed that improve the efficiency of construction processes and the quality of structures. New technology plus the growing need for efficiency have resulted in the development of new construction methods and new designs. Biotechnology is being increasingly used to make sure structures are designed to be comfortable, easily accessible, and safe. A new world of possibilities opens up as we begin construction in space.

Apply Your Thinking Skills

1. Think about a construction project you've seen recently in your community. Were the techniques being used at this construction site any different from those that have been in use for hundreds of years?

Tools and Equipment

The purpose of any tool is to help a person do work or perform a task. In the last twenty years or so, several important tools have been developed. The computer and the laser are among these tools. Various new items of equipment have also been developed.

Computers

Curiously enough, the most significant innovation in construction is not a "tool" in the traditional sense, nor is it a new piece of equipment. It is the computer.

The computer is an invaluable tool to architects and design engineers. They can design, test, modify, and revise design and engineering ideas on a computer screen, without cutting a single board or hammering a single nail. Fig. 25-1.

Using a computer, detailed architectural and engineering drawings can be prepared rapidly. If necessary, they can be revised. Some computers can prepare accurate three-dimensional drawings of objects from any angle. Even the most complex and fantastic architectural and engineering concepts and designs can be "built" on a computer. Drawings can be "printed out" almost instantly. They can be printed in color on new high-speed plotters. They can also be transmitted electronically. A combination of telephone lines, satellites, and microwave relay stations can be used to send the drawings almost anywhere in the world.

Fig. 25-1. New architectural and engineering concepts can be developed and tested on a computer.

Using mathematics on a computer, construction engineers and materials specialists can pretest new and experimental materials. They can determine, for example, exactly how well materials will perform or "hold up" under actual use.

Computers are also a useful tool in managing construction. They can be used when preparing estimates and specifications for projects. They can be used to plan schedules and keep track of work being done. They can be used to monitor costs and keep records. Contractors can use computers to accumulate knowledge and apply it from project to project.

Lasers

As you learned in Chapter 11, lasers strengthen and direct light to produce a narrow, high-energy beam. This concentrated, powerful beam of light can be used in construction in many ways.

Just as in manufacturing, laser cutting tools can be used to cut metals. Fig. 25-2. There are also laser tools that are used to weld metals. Concentrated laser beams can reach very high temperatures, so they are especially good for cutting and welding heat-resistant metals.

Laser equipment is also used in surveying and leveling construction sites and leveling interior walls. In surveying, a device called a construction laser flashes a narrow, accurate beam of light. Workers use this as a baseline for making measurements. A laser beam can help earthmoving equipment operators level the ground to the right height, or elevation. Fig. 25-3. There are laser levels that can be used to increase the accuracy of interior wall construction. Fig. 25-4 (p. 522).

Fig. 25-2. A laser beam is powerful enough to cut through metal quickly.

Fig. 25-3. The tripod holds a laser-emitting device. Electronic receivers attached to the grader blade receive a signal from the laser. This allows the laser to control the height of the grader blade and enables the grader to level the earth to an exact elevation.

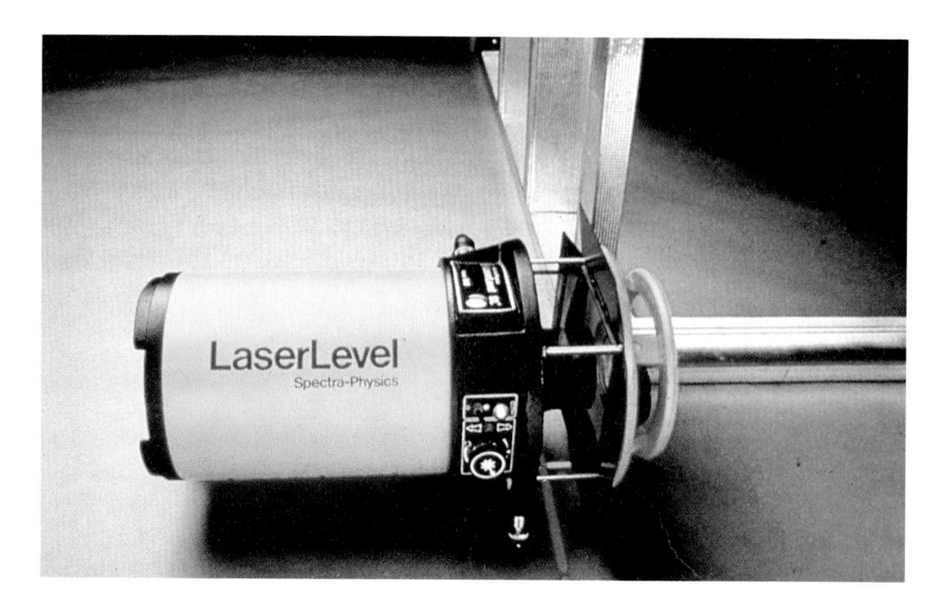

Fig. 25-4. This laser level helps construction workers accurately align interior walls.

Apply Your Thinking Skills

1. Can you think of any ways, other than those mentioned in this text, that computers are used in construction?

Materials

The materials used in today's construction must be strong, durable, economical, easy to use, and, in many cases, lightweight. New metals, plastics, adhesives, and wood materials have been developed to meet these construction needs.

Lightweight, Ultrastrong Metals

A variety of new, lightweight, ultrastrong steel and aluminum materials have been developed. These unique materials give architects and design engineers greater freedom to experiment with new design and construction techniques. Examples can be seen in the steel and glass towers in many major cities. Fig. 25-5. Other examples of new techniques can be seen in the increasing number of clear-span structures. Fig. 25-6.

Fig. 25-5. New lightweight, yet strong, materials are often used in building skyscrapers.

Fig. 25-6. Lightweight materials and new design techniques are creatively combined in this clear-span structure. The curved roof of the structure requires no vertical support members inside the structure.

Plastics

Plastics have been used in many areas of construction for several years. Plastic pipe is now used in most plumbing systems. Plastic is used to make many insulating materials. Some bathtubs and sinks are made from fiberglass, which is made from plastic resin and spun glass fibers. New applications for plastics are constantly being developed. Because they are lightweight, strong, waterproof, economical, and resistant to corrosion and rust and can be formed into any desired shape, plastics may someday replace many of the materials with which we now build.

Plastics are now being used for roofing materials, liquid storage tanks, exterior siding, protective coatings, and fasteners. Fig. 25-7. A new type of "lumber" made from recycled plastics is being used to make outdoor items such as picnic tables and decks.

Another new application of plastics is geotextiles. **Geotextiles**, also called engineering fabrics, are like large pieces of plastic cloth. Geotextile materials are spread

Fig. 25-7. Because of their resistance to rust and corrosion from salt water, fiberglass bolts are used on this tower near the ocean.

on the ground as an underlayment, or bottom layer, for roadbeds. They are also used on slopes along highways to help keep soil in place and prevent erosion. Fig. 25-8.

Adhesives

Adhesives are materials that are used to bond together, or adhere, two objects. Glue is a common type of adhesive. New, stronger adhesives that can bond almost any kind of material to almost any other kind of material are being used and perfected. Fig. 25-9. Adhesives are used to bond many of the new plastic construction materials. Adhesives are also being used more and more in place of nails in today's construction. Adhesives help save time on a construction job. For example, drywall can be fastened to studs more quickly with adhesives than with nails.

Fig. 25-9. These concrete bridge segments are bonded together with an adhesive.

Fig. 25-8. This geotextile material is being used to help reinforce this levee (an embankment for preventing flooding).

Wood Materials

Our timber supply is gradually being used up, in spite of the fact that most lumber companies plant seedlings to replace trees they cut down. It is becoming increasingly important to stretch the supply of our valuable wood resources. New construction materials, such as oriented-strand board and Micro-lam®, have been developed that make the best use of our wood supply.

Oriented-strand board is a wood panel made by mixing woods from small, crooked trees that would otherwise be unprofitable to harvest. *Micro-lam®* is made of pieces of *veneer* (thin, even layer of wood) glued together with waterproof adhesive and bonded under heat and pressure. Micro-lam® is up to 30 percent stronger than comparable lumber. In addition, Micro-lam® is a more efficient way of using our wood resources. About 75 percent of a tree can be used when making this wood composite. Only about 40 percent of a tree is used when it is harvested for conventional lumber.

Apply Your Thinking Skills

1. Some people have a hard time accepting the use of plastics as construction materials. Why do you suppose this is so?
2. Since wood is a renewable resource, why do you think it is so important to stretch our supply of this construction material?

Fig. 25-10. These prefabricated sections can be quickly assembled into a complete house.

New Methods and Designs

New construction methods have been developed that help save time and money. They make construction more efficient. New electronic technology has made it possible to design and construct "smart" buildings.

Prefabrication

Components of structures and even whole structures are now being built in factories. Another term for these prebuilt components is **prefabricated**. These prefabricated components may be as simple as roof trusses or may be entire sections of houses. Fig. 25-10. The prefabricated parts or sections are shipped to the construction site, where they are assembled. Two types of prefabrication are panelized construction and modular construction.

In **panelized construction**, the floors, walls, and roof all consist of prefabricated panels that have been made in factories. These panels are then shipped to the construction site, where they are assembled to produce the framed and sheathed shell of the structure. Prefabricated roof trusses and floor joists are usually used with the panels. Such things as insulation, utilities, and siding are added at the site in the traditional manner.

In **modular construction**, entire units, or modules, of structures are built at factories and shipped to the site. The modules are assembled at the site to produce a finished structure. This construction method has been used to build motels. Identical modules—complete with plumbing, electricity, heating, and air conditioning—are delivered to the site and assembled into a motel.

When prefabrication methods are used, site preparation and the building of the foundation can proceed while the structure itself is being constructed.

Smart Buildings

New electronic technology has made it possible to design and construct buildings that are intelligent, or smart. **Smart buildings** use computers to control day-to-day operations. Among other things, these computerized controls can operate the mechanical and electrical systems in a building. For example, some controls can turn furnaces and air conditioners on or off as needed. Others can sense when no one is in a room and turn off lights, and then turn them back on when someone enters. These control systems improve the efficiency of building operations and reduce costs.

Other computerized control systems can help make buildings safer. For example, they can sense when an intruder is in a building and notify the police.

Fascinating Facts

Have you ever thought of ordering a house through the Sears Catalog? Back in 1908, Sears introduced its first mail-order house kits. They sold more than 100,000 kits before they discontinued the line in 1937.

Back in those days, you could order a kit containing everything you needed to complete a six-room house, including paint and varnish, for just $645! Add local labor costs, if needed, and costs of other desired materials and you could still complete the house for under $1,600!

Smart buildings are becoming increasingly popular. Research is now being done on other electronic systems that will make buildings even "smarter" in the future.

Apply Your Thinking Skills

1. Sears no longer sells house kits. However, some manufacturers do sell certain types of houses in kit form. What kinds of skills do you think a person interested in building a house should have? (*Note*: Specialists are usually hired for electrical and most plumbing work.)

Construction and Biotechnology

Biotechnology has always been a part of construction. For example, people needed to understand the many properties of wood in order to use it for building. Native Americans would sometimes use animal skins in constructing housing, so they needed to know how to treat and cure such skins to make them usable for construction.

As new construction materials, designs, and methods are being developed, researchers must make sure that they are safe for and well-suited to construction workers and to the people who will be using the finished structures. For example, for a long time, a material called asbestos was used to insulate and fireproof parts of many structures. It was finally discovered, however, that asbestos can seriously damage the health of people who must work with the materials, such as construction workers, as well as the people who live and work in structures in which asbestos was applied. Inhaling air that has come in contact with asbestos can cause serious lung ailments, including lung cancer. Biotechnologists now test many new construction materials before they come into widespread use in order to prevent similar health hazards.

Another way biotechnology is related to construction is the area of ergonomics. As you read in Chapter 11, **ergonomics** involves designing products and structures to match human needs. People who are concerned with ergonomics try to create technologies that take human characteristics into account. For example, computer display designers try to make computer screens that are easy for people to use. Furniture designers make chairs with the contours of the human body in mind. Even something as simple as the design of a water faucet depends on ergonomics, because certain designs are better suited to the human hand.

Most construction projects are intended for human use, so naturally information about the humans who will be using a structure is important to the structure's design. For example, very tall people have trouble moving around in homes that have low ceilings. Architects can design homes that have higher kitchen counters and higher doorway openings and ceilings to accommodate additional height. Suppose someone must use a wheelchair to move around. Homes can be designed to have extra-wide doorways so wheelchairs can pass through easily. Kitchens can be designed to have low countertops and a low sink with an open space underneath for the seated person's legs. Fig. 25-11.

Many public places are making their facilities more easily accessible to all people, regardless of their physical capabilities. Because such places were often originally designed for people who could walk and get around easily, extensive alterations sometimes need to be made. The changes

Fig. 25-11. Using ergonomics, structures can be designed to fit any special needs of those people who will be using them.

made are based on information about the human body. For example, wider doorways were built to allow easy passage of wheelchairs and walkers. Entrance ramps were built so that people could propel their own wheelchairs into buildings instead of having to be lifted up steps. Public places also include specially designed restrooms with larger stalls and gripbars for the physically challenged.

Apply Your Thinking Skills

1. Think about the shopping and recreational (movie theaters, restaurants, civic centers, zoos, etc.) areas in your community. Are they designed to be easily accessible and usable for people with special needs? If so, what are some of these special design features? If not, can you suggest some possible design improvements?

Construction in Space

It's likely that our future will involve outer space. How will construction and the construction industry be involved? Space structures will be needed to provide areas for living and working.

Techniques for Building in Space

New technology must be developed to build space structures. Parts of the structures will probably be "built" on earth and assembled in space. Workers in space will have to do this assembly work in near-zero gravity conditions. What does this mean? In space, everything is relative. "Up" and "down," "top" and "bottom," "right" and "left" have no fixed meanings. Tools must be tethered (attached with a cord or line) to the workers' suits, or the structure itself, so they will not float away.

Fig. 25-12. Space construction presents a challenge to the construction industry. Workers would have to wear jet-powered backpacks to get around in the near-zero gravity of outer space.

In addition to these problems, workers will have to wear bulky suits for protection. They will also have to carry their own supply of oxygen for use while they are building. It won't be convenient to build in space.

To help the workers get around in space while they are building, they may use *manned maneuvering units (MMUs)*. These are jet-powered backpacks. Fig. 25-12. Once the larger, stationary parts of the space structures are in place, construction workers could use vehicles attached to the structures to move along the structures and continue working.

Special kinds of tools, materials, and techniques will be required for space construction. Many have yet to be invented. One kind of system scientists are working on is a *telerobotic system*. The word *tele* means from a distance. In this system, a robotic arm mounted on a special base performs the construction operations in space. The robotic arm is controlled from a space shuttle or from ground units back on earth. Fig. 25-13.

Deployable methods may also be used to build space structures. This means that structures would be designed to erect themselves by unfolding automatically.

Space Structures

What will space structures be like? The National Aeronautics and Space Administration (NASA) has developed a set of requirements for the first one. This space station will have to provide servicing for satellites. It will need research and development facilities and living quarters for workers. The structure will have to be

Fig. 25-13. Telerobotic systems such as this one do some of the assembly and repair work on space structures.

expandable. Facilities such as special laboratories may be added. The area for living quarters may need to be increased.

Of special concern are the environmental and life-support systems. Workers will need air, food, and water. They must be able to live and work safely and comfortably within the structure.

The space station has been redesigned several times since it was first conceived. Building a structure in space is not like building one on earth, where years of history and experience have made design problems relatively simple to solve. There are many factors in space construction that make space station *Freedom* a difficult project.

Space station *Freedom* will be built in stages. The first step will be building an aluminum framework. This framework will be assembled in space. Modules that have been built on earth and transported to space will be attached to the framework. The modules are airtight structures in which human beings can live. Each module is designed so that it fits into the space shuttle, which will carry the modules to the space station. Once the station is assembled, astronauts will begin living and working on *Freedom*. Fig. 25-14.

Freedom will be powered by solar cells. Two huge panels of cells will use solar energy to power the station.

Many nations are cooperating to build this space station. Japan, Canada, and the European Space Agency are partners with the United States in this ambitious project. Recent cutbacks in NASA funding and other problems have delayed the progress of the space station. However, current hopes are that it may be ready for use by the year 2000.

Apply Your Thinking Skills

1. You just read that several countries are and will be working together to develop and use the space station *Freedom*. What might be some advantages and disadvantages of this joint effort?

Fig. 25-14. Astronauts from several countries will live and work together for extended periods on the space station *Freedom*.

Global Perspective

Swifter, Higher, Stronger

"Citius, Altius, Fortius"—this is the Olympic motto. It's Latin for "swifter, higher, stronger''—words that express the drive of Olympic athletes to constantly improve their athletic performances. The Olympics brings out the best, not only in athletics, but in areas such as architecture as well. Structures built by host cities for the Olympic games are usually designed and constructed using the latest—and often innovative—techniques.

The Olympic games originated in the city of Olympia, Greece, more than 2600 years ago. The name "Olympics" was derived from this tradition. The first modern organized games were held in 1896, in Athens, Greece. The stadium built there was designed to be like the first stadium in ancient Olympia.

The 1932 Olympics were held in Los Angeles, California. The main reason they were held there was that the organizers there offered two appealing innovations. First, were plans for an impressive stadium. The resulting structure, the Los Angeles Coliseum, was, at the time, one of the finest stadiums in the world. The second innovation was the idea of building a group of living quarters for the participating athletes—the first Olympic Village.

Probably the first architectural structure to become a symbol of the Olympic games was the Sports Palace, which was constructed in Mexico City, Mexico, for the 1968 games. The Sports Palace has a thick masonry base and a very impressive copper-covered dome roof.

Another unique building was designed for the 1980 games in Moscow, U.S.S.R. The Velodrome, designed to house cycling events, is covered by a roof that appears to be in two parts, resembling the wings of a butterfly.

Fig. GP-25. The 1992 Olympics in Barcelona, Spain, continues the tradition of innovative architecture.

The Olympics encourage athletes to be "swifter, higher, stronger.'' The structures built to showcase the Olympic games contribute in their own ways to the Olympic spirit.

Extend Your Knowledge

1. The moon and other planets will not themselves provide a hospitable living environment for human space travelers. One will need to be constructed. Do research to find out about experiments and studies presently being done. Then work with other members of your class to design a biosphere for a colony on the moon.

Chapter 25 Review

Looking Back

The development of new technology is changing construction. Ideas about construction are changing along with the tools and techniques.

Innovative construction tools and equipment include computers and lasers. Computers play an ever-increasing role in practically all aspects of construction. Laser devices are being used increasingly in a variety of construction operations.

New materials are being developed to meet today's construction needs. These include new or improved metals, plastics, adhesives, and wood materials.

Advancements are being made in design and construction techniques. More and more components of structures will be built in factories, then shipped to construction sites to be assembled. New construction technology has made it possible to construct "smart" buildings.

Biotechnology is becoming increasingly important in construction. It is used to help design buildings that are comfortable and easily accessible for everyone, regardless of their physical capabilities. Biotechnology is also used to make sure the materials used in construction do not create an unsafe environment in which to live or work.

In the future, structures will be needed in space. New construction technology is being developed to meet this challenge.

Review Questions

1. Briefly describe three ways in which computers are useful in construction.
2. What is a laser? Describe three ways laser devices can be used in construction.
3. List five qualities of plastics that make them desirable as construction materials.
4. What are geotextiles? How are they used in construction?
5. Describe new wood material and tell how it helps us stretch our wood supply.
6. Describe panelized construction.
7. Describe modular construction.
8. Briefly describe a smart building.
9. Define ergonomics. Give an example of how a structure can be ergonomically designed.
10. What special problems will the near-zero gravity conditions present for construction in space?

Discussion Questions

1. Discuss some advantages and disadvantages of using prefabricated construction for homes.
2. Imagine you are turning your house into a smart house. What would you program your control center to do?
3. What design features might be included in a home inhabited by someone who is blind or has very limited vision?
4. Describe some design features you would like to have in the home you will be living in twenty years from now.
5. What personality characteristics do you think someone would need to be a good member of a space station crew?

Chapter 25 Review

Cross-Curricular Activities

Language Arts

1. Identify a new building or other structure in your city or state that you feel is especially interesting or beautiful. Describe its appearance. Were any new construction methods or materials used in its construction? If so, describe them.
2. The "Technology Focus" at the beginning of this chapter gives some of the author's ideas of what life might be like within the next twenty-five to thirty years. Write a brief essay describing what you think life will be like 50 or even 100 years from now. Include descriptions of construction as well as some ideas about future transportation systems and new products.

Social Studies

1. Do some research to learn what tools and equipment were needed to construct a home in the mid-1800s. Compare those tools with the tools and equipment most commonly used in today's home construction. What inventions or developments had to precede the creation of these newer, more innovative tools?
2. Do some research to learn under what presidential administration the space program was established. What significant accomplishments have occurred within the space program since it began?

Science

1. In the microgravity of an orbiting space station, plants will grow differently than they do on earth. On earth plants respond to gravity during growth, sending roots downward. Use a moistened piece of paper towel inside a glass to sprout a bean seed. This will allow you to see the orientation of the plant. The leaves will be up; the roots down. Now turn the glass over. What happens in the next few days? Do some research to find out what happens to plants grown in a microgravity environment.

Math

R-value is a number used to rate the insulating capacity of various materials. R-value depends on both the composition and thickness of the material used to insulate. For example, fiberglass has an R-value of 3.33 per inch. The proper R-values for a home will vary according to the climate of the region.

A family living in Arkansas needs an R-value of R-26 in their ceiling below the roof. They measure their existing fiberglass insulation and find that it is 3 1/2" thick.

1. What is the current R-value of the fiberglass insulation in this family's ceiling below the roof? (Round to the nearest whole number.)
2. Approximately how much fiberglass insulation must be added to reach the recommended protection of R-26? (Round to the nearest 1/2".)

Chapter 25 Technology Activity

Building Model Space Structures

Overview

A *satellite* is an object that orbits the earth. Today, the satellites that people manufacture and send into orbit are communication tools. They transmit information. Some detect weather patterns. Others relay telecommunication messages and transmit television signals from around the world. Plans are now being made for a new type of satellite. It will be a space station in which people will live and work—an actual space structure! Figure A25-1 shows designs presently being developed.

In this activity, you will research the needs that must be met to make a space station livable and functional. You will also design and build a three-dimensional scale model. In the process, you will develop your skills in research, design, and development.

Goal

To research space stations, and to design and build a three-dimensional scale model that would be livable and functional.

Equipment and Materials

paper
poster board
drafting instruments
scissors
colored pencils
glue
tape
mylar film
aluminum foil
paints
contact cement
string

Procedure

1. *Safety Note:* Before doing this activity, make sure you understand how to use the tools and materials safely. Have your teacher demonstrate their proper use. Follow all safety rules.
2. Research books, magazines, scientific journals, and other sources for information about the design and possible uses of space stations. Make a list of needs that must be met by the design of the structure. For example, areas must be planned for living and working. How about a workshop for satellite repair? What shape should the station be? Use your imagination as well as the information you find in your research. Remember, this will be a low-gravity environment. People will be able to stand, work, or sleep on the "ceiling" as easily as on the "floor."
3. Prepare three or four thumbnail sketches of possible exterior designs. Keep the designs fairly simple.
4. Select one design. Make a rendering of your space station.

Chapter 25 Technology Activity

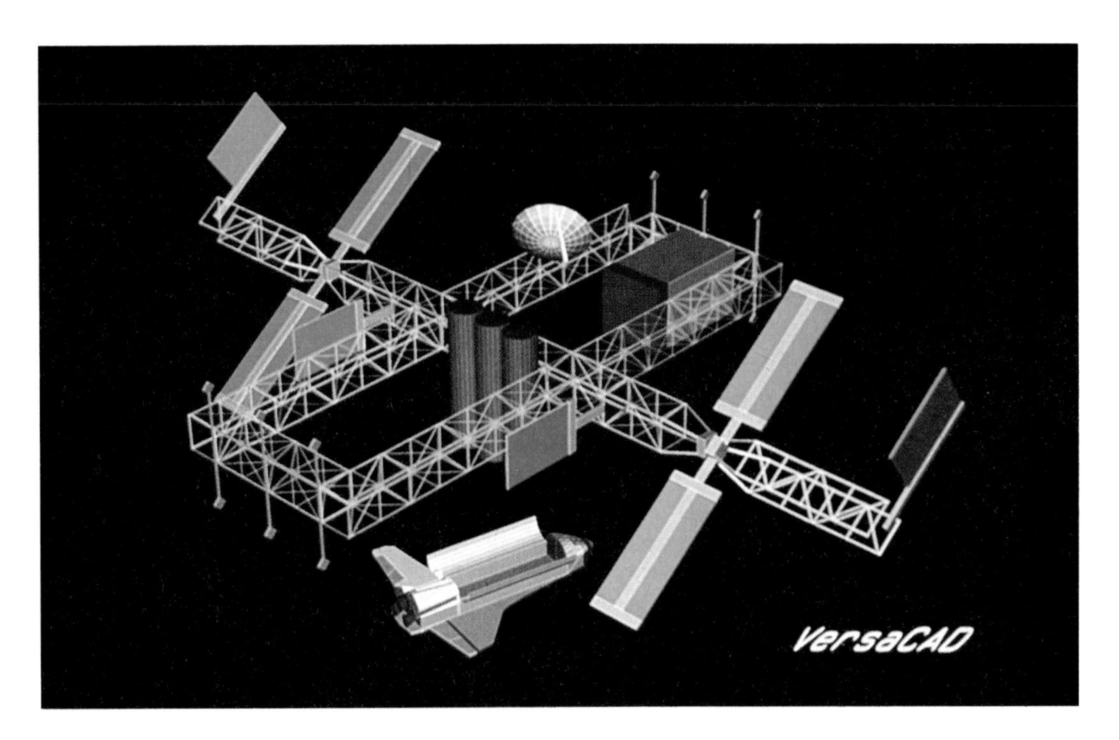

Fig. A25-1. Possible space station designs.

Fig. A25-1. (continued)

5. Develop and draw a floor plan for each level of your station.
6. Plan how to build your model. Determine the best view to use to show both the inside and the outside of your space station. See Fig. A25-2 for two ideas, but try to create your own design.
7. Construct your model. One portion of the exterior must open or be removable to show the interior.
8. Prepare a presentation of your ideas for a space station, using your rendering, drawings, and model to show your ideas as you explain them.

Chapter 25 Technology Activity

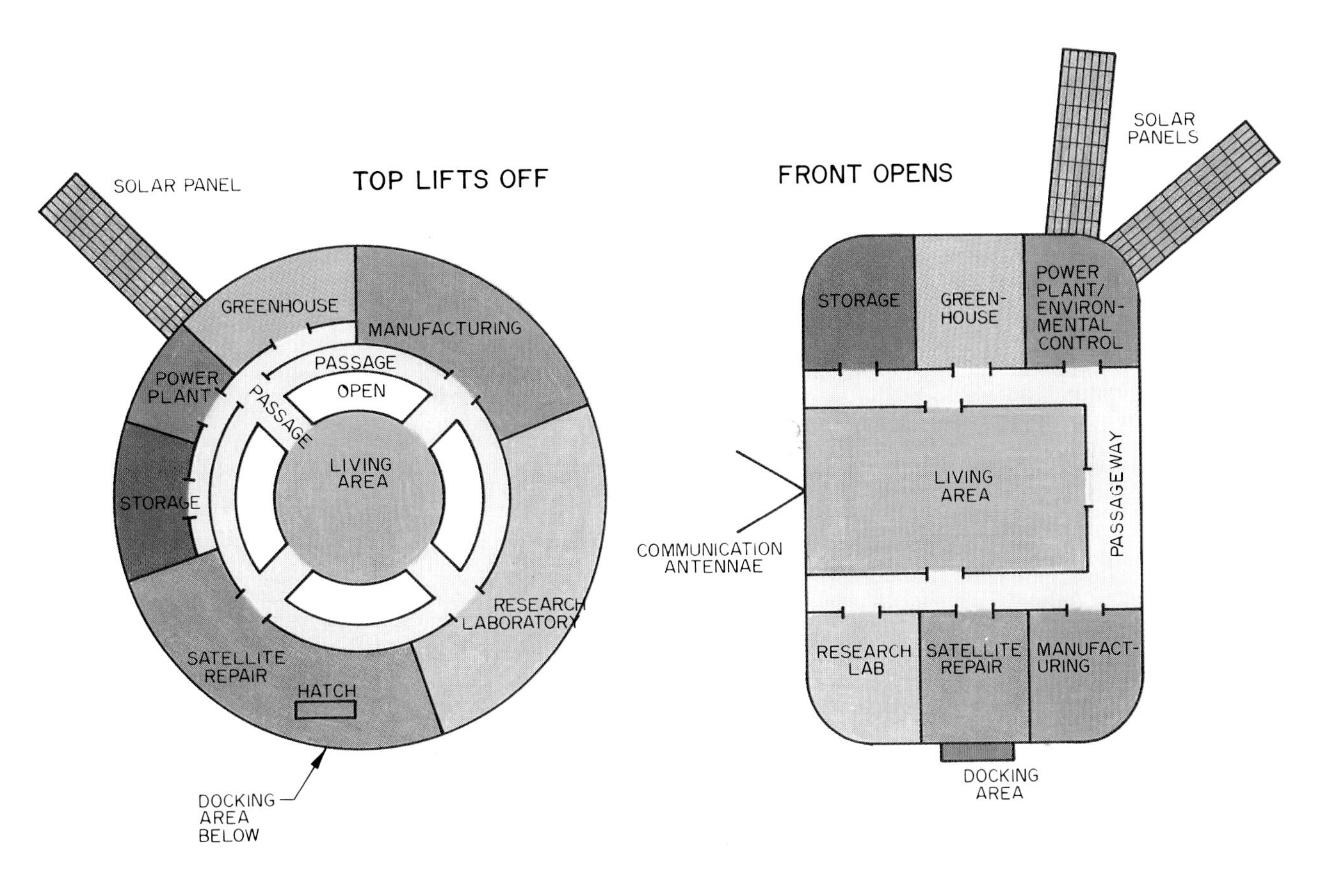

Fig. A25-2. Examples of possible model designs.

Evaluation

1. Review the research you did on space stations. How have the possible designs of space stations changed since they were first planned?
2. How livable and functional is your space station? Look at your drawings and finished model. Are there any parts that you have forgotten to include?
3. If you could design and build another space station, what would you do differently?

Section V Activity

Designing a Park

Overview

Because of several new industries that moved into town, a community that was fairly small just four years ago has grown into a rather large city. There have been several thousand new houses built in this community during this time. One of the housing developments was planned by a conscientious architect who wanted to be sure that there was space set aside for a park for the homeowners.

From the landscape architectural firms that submitted proposals, your firm has been chosen to design this neighborhood park.

STEP 1: State the Problem (Design Brief)

Design a neighborhood park that would meet the needs of a young community whose average family has two children, one in grade school and the other in middle school. The land allocated for this park is one city block (660 ft. by 660 ft.).

On a clean sheet of paper, type or write neatly a design brief in which you state your problem and the major points you must consider in solving it. (Here, that includes such things as the age of the neighborhood children and the amount of land allocated to the park.)

STEP 2: Collect Information

Here are ways you might begin designing the park:

- Visit your favorite park and list the various pieces of equipment that you think would be needed in the park you are to design. Fig. SCV-1.
- Visit your local park district office and ask to look at some of their magazines devoted to park design and equipment.
- Go to the library and look for the reference book entitled *Architectural Graphic Standards*. This book will help you with dimensions for athletic fields and playground equipment. Fig. SCV-2.

STEP 3: Develop Alternative Solutions

On the same sheet of paper as your design brief, write down several ideas. For example, what kind of sports are the children in the neighborhood most likely to be interested in? Give the planning board several options of athletic field layouts. Give them options in soccer, basketball, baseball, swimming, etc.

STEP 4: Select the Best Solution

Consider the pros and cons of each design.

- Which design is the least expensive?
- What do other parks in town offer?
- Which design would take the least amount of time to build?
- Could the design be built in stages?
- Could there be multiple uses for the various areas?

STEP 5: Implement the Solution

Prepare sketches and drawings of the park. Make a model based on your drawings.

STEP 6: Evaluate the Solution

Ask your teacher and classmates to evaluate your park design. Changes may be made based on what the rest of the class thinks is important. Evaluate the suggestions and incorporate those that you think are important into your final drawings.

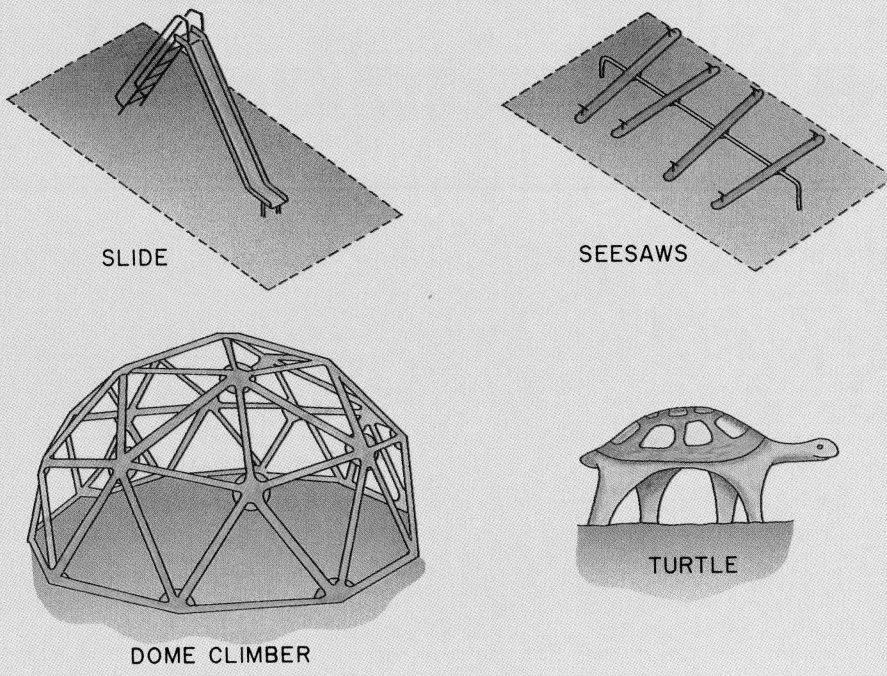

Fig. SCV-1. Different types of playground equipment.

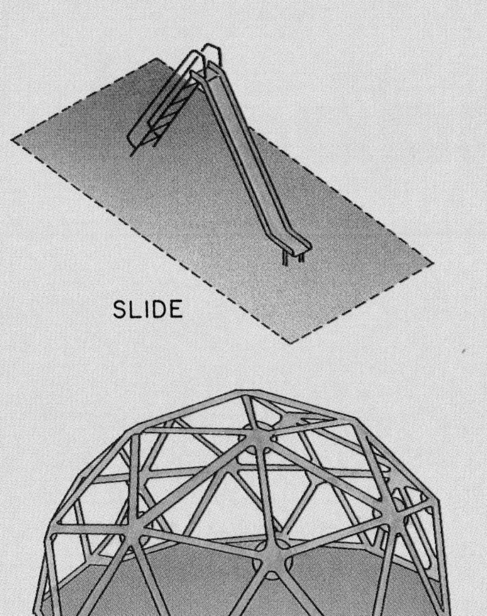

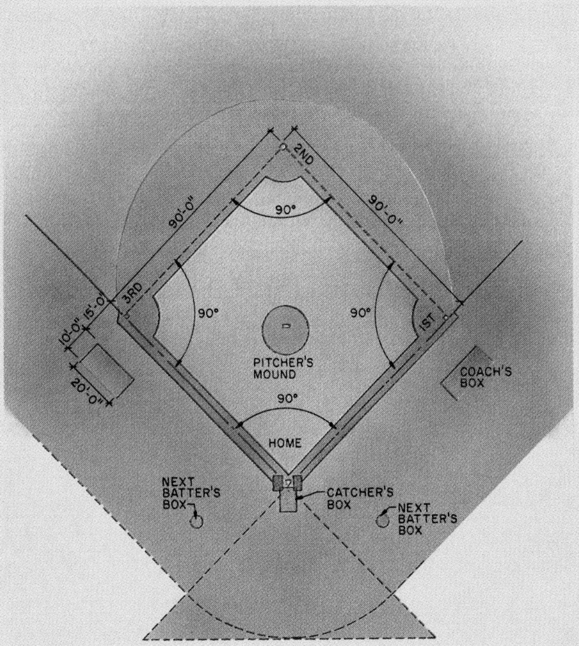

SEESAWS

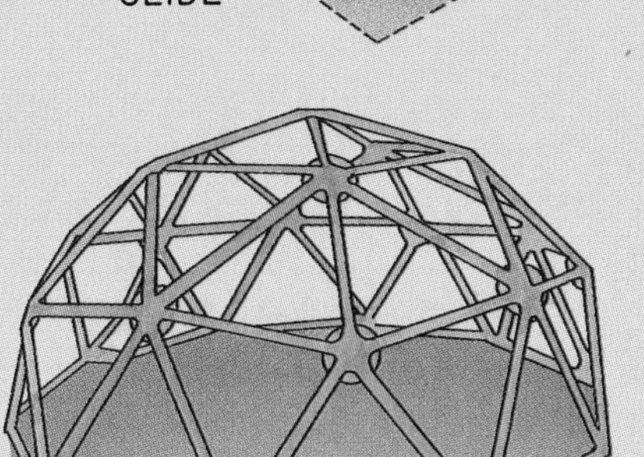

Fig. SCV-2. Some playing fields may take up more space than you expected.

Credit:

Developed by Patti Farrar for the Center for Implementing Technology Education
Ball State University
Muncie, Indiana 47306

Careers Handbook

The articles in this handbook give you information about jobs available in the different areas of technology (communication, manufacturing, transportation, construction, and biotechnology). These articles describe only one career that is available in each area of technology. Since many other careers exist within these five areas, these articles should just be the start of your research into a possible career in technology.

For each of the careers discussed, the article provides basic current information on job qualifications, required training and education, job responsibilities, working conditions, and sources of more information.

The following career possibilities are included in this handbook:

- Industrial Designer (Communication)
- Numerical-Control Machine-Tool Operator (Manufacturing)
- Automotive Mechanic (Transportation)
- Architect (Construction)
- Radiologic Technologist (Biotechnology)

Industrial Designer (Communication)

Industrial designers combine artistic talent with knowledge of marketing, materials, and methods of production to improve the appearance and functional design of products.

As the first step in their work, industrial designers gather information on how the product compares with competing products, the needs of the user of the product, fashion trends, and effects of the product on its environment. After the initial research, industrial designers sketch different designs and consult with managers, engineers, production specialists, and sales and market research personnel about the feasibility of each idea. This development team considers such factors as visual appeal, convenience, utility, safety, maintenance, and cost to the manufacturer, distributor, and consumer.

After company officials select the most suitable design, the industrial designer or a professional modeler makes a model. After any necessary revisions, a final or working model is made. The approved model then is put into production.

Although most industrial designers are product designers, many others are involved in different facets of design. To create favorable public images for companies and for government services, some designers develop trademarks or symbols that appear on the firm's product, advertising, brochures, and stationery. Some design containers and packages that both protect and promote their contents. Others prepare small display exhibits or the entire layout for industrial fairs. Some design the interior layout of special purpose commercial buildings such as restaurants and supermarkets.

Industrial designers generally work in clean, well-lighted, and well-ventilated rooms. They normally work 5 days, 35-40 hours a week, but occasionally, they work overtime to meet deadlines.

Completing a course of study in industrial design in an art school, in a university, or in a technical college is the usual requirement for entering this field of work. Persons majoring in engineering, architecture, and fine arts may qualify as industrial designers if they have appropriate experience and artistic talent. Most large manufacturing firms hire only industrial designers who have a bachelor's degree in the field.

Industrial designers must have creative talent, drawing skills, and the ability to translate abstract ideas into tangible designs. They must understand and meet the needs and tastes of the public, rather than design only to suit their own artistic sensitivity. Designers should not be discouraged when their ideas are rejected—often designs must be resubmitted many times before one is accepted. Since industrial designers must cooperate with engineers and other staff members, the ability to work and communicate with others is essential. A sound understanding of marketing, sales work, and other business practices also is important.

For further information, write to:
Industrial Designers Society of America, 1142-E Walker Rd., Great Falls, VA 22066.

Numerical-Control Machine-Tool Operator (Manufacturing)

Numerical control machine tools have two major components: an electronic controller (a type of computer) and a machine tool. The controller directs the mechanisms of the machine tool through the positioning and machining described in the computer program for the job. A program, for example, could contain commands that cause the controller to move a drill bit to certain spots on a workpiece and drill a hole at each spot. Many types of machine tools—milling machines, lathes, punch presses, and others—can be numerically controlled.

The duties of these operators vary. In some shops, operators merely tend one machine. In others, they might program and tend machines, operate more than one machine at a time, or operate more than one type of machine.

Working from written instructions or directions from supervisors, operators must position the workpiece, attach the necessary tools, and load the program into the controller. The machine tool cannot "see" the workpiece, but moves and operates in relation to a fixed starting point on the piece. It is, therefore, critical that operators position the workpiece correctly or all subsequent machining will be wrong. Operators also must secure the workpiece to the worktable correctly, so the piece does not move while it is machined. During the setup and running of a job, operators must install the proper tools in the machine. Many numerically controlled machines are equipped with automatic tool changers, so operators have to load several tools in the proper sequence. The time an operator needs to position and secure the workpiece and load the tools may be only a few minutes or several hours, depending on the size of the workpiece and complexity of the job.

The way a program is loaded into a controller depends on how it is stored. If the program is stored on a paper or magnetic tape, the tape must be run through a tape reader that transmits the program to the controller. Increasingly, machine-tool controllers are connected to minicomputers. Operators load programs that are stored on disks or tapes directly into the controller via the computer.

When a job is properly set up and the program has been checked, the operator merely has to monitor the machine as it operates.

The job requires stamina because operators stand most of the day and may lift moderately heavy workpieces onto the worktable. In some shops, operators may have to work evening or night shifts.

This generally is not an entry-level job. Employers prefer to fill operator jobs in-house. They select machine-tool operators or shop helpers who have some experience in machine-tool operation and have demonstrated good work habits and mechanical aptitude. Courses in shop math and blueprint reading may improve an employee's chances of getting selected for an operator job.

Working under a supervisor or an experienced operator, trainees learn to set up and run one or more kinds of numerically controlled machine tools. Trainees usually learn the basics of their job within a few weeks. However, the length of the training period varies with the number and complexity of the machine tools the operator will run and the individual's ability. If the employer expects operators to write programs, trainees may attend programming courses offered by machine-tool manufacturers. These courses usually last 1 to 2 weeks.

For general information about this occupation, contact:
The National Machine Tool Builders, 7901 Westpark Dr., McLean, VA 22103.

Automotive Mechanic (Transportation)

Automotive mechanics, often called service technicians, repair and service automobiles and occasionally small trucks, such as vans and pickups, with gasoline engines.

The ability to make a quick and accurate diagnosis, one of the mechanic's most valuable skills, requires good reasoning ability and a thorough knowledge of automobiles.

When mechanical or electrical troubles occur, mechanics first get a description of the symptoms from the owner. The mechanic may have to test drive the vehicle or use testing equipment, such as engine analyzers, spark plug testers, or compression gages, to locate the problem. Once the cause of the problem is found, mechanics make adjustments or repairs.

Automatic transmission mechanics work on gear trains, couplings, hydraulic pumps, and other parts of automatic transmissions. Because these are complex mechanisms, their repair requires considerable training and experience, including a knowledge of hydraulics. *Tune-up mechanics* adjust the ignition timing and valves, and adjust or replace spark plugs and other parts to ensure efficient engine performance. They often use electronic test equipment to help them adjust and locate malfunctions in fuel, ignition, and emissions control systems.

Automotive air-conditioning mechanics install air-conditioners and service components such as compressors and condensers. *Front-end mechanics* align and balance wheels and repair steering mechanisms and suspension systems. They frequently use special alignment equipment and wheel-balancing machines. *Brake repairers* adjust brakes, replace brake linings and pads, repair hydraulic cylinders, turn discs and drums, and make other repairs on brake systems. Some mechanics specialize in both brake and front-end work.

Automotive-radiator mechanics clean radiators with caustic solutions, locate and solder leaks, and install new radiator cores or complete replacement radiators. They also may repair heaters and air-conditioners and solder leaks in gasoline tanks.

Generally, automotive mechanics work indoors. Most repair shops are well ventilated and lighted, but some are drafty and noisy. Mechanics frequently work with dirty and greasy parts, and in awkward positions. They often must lift heavy parts and tools. Minor cuts, burns, and bruises are common, but serious accidents may be avoided when the shop is kept clean and orderly and safety practices are observed.

Many automotive mechanics still learn the trade by assisting and working with experienced mechanics. However, automotive technology is rapidly increasing in sophistication, and most training authorities recommend that persons seeking trainee automotive mechanic jobs complete a formal training program. Programs in automotive mechanics are offered in high schools, community colleges, and public and private vocational and technical schools. High school programs, particularly, vary greatly in quality. Post-secondary automotive mechanic training programs vary greatly in format. Some concentrate the instruction in only 6 months or a year, depending on how many hours the student must attend each week. Some community college programs spread the training out over 2 years, supplement the automotive training with instruction in academic subjects, and award an associate degree.

Knowledge of electronics is increasingly desirable for automotive mechanics.

Most mechanics work between 40 and 48 hours a week, but many work even longer hours during busy periods.

For general information about the work of automotive mechanics, write to:
Automotive Service Association, Inc., P.O. Box 929, Bedford, TX 76021-0929.
Automotive Service Industry Association, 444 North Michigan Ave., Chicago, IL 60611.

Architect (Construction)

Designing a building involves far more than planning an attractive shape and exterior. Buildings must also be functional, safe, and economical and must suit the needs of the people who use them. Architects take all these things into consideration when they design buildings.

Architects provide a wide variety of professional services to individuals and organizations planning a building project. They are involved in all phases of development. Their duties require a variety of skills—design, engineering, managerial, and supervisory.

The architect and client first discuss the purposes, requirements, and cost of a project. Based on the discussions, the architect prepares a program—a report specifying the requirements the design must meet. The architect then prepares carefully scaled drawings presenting ideas for meeting the client's needs.

After the initial proposals are discussed and accepted, the architect develops final construction documents that incorporate changes required by the client. These documents show the floor plans, elevations, building sections, and other construction details. Accompanying these are drawings of the structural system, air-conditioning, heating, and ventilating systems, electrical systems, plumbing, and landscape plans. Architects also specify the building materials and, in some cases, the interior furnishings. In developing designs, architects follow building codes, zoning laws, fire regulations, and other ordinances, such as those that require easy access by physically challenged persons.

Throughout the planning stage, the architect may make changes to satisfy the client. The architect may also assist the client in obtaining bids, selecting a contractor, and negotiating the construction contract. As construction proceeds, the architect visits the building site to ensure that the contractor is following the design, using the specified materials, and that the quality of work meets the specified standards.

Architects design a wide variety of structures, such as office buildings, churches, hospitals, houses, and airports. Architects generally work in a comfortable environment. Most of their time is spent in offices advising clients, developing reports and drawings, and working with other architects and engineers. However, they also often work at construction sites reviewing the progress of projects.

All states and the District of Columbia require individuals to be registered (licensed) before they may call themselves architects or contract for providing architectural services. To qualify for the registration examination, a person generally must have at least a Bachelor of Architecture degree from a program accredited by the National Architectural Accrediting Board and 3 years of acceptable experience in an architect's office.

Many architecture school graduates work in the field even though they are not registered. However, a registered architect is required to take legal responsibility for all work.

Persons planning a career in architecture should have some artistic ability, at least to the extent of being able to make reasonable freehand sketches. They should have a capacity for solving technical problems and should be able to work independently. General information about careers in architecture can be obtained from:

Director, Education Programs. The American Institute of Architects, 1735 New York Ave. N.W., Washington, DC 20006.

Radiologic Technologist (Biotechnology)

No hard and fast rules about job titles exist in this field. However, operators of radiologic equipment should not be confused with radiologists—physicians who specialize in the interpretation of radiographs. Radiologic personnel may be called radiologic technologists in one hospital, radiographers in another, and X-ray technicians in yet a third. The size of the facility, amount of specialization, and organizational policy are among the factors that determine which job titles are used. Another reason for inconsistency in job titles is the rapidity with which new medical technologies have emerged and practice patterns have changed. When new equipment is introduced, existing staff are taught to operate it, and it may be some time before job titles are changed.

Radiographers take X-ray films (radiographs) of all parts of the human body for use in diagnosing medical problems. They prepare patients for radiologic examinations, assuring that they remove any articles, such as belt buckles or jewelry, through which X-rays cannot pass. Then they position the patients, who either lie on a table, sit, or stand, so that the correct parts of the body can be radiographed, always taking care not to aggravate injuries or make the patients uncomfortable. To prevent unnecessary radiation exposure, the technologist surrounds the exposed area with radiation protection devices, such as lead shields, or in some way limits the size of the X-ray beam.

After the necessary preparations, the technologist positions the radiation equipment at the correct angle and height over the appropriate area of a patient's body. Using instruments similar to a measuring tape, the technologist measures the thickness of the section to be radiographed and then sets the controls on the machine to produce radiographs of the appropriate density, detail, and contrast. The technologist then places a properly identified X-ray film of the correct size under the part of the patient's body to be examined and makes the exposure. Afterward, the technologist removes the film and develops it. Throughout the procedure, the technologist is careful to use only as much radiation as is necessary to obtain a good diagnostic examination.

With the successful use of radiation as a cancer treatment, radiation therapy technology has developed into a separate specialty. *Radiation therapy technologists* prepare cancer patients for treatment and administer prescribed doses of ionizing radiation to specific body parts. Technologists operate many kinds of equipment, including high-energy linear accelerators with electron capabilities. They must position patients under the equipment with absolute accuracy in order to expose affected body parts to treatment while protecting the rest of the body from radiation.

Sonographers select equipment appropriate for use in ultrasound tests ordered by physicians. They also check the patient's other diagnostic studies for information. Sonographers explain the procedure, record any additional medical history considered necessary, and then position the patient for testing. Viewing the screen as the scan takes place, the sonographer must be able to recognize subtle differences between healthy and pathological areas; to check for factors such as position, obstruction, or change of shape; and to judge if the images are satisfactory for diagnostic purposes. A high degree of technical skill and knowledge of anatomy and physiology are essential to recognize the significance of all body structures present in the ultrasound image.

Radiologic technologists generally work a 40-hour week that may include evening and weekend or on-call hours. Some hospitals offer extremely flexible work schedules. A technologist may choose to work three 13-hour days a week, for example. Part-time work is widely available.

There are potential radiation hazards in this field; however, these hazards have been reduced by the use of safety devices such as instruments that measure radiation exposure, lead aprons, gloves, and other shielding. Because of the presence of radiation and radioactive materials, technologists wear special badges that measure radiation levels while they are in the radiation area, as well as the cumulative lifetime dose.

Preparation for this field is offered at the postsecondary school level in hospitals, medical centers, colleges and universities, trade schools, vocational-technical institutes, and the armed forces. Hospitals, which employ most radiologic technologists, prefer to hire individuals who have completed a formal training program.

Formal training programs are offered in radiography, radiation therapy technology, and diagnostic medical sonography (ultrasound). These programs vary in a number of respects: length of training, prerequisites, class size, and cost. Programs range in length from 1 to 4 years and lead to a certificate, associate degree, or bachelor's degree. Two-year programs are most prevalent.

Some of the 1-year certificate programs are designed for individuals from other health professions who wish to change fields—medical technologists, registered nurses, and respiratory therapists, for example. Certificate programs also attract experienced radiologic technologists who want to specialize in radiation therapy technology or sonography. A bachelor's or master's degree in one of the radiologic technologies is desirable for supervisory, administrative, or teaching positions.

Although hospitals will remain the principal employer of radiologic technologists, opportunities in nonhospital settings are increasing rapidly. Many technologists will find jobs with walk-in clinics, freestanding imaging centers, medical group practices, and health maintenance organizations. Radiology groups will constitute a particularly important employer of technologists. Health facilities such as these are expected to grow very rapidly through the year 2000 due to the strong shift toward outpatient care.

For career information about radiologic technologists, write to:

American Society of Radiologic Technologists, 15000 Central Ave. S.E., Albuquerque, NM 87123.

Society of Diagnostic Medical Sonographers, 12225 Greenville Ave., Suite 435, Dallas, TX 75231.

Safety Handbook

Safety Is the First Priority!

One of the first rules we learned as children had nothing to do with others around us but had to do with our own safety. Our parents removed any objects around the house that might harm us, and then they watched over us as if we were delicate soap bubbles. Our own safety is one of the basic needs of life. We have been taught from early childhood to "be careful" doing this or that. We must all feel safe in our endeavors or we will fail to endeavor.

This Industrial Technology course is no exception when it comes to making you feel safe. During this course, you will be asked to use some of the tools and materials of industry. The tools may be "industrial" types such as stationary table saws or planers, or they may be of a "chemical" type such as photographic developers or printing inks and solvents. Even drafting instruments are dangerous if used incorrectly. In all cases, there is a certain amount of respect you must give to the tools and materials you are using. There will always be certain rules you must follow to ensure your safety and the safety of those around you. Unless you are working with some experimental equipment (which you probably won't be), the rules for the material or equipment you are using have been tried and tested before it was put on the market. While working with the materials and equipment in this course, you will be required to follow *all* of the safety rules associated with such materials and equipment. Never use any materials or equipment without prior instruction and the instructor's permission. Failure to abide by all safety rules could have devastating results and may require expulsion from the labs. We all want to have a good time while learning technology, but we must all be safe in the process.

Safety starts with an attitude that says, "I will practice *all* safety rules at *all* times." The key word here is "I." You, not your neighbor or your friend, but you and only you are responsible for your actions. It only takes one time of disobeying or forgetting a safety rule for you to be seriously injured. You could hurt yourself or you could hurt some of your classmates and maybe even permanently disable someone if you have a poor attitude about safety.

With safety being "First Priority" in this class, you and your classmates will learn a lifelong skill which time and technology will never overtake. A proper attitude about safety will help you become a more productive citizen and worker in whatever field of employment you choose as an adult. If we all work together in this class, we will be safe in our endeavors and we may endeavor great things.

Think Safety!

Industrial Technology

General Safety Study Sheet

Note: Many of the general safety rules listed here also apply to other areas of safety such as electrical, chemical, power, or hand tool safety.

- Never wear jewelry or loose clothes in the lab.
- Wear an apron while using tools or machines.
- Long hair should be tied back or worn under a cap to avoid getting it caught in the machinery.
- Don't use any equipment unless the instructor is in the lab.
- Before you use any equipment, whether you have used it before or not, you must get the instructor's permission.
- You should wear hard shoes or boots with rubber soles.
- The most important reason to use the right tool for the job is safety.
- If a machine doesn't sound right, or if you can see that something is wrong, tell the instructor immediately.
- Keep the floor and aisles clean at all times.
- Always use a brush to clean a table or piece of equipment.
- Before using any equipment, you should know where the fire extinguishers are located and how to use them.
- Oily rags must be kept in a closed metal container to prevent spontaneous combustion.
- See to it that you and others stay out of the danger zones marked by red or red/white striped tape around machines.
- If you have to leave a machine, turn it off and wait till it stops.
- Lift with your legs, not your back.
- Get help to lift or move long or heavy objects or materials.
- Pull the plug, not the cord, when you unplug a machine.
- Machine guards must be used at all times to prevent injury.
- Sharp or pointed tools must be carried with the point down.
- Horseplay will not be tolerated. This is evidence of a poor safety attitude.
- When you approach a machine, be sure it is off and that it is not coasting.
- If you do not feel well, tell the instructor and do not work in the labs.
- When you finish using a machine, turn it off, wait till it stops, and clean the machine and the area around it.
- The safe way is always the best way.
- Your safety is everyone's responsibility.
- Others' safety is everyone's responsibility.
- Store all materials properly.
- The shortest piece of lumber that can safely be run through most equipment is 12 inches long.
- A safety rule which applies to all power equipment is to never talk to a person while he/she is operating a machine.
- While you are learning to operate power machinery, have the instructor check the setup before turning the power on.
- Mistakes cause accidents.
- Check all stock for cracks, loose knots, and nails.
- Long stock must be supported when cutting.
- If there is an accident, even a minor one, it should be reported immediately to the instructor.
- Always wear proper eye protection when using or watching any power equipment.
- Wait until the blade stops before removing any scraps.
- Make all adjustments with the power off and the blade stopped.
- Machines should be allowed to reach full speed before starting to cut.

- Avoid standing in danger zones.
- Never distract the machine operator.
- Never place your hand or fingers in line with any moving parts.
- When operating any machine, give it all of your attention.
- Don't use any equipment until the instructor has shown you how to use it.
- The cord must be disconnected from the power source before changing bits, belts, or blades.
- Sometimes the equipment is loud, and you should use ear protection.
- Damaging the cord of electrical hand-held tools may cause an electrical shock.
- Never set a hand-held power tool down while it is running or coasting.
- Be sure blades, belts, and bits are installed properly before using.
- Your hand should never be placed directly in line with any moving parts.
- You must be able to measure before you can use most power equipment.
- The number one cause of accidents is a poor attitude.
- Always secure your work with a clamp or vise.
- Do not overreach; keep your balance.
- Use only properly insulated or grounded tools.
- Take your time when working with tools.
- Plan your work; measure twice and cut once.
- Use only tools that are sharp and in good condition.
- Return all tools to their proper places when you are finished using them.
- Stand to one side when using power tools.
- Use the right tool for the job.
- Read and follow all posted safety rules.
- Keep your mind on your work.
- Use compressed air with caution and only with the instructor's permission.

Industrial Technology

Electrical Safety Study Sheet

Note: Many general safety rules also apply to electrical safety.

- Always work in a dry area.
- Always short out large-value capacitors with the proper equipment.
- Never work alone.
- Always get the instructor's permission before working on any electrical components.
- If you have questions about a circuit, ask the instructor.
- When checking voltage, always keep one hand behind you or in your pocket.
- Know where the fire extinguishers are located and how to use them.
- Never use equipment that has had the ground prong removed from the plug.
- Never work on an electrical circuit with the power on.
- Never use any equipment without proper instruction first.
- Use caution when dealing with hot tubes and resistors; let them cool first.
- Less than one (1) ampere is enough electrical current to cause death.
- Always check a circuit for power before working on it.
- Use the right tool for the job.
- Never bypass or defeat a ground circuit.
- Use only approved chemicals or solvents and follow all label safety precautions.
- Use only approved electrical fire extinguishers for electrical fires.
- Wear proper clothing and eye protection.
- Read and follow the manufacturer's safety literature when available.
- Work only in a well-lighted area.
- Never activate a circuit without first checking with the instructor.

- Never assume anything around electricity; check it for yourself.
- Wear rubber-soled shoes when working with electricity.
- Stand on a rubber mat when working with electrical tools and machines.
- Know where the circuit breakers are located and see that they are properly labeled.
- Consider all circuits *live* until confirmed personally.
- Never overload a circuit.

Industrial Technology

Chemical Safety Study Sheet

Note: Many general safety rules also apply to chemical safety.

- Always wear protective clothing and approved eye protection.
- Always pour acid into water when mixing or diluting.
- Read all label precautions.
- Work only in well-ventilated areas.
- Use only the proper fire extinguisher for chemical fires.
- Use appropriate gloves or tongs when needed.
- Store all chemicals according to label directions.
- Know where the eyewash station is and how to use it.
- Know where the "Material Safety Data Sheets" (MSDS) are located in your lab.
- Know where the poison-control phone number is located.
- Clean all tools and equipment properly after using.
- Label all chemicals properly.
- Pay attention to others around you when working with chemicals and note any unusual reactions.
- Be aware of skin contact with certain chemicals and wash thoroughly before leaving the area.
- Store chemical-soaked rags in a proper container.
- Be careful not to mix certain chemicals, as they may become toxic.
- Dispose of chemicals properly.
- Use a respirator when working with certain chemicals.
- Mix chemicals only as directed.
- Clean spills immediately.
- Use only approved solvents for cleaning.
- Store all chemicals in properly labeled containers.

Glossary

A

abutments The supports at the ends of a bridge that support the bridge itself as well as the earth at the ends of the bridge.

acceptance sampling Randomly selecting a few typical products from a production run, or lot, and inspecting them to see whether they meet specified standards.

acoustics Refers to how clearly sounds can be heard in a room.

adhesives Materials that are used to bond together, or adhere, two objects. Glue is a common type of adhesive.

advertising The method or methods a company uses to persuade, inform, or influence consumers to buy a certain product.

aerodynamic drag The strong, power-robbing force created as a vehicle moves through air; causes a slowing of the vehicle's movement.

aerodynamics The science that deals with the interaction of air and moving objects.

airspace The area above the earth through which aircraft move.

alloy The new metal created when two or more metals or a metal and a nonmetal are combined.

amplitude The measurement of the intensity or strength of an electromagnetic wave.

AMTRAK The AMerican TRavel trAcK system, which is owned by the federal government and operated by the railroad lines that own the track the trains are using. This system provides all long-distance rail passenger service in the U.S.

arch bridge Bridge in which the curved portion or arch carries the weight of the load.

architects Professionals who design structures and develop the plans for building them.

artificial intelligence The process computers use to solve problems and make decisions that are commonly solved or made by humans. These computers have to be programmed with instructions in the form of human reasoning processes.

assemblies Components or parts of a product that have been assembled or put together in a planned way.

assembly drawings Drawings that show how all the parts fit together to make a finished product.

assembly line In factories, an arrangement in which the product being made moves from one work station to the next while parts are added.

assets Term used to describe anything a company owns that has value.

automated factory Factory in which many of the processes are directed and controlled by computers.

automated storage and retrieval system (AS/RS) Type of materials-handling system used in manufacturing in which a computer-controlled crane travels between a set of tall storage racks to store and retrieve materials.

automatic factory Factory in which everything is done automatically by machines; there are no people working.

automatic guided vehicle system (AGVS) Type of materials-handling system used in manufacturing in which specially built computer-controlled driverless carts (AGVs) that carry materials follow a wire "path" installed in the floor.

automatic processing Computer-controlled methods for processing materials in manufacturing.

B

bar codes The striped code printed on most products, with each product having its own pattern of stripes. A laser scanner reads light/no light from the stripes and sends this information to a computer.

barge A large, flat-bottomed water vehicle used to carry bulk cargo such as petroleum products and grain on rivers, canals, and lakes.

batter boards Boards held horizontally by stakes driven into the ground to mark the boundaries of a proposed building.

beam bridge Bridge in which a steel-reinforced concrete roadway is supported by steel or concrete girders (beams).

bid A price quote for how much a contractor will charge for a proposed project. The amount equals the total of the contractor's best estimate of what it will cost to build the project according to the owner's plans and specifications, plus the amount of profit the contractor hopes to make.

bill of materials (BOM) A complete list of the materials or parts, as well as the amounts of each, needed to make one product.

binary digital code Coded information consisting of varying series of the two digits, 1 and 0. Information is changed electronically into this code so a computer can "understand" it.

biomaterials Human-made materials designed to be placed within the human body. Examples include artificial joints and blood vessels.

biotechnology All the technology connected with plant, animal, and human life. Examples include medical technologies like X-ray machines and NMR machines and genetic engineering, which uses special techniques to create improved plant varieties in order to provide more food.

bit Each on or off pulse of the binary digital code; a tiny amount of computer data.

booster rockets Rockets used to push a payload into space.

brainstorming Problem-solving process in which a group of people tries to think of as many possible solutions to the problem as they can without stopping to evaluate the suggestions.

break bulk cargo Single units or cartons of freight.

building codes Regulations that specify the methods and materials that can or must be used for each aspect of construction.

bulk cargo Loose cargo such as sand, oil, or a gas.

byte Eight bits of information that are combined and processed as a group in a computer.

C

cable-stayed bridge Bridge supported by cables that are connected directly to the roadway.

cantilever bridge Bridge consisting of two beams, or cantilevers, that extend from each end and are joined in the middle by a connecting section called a suspended span.

capital Includes the money, land, and equipment needed to set up a technological system.

cargo Everything other than people that is transported in vehicles. Also called *freight*.

ceramics Materials, such as earthenware and porcelain, made from nonmetallic minerals that have been fired at high temperatures. Ceramics can withstand extreme temperature and are good insulators.

chain of distribution The "path" that goods take in moving from the manufacturer to the consumer.

circuit A path over which electric current or impulses flow.

city planner Person hired to create plans for community development.

classification yards Places where trains are disconnected, cars are sorted according to their destinations, and then new trains are made up of cars going in the same direction. Also called *switch yards*.

coaxial cable Cable for carrying electrical signals that consists of an outer tube made of electrical-conducting material (usually copper) that surrounds an insulated central conductor (also copper).

cofferdams Temporary watertight walls built to shift the flow of water around the construction area as a permanent dam is being built.

combining Process of joining or adding two or more materials together, such as welding two pieces of metal together or mixing chemicals to make a product.

combustion The process of burning, such as the burning of fuel by engines.

commercial construction Building structures used for business. Examples include supermarkets, shopping malls, restaurants, and office buildings.

commission A percentage of the selling price that some salespersons earn.

communication The process of sending and receiving messages.

communication channel The path over which a message must travel to get from the sender to the receiver.

communication satellite A device placed into orbit above the earth to receive messages from one location and transmit them to another.

communication technology All the ways people have developed to send and receive messages.

commuter service Regular back-and-forth passenger transportation service, such as commuter trains and subways.

component Each individual part of a product.

composite A new material made by combining two or more materials. The resulting material has more desirable qualities.

computer-assisted design (CAD) Using a computer to assist in the creation or changing of a design.

computer-assisted engineering (CAE) Computers are used to perform needed math calculations as well as to generate (produce) working drawings and analyze parts.

computer-assisted production planning (CAPP) Using a computer to determine the best processes for production flow and manufacturing times.

computer-integrated manufacturing (CIM) Using computers to help tie all the phases (planning, production, and control) of manufacturing together to make a unified whole.

computerized tomography Digital image processing used to diagnose medical problems. The patient's body is scanned by several X-ray machines; then this data is collected and the image is graphically reconstructed on a computer screen.

computer numerical control (CNC) Manufacturing process in which numerical directions contained in a computer program control and monitor machines' operations.

construction The process of building structures such as houses, highways, and dams.

construction industry The production system that produces constructed products such as buildings, highways, and dams.

construction technology All the technology used in designing and building structures.

consumers People who buy products for their own personal use.

container on flatcar (COFC) Method of transporting containers by rail in which containers are loaded directly upon flatcars.

containerization Containerized shipping; cargo is loaded into large containers before it is transported.

containerships Ocean-going ships specially designed and built to carry containers.

continuous production Type of production in which a large quantity of the same product is made in one steady process using an assembly line.

contractor Person who owns and operates a construction company.

coordinate measuring machine (CMM) A very accurate computer-controlled measuring device used to measure "hard-to-measure" parts, like rounded or spherical parts.

custom production Type of production in which products are made one at a time according to the customer's specifications.

D

data Information.

data bank A central computer that stores the information from many smaller computers.

demolition Destroying a structure by tearing it down (wrecking) or blasting it down with explosives; done to clear a site of old structure so new structure can be built in its place.

detail drawing Drawing that specifies the details (such as shape, materials, dimensions, tolerances) of a particular part.

die A piece of metal with a cut-out or raised area of the desired finished shape of a part; the material is then forced through or against the die to take on the shape of the die.

diesel-electric locomotives Train engines in which diesel engines turn electrical generators to produce the electricity that powers the traction motors that turn the wheels of the train.

digital image processing Using computers to change electronic data into pictures and manipulate these pictures in certain ways to make them more useful. For example, contrast can be enhanced, certain features can be emphasized, and colors can be changed.

digitize The process of changing information into a number code.

direct sales Type of sales in which a manufacturer sells its product directly to the customer.

distribution The methods used to get manufactured goods to the purchaser.

downlink Process in which a satellite transmits signals it has received from one earth station back to another location on earth.

downsizing Designing and producing vehicles that are smaller and lighter-weight so they will be more fuel-efficient.

drafting The process of accurately representing three-dimensional (having height, width, and depth) objects and structures on a two-dimensional surface, usually paper.

E

earthmoving Using heavy and powerful equipment such as bulldozers and graders to clear away excess earth and rock and to level and smooth remaining earth.

earth station A large, pie-shaped antenna that transmits signals to and receives signals from satellites; sometimes called a *ground station*.

economy A system for producing and distributing products and services; a money system.

efficiency The ability to bring about the desired effect or result with the least waste of time, energy, or materials. In production—producing as many quality products as possible in the least amount of time.

electromagnet A soft iron core surrounded by a coil of wire that temporarily becomes a magnet when electric current flows through the wire.

electromagnetic radiation An invisible source of energy given off by the movement of electrons when electrical devices are in operation.

electromagnetic waves Waves created by a magnetic field; they travel through the atmosphere and make communication without a connecting wire possible.

electronic mail (E-mail) Refers to sending such things as letters, messages, and documents using computers and telephone lines instead of paper.

electrostatic printing A printing process that relies upon a charge of static electricity to transfer the message from the plate to the paper. An example of this process is a photocopy machine.

elevations Drawings that show the finished appearance of the outside of the structure, as viewed from the ground level. A separate elevation is made for each side of a structure.

emulsion One or more thin layers of a light-sensitive material on photographic film that captures the image.

energy The ability to do work.

engine A machine that produces its own energy from fuel.

engineers People who figure out how a structure will be built or a product will be made and what materials will be used.

entrepreneur A person who starts his or her own business.

environment All the conditions that surround a person, animal, or plant and affect development and growth.

ergonomics Designing products and structures to match human needs.

escalator A moving stairway.

excavating Digging.

external-combustion engine Engine powered by fuel that is burned outside the engine.

F

facsimile (fax) system System for sending data in which a fax machine turns a picture or words into a number code and sends this data over telephone lines to another fax machine. The fax machine at the receiving end converts this information back into pictures or words.

Federal Aviation Administration (FAA) A government agency that controls all air traffic above the United States.

feedback The information about the outputs of the system that is sent back to the system to help determine whether the system is doing what it is supposed to do.

fifth wheel A hook-up arrangement between a tractor and a semi-trailer; a kingpin joins the two vehicles securely, while allowing the semi-trailer to be swiveled or maneuvered.

fixture A special device that holds a part in place during processing.

flexible machining center (FMC) A computer-controlled combination machine tool capable of drilling, turning, milling, and doing other processing.

floor plan Drawing that shows the locations of rooms, walls, windows, doors, stairs, and other features of a structure.

footing The part of the structure below the foundation wall that distributes the structure's weight. Footings are made of reinforced concrete.

forming Changing the shape of a material without adding or taking away any part of the material; for example, casting molten aluminum in a mold.

fossil fuels Fuels that were formed from decayed plants and animals that lived millions of years ago. Coal, petroleum, and natural gas are fossil fuels.

foundation The part of the structure that rests upon the earth and supports the superstructure. Foundations are usually made from concrete.

four-stroke cycle engine Internal-combustion engine that has four strokes of the piston for one combustion cycle.

freight Everything other than people that is transported in vehicles. Also called *cargo.*

frequency The number of electromagnetic waves that pass a given point in one second.

friction Resistance to motion; often caused by the rubbing together of parts.

functional design Design elements that ensure that a product will work properly or serve its planned purpose.

G

gages Inspection tools used to compare or measure sizes of parts and depths of holes.

gasohol A fuel made by combining gasoline with methanol, a chemical derived from distilling corn.

geotextiles Large pieces of material similar to a plastic cloth that are spread on the ground as an underlayment, or bottom layer, for roadbeds and on slopes to help prevent erosion. Also called *engineering fabrics.*

graphic communication Methods of sending messages using primarily visual means. The basic methods are printing, photography, and drafting.

gravure printing Printing process in which images are transferred from plates that have recessed (sunken) areas. Also called *intaglio printing.*

gridlock Traffic jam in which no vehicles can move in any direction.

group technology Computerized method of keeping track of similar parts that a company manufactures so that information can quickly be retrieved and revised to develop production plans for a new similar part.

H

hologram The printed three-dimensional image produced by holography, which uses lasers to record realistic images of three-dimensional objects. You see the subject from different angles as you change your view of the hologram.

horsepower A unit for measuring the power supplied by a motor or engine. One horsepower equals the force needed to raise 550 pounds at the rate of one foot per second.

hydraulics Using the reactions of fluid under pressure to develop motion. Force is exerted on one part and the fluid transfers the force to another part.

I

industrial construction The building and remodeling of factories and other industrial structures.

Industrial Revolution Refers to the great changes in society and the economy caused by the switch from products being made by hand at home to being made by machines in factories.

ink-jet printing Computer-controlled printing process in which tiny nozzles spray ink droplets onto the material being printed upon.

input Anything that is put into the system. Input comes from the system's seven resources: people, information, materials, tools and machines, energy, capital, and time.

insulation Material that is applied to walls and ceilings of a building to help keep heat from penetrating in the summer and cold from penetrating in the winter.

integrated circuit (IC) A tiny piece of silicon that contains thousands of tiny, interconnected electrical circuits that work together to receive and send information. Also called a *microchip.*

interchangeability of parts Parts are made exactly like one another so that any one of them will fit the product.

intermodal transportation The process of combining two or more of the five modes of transportation (air, rail, water, highway, and sometimes pipeline) to efficiently transport passengers and/or cargo.

internal-combustion engine Engine in which the fuel is burned within the engine.

inventory The quantity of items a company has on hand.

inventory control Keeping track of materials, parts, supplies, and finished products on hand.

J

jig A special device that holds a part being processed and guides the tool performing the process.

job-lot production Production method in which a certain quantity of a product or part, called a "lot," is made; then any necessary retooling or other changeovers are made so another job of a different part or product can be produced.

joists Evenly spaced horizontal boards that form the frames that support floors and ceilings.

just-in-time (JIT) Computerized system of scheduling parts and materials to be delivered just as they are needed for use in production.

L

landscaping Changing the natural features around a structure to make it more attractive. Landscaping includes shaping and smoothing the earth as well as planting trees, shrubs, grass, and other vegetation.

laser A narrow, high-energy beam of light that concentrates a lot of power in a very small space.

laser curtain A quality-control machine that uses a moving laser beam to record the measurements of a part.

laser cutting Using a laser (a concentrated, high-energy beam of light) to cut or engrave materials.

layout Graphic version of how words and pictures will be arranged on a printed page.

linear motion Motion in a straight line, such as in a rocket.

lithography Printing process in which the image areas of the printing plate have an ink-attracting grease surface and the nonimage areas are covered by an ink-repelling coating of water.

load-bearing wall structures Structures in which the heavy walls support the weight of the structure; there is no frame. Usually made of concrete blocks, poured concrete, or precast concrete panels.

lock An enclosed part of a canal or other waterway that is equipped with a gate; the level of water within the lock can be changed in order to raise or lower ships from one water level to another.

logos Symbols with which companies or organizations are quickly and easily identified.

M

maglev systems Rail systems that operate on the scientific principle that like poles of a magnet repel each other. Magnets are used to both levitate and propel the train.

manufacturing The process of making products.

manufacturing resource planning (MRP II) Involves using computers to plan for material, people, time, and money requirements.

manufacturing technology All the technologies people use to make the things they want and need.

market A specific group of people who might buy a product.

marketing All the activities involved in selling products.

market research All the activities used to determine what people want to buy and how much they will pay for it. The results indicate how well the company can expect the product to sell.

mass communication Communication with large groups of people.

methane gas A waste product given off when plant or animal waste decays.

methanol A chemical derived from distilling corn that can be used as fuel in internal-combustion engines. Usually mixed with gasoline to make a mixture called gasohol.

microchip A tiny piece of silicon that contains thousands of tiny, interconnected electrical circuits that work together to receive and send information. Also called *integrated circuit.*

microgravity The very low pull of gravity and resulting near-weightlessness that occurs in space.

microwaves Very short electromagnetic waves that can be used to carry telephone messages through the atmosphere.

mock-up A three-dimensional model of the proposed product; it looks real, but has no working parts.

modem A device that sends computer data over telephone lines to other computers.

modular construction Type of construction in which entire units, or modules, of structures are built at factories and shipped to the site, where they are assembled to produce a finished structure.

modular design Designing a product that will use basic standard units called modules or modular parts. Makes it possible to make a variety of products by using basic parts and adding different modules.

monitor To watch over and inspect to make sure a product is being made or a structure is being built according to the plans.

motor A machine that uses energy supplied by another source.

movable bridge Bridge designed so that a portion of the roadway can be moved to allow large water vessels to pass underneath.

myoelectric hand A realistic-looking prosthetic (artificial) hand that has electronic sensors that detect signals from nerve endings in the remaining portion of the arm and relay these signals to activate the wrist, hand, and fingers.

N

National Aeronautics and Space Administration (NASA) Plans and operates the U.S. space program.

navigable Refers to a waterway that is wide enough and deep enough to allow ships or boats to pass through.

nonrenewable energy sources Sources of energy that cannot be quickly renewed or replenished. Fossil fuels, such as coal and petroleum, and nuclear energy are nonrenewable energy sources.

O

Occupational Safety and Health Administration (OSHA) Agency of the federal government that sets standards for safety and health in the workplace and sees that these standards are met.

on-site transportation Transportation within a building or a group of buildings. Moving from one story of a building to another in an elevator and loading freight are examples of on-site transportation.

optical fibers Thin, flexible fibers of pure glass used to carry signals in the form of pulses of light.

output Includes everything that results when the input and process parts of the system go into effect.

P

panelized construction Type of construction in which floors, walls, and roof all consist of prefabricated panels that have been made in factories and then shipped to the construction site, where they are assembled to produce the framed and sheathed shell of the structure.

part print analysis When a process planner studies the working drawings of a part to get ideas about how the part could be made.

passengers People who are moved from one place to another in vehicles.

patent A government document granting the exclusive right to produce or sell an invented object or process for a specified period of time.

payload Anything transported into space.

photographic printing Printing technique in which light is projected through a plate (usually called a negative) onto a light-sensitive material.

photovoltaic cells A device that can change the energy from the sun's light (solar energy) into electricity.

piggyback Intermodal method of transporting cargo in which semi-trailers full of cargo are carried on flatcars. Also called *trailer on flatcar (TOFC).*

pilot run A practice production run done to find and correct problems before actual production begins.

pixels Glowing dots that make up the image on a TV screen.

planographic printing Any process that involves the transfer of a message from a flat surface. Lithography is the most popular planographic printing method.

plant layout The arrangement of machinery, equipment, materials, and traffic flow in a manufacturing plant as determined by production flow, related activities, and space requirements.

power The use of energy to create movement.

power of eminent domain A law that states the government has the right to take private property for public use.

prefabricated Refers to components of structures that have been built in factories. These parts or sections are shipped to the construction site, where they are assembled.

prime mover A basic engine or motor that supplies the power in transportation.

principles of design Factors that help to determine the effectiveness of the design. These factors are balance, proportion, contrast, harmony, and unity.

private sector Ordinary people; the general public.

problem-solving process A multi-step process used to develop workable solutions to problems.

process All of the activities that need to take place for a system to give the expected result.

process chart A chart that shows the sequence (order) of manufacturing steps.

product engineering Planning and designing to make sure the product will work properly, will withstand extensive use, and can be manufactured with a minimum of problems.

production control Methods used to control what is made and when it is made.

productivity The measure of the amount of goods produced (output) and the amount of resources (input) that produced them.

profit The amount of money a business makes after all expenses have been paid.

program Set of instructions that control the operations of a computer.

programmable controller A small self-contained computer used to run machines and equipment; workers can reprogram it to change the way it functions.

prototype A full-size working model of the actual product. It is the first of its kind, and is usually built by hand, part by part.

public sector Includes municipal (city), county, state, and federal governments; responsible for initiating (beginning) construction projects such as highways, post offices, and fire stations.

public works construction Building structures intended for public use or benefit. Examples include dams, highways, bridges, sewer systems, and schools.

Q

quality assurance Methods used to make sure the product is produced according to plans and meets all specifications. Also called *quality control.*

quality circles Workers who perform similar tasks meet periodically with managers to discuss production problems and offer suggestions for improvement.

R

raw materials Materials as they occur in nature, before any processing has been done. Examples include trees, metal ore, and raw cotton.

reciprocating motion Up-and-down or back-and-forth motion.

relief printing processes Methods that print from a raised surface.

renewable energy sources Sources of energy that can replenish (restore) themselves regularly. Examples include sun, wind, water, wood, humans, and animals.

reservoir A lake in which water is stored for use. The water that collects behind a dam creates a reservoir.

residential construction Building structures in which people live.

resource Anything that provides support or supplies for the system.

retailers People or companies who buy products from manufacturers or wholesalers and then sell them directly to consumers.

robotics Using robots to perform tasks.

robots Special machines that are programmed to move things or do certain tasks automatically.

rolling stock Rail transportation vehicles.

rotary motion Circular motion.

routes Particular courses over which vehicles travel. Examples include shipping lanes and air routes.

RPM Revolutions of the crankshaft per minute.

S

sales forecast A prediction of how many products the company will sell.

scale drawings Drawings in which a small measurement is used to represent a large measurement. For example, one-fourth inch (1/4") may represent one foot (1'). Correct size relationships are maintained.

schedule A plan of action that lists what must be done, in what order it must be done, when it must be done, and, often, who must do it.

schematic drawing A type of drawing that uses symbols to show the relative positions of parts in a product or system.

screen printing Printing process in which ink is transferred through a stencil held in place by a porous screen.

sea-lanes Shipping routes across oceans.

sensors Devices that sense things such as light, temperature, sound, or pressure.

separating Changing the shape of material by removing some of it.

sheathing A layer of material, such as plywood or insulating board, that is placed between the framing and the finished exterior of a building.

simulation Involves using a computer or equipment to imitate as closely as possible the real-life circumstances for which a solution or product is designed to be used.

site The land on which a project will be constructed.

site plan Plan that shows where the structure will be located on the lot; includes boundaries, roads, landscaping, and utilities.

slurry A rough solution made by mixing liquid with solids that have been ground into small particles so that the material can be sent through a pipeline.

smart building Building in which computer systems, aided by sensors, control mechanical, electrical, and security systems.

space observatories Spacecraft that send information back to earth for scientists to study and interpret.

specifications *In architecture:* the written details about what materials are to be used for a project, as well as the standards and government regulations that must be followed. *In manufacturing:* the detailed descriptions of the design standards for a part or product; these standards include rules about the type and amount of materials, size, shape, function, and performance.

standardization Setting a uniform or common size for certain parts.

standard stock Materials that are formed or packaged in a widely used (standard) size, shape, or amount that is easy to ship and to use.

statistical quality control (SQC) Quality control method in which a computer uses a sampling system to determine how well parts are being made.

structural materials Those materials used to support heavy loads or to hold the structure rigid.

studs Evenly spaced vertical boards that form the frame of exterior and interior walls.

subassembly An assembly of components that is used as part of another assembly.

subcontractors Companies or people who specialize in certain types of construction work, such as plumbing or electrical work.

subfloor Sheets of plywood nailed or glued to the floor joists to serve as a base for the finish flooring.

subsidiaries Companies owned and controlled by a larger "parent" company.

subsystems Many smaller systems that make up a larger system.

superconductor A material that will carry electrical current with virtually no loss of energy.

superstructure The part of the building or other structure that rests on the foundation. In buildings, the superstructure consists of everything from the first floor up.

surveying Measuring and marking the site in order to identify the boundaries of the property.

suspension bridge Type of bridge that has two tall towers that support main cables that run the entire length of the bridge and are secured by heavy concrete anchorages at each end. Suspender cables dropped from the main cables are attached to the roadway.

system A group of parts that work together to achieve a goal.

T

target market The group of consumers a company's marketing department determines is most likely to use and want a particular product. Advertising is then aimed at this target market.

technology The way people use resources to meet their wants and needs.

telecommunication Communication over a long distance.

teleconference Conference held simultaneously among participants who are in different locations but are connected by telecommunication lines.

telemarketing The use of the telephone to sell goods and services.

test marketing Trying out a new product in a limited area to get consumers' reactions and opinions.

time and place utility Change in value caused by transportation.

tolerance The amount that a part can vary from the specified design size and still be used.

tooling-up Getting tools and equipment ready for production.

topography A construction site's surface features, such as lakes, hills, gullies, large rocks, trees, and soil type.

tractor-semi-trailer rig A combination of a tractor and a semi-trailer; the *tractor* is the base unit that pulls the *semi-trailer,* which contains the cargo.

trailer on flatcar (TOFC) Intermodal method of transporting cargo in which semi-trailers full of cargo are carried on flatcars. Also called *piggyback.*

transit Surveyor's tool used to measure horizontal and vertical angles.

transportation The movement of people, animals, or things from one place to another using vehicles.

transportation technology Includes all the means we use to help us move ourselves or cargo through the air, in water, or over land.

truss bridge Type of bridge supported by steel or wooden trusses, which are beams put together to form triangular shapes.

trusses Preassembled triangular frames; often used to frame the roof.

two-stroke cycle engines Internal-combustion engine that has two strokes of the piston for one combustion cycle. Also called *two-cycle* or *two-stroke* engine.

U

unit train A type of train that carries the same type of freight in the same type of car to the same place time after time.

uplink Process in which an earth station transmits signals to a satellite.

utilities Services such as electricity, natural gas, and telephone.

V

value added Increased worth of a material after it has been changed into a more useful form by one or more manufacturing processes.

value analysis Analyzing each part of a product to determine whether the best (most functional, yet lowest cost) material has been used to make the part.

vehicle navigation system Computerized system in which area maps stored on compact discs can be displayed on a small screen on a car's dashboard to help the driver find the desired location quickly and efficiently.

vehicles Any means or device used to transport people, animals, or things.

videotex A system that allows one to receive computer text and graphics via telephone lines.

W

warehouse A building where products are temporarily stored.

waterjet cutting Computer-controlled process of using a highly pressurized jet of water to cut a material.

ways The spaces set aside for use by a particular mode of transportation. Examples include pipeline right-of-ways, highways, and railways.

wholesalers People or companies who buy large quantities of products from manufacturers, then sell the products to commercial, professional, retail, or other types of institutions that also purchase in quantity.

working drawings Drawings that provide the information needed to make a product or construct a project.

Z

zoning laws Regulations that tell what kinds of structures can be built in specific parts of the community; may also specify such things as minimum property size, maximum building height, the number of families that can occupy a building, and the number of parking spaces a business must provide.

Index

A

B

C

D

E

F

N

O

P

CREDIT LIST

We wish to thank the following people and corporations for their cooperation with the production of this text:

W.D. Adams/Superstock (Four By Five, Inc.), 469. Advanced Environment Recycling Technologies, Inc./Arnold & Brown, 439. American Airlines, 27, 319. American Petroleum Institute, 367. Ametek, Incorporated, 85, 219, 328. Amoco Corporation, 200. AMP Incorporated, 199, 223, 245. Apple Computer, Incorporated, 176, 217. Archer Daniels Midland Company, 221. Arco Solar, Inc., 299. Chris Arend/Alaska Stock Images, 17. Arnold & Brown, 23, 37, 44, 91, 93, 131, 134, 139, 260, 324, 330, 378, 420, 421. Ashland Oil Incorporated, 177, 499. Association of American Railroads, 368. AT&T Bell Laboratories, 79, 105, 106. Craig Aurness/West Light, 318, 344, 430. Autodesk, Inc., 291. David Barnes (1991)/The Stock Market, 14. Steve Barnett/Tony Stone Worldwide, 171. David Bartruff (1991)/FPG International, 16. Baxter Health Care Corporation/David Joel, 271. Roger B. Bean, 37, 40, 261, 296, 429, 436, 437, 442, 466, 476, 485. Brent Bear/West Light, 342. Keith M. Berry, 120, 256, 258, 288, 444. The Bettmann Archive, 175, 196, 331, 422. Blaupunkt, 151. Boeing, 27. Bradley University/Roger B. Bean, 277. Matt Brown (1991), 424. Bureau of Reclamation, 497. Burton Associates, 445. Calcomp, 208. California Energy Commission, 326. Curt Campbell, 67. Carr Lane Manufacturing Company, 226. Caterpillar, Incorporated, 122, 224, 281, 377. Caterpillar, Incorporated/Arnold & Brown, 124. Claude Charlier, Science Source/Photo Researchers, Inc., 158. Chevrolet Motor Division, 63. Chicago Convention and Tourism Bureau, 25. Chrysler Motor Corporation, 20, 41, 122, 208. Cincinnati Milacron, 226, 238, 276, 280. City of Peoria, IL/Roger B. Bean, 446. Colonial Williamsburg Foundation, 175. Comstock, 15, 17, 59, 312, 425. The Cousteau Society, 406. Thomas Craig (1991), 422. Creative Sources Photography, Atlanta (1986)/FPG International, 57. CS&A, 23, 43, 44, 45, 47, 69, 81, 106, 109, 179, 183, 194,195, 242, 278, 290, 319, 322, 379, 431. CSX Corporation, 363, 364. Daemmrick/Stock Boston, 313. Dallas/Fort Worth International Airport, 370. Howard T. Davis, 21, 52, 53, 77, 103, 114, 127, 128, 129, 130, 132, 133, 142, 143, 144, 145, 164, 190, 191, 211, 213, 214, 220, 232, 233, 245, 249, 250, 251, 252, 253, 265, 267, 275, 283, 306, 308, 309, 323, 334, 335, 341, 351, 356, 357, 358, 359, 360, 379, 381, 382, 383, 384, 385, 386, 387, 391, 398, 399, 407, 410, 415, 418, 419, 438, 461, 473, 474, 480, 482, 486, 488, 493, 494, 495, 500, 501, 503, 505, 511, 513, 515, 516, 536, 537, Deere & Company, 201, 216. Delta International Machinery Corporation, 181. Department of Treasury, 227. DeVilbiss, 292. Diamond-Star Motor Corporation/Jim Gaffney, 434. Dresser Industries, Inc., 180. Drexler Technology/Chris Dyball, 61. Duchossors Industries, Incorporated, 361. Steve Dunwell/The Image Bank, 171. Emerson Electric Company/Obata Design Incorporated, 236. Robert Everts/Tony Stone Worldwide, 15. Donna Faull, 135. FEC Incorporated, 239. Scott Fishel, 40. Flow Systems Incorporated, 239. Ford Motor Company, 244, 380. FPG, 315, 423. Lois Ellen Frank/West Light, 80. David R. Frazier Photolibrary, 26, 27, 30, 78, 88, 149, 183, 187, 342, 345, 347, 352, 363, 365, 453, 479, 525, 531. Robert Frerck/Odyssey Productions, 312, 313. G. Fritz/Superstock (Four By Five, Inc.), 36. Gerald Fritz/FPG International, 57. Jim Gaffney, 91, 197, 433. Bill Gallery/Stock Boston, 425. Bob Gangloff, 81, 104, 105, 112, 113, 123, 125, 128. Garrett Auxilary Power Division/Allied Signal Aerospace Company, 403. Kenneth Garrett/West Light, 320. Ann Garvin, 24, 29, 71, 196, 258, 263, 435, 450,487. General Electric, 152. General Motors Corporation, 42, 207, 235, 236, 302, 402, 405. Giddings & Lewis, AMCA International, 279, 293. Gary Gladstone/The Image Bank, 314. Goodyear Tire and Rubber Company, 180. Spencer Grant/Stock Boston, 423. Rainier Grosskopf/Tony Stone Worldwide, 14. Jeff Isaac Greenberg, 224, 228, 323, 326, 344, 364, 388, 483. Grumman Corporation, 198. Jim Gund/Allsport USA, 121. Harnischfeger Industries, Incorporated, 225. Brownie Harris/The Stock Market, 57, 59. Robert Herko (1985)/The Image Bank, 58. Hewlett Packard Company, 66, 150, 198. Walter Hodges/West Light, 349, 432. Charles Hofer, 119. Doug Hoke (1991), 424. Robert Houser (1987)/Comstock, 314. Huffy Bicycles, 201. Illinois Department of Energy Resources/Community Recycling Center, Inc./Department of Mechanical and Industrial Engineering, University of Illinois/Champaign-Urbana News Gazette, 193. Impact Communications, 447. Ingersoll-Rand Company, 2, 201. International Business Machines Corporation, 31. Shinichi Kanno (1991)/FPG, 312. Kansai International Airport Company, 455. Kellogg Company, 122. Russ Kinne/Comstock, 312, 315, 425. Kent Knudson (1982)/West Stock, 423. R. Kord/H. Armstrong Roberts, 16. K's Merchandise, Peoria, IL/Arnold & Brown, 154. Knight-Ridder, Incorporated, 237. Kohler, Company, 378. KPMG Peat Marwick, 173. Robert Kusel (1989), 314. Laser Alignment, Inc., 521. LA-Z-Boy Chair Company, 207. Larry Lee/West Light, 17, 347, 433, 464, 506, 522. Lemark International Incorporated, Lexington, KY, 204. Litton, 221. R. Llewellyn/Superstock, 423. R. Ian Lloyd/West Light, 225, 234. Lockheed Corporation, 275. Lawrence E. Manning/West Light, 18, 520. James Marshall, 2, 376, 518. Sven Martson/Comstock, 425. Tom McCarthy/West Stock, 314. Kathy L. McCloskey, 28. Dan McCoy, Rainbow, 295. McDonnell Douglas Corporation, 206, 273, 274. McDonnell Douglas Corporation/NASA, 294. Bob McElwee, 38, 464. Fred McKinney/FPG, 314. MEGA CADD, Inc., 459. Method Research Corporation, 240. Minolta Corporation, 66. Mishima, 19, 20, 25, 30, 41, 42, 46, 65, 101, 196, 316, 393. MMFG Company, Bristol, Virginia, 523. Moller International, Inc., 401. Mason Morfit/FPG, 425. Mason Morfit/Lone Star Industry, 259. E. Nagele/FPG, 425. NASA, 6, 155, 156, 294, 298, 341, 390, 529, 531, 536. NASA, Science Source/Photo Researchers, 400, 410, 411. National Association of Women in Construction/John Swisher, 477. National Park Service, 428. National Semiconductor, 72. New United Motors Manufacturing, Incorporated, 301. New York State Department of Economic Development, 463. Nike, Inc., 122. Nippon Television Network Corporation, Tokyo (1991), 73. Noah Herman Sons, Peoria, IL/Roger B. Bean, 40. North American Philips Corporation, 89. North Wind Picture Archives, 159. C. Orrico/Superstock, 437. Ontario Industry & Tourism, 427. Optical Artists/West Light, 350. Chuck O'Rear/West Light, 60, 100, 146, 327, 338, 343, 344, 346, 364, 403, 408, 521. Otis Elevator, 318, 369. Packard Electric Division of General Motors, 405. Palmer & Kane (1986, 1979)/The Stock Market, 58, 168. Panasonic Company, 62. Pelton & Associates/West Light, 426. Pennsylvania Bureau of Travel Marketing, 449. Pentain, Incorporated, 218. Brent Phelps, 178, 264, 303, 432, 467, 484, 487. Philips International B.V., 115. Jim Pickerell/West Light, 158, 276, 422, 465. Philippe Plailly/Science Photo Library/Photo Researchers, Inc., 168, 171. Porter-Cable Corporation, 181. Potlatch Corporation/Toni Tracy, 126. Allen Prier (1989)/West Stock, 315. Principle Financial Group, 49. Proctor Community Hospital/Charles Hofer, 83. Public Safety Dispatch Center of Peoria County/Roger B. Bean, 83. Liz Purcell, 119, 136, 152, 269, 321, 482. P & W Builders, Incorporated/Roger B. Bean, 435, 478. Ed Pritchard/Tony Stone Worldwide, 15. Questar Corporation, 351, 496, 507. Randolph & Associates/Arnold & Brown, 272. Raychem, 195. Cloyd Richards, 98, 137, 138, 203, 222, 243, 285, 286, 287, 347, 375, 397, 398, 417, 418, 454, 457, 514. Rockwell International Company/Allen-Bradley, 278. Ken Rogers/West Light, 371. Rohr Industries, Incorporated, 381. Mark Romine, 58, 59. Bill Ross/West Light, 370. Saint Francis Medical Center, Peoria, IL, 299. Lucille Sardegna/West Light, 484. Scitex America Corporation/Brian Smith, 126. Mark Segal/Panoramic Stock Image, 425. Smith Floor Covering Company/Ann Garvin, 487. Smithsonian Institute, 110. SNCF-CAN-Jean Marc FABRO, 413. Solar Electric, 406. Sonoco Products, 185. Spectra Physics, 522. William P. Spence, 205. Spencer Industries, 289. E. Dale Spillman, 339. Stanley Continental, Inc., 2. Stanley Tools, 181. Ralph Starkweather/West Light, 1. Starnet Structures, Incorporated, 523, 528. Gene Stein/West Light, 329. Mark Stephenson/West Light, 329. Dennis Stock/Magnum Stock (1969), 313. Stryker Corporation, 201. Superstock (Four By Five, Inc.), 36, 313, 315, 424, 437, 469. The Telegraph Colour Library/FPG International, 16, 169, 170. Technicraft Supply Store/Duane R. Zehr, 85. Technics Audio Group, a division of Panasonic Company, 44, 148, 154. The Tensar Corporation, 498, 524. Tensar Structures, 508. Terra Mar Resources Information Services, Inc., Mountain View, CA, 147. Texas Instruments, Incorporated, 196. Charles Thatcher/Tony Stone Worldwide, 169. Michael S. Thompson (1989)/Comstock, 424. 3M Company, 70, 229, 257. Tom Tracy/FPG International, 170, 424. Transitions Research Corporation, 299. Union Pacific Railroad, 320, 364. Unity Systems, 519. Walter Urie/West Light, 270. U. S. Air Force, 368. U. S. Army Corps of Engineers, 509. U. S. Fish and Wildlife Service, 353. Versa CAD, 535. Videojet Systems International, Inc., 130. Terry Vini (1991), 422. Visual Horizons/FPG (1990), 422. Volvo Cars of North America, 247. Vortek Industries Limited, 209. Wagner Spray Technology Corporation, 437. Nicholas Waring, 423. Warner Brothers, Inc. (1988), 87. William James Warren/West Light, 69, 107, 328, 340. Ron Watts/West Light, 172. Jervis B. Webb Company, 223. Westmoreland Coal Company, 367. West Stock, 315. Whirlpool Corporation, 527. Ken Whitmore/Tony Stone Worldwide, 56. Wide World Photos, 317, 438, 489. Reg Wilkins/Tony Stone Worldwide, 170. The Williams Corporation, 352. T. D. Williamson, Inc., 352. Doug Wilson/West Light, 325, 360, 392, 502, 504. Don Wilson/West Stock, 314. WMBD TV 31, Peoria, IL/Roger B. Bean, 83. World Trade Center, St. Louis, MO, 452. Mike Yamashita/West Light, 362. Duane R. Zehr, 186. Frank Zosky, 259. Section Openers I, II, and III designed by 10 Adams. Section Openers IV and V designed by Design Associates.